Applied Bioinformatics, Statistics & Economics in Fisheries Research

AF694154

books for everyone
PAPERBACKS

NIPA® GENX ELECTRONIC RESOURCES & SOLUTIONS P. LTD.

New Delhi-110 034

About the Editors

Dr. Ajit Kumar Roy born in 1946, obtained his M.Sc. degree in Statistics. Dr. Roy joined Agricultural Research Service (ARS) of Indian Council of Agricultural Research (ICAR) as Scientist in 1976. Prior to that, he has served in Damodar Valley Corporation (DVC) and Ministry of Home Affairs, Govt. of India.

Dr. Roy has been engaged in R & D activities and successfully handled various research projects in fisheries sector for the last 36 years. He was deputed by Govt. of India to take up the assignment of Computer Specialist at SAARC Agricultural Information Centre (SAIC), Dhaka, Bangladesh for over 3 years. He has to his credit ten books, eight technical manuals, many e-publications, (published by SAIC and CIFA), and over eighty research publications covering the areas of statistics, economics, bioinformatics and ICT applications in fisheries research of National & International repute. *Presently*, as a co-ordinator of BTIS Project he developed the informatics based infrastructure to establish Bioinformatics Centre on Aquaculture at CIFA, Bhubaneswar. He has organized workshops on application of ICT in Fisheries Research.

Dr. A. K. Roy is principal scientist and head, social science section, Central Institute of Freshwater Aquaculture, Bhubaneswar, Orissa. He is involved in research in the area of application of bioinformatics and statistics in fisheries research.

Dr. N. Sarangi, born on 2nd July, 1946, is the Director, Central Institute of Freshwater Aquaculture, Bhubaneswar, India. He did M.Sc. (Zoology) from Andhra University, India in 1976 and obtained Ph.D. degree from Calcutta University, West Bengal in 1984. His major research contributions are genetic improvement of Indian major carps through systematic selective back crossing, gene transfer in carps and standardization of induced breeding of carps through various hormonal applications and freshwater aquaculture in the rain-fed ponds. He is the recipient of several honours and awards including ICAR Team Award for outstanding Team Research for the biennium 1994-1996 and ICAR Team Award for outstanding Team Research for the biennium 2003-2004. He is member of several policy making bodies in India and abroad. He has more than 100 research publications in reputed international and national journals.

Applied Bioinformatics, Statistics & Economics in Fisheries Research

Editors

Ajit Kumar Roy
Principal Scientist and Head
Social Science Section, Central Institute of Freshwater Aquaculture (CIFA)
Bhubaneswar, Orissa

Niranjan Sarangi
Director
Central Institute of Freshwater Aquaculture (CIFA)
Bhubaneswar, Orissa

nipa books for everyone PAPERBACKS

NIPA® GENX ELECTRONIC RESOURCES & SOLUTIONS P. LTD.
New Delhi-110 034

books for everyone
PAPERBACKS

NIPA® GENX ELECTRONIC RESOURCES & SOLUTIONS P. LTD.

101,103, Vikas Surya Plaza, CU Block
L.S.C. Market, Pitam Pura, New Delhi-110 034
Ph : +91 11 27341616, 27341717, 27341718
E-mail: newindiapublishingagency@gmail.com
www: www.nipabooks.com

For customer assistance, please contact
Phone: + 91-11-27 34 17 17
Fax: + 91-11- 27 34 16 16
E-Mail: feedbacks@nipabooks.com

ISBN:978-81-19002-10-8

Composed and Designed by NIPA®.

डा0 एम0 के0 भान
Dr. M.K. Bhan

सचिव
भारत सरकार
विज्ञान और प्रौद्योगिकी मंत्रालय
बायोटेक्नोलॉजी विभाग
ब्लाक–2, 7 वां तल, सी0जी0ओ0 कम्पलेक्स
लोदी रोड, नई दिल्ली–110003

Secretary
Government of India
Ministry of Science & Technology
Department of Biotechnology
Block-2, (7th Floor) C.G.O. Complex
Lodi Road, New Delhi-110003

Foreword

Aquaculture is one of the fastest growing food producing sectors in India. It contributes about 4.2% of global aquaculture production of 45.5 million tones. Per capita food supply from aquaculture has increased manifold compared to that obtained from terrestrial meat production during the last decade. Apart from supplying cheap protein, this sector is contributing significantly in poverty alleviation, food security and income generation for rural population. Demand from aquaculture sector is increasing because production from capture fisheries has leveled off. Government of India has given due attention for all round development of aquaculture sector through establishment of National Fisheries Development Board in the recent past. To augment the productivity further on a sustained basis, capacity building through information resource generation and training of manpower in application of modern methods of Information and Communication Technologies (ICT) in research and development activities is essential. Bioinformatics Division of Department of Biotechnology, Ministry of Science and Technology has been strengthening the infrastructure for Bioinformatics application countrywide. The need for training and developing manpower resources in this area is great as trained human resources are limited particularly recognizing the importance of IT in pursuing advanced research in modern biology for translating information into actionable knowledge. Moreover micro level knowledge management oriented research is to empower computer system in such a way that they can handle knowledge like content *i.e.* they become knowledge intensive transition from conventional database technology to knowledge intensive systems.

I am extremely happy to learn that a book entitled *"Applied Bioinformatics, Statistics and Economics in Fisheries Research"* has been compiled and edited with up-to-date and interesting articles contributed by the eminent professionals of the national institutes. The book is broadly divided into three disciplines like Bioinformatics, Statistics and Economics. Within bioinformatics many important topics such as data mart, databases, artificial neural networking and information management are covered besides basics of bioinformatics. Methodological aspects of estimation of fish stock, modeling, productivity analysis, impact assessment survey, forecasting, market intelligence, geographical information system are new introductions in aquaculture research along with statistical design of experiments, sample survey and econometric analysis using statistical and econometric packages. Further application of statistics in bioinformatics related to fish sequence data with illustrated data is unique.

It is expected that this publication will be useful for research workers and planners engaged in fisheries research equally that may serve as a unique source of reference material on the application of bioinformatics and statistics tools in fisheries research for the future generations in the era of knowledge based society.

I wish this endeavor a grand success.

(Dr. M.K. Bhan)

डा0 एस0 अय्यप्पन
उप महानिदेशक (मत्स्य)
Dr. S. Ayyappan
Deputy Director General (Fisheries)

भाकृ अनुप
ICAR

भारतीय कृषि अनुसंधान परिषद
कृषि अनुसंधान भवन–II
पूसा, नई दिल्ली–110012
Indian Council of Agricultural Research
Krishi Anusandhan Bhavan-II
Pusa, New Delhi-110003

Foreword

Knowledge is increasingly becoming more valuable asset than the traditional factors like labour and capital. Knowledge asset is considered as new national wealth. Therefore, major initiatives have been taken by Government of India through formation of National knowledge Commission with the objective of transforming India into a knowledge society. Indian Council of Agricultural Research (ICAR) has initiated appropriate steps to upgrade the information system for effective information and knowledge management to enhance the national competitiveness in the emerging Indian knowledge economy. Keeping in view the objectives of ICAR, CIFA is engaged in the task of modernizing the hardware and software infrastructure involving computers, communication facilities and databases for effective knowledge management. Various knowledge products in the form of bibliographic database and software like Aquatic Science and Fisheries Abstract (ASFA), FISHSTAT Plus, FISHBASE. The FAO Fisheries Atlas and Diagnosis of Shrimp Diseases have been acquired for online and offline information search. Statistical packages like SAS, SPSS, and econometric packages like DEAP & FRONTIER facilitate advanced and complex qualitative and quantitative analysis of fisheries research data. Software like CodonW, BLAST, and Codon Manipulation Suite that are essential bioinformatics tool in computational biology has been acquired. Therefore, thorough awareness regarding tools and techniques for exploring, adopting and applying new technologies in knowledge and information management in the field of fisheries research for relating information-centric network to action-centric networks of knowledge ecology is essential.

The book entitled "*Applied Boinformatics, Statistics and Economics in Fisheries Research*" has been compiled and edited with up-to-date and interesting articles contributed by the eminent professionals of the national institutes of India is timely. Chapters on statistical approach of genomic sequence and microarray expression data analysis and biotechnological tools in fish disease diagnosis are new in the field of aquaculture. Besides, design of experiments, sample survey, statistical tests, biometric techniques and quantitative genetics are also added with illustrated examples. Many important articles like cryoconservation, probiotics, disease diagnostic tool, nanotechnology, datamart and information management as applied to fisheries are also covered with latest development. Articles on, econometric analysis, market intelligence, impact assessment, IPR and policy issues and as applied to fisheries research are covered.

I hope the book will be a very good source of reference material for all those seeking knowledge from conventional information & communication technology to knowledge intensive tools and techniques with focus on information ecology.

(S. Ayyappan)

Preface

The idea of compiling the book was conceived while conducting National workshops on 'Bioinformatics and Statistics in Aquaculture Research at the Bioinformatics center of Central Institute of Freshwater Aquaculture, Kausalyaganga during past ten years. During the workshops the course content in the form of lecture, hands on training and demonstration of various aspects of Internet, CD-ROM search, data analysis using statistical and bioinformatics soft wares, web page & database development and many other frontier areas of science and technology were covered. Many participants expressed their keen interest in the emerging areas aquaculture and expressed for a book which will cater to the need of understanding the basics of computers, communications and micro electronics, analytical tools and techniques. The book is on out growth of the workshops. The book is an effort to compile and document the latest trend and technologies especially ICT application in fisheries and aquaculture besides statistics and economics. The book spread over broadly in three sections like ***Bioinformatics, Statistics*** and ***Economics***. Each section consists of a number of chapters. ***Bioinformatics section*** covers computational aspect of genomics and microarray data using latest tools and techniques apart from fundamental issues of carp, prawn and ornamental fish breeding, culture and management. Chapters on probiotics, immunology, cryo-conservation and disease that are included in this section are new and of paramount importance. Chapters like non-evasive methods of fish freshness determination by electronic olfaction, fisheries data mart, nano-technology, mapping fisheries research in India are altogether new approaches in fisheries and aquaculture research in Indian context. ***The statistics section*** covers chapters on application of designs of experiment, sample survey, biometrical techniques, quantitative genetics and selective breeding techniques that are noteworthy. ***Economics section*** is embodied with interesting chapters like quantitative techniques in econometric analysis and impact assessment in addition to fundamental computational aspects of economic indicators along with illustrated examples. Chapters on estimation of technical efficiency, project evaluation, IPR issues and fisheries legislation and policy in India are new introductions.

The endeavour could not have been materialized without the active support in terms of contributing articles by the dedicated subject matter specialists, professionals from various National Institutes of repute like DBT, IASRI, C-DAC, CIBA, NBFGR, B.C.K.V.V, KIITS, IGAU, CRIDA, NAARM, NCAP, CARI, WTCER, OUAT, RRL, MANAGE,ILS, CIFRI, CIFA, ZCU-VI, KVIC, Essar, China, Purdue University USA, IMS Engineering College, Utkal, Jadavpur and Delhi University. All out support, guidance, suggestion and encouragements from Dr. M.K. Bhan, Secretary, DBT and Dr. T.Madhan Mohan, Advisor, DBT are thankfully acknowledged. We express our gratitude to Dr. S.Ayyappan, DDG (Fy.), ICAR for constant encouragement and guidance at all stages of the programme.

Invitation to resource persons, conducting the workshop and subsequent publication of this book would not have been possible without the financial support of Bioinformatics Division of Dept. of Biotechnology, Ministry of Science & Technology, Government of India.

Mrs. N. Panda, Sri. D.P.Rath, Mrs. Nibedita Jena, Miss Sushma Martha, Sri Surya Narayan Rath deserve special appreciation for rendering help in processing the manuscript.

It is expected that the book will of immense help to all section of users like students, researchers, planners all those engaged in R/D activities in the field of fisheries and aquaculture.

Ajit Kumar Roy
Niranjan Sarangi

Contents

STATISTICS

ECONOMICS

Contributors

** denotes the chapter numbers*

A. Antony *[21,31]
Ph.D. Scholar
Perdune University, USA

A. Barat *[20]
Senior Scientist
NRC on Cold Water Fisheries
Bhimtal-263 136, Uttaranchal, India

A. Girija *[25]
Technical Officer (Statistics)
CRIDA, ICAR, Hyderabad-500 059
Andhra Pradesh, India

A. K. Roy *[2,4,5,21,35]
Principal Scientist & Head
Social Science Section
CIFA, Bhubaneswar-751 002
Orrisa, India

A. K. Vasisht *[37]
Principal Scientist & Head
Division of Econometrics, IASRI
Library Avenue, Pusa Campus
New Delhi – 110 012, India

A.G. Ponniah *[36]
Director
Central Institute of Brackishwater
Aquaculture, ICAR, Chennai – 600 028
Tamil Nadu, India

Abhinav Pragya *[41]
Deshbandhu College
University of Delhi
New Delhi-110 019, India

Ajit Kumar Das *[27]
Prof. & Head
B.C.K.V., Mohanpur, Nadia
West Bengal, India

Anil Rai *[19]
Senior Scientist
IASRI, Library Avenue, Pusa
New Delhi-110 012, India

Arun Jana *[17]
Technical Officer
(C-DAC), E-2/1, Block GP, Sector–V
Salt Lake, Kolkata-700 091
West Bengal, India

B.S.Sontakki *[14,38]
Senior Scientist
NAARM, Hyderabad
Andhra Pradesh, India

Bibhudatta Mishra *[16]
Technical Officer
Aquaculture Production & Environment
Division, CIFA, Bhubaneswar-751 002
Orissa, India

Bindu R. Pillai *[10]
Senior Scientist
Aquaculture Production & Environment
Division, CIFA, Bhubaneswar-751 002
Orissa, India

Bipin Tudu *[17]
Lecturer
Department of Instrumentation &
Electronics Engineering
Jadavpur University, Salt Lake Campus
Kolkata-700098, West Bengal, India

Chinmayee Sahoo *5
Researcher
OUAT, Bhubaneswar, Orissa, India

Chinmoy Chakraborty *34
Assistant Director
Office of Jt. Director of Fisheries
Government of West Bengal
60a, Kolkata - 700 073
West Bengal, India

D. B. Ramesh *23
Head, Library & Documentation Division
Regional Research Laboratory (CSIR)
Bhubaneswar - 751 013, India

D.P. Rath *40
Technical Officer (T-4)
Social Science Section, CIFA
Bhubneshwar, Orissa, India

Devdulal Ghosh *17
Technical Officer
(C-DAC), E-2/1, Block GP, Sector–V
Salt Lake, Kolkata-700 091
West Bengal, India

Fahima Dilnawaz *18
Research Associate
Institute of Life Science
Bhubaneswar, Orissa, India

G.P.Reddy *14,38
Principal Scientist
NAARM, Hyderabad
Andhra Pradesh, India

G.R.Maruthi Sankar *25
Principal Scientist & Head
Department of Computer Science &
Application, CRIDA, Hyderabad-500 059
Andhra Pradesh, India

K.K.Chaturvedi *19
Scientist
IASRI, Library Avenue, Pusa
New Delhi-110 012, India

K.N.Sahoo *33
Country Manager, Essar China
Sanghai, China

Kanta Das Mahapatra *29
Senior Scientist
Fish Genetics & Biotechnology Division
CIFA, Bhubaneshwar-751 002
Orissa, India

Kumar Das *32
Prof. & Head
Department of Economics
Utkal University, Vani Vihar
Bhubaneswar, Orissa, India

Lakshmi Murthy *22
Assistant Director (Documentation)
National Institute of Agricultural
Extension Management (MANAGE)
Rajendranagar, Hyderabad
Andhra Pradesh, India

M. Krishnan *36
Principal Scientist & Head
Central Institute of Brackishwater
Aquaculture, ICAR, Chennai – 600 028
Tamil Nadu, India

M. N. Reddy *14,38
Principal Scientist & Head
Computer Section, NAARM
Rajendranagar, Hyderabad
Tamil Nadu, India

Meenakshi Prajneshu *41
Reader
Deshbandhu College, University of Delhi
New Delhi-110 019, India

N. K. Maiti *7,12
Senior Scientist
Fish Health Management Division, CIFA
Bhubaneswar-751 002, Orissa, India

Nabarun Bhattacharya *[17]
Additional Director
(C-DAC), Ministry of Communication & Information Technology, E-2/1
Block GP, Sector–V Salt Lake
Kolkata-700 091, West Bengal, India

Nagesh K. Barik *[39]
Scientist
CIFRI, Barrackpore, Kolkata – 700 120
West Bengal, India

Nibedita Jena *[35,40]
Economic Investigator
Directorate of Economic Research
Khadi & Village Industry Commission
Gramodaya, 3, Irla Road, Vile Parle (West)
Mumbai – 400 056, Maharashtra, India

Nirupama Panda *[2,40]
Technical Officer
Social Science Section, CIFA
Bhubaneswar, Orrisa, India

P. Das *[3,20]
Senior Scientist
Fish Genetics & Biotechnology Division
CIFA, Bhubaneswar-751 002, India

P. K. Katiha *[34]
Senior Scientist
CIFRI, Barrackpore, Kolkata - 700 120
West Bengal, India

P. K. Meher *[20]
Scientist (Sr. Scale)
Fish Genetics and Biotechnology Division, CIFA, Bhubaneswar-751 002
Orissa, India

P. K. Nanda *[8]
Scientist (Sr. Scale)
Eastern Regional Campus
Indian Veterinary Research Institute
Belgachia, Kolkata, West Bengal, India

P. K. Satapathy *[31]
Technical Officer (T - 6)
Zonal Coordinating Unit VI
ICAR, CAZRI Campus, Jodhpur
Rajasthan, India

P. Routray *[11]
Senior Scientist
Aquaculture Production & Environment Division, CIFA, Bhubaneswar-751 002
Orissa, India

P. Swain *[8]
Senior Scientist
Fish Health Management Division
CIFA, Bhubaneswar-751 002
Orissa, India

P. K. Sahoo *[6]
Senior Scientist
Fish Health Management Division
CIFA, Bhubaneswar-751 002
Orissa, India

P. K. Malhotra *[19]
Principal Scientist & Head
Computer Division, IASRI
Library Avenue, Pusa
New Delhi-110 012, India

R. N. Mandal *[12]
Scientist (Sr. Scale)
Aquaculture Production & Environment Division, CIFA, Bhubaneswar-751 002
Orissa, India

Rabi N. Subudhi *[24]
Professor
School of Rural Management (KSRM)
KIIT University, Bhubaneswar - 751 024
Orrisa, India

Rajendra Nath Das *[27]
Technical Officer
B.C.K.V., Mohanpur, Nadia
West Bengal, India

Rajib Bandhopadhyaya *17
Professor
Department of Instrumentation &
Electronics Engineering
Jadavpur University, Salt Lake Campus
Kolkata-700098, West Bengal, India

Ravi R Saxena *26,28
Reader
Indira Gandhi Agricultural University
Raipur- 492 006, Chattisgarh, India

Ritu R. Saxena *28
Lecturer
Department of Biotechnology
Indira Gandhi Agricultural University
Raipur- 492 006, Chattishgarh, India

S. C. Rath *9
Senior Scientist
Aquaculture Production & Environment
Division, CIFA, Bhubaneswar-751 002
Orissa, India

S. D. Gupta *9
Retd. Principal Scientist,
CIFA, Bhubaneswar-751 002
Orissa, India

S. Dash *8
Senior Research Fellow
Fish Health Management Division
CIFA, Bhubaneswar, Orissa, India

S.Mohanty *7,12,23
Senior Scientist
Fish Health Management Division
CIFA, Bhubaneswar-751 002
Orissa, India

S. P. Singh *42
Senior Scientist
National Bureau of Fish Genetics
Resources (NBFGR), Telibagh, Dilkusha
Lucknow-226002, Uttar Pradesh, India

S.T. Rout *32
Department of Economics
Utkal University, Vani Vihar
Bhubaneswar, Orissa, India

Sanjeeb K. Sahoo *18
Scientist
Institute of Life Science
Bhubaneswar, Orrisa, India

Sanjeev Kumar Sahu *15
Scientist
CIFRI, Barrackpore, Kolkata-700 120
West Bengal, India

Saroj K. Swain *13
Senior Scientist
Ornamental Fish Culture Unit
Aquaculture Production & Environment
Division, CIFA, Bhubaneswar-751 002
Orissa, India

Snehasish Mishra *30
Chief Editor
Bioimpulse, Noida
Uttar Pradesh, India

Subhendu Adhikari *16
Senior Scientist
Soil Chemistry Section, Aquaculture
Production & Environment Division
CIFA, Bhubaneswar-751 002
Orissa, India

Suphiya Parveen *18
Researcher
Institute of Life Science, Bhubaneswar
Orrisa, India

Sushma Martha *4,5
Researcher
Bioinformatic Centre, CIFA
Bhubaneswar - 751 002
Orissa, India

T. Madhan Mohan *[1]
Advisor
Department of Biotechnology
Ministry of Science & Technology
Govt. of India, C.G.O. Complex
New Delhi, India

T. Ravisankar *[36]
Senior Scientist
Central Institute of Brackishwater
Aquaculture, ICAR, Chennai - 600 028
Tamil Nadu, India

BIOINFORMATICS

Chapter 1

Growth of Biotechnology Information System Network (BTISnet) in India

T. Madhan Mohan

Need of Bioinformatics

Bioinformatics, an emerging area offering a fundamental tool to the scientific community, to speed up the research, application and commercialization of biotechnology. In fact the best thing, which has happened towards the end of the 20th century and in the 21st century, is the marriage between biotechnologists and information technologists leading to the growth and development of this field. India took a lead as early as in the 1980s but more precisely from 1986 onwards by establishing a strong base of Bioinformatics with the necessary infrastructure. As a result one has seen major breakthroughs in biology and the growth of biotechnology has been phenomenal in the last decade, especially with the most outstanding technological breakthrough of the 20th century wherein the draft human genome sequence was completed in 2000 and now the genetic code is completely sequenced. This has provided the worlds scientific community information on the vast sequence and structure of the Genomes, the crystal structures and there is an increasing dependence on computational approaches. Biotechnology has emerged as a front-line area with vital significance in unraveling secrets of life, particularly in the studies of new biology and biotechnology. The genomic revolution has underscored the central role of Bioinformatics in understanding the very basic of life processes.

Applied Bioinformatics, Statistics and Economics in Fisheries Research: pp. 1-7

New India Publishing Agency, New Delhi (India)

Let us start with the issue of the genetic resources and precious biodiversity of India as also of the world. The need to conserve, protect and utilize them on a sustainable basis is the need of the hour. Genetic resources are made to improve the agricultural productivity and are the vital components of the global environment. Gene Bank concept has become very important, both for ex-situ conservation and the utilization of the valuable germplasm on a long -term basis.

In fact, world over the scientists are associating the computers and the new biology looking at a speed with which data collection is done, automated high throughput data gathering, i.e. sequencing the whole genome and the web based data "Warehouses" integrate the information from the research efforts from all over the world. We know that there will be more than 3-27 million undiscovered species of plants and animals on the planet earth. However, through the endeavor of the scientists, hundreds of new species are discovered and classified every day. The accelerated progress of information technology has completely transformed the research in life sciences. The focus on genome technologies has made it still more relevant to establish very close linkages between computers and experimental biologists. There are many developments, for e.g. Enterprise Bioinformatics, high system computing, genomic data and data base integration and DNA amplification and sequencing, genome assembly and gene types, web based Bioinformatics, biochips and gene expression analysis, Proteomics/and mass spectrometry.

Recent studies have shown that even the implication of mathematical theory in cell biology have become very important. In fact Leroy Hood from the institute for systems biology, in Seattle says "The future will be the study of genes and proteins of organisms in the context of their informational pathways and networks". Similarly the areas of modular biology, synthetic biology is at the forefront of scientific excitement.

The announcement of the remarkable success of the human genome project supported in 1986 and making the draft sequence available to the people was a landmark in the history of modern biology and science. This has generated tools to produce the whole gene catalogues for many microbes, the plant *Arabidopsis thalina*, the fruit fly *Drosophila melanogaster*, the roundworm *Caenorhabditis elegans*, and soon the puffer fish *Fugu rubripes*. The complete genome sequences have almost served as recipes for life and the foundation of modern biology in the 21st century aimed at the biological goals as defined by the scientific community world over as "Achieve a fundamental, comprehensive and systematic understanding of life".

It is important now that scientists work towards characterization of the full repertoire of molecular machines of the living systems. Also they must understand how these machines are orchestrated with the life system of single cell and complex multicellular organisms.

Concept of Bioinformatics: Bioinformatics or Life Science Informatics is a young, fast developing science branch, dealing with storage, analysis integration and simulation of molecular biological data. Bioinformatics is an applied discipline that utilizes computational tools to conceptualize biology. It has emerged as a cutting edge technology and a knowledge revolution. Extraordinary growth of information technology and unprecedented advances in molecular biology and recombinant *DNA* techniques have ushered in the age of Bioinformatics.

Human kind is on the brink of another revolution. There is no doubt that the mapping of the human genome, completed in June 2000, is one of the greatest scientific advancements in history. This breakthrough in biological research was made possible by advancements in using Bioinformatics and Computational Biology. Areas such as proteomics, genomics, combinatorial chemistry, statistics, nanotechnology, spectroscopy and structural and computational Biology will have increasing applications of Bioinformatics in days to come.

Goal of Bioinformatics: The major goal of bioinformatics is to obtain the complete sequences of as many different genomes as possible. By having that sequence information, companies and research organizations can start to do what is commonly referred to as "sequence-based biology." They take that sequence information, and use it to give scientists more direction as to how they should design experiments, and how these scientists should analyze the experiments. So bioinformatics is causing a fundamental shift in how scientists actually approach molecular biology. The "Holy Grail" for bioinformatics is to map all the genes in the human body and decipher what the role each gene and its associated base pair plays in the expression of the gene.

Role and Relevance in Biotechnology: Bioinformatics is becoming increasingly important due to the interest of the pharmaceutical industry in genome sequencing projects. There is a vital need to harness this information for medical diagnostic and therapeutic uses, and there are opportunities for other industrial applications. This field is evolving rapidly, which makes it challenging for biotechnology professionals to keep up with recent advancements. The area has evolved to deal with four distinct problems viz. (i) Handling and management of biological data, including its organization, control, linkages, analysis, and so forth. (ii) Communication among people, projects, and institutions engaged in the biological research and applications. The communication, which includes e-mail, file transfer, remote login, video conferencing, electronic bulletin boards and establishment of web-based information resources; (iii) organization, access, search and retrieval of biological information, documents, and literature; and (iv) analysis and interpretation of the biological data through the computational approaches including visualization, modeling & simulation, and development of algorithms for highly parallel processing of complex biological structures. The following are the thrust areas of Biotechnology, which are fully dependent on Bioinformatics:

Comparative Genomics: The assumption that the similarity of two sequences whether it is DNA, RNA or protein implies functional correlation. Some of the most successful bioinformatics applications deal with this kind of analysis. Several software tools have been developed like BLAST that efficiently performs sequence-alignments against large databases of known sequences. It became a routine task to compare a new sequence against several databases.

Functional Genomics: To investigate genes in their cellular context, expression analysis via Microarray and DNA-chips takes place. The comparison of expression patters of well-defined metabolic states allows identifications of pathological phenotypes on a molecular level. Based on the identifications a reverse engineering of pathways and localization of pharmaceutical targets is possible.

Proteomics: The publication of entire genome sequences led to a shift of interests from pure DNA sequencing to Protein localization and characterization within their cellular context. This became necessary as one gene can give rise to a number of products via mechanisms such as splicing or post-translational modifications. The proteome refers to the identification and analysis of all proteins of a cell. This involves the determination of structures as well as the identifications of protein interactions and function in biological pathways.

Structural Genomics: Structural Genomics covers the calculation of three-dimensional structures based on the sequence of a macromolecule. The theoretical basis of the relationship between sequences and structure is the most fundamental problem of in silico biology. From a computational point of view the so-called folding problem is the most demanding objection of computational biology. Nevertheless only the knowledge of the structure of a protein can provide a deeper understanding of its function.

Pharmaco Genomics: The development of drugs aims to maximize effect and minimize side effects. It would therefore be very convenient to personalize drugs for each patient. The genetic variations among all human is only 0.1% of the total DNA. The differences are mostly point mutations, having phenotypic impact. These so-called SNPs (Single Nucleotide Polymorphisms) become good candidates for drug development and diagnosis. With the investigation of SNPs using Microarrays the investigation of entire populations becomes feasible. Databases of SNPs will reveal the patterns that cause cancer or Alzheimer.

Cellomics or System Biology: If sufficient data is available and all relevant components for life are identified more complex interactions can be investigated. For a holistic biological understanding of cell, simulations of cells, entire organisms and populations provide new insights. The simulation of life *in silico* is a future directive for bioinformatics that started now.

Bioinformatics Network in India : India's achievements and strategic advantage in information technology and biotechnology have positioned the country to play a crucial role in the post-genomic era. This role would range from analysis of genes and structures, defining metabolic pathways to identification and validation of drug targets and classification of the natural resources based on genetic profiles. Biotechnological applications in agriculture, medicine and environment protection and biodiversity conservation have not only scientific relevance but also offer opportunities for overall sustainable development, most desirable for 21st Century. The countries rich biodiversity and bioresources can reap the harvest from these scientific breaks through.

With the objective of integrating the country's vast scientific expertise in the different areas in biotechnology and computational sciences, the Department of Biotechnology envisioned in 1986 the establishment of a distributed information network. The network was to provide a common platform for exchange of information among the scientific community and in the process speed up the scientific discoveries. The network, referred to as the Biotechnology Information System Network, (BTISnet) presently consists of sixty-one centres spread all over the country. Over the years, the network has created a resource base including computational infrastructure, databases and software. Human resource development through short and long-term programmes, research and applications in frontier areas of Bioinformatics and biotechnology, are the main objectives of the Bioinformatics programme.

The BTISnet has established a link among scientists and organizations involved in research and development activities in Biotechnology. The network today offers a single window information resource in the country covering inter-disciplinary areas of biotechnology and molecular biology. Six National Facilities on Interactive Graphics are dedicated to the promotion of molecular modeling and other related activates. More than 50 subject specific databases have been developed through the BTISnet. Several major International databases for application to genomics and proteomics have been established in the form of Mirror sites as part of the network. These databases are being linked through high speed and large bandwidth network in the form of VPN called BIOGRID INDIA to promote faster sharing of information as well as to encourage tandem research by various R&D labs and industrial units. These sites are designed to act as knowledge pathways for discoveries in biotechnology. At the micro level, the programme deals with the various issues related to the biological data. It covers the development of data analysis tools, modeling of biological macromolecules and their complexes, metabolic pathways, designing of new molecules such as drugs, peptide vaccines, proteins etc.

Human Resource Development

Advance Diploma Course: The skilled human resource in bioinformatics is of utmost importance for effective implementation of the high throughput programmes in biotechnology. The BTISnet is running five long-term advanced diploma courses in bioinformatics at the post-M.Sc. level through leading universities of the country: Madurai Kamaraj University, Madurai; University of Pune, Pune; University of Calcutta, Kolkata; Jawaharlal Nehru University and Pondicherry University. The number of seats at each University has been increased based on the demand from the industrial sector. Each course is now producing approximately 15-20 candidates per year. Every year more than 60 trained professionals are being generated through this programme.

Post-Graduate Degree Courses: Post-Graduate courses such as M.Sc., M.Tech and PhD in bioinformatics have been introduced through Pune University, IIIT-Allahabad and JNU, New Delhi respectively to promote leadership quality human resource in the area of bioinformatics.

Short-Term Training: The BTISnet is also supporting and organising several short-term training and workshops in front-line areas of bioinformatics and Biocomputing all over the country. A large number of scientists and research scholars from all over the country are making use of these training programmes of which more than 70 have been conducted this year.

R & D on Bioinformatics: The Bioinformatics Centres are being extensively used for intensive research by the hosts and neighboring institutions. The thrust areas provided are sequence analysis and molecular modeling. Apart from supporting research, the centres are also actively engaged in research on frontier areas of bioinformatics. Intrinsic research in the bioinformatics centres includes gene analysis, protein structure prediction & engineering, modeling macromolecules assembly, evolutionary biology and mechanisms of disease. Under the programme several R&D projects have been supported and the increase of R&D proposals in Bioinformatics has been witnessed in this year. The Department is considering this area as a thrust area to extend more support towards development of Bioinformatics tools.

According to the views of the scientific community world over, there are four goals before science, namely,

i. Identify and characterize the molecular machines of life- the multiprotein complexes that execute cellular functions and govern cell form.

ii. Characterize gene regulatory networks.

iii. Characterize the functional repertoire of complex microbial communities in their natural environments at the molecular level.

iv. Develop the computational methods and capabilities to advance understanding of complex biological systems and predict their behavior (Genes, proteins and molecular machines).

While it is very fascinating and exciting to link the new development in biology emerging out of the tools of Bioinformatics, all aspects relating to ELSI (Ethical Legal Social Implications) of modern research and its application are equally important. Biosafety of genetically modified products, aspects of environmental risks and assessment, Intellectual property rights etc. have become an integral part of biology today.

Chapter **2**

Capacity Building for Knowledge Management in Aquaculture Research

A.K. Roy & Nirupama Panda

Introduction

In this era of international competitiveness of knowledge economy, there is an emergent need for identifying new knowledge streams likely to emerge for mapping the directions for transition of knowledge from industry to information technology. National Knowledge Commission, Government of India has undertaken a survey of agricultural sector for exploring and discovering the dynamic role of innovation of knowledge for growth and competitiveness and recommended for ascertaining the role of the Governmental factors as innovation enablers in development of Indian economy (Anon, 2006). The knowledge network and development of knowledge infrastructure in the field of information and communication technology lead for knowledge generation, knowledge transfer and it's preservation as intellectual property right which will support for adoption of effective knowledge management practices and provide guidelines for envisioning the policy makers to formulate appropriate plans in the filed of research and development.

Knowledge Management caters to the critical issues of organizational adoption, survival and competence in face of increasingly discontinuous

Applied Bioinformatics, Statistics and Economics in Fisheries Research: pp. 9-22

New India Publishing Agency, New Delhi (India)

environmental change. Essentially, it embodies organizational processes that seek synergistic combination of data and information processing capacity of information technologies, and the creative and innovative capacity of human beings (Malhotra 1997).

Central Institute of Freshwater Aquaculture (CIFA), Bhubaneswar, Orissa is a leading research institute under Indian Council Agricultural Research (ICAR) system is involved in research and development in freshwater aquaculture. Embodied with rich potentials of freshwater resources, there is an ample scope for production of freshwater fish in India. Hence, the emergent research needs for knowledge management of organizational processes that seek synergistic combination of information processing, simulation of data, generation of technologies and building of creative & innovative capacity for human resource development.

The CIFA regularly conducts research, development and capacity building activities related to freshwater aquaculture with the goal of improving fish production through natural resource management and various multidisciplinary fields, *viz.* physiology, nutrition, pathology, breeding and genetics. The institute is creating innovative approaches to translate natural resource management research into actionable recommendations for policy makers, resource managers, fish farmers and poor rural communities. The Institutes constantly endeavors for ensuring the organization to achieve its strategic goals and objectives by integrating communications and knowledge sharing, e-publishing, e-library and information technology. The institute is involved in monitoring and sharing the information through electronic media technologies within the organization and between the regional research centers, fisheries research institutes and fisheries colleges. The institute is also trying to mobilize knowledge management across the organization with the aim of providing rapid access to useful information for researchers throughout the country.

Bioinformatics Network in India

India's achievement and strategic advantages in Information Technology and Biotechnology have positioned the country to play a crucial role in the Bioinformatics Network System. With the objective of integrating the country's vast scientific expertise in different areas of Biotechnology and Computational Science, the Department of Biotechnology envisioned in 1986 the establishment of a distributed information network. The network is to provide common platform for exchange of information among the scientific community and in the process to speed up the scientific discoveries. The network referred to as the Biotechnology Information System Network (BTIS net), presently consists of sixty-five BTIS centers and fifty Bioinformatics Facility (BIF) spread all over the country and these are attached with various national institutes of Indian Council of Agricultural Research, Council of Scientific and Industrial Research, State Agricultural Universities etc. Over the years, the network has created a resource base including computational infrastructure, databases and software. Human resource development through short and long-term training programmes, research and applications in frontier areas of Biotechnology and Bioinformatics are the main objectives of the Bioinformatics center. BTIS center on aquaculture is the only center in the country located at Bhubaneswar. Over the years the center created the following infrastructure.

Knowledge Infrastructure

Hardware: The institute is equipped with Informatics-based infrastructure that includes the latest P-4 Multimedia Computers, 1, 3 & 5 KVA online UPS, Dot Matrix/ DeskJet, HP LaserJet 1320 N (Network) Printers, LCD Projector, Colour Scanner, Modem, 5.5 KVA Generator, network switches, routers etc.

Software: Presently software like MS Office 97, 2000, Novel Netware 4.1, SPAR1, FoxPro, QPRO, Oracle, Visual Basic, M.S. Project, Visual Studio are available in the Institute. The Operating Systems are: DOS, Windows 95, 98, UNIX, Linux, CD Mirror Server for bibliographic search services through LAN.

Communication facilities: BSNL Lease Line with 512 kbps bandwidth is installed and is now under operation for faster Internet browsing and e-mail uploading and downloading. The centre is now operating a VSAT sky blaster with 256 kbps bandwidth for Internet and e-mail uploading and downloading. BSNL broadband connectivity and SLIP/PPP connectivity from NIC for Internet browsing is maintained at the BTIS center of CIFA. More than one hundred computers are distributed to different divisions/sections of the Institute that are connected through Local Area Networking (LAN) for sharing of information resources.

Scientific Software Packages: Quantity one (gel analysis), DNA Star, Reference Manager and UVI - Software (gel analysis) are available in the institute.

Statistical Packages: Statistical Packages (Statistical Package for Social Science (SPSS), Version 14.0 and Statistical Analysis System (SAS), Version 2.31) are available for computational work and statistical analysis of vast array of data generated from research activities involving various disciplines.

Databases subscribed: (1) ASFA (Aquatic Sciences and Fisheries Abstracts) online version covering 1971 to present and offline CD ROM from 1971 to

2003 are provided to scientists and research workers of the Institute for online/ offline retrieval of bibliographic information resources. Fish Base, The FAO Fisheries Atlas, 2001, Diagnosis of Shrimp Disease, 2000, FAO - ICLARM (Stock assessment tool), FishstatPlus (V 2.30) – FAO, Rome, 2004, Archives of the Bay of Bengal Programme 1979-2000 are also available for retrival of information for carrying out research work.

Databases Created

(i) *Database on socio-economic status of fish farmers at Kolleru Lake Area:* This database has been developed inhouse. This covers the socio-economic parameters like fish farmers' personal details, assets required, loans available, assistance provided, insurance offered, types of species used for culture, feeds, chemicals, fertilizers used, disease occurred, drugs used, technology used, training, water management etc.

(ii) *Database on socio-economic status of fish farmers at Kolleru Lake Area (Pictorial):* The database provides in-depth details of socio-economic conditions of fish farmers, which will support in boosting up the scope of result oriented research and development of fish farmers. It would support the researchers and the policy makers in decision-making.

(iii) *Fish Disease Database Information System*: The CD-ROM being the first of its kind was made in collaboration with IVLP- TAR-NATP of CIFA, which contains Fish Disease Database Information System for aquaculture farms. It includes standardized definitions of disease events, systematic classification of diseases, standard disease indices for recording and processing of epidemiological data.

Fish related database available free (online): Listed below are some of the online databases containing a wide range of information resources related to fisheries. Browsing of the databases can help research workers immensely. (i) Global information system on fishes (*http://www.fishbase.org*), (ii) Online information for medaka rice fish (*www.bio1.bio.nagoyau.ac.jp.8000*), (iii) Information on the pocciliid fish xiphophorus (*www.xiphorus.org*).

Learning and Knowledge Sharing

Manpower Development in Bioinformatics through workshop/ training/ teaching/ offering studentship and traineeship: National Workshop cum Training Programs on Bioinformatics and Statistics in Aquaculture Research is being organized every year, which is a boost for human resource development. The students and trainees are also appointed under the BTIS project for providing training and professional knowledge in the field of Information Technology and Bioinformatics. Many state and central government officials engaged in various R&D activities in the field of aquaculture are being trained.

listservs and discussion forums : A survey of listservs and discussion forums related to aquaculture was conducted to facilitate active participation and contribution to the world's growing digital knowledge bank that has the potential for knowledge sharing in the area of aquaculture. One way to facilitate for discussion is through listservs or electronic discussion forums. Details of email addresses of various listserves or electronic discussion forums are noted below (Murthy, 2006).

Listservs/Discussion Forums (clearinghouse@seaweb.org), AQUANIC: Aquaculture Network Information Center (http://aquanic.org), Aquaponics: (http://groups.yahoo. com/group/Aquaponics/), AquaBiz: The Aqua Business Club (http://groups.yahoo.com/ group/AquaBiz/), Aqua-culture (http://groups.yahoo.com/group/Aqua-culture/), Aquaculture (http://finance.groups.yahoo.com/group/aquaculture/) Network of Aquaculture Centres in Asia-Pacific Forums (http://www.enaca.org/modules/ newbb/index.php), WATER-FORUM(waterforum-subscribe@egroups.com), AQUA-L (majordomo@killick.mi.mun.ca,FISHFOLK(listserv@mitvma.mit.edu), FISH-JUNIOR (listserv@searn.sunet.se), OCEANF-L(listserv@mitvma.mit.edu). Virtual Library on Knowledge Management Network: World's most reputed knowledge management resource and it holds award winning content, community, research and practices (http://www.brint.com/km)

Web page: Keeping in pace with the latest technologies designed, developed and hosted the content created with independent domain [http://www.cifa.in] covering broadly different aspects like Profile, Director's Office, Staff, Division, Research Projects, Collaborators, Research Contribution, Post Harvest Technology, Transfer of Technology, Publications, Consultancy, BTIS, Training Programme, Statistics, Products, Recognition alongwith Useful Links.

A brief information about the institute based and externally funded projects both ongoing and completed covering the areas of fish breeding, genetics, nutrition, physiology, biotechnology microbiology, pathology, integrated nutrient management, immunology, aquaculture diversification and wastewater management, network project on germplasm exploration & conservation of fish and shellfish resources of India, post harvest processing and value addition of freshwater fish and shellfish, ornamental fish culture and pearl culture, cryopreservation and broodstock characterization and socio-economic research in freshwater aquaculture is presented in the web page. In the list, some of the important externally funded (national and international) projects are Biotechnology Information System on Aquaculture (DBT funded), Application of plastics in agriculture-plant environment control and agricultural processing, Development of protocol for organic farming of carps and freshwater prawns, Sustainable aqua feeds to maximize the health benefits of farmed fish for consumers (European Commission), Economic and livelihood development of SC/ST population through freshwater aquaculture technologies (DBT), Application of microorganisms in agriculture and allied sectors (NBAIM), Impact assessment of fisheries research in India (AP Cess), Studies on association of viral infections in diseases of Koi carps by immunological and molecular techniques (AP Cess). Publication includes list of books, annual reports and newsletters incorporating research achievement.

Knowledge Creation

As a nodal research center, the institute generates state-of the art knowledge that emphasizes responsiveness and relevance by applying knowledge and distilling the lessons of experiments into knowledge that can serve as a platform for policy recommendations for implementation in the form of technologies throughout the whole country. Brief account of the following technologies generated in aquaculture are presented in the website (www.cifa.in) for knowledge sharing and dissemination to the interested users.

1. Carp breeding and hatchery management
2. Production of carp fry and fingerlings
3. Grow-out carp culture
4. Selective breeding of rohu (Labeo rohita)
5. Cryopreservation of carp milt
6. Portable FRP carp hatchery
7. Breeding and culture of magur (Clarias batrachus)
8. Hatchery technology of freshwater prawn, *Macrobrachium rosenbergii*
9. Culture technology of freshwater prawn, *Macrobrachium rosenbergii*
10. Seed production and grow-out culture of Indian river prawn, *Macrobrachium malcolmsonii*
11. Freshwater pearl culture
12. Ornamental fish farming
13. Wastewater treatment through aquaculture
14. Commercial feeds for carps, magur and freshwater prawns
15. Control of Epizootic Ulcerative Syndrome with 'CIFAX' formulation
16. Diagnostic kits, immunodiagnostic aids and therapeutics for diagnosis and treatment of fish diseases

Transfer of Technology

The Institute has been actively engaged in transfer of the developed techniques and technologies in different aspects of freshwater aquaculture through research publications in scientific journals, publication of annual reports, technology manuals and pamphlets, organisation of national and international training programmes, organisation of national symposia and workshops, participation of the Institute and its staff members in national and international symposia and exhibitions, demonstration in farmers' ponds, etc., apart from on-campus and off-campus training programmes of Krishi Vigyan Kendra (KVK). The Agricultural Technology Information Centre (ATIC) as a sub-component of the Innovations in Technology Dissemination under National Agricultural Technology Project (NATP) is being operated at the Institute to provide services viz., water and soil testing, diagnostic services, planting materials, fish seed, live stocks etc. through a single window system to farmers and entrepreneurs throughout the country. Technology was transferred to the farmers and entrepreneurs of the tribal and hilly areas of Chhattishgarh and Orissa through Jay Vigyan National Science and Technology Mission on "Household food and nutrition security" and Institute Village Linkage Project for technology assessment and refinement.

Knowledge Management Applications

The Bioinformatics Center of CIFA supports the system of development of knowledge learning and knowledge management applications, document storage and retrieval system of databases. In cooperation and collaboration with all the offices of its regional centers, the institute is making constant effort for alround development in knowledge management. It monitors compliance with knowledge management practices and progress towards Institute's work plan and mandate for disseminating the technologies at the grassroot level.

Knowledge products: Major initiative has been taken for digitization of the publication aiming at e-governance in future activities to keep pace with latest trends.

e-library: The computerized system in the library facilitates the researchers as well as the students for ready reference. CDS/ISIS package is used for author-wise/ title-wise/discipline-wise search of books/journals available in the library.

e-publishing : The following titles of publications in the form of CD-ROM are created.

- *Book on Bioinformatics and Statistics in Fisheries Research, Vol-3, 2007.* Central Institute of Freshwater Aquaculture (CIFA), Kausalyaganga, Bhubaneswar.
- *Book on Bioinformatics and Statistics in Fisheries Research, Vol-2, 2006.* Central Institute of Freshwater Aquaculture (CIFA), Kausalyaganga, Bhubaneswar.
- *Book on Bioinformatics and Statistics in Fisheries Research, Vol-1, 2004.* Central Institute of Freshwater Aquaculture (CIFA), Kausalyaganga, Bhubaneswar.
- *Compendium of Lectures for National Workshop Cum Training on Bioinformatics and Statistics in Aquaculture Research* held during February 7-10, 2006 at CIFA, Bhubaneswar.
- *Compendium of Lectures on Training Programme on Recent Advances in Freshwater Aquafarming* held at CIFA, Kausalyaganga, Bhubaneswar.
- *National Workshop cum Training on Recent Developments in Bioinformatics and Statistics in Aquaculture* (Proc. of workshop held during Feb. 10-13, 2004 at CIFA, Kausalyaganga).
- *Aquaplosion* – Multimedia presentation of the activities of CIFA, 2001
- *National workshop cum training on Application of Bioinformatics and Statistics in Aquaculture* (Proc. of workshop held during Feb. 4-7, 2003 at CIFA, Kausalyaganga)
- *National workshop cum training on Bioinformatics and Statistics in Aquaculture Research* (Proc. of workshop held during Jan.8-11, 2002 at CIFA, Kausalyaganga)
- *National workshop and training on Bioinformatics and Statistics in Aquaculture Research* (Proc. of workshop held during Jan. 16-19, 2001)
- *National workshop and training on Bioinformatics and Statistics in Aquaculture Research* (Proc. of workshop held during Feb. 8-11, 2000)
- *National workshop cum training on Bioinformatics and Statistics in Aquaculture Research* (Proc. of workshop held during Feb. 2-5, 1999 at CIFA, Kausalyaganga).
- *CIFA at a Glance, 2000, CIFA, Kausalyaganga.* The CD contains all round activities of CIFA in research, training & extension in the field of freshwater aquaculture development in the country. There are about 600 self explanatory photographs depicting infrastructure, cultivable species, breeding, culture, harvest, marketing, training, extension, etc.

Publications

Books/Compendium of lectures

- Book entitled "Bioinformatics and Statistics in Fisheries Research", Vol-3, 2007;CIFA (ICAR), Kausalyaganga, Bhubaneswar, pp.556
- Book entitled "Bioinformatics and Statistics in Fisheries Research", Vol-2, 2006; CIFA (ICAR), Kausalyaganga, Bhubaneswar, pp.516
- Book entitled "Bioinformatics and Statistics in Fisheries Research", Vol-1, 2004; CIFA (ICAR), Kausalyaganga, Bhubaneswar, pp.473
- A compendium of lectures of National Workshop cum Training on Bioinformatics and Statistics in Aquaculture Research held during February 7-10, 2006, pp.399.
- Proc. National Workshop cum Training Programme on Recent Developments in Bioinformatics and Statistics in Aquaculture" held at CIFA, Kausalyaganga during –10-13, Feb., 2004. pp.270.
- Proc. National Workshop cum Training Programme on Application of Bioinformatics and Statistics in Aquaculture" held at CIFA, Kausalyaganga during 04-07, Feb., 2003. pp.270.
- Proc. National Workshop cum Training Programme on Bioinformatics and Statistics in Aquaculture Research held at CIFA, Kausalyaganga during 08-11, Jan., 2002. pp.185.
- Proc. National Workshop cum Training Programme on Bioinformatics and Statistics in Aquaculture Research held at CIFA, Kausalyaganga during 16-19, Jan., 2001. pp.258.
- Proc. National Workshop cum Training Programme on Bioinformatics and Statistics in Aquaculture Research held at CIFA, Kausalyaganga during 8-11, Feb., 2000. pp.200.
- Proc. of 'National Workshop Cum Training on Bioinformatics and Statistics in Aquaculture Research' held during February 2-5, 1999 at CIFA, Bhubaneswar. pp.175.
- Proc. National Workshop on Information Technology in Aquaculture Research held during February 10-13, 1998 at CIFA, Bhubaneswar. Pp.88.
- Proc. Networking and Biological Data Analysis held during February 4-6, 1997 at CIFA, Bhubaneswar. pp.137
- Proc. Perspectives in Bioinformatics and Its Application to Aquaculture held during 22-26 February 1994 at CIFA, Kausalyaganga, and Bhubaneswar. pp.132.

Research Management Information System

ARIS : Agricultural Research Information System (ARIS) cell is being established at CIFA aiming to bring information management culture so that fishery scientist can carry out research more effectively by having systematic access to research information available in India as well as in other countries, better project management of agricultural research, and modernization of the office tools. The basic infrastructure like hardware, software, databases and communication facilities required for linking all ICAR institutes and SAU's has already been created. The E-mail connectivity has been established at the institutes by linking through VSAT connectivity using ERNET and BSNL lease line services. Efforts are on to implement all the four information modules namely Agricultural Research Personnel Information System (ARPIS); Agricultural Research Financial Information System (ARFIS); Agricultural Research Library Information System (ARLIS) and Agricultural Research Management Information System (ARMIS) to maintain uniformity in functioning in administrative and financial aspects.

PERMISnet: Personnel Management Information System Network (PERMISnet) has been developed by IASRI for ICAR set up which provides information for systematic planning and development of human resources in ICAR. This system is maintained at the Headquarter of ICAR and information is updated regularly at the institute level on various attributes of different categories of personnel. The information is collected on scientists, technical personnel, administrative and supporting staffs which documents the details about personal information, family details, qualification, experience, deputation, visit abroad, projects undertaken by the individual scientist, awards received and submission of ACR and property statement, languages known, study and sabbatical leaves availed etc. The system also maintains various reports on cadre strength (sanctioned and in position), Institute's report (Division wise), list of person's who have retired etc.

Scientist's Progress Monitoring System: Scientist's Progress Monitoring System was developed by the ICAR, New Delhi and is maintained by the ARIS cell of CIFA for regular updating and communicating half yearly target and progress report of the scientists to the ICAR Hqrs., New Delhi.

Survey of software related to aquaculture/fisheries

A study was undertaken to update knowledge about different areas of automation in the field of aquaculture and fisheries research as well as to utilize available software and to develop software/databases specifically needed for aquaculture research. Different available software packages available globally were broadly classified into following categories: *Disease diagnosis and treatment:* Hawaii Aquaculture Module Expert System (HAMES), Fish Vet, Samakia, Diagnosis of shrimp diseases, etc. *Management decision making and record keeping:* Shrimp farm management, AbaSim, MICROFISH, CAGEAN, GENMOD, FISHPARM, FISHCALC89-DISBCAL89, AquaCalc 1, Promanager, Harvest Optimizer, Water Wizard (for aquaculture farm management), Aquadab, POPDYN, Fishy, Pond Class, Pond 3, ECOPATH I & II, CARP Plus and CARP plus II, *SAMCALRE 2.1,* FISHY 3.2, Fisher's associate, Fastfish, Fisher's Guide, FishTrek, FISHING EXPERT, Trout Dynamics, Fisher's Associate. Environmental regulation: DSS for Mari culture Regulation, Modeling the Trophodynamics of the Black Azor Sea Ecosystem, Hydrotech, Fishing Net, SURPH Nutrition: MAXIMS, PREFER, Sam Calc, Fishbase (Frocse and Pauly, 1997) Climprod - Plus, Nutribase etc. *Facility design and management planning:* Aquafarm, Pond 4.0, NID-Tech, ASPIC, etc. *Facility construction:* Aqua CAD, Levee. *Development planning and project appraisal*: The aquaculture-planning simulator (APS), Aquaculture Development Decision Support System (ADDSS). *Financial analysis & economic analysis*: A guide to the financial analysis of shrimp farming, Aqua Sim PC, Soft crab Education program, Shellfish farm Spread sheet and Economic Analysis Software Package, Beam 1 and 2, Beam 3, Beam 4. *Statistical software (FAO)*: Fishstat Plus, Conversion factor, FAO/ICLARM stock assessment tool, Globe fish databases, NASIS, Species dab, Spatial, Thompson & Bell yield analysis, Fishers, Fishery Fleet, Importer, Listal, Popdyn, etc. *Bibliographical Databases*: ASFA, The FAO Fisheries Atlas, Fishbase, Cities under the sea, DIAS, Encyclopedia of Hydroponics gardening. *Aquarium Maintenance*: Tankminder, AquaScape, Aquarix, etc.

Data Warehouse, Data Mining, Exploratory Data Analysis and Visualisation

A data warehouse is a subject-oriented, integrated, time-variant and nonvolatile collection of data in support of management's decision-making process. Data mining is an interdisciplinary field, the confluence of a set of disciplines including database systems, statistics, machine learning, visualization and information science. Data mining is the process of extracting meaningful relationships from usually very large quantities of seemingly unrelated data. Specialized data mining tools allow researchers to perform complex analyses and predictions on data (Thainese, 2006). Knowledge Discovery in Databases (KDD) includes data cleaning, data integration, data selection, data transformation, data mining, pattern evaluation and visualization.

Realizing the importance of Knowledge Discovery in Databases, Data Warehouse and Data Mining technologies, the database on genomic sequence and protein analysis is in major operations with the systems of "Online Transaction Processing Systems"(OLTP) and "Online Analytical Processing"(OLAP) for efficient management and decision-making process. In the field of Bioinformatics, searching nucleotide sequences and visualizing protein-folding patterns to simulating complex 3-D protein-protein interactions, robotic gene sequencing machines are helpful in processing of numeric computation (Rai, 2006). In the area of exploratory data analysis the following areas are covered at CIFA.

(i) Development of Biophysical Models for *Catla catla L. rohita, H. molitrix, C. idella* and *C. mrigala:* Developed biophysical models identifying and quantifying various factors responsible for the growth of different species of carps reared in sewage fed ponds through multivariable exploratory data analysis (Anon, 2005).

(ii) Factor analysis was performed on the data set of sewage-fed fish culture pertaining to the growth of *Catla catla* in order to bring down the factors capable of explaining most of the variances. (Anon, 2005)

(iii) Path analysis was done to identify and visualize the direct and indirect effect of various biotic and abiotic factors on the Specific Growth Rate (SGR) of silver carp (Anon, 2005).

(iv) Statistical Analysis of breeding experiments on *L. rohita:* Data on breeding experiments (421 sets) of *L. rohita* conducted during 1970-1982 were computerized and analyzed statistically to study the breeding performance of the species in relation to biotic and abiotic factors. (Anon, 2005)

(v) Multivariate analysis of breeding data: In order to assess the possible effect of different independent variables like weight of female, doses of hormone, temperature, quantity of eggs released and percentage fertilization on the dependent variable like spawn recovery; a multivariate approach of data analysis was taken using SPSS package (Anon, 2005).

(vi) Application of Statistics in Bioinformatics: For the first time initiated research work on application of statistics in Bioinformatics related to cultivable fresh water fish species. So far the following studies are made based on secondary data downloaded from various sequence databases using statistical package, SPSS (14.0) and Bioinformatics tools like Codon Usage Software of Sequence Manipulation Suite.

a) Correspondence Analysis in study of Genomic Sequence of *Labeo rohita* to study the association between gene and codon

b) Study of Codon Usage variation in *Labeo rohita* taking into consideration fully expressed genes

c) Cluster analysis of Microarray data of *C. carpio* using K-means clustering approach

Knowledge Networks

Aquatic Sciences and Fisheries Abstracts (ASFA): Aquatic Sciences and Fisheries abstracts (ASFA) is an International Cooperative Information System which comprises an abstracting and indexing service covering the world's literature on the science, technology, management and conservation of marine, brackish water and freshwater resources and environments including their socio-economic and legal aspects.

FAO GLOBEFISH: FAO GLOBEFISH is the unit in the FAO Fisheries department responsible for information on international fish trade

FAOSTAT: It is an online multilingual database currently containing over 3 million time-series records from over 210 countries and territories covering agriculture, nutrition, fisheries, forestry and food aid.

Fishery Resources Monitoring System (FIRMS): FIRMS, a global monitoring system, provides access to a wide range of high quality information on the status and trends of marine fishery resources including their management.

Knowledge dissemination : Knowledge creation activities of the institute over the plan period have an associated dissemination strategy, with knowledge products increasingly tailored to the needs and interests of different product users. For effective dissemination and outreach, the institute has made considerable progress in reaching a wider spectrum of users with its knowledge products. In the form of books/compendium of lectures and electronic publications; the institute has rendered its help in disseminating proper, clear and relevant knowledge in various technologies and skills to the researchers working inside and outside the state.

REFERENCES

Anon, 2005. Final Report (RPF-III), Social Science Research in Aquaculture (1996-2005), CIFA, Kausalyaganga, Bhubaneswar.

Anon, 2005-06. Annual Report, Central Institute of Freshwater Aquaculture, Kausalyaganga, Bhubaneswar, Orissa, India.

Fr. Thainese, S. J. and V.Kannon, 2007. Design and Techniques of Data Mining in Biotech Research. *In:* Bioinformatics and Statistics in Fisheries Research, Vol.-III, A. K. Roy & N. Sarangi (Eds.), pp: 74-89

Anon, 2006. http://knowledgecommission.gov.in/consultations/sy.innovation.asp

http://www.cifa.in

http://www.food.gov.uk/science/ouradvisors/toxicity/cotmeets/49737/49750/49831

http://www.iasri.res.in/permisnet

Malhotra, Yogesh. (1998). Knowledge Management, Knowledge Organizations & Knowledge Workers: A View from the Front Lines [WWW document]. URL: http://www.brint.com/interview/maeil.htm

Pascale Baelde, 2003. Putting Fishers Knowledge to work. *In:* Haggan, N., Brignall, C. and Wood, L. (eds.). Proc. of a conference held at the Fisheries Centre, University of British Columbia during august 27-30, 2001. pp. 78-86.

Rai, Anil (2006). Data Warehousing – An Introduction. *In:* Bioinformatics and Statistics in Fisherics Research Vol.-II, A. K. Roy & N. Sarangi (Eds.), pp.63-76.

Roy, A. K. and N. Sarangi, 2007. Status of Bioinformatics Centre on Aquaculture. *In:* Bioinformatics and Statistics in Fisheries Research Vol.-III, A. K. Roy & N. Sarangi (Eds.), pp.9-25.

Chapter 3

Data Generation Resources for Bioinformatics

P. Das

Introduction

The biological information that are being generated by research in modern biology worldwide is so huge that it is not possible to manage through conventional file keeping or simple computing. Secondly, analysis of these data certainly needs specific interfaces, software and management facilities. Therefore, such requirements to store, analyze and manage the biological information/data in an appropriate way has led to the origin of a new branch of science called "Bioinformatics". Here, emphasis will be given on the disciplines of research mostly from which generation of data is made. Broadly, they can be classified as follows.

Genomics

Genomics deals with structural and functional analysis of genes in the genome. Often it is described under two sub-disciplines namely structural and functional genomics. Human genome was being expected to contain about 20000 - 30000 genes. The 15-year human genome project that got completed recently has elucidated that there are about 29,000 genes. The objective of this project was to map and sequence all the genes in 23 pairs of chromosomes. The goals of the project have been divided into smaller tasks, one of which is

Applied Bioinformatics, Statistics and Economics in Fisheries Research: pp. 23-28

New India Publishing Agency, New Delhi (India)

to develop genetic linkage maps to establish markers on each of the 46 human chromosomes in order to work on smaller and more clearly defined sectors in the search for genes. A second task is to get a physical map of overlapping clones of large DNA fragments that will span all the chromosomes. The map with the highest resolution will show the location of genes or markers on the chromosomes as measured by the number of DNA bases between them. The ultimate goal is to get the entire DNA sequence of the human genome. The next challenge is to identify the functions of these genes.

Now we can have access to individual genes with its structural details much faster than ever before. Soon we will have an accurate function of all the genes. Likewise, the genome sequences of more than fifty prokaryotic and eukaryotic organisms have been completed. As said earlier, the primary goal of a mapping program is to construct genetic linkage maps sufficient to identify either a disease gene or chromosomal segments that harbor quantitative trait loci (QTL) affecting traits of economic importance. The gene maps or genetic linkage maps of almost all the domestic animals and several model organisms have been completed. Further, comparison of this information with closely related phyletic groups builds the comparative genomics data. Ultimately, the information those are generated in the internet are in the form of database and these include enormous DNA sequence, genetic-, physical- and comparative maps and so on for bioinformatics application. The databases are operated jointly by three groups *viz.* European Molecular Biology Laboratory (EMBL) at European Bioinformatics Institute (EBI), The DNA Database of Japan (DDBJ) and The GenBank at the National Center for Biotechnology Information (NCBI, USA). These three banks exchange and update data between themselves on a daily basis so that one can deposit information only at one place.

Internet database sites: (Exemplary)

http://www.ncbi.nlm.nih.gov

http://www.ebi.ac.uk

Transcriptomics

Expression of a biological trait in an organism is linked to the function of one or combination of few genes. Alternatively, it may be said that genes express the biological traits of humans, animals, plants and microorganisms. The acquisition of new knowledge about genes will make it possible to identify, prevent or treat disease in humans, animals and plants. Using new technology, it is possible to study tens of thousands of genes at once. Thus it has become a realistic objective to determine how the genes in an organism function. At the messenger RNA level, almost 4 million ESTs (expressed sequence tags) of the complementary DNAs (cDNAs) of messenger RNAs have been sequenced in humans. In plants, enormous ESTs have been sequenced in soyabean, maize, tomato and *Arabidopsis.* These sequences can be used to create high-density filters, DNA chips or microarrays, where thousands of cDNAs are fixed on a small surface, which is then hybridized with the mRNAs from a given stage or organ. It can thus be answered which genes among thousands are expressed in a particular cell type of an organism, at a particular time and stage. Such transcriptomics tools will soon permit extensive study of transcription and its regulation.

Internet sites: (Exemplary)

1. http://www.ncbi.nlm.nih.gov/dbEST
2. http://www.ebi.ac.uk/2can/databases/microarray.html

Proteomics

Most genes code for and work through proteins. Proteins may have a wide variety of functions in an organism. Thus it is of great significance to know how proteins function. The complete profile of proteins expressed in a cell is called the cell's proteome. There are thousands of different proteins in each individual cell, and different types of cells contain different sets of proteins. The objective of proteomics is to discover how these proteins function and interact with one another. In order to determine the effect of proteins, it is necessary to know their composition; i.e., the way in which their components are arranged. It is also necessary to know the form that they take, i.e., their three-dimensional structure. Most proteins are coiled in a particular way and can only fulfill their function if they assume the correct three-dimensional structure. Techniques have been developed for visual imaging of proteins, but determining protein structures is technically demanding. Data have been stored for a couple of thousand different protein structures in international databases, but the three-dimensional structure for most of the proteins has not yet been determined. In many cases only parts of these proteins' structures have been determined, perhaps only the biologically active part.

At the protein level, the method used today is still two-dimensional (2D) gel electrophoresis. This separates the denatured proteins according to two independent criteria, the isoelectric point in the first dimension (isoelectrofocusing) and the apparent molecular mass in the second one (electrophoresis in the presence of SDS). On the 2 D gels obtained, several hundreds to several thousands of protein spots (i.e. of gene products) are revealed. These proteins can then be identified or characterized by different methods. Immunological characterization can be done when the identification of a spot is *a priori* suspected or when we are looking for an already known protein on a 2 D gel. The classical protein sequencing of the N-terminal first 10 or 15 amino acids, or of internal sequences, is a very reliable but expensive and relatively slow running method (one spot a day) and requires reasonable amounts of protein. The combination of different criteria can be adopted to identify a protein spot faster and at a lower cost. It permits one to unambiguously identify a protein, if its gene is already present in the databanks. More samples can be analysed for a lower cost than with Edman sequencing. However, protein identifications are performed today by mass spectrometry. The MALDI-TOF (Matrix Assisted Laser Desorption Ionisation – Time of Flight) mass spectrometers give masses of peptides obtained after trypsin digestion of the spots. These high-throughput apparatuses permit a fast identification of numerous proteins when detailed genomic information is available on the studied species. The data obtained are compared to those generated from the databanks such as SwissProt or EMBL. Another type of mass spectrometer, the ESI-MS/MS (Electro Spray Ionisation - Mass Spectrometer / Mass Spectrometer), gives, by fragmentation of the peptides, values that are diagnostic of amino acid sequences. Then, as with Edman sequencing, the sequence itself can be compared to homologous ones from other species and is thus preferred when the genome of the studied species is not well represented in databanks.

Internet sites: (Exemplary)

http://www.ebi.ac.uk/2can/databases/protein.html

http://www.expasy.ch/enzyme/

Metabolomics

Metabolomics is a powerful emerging technology, whereby the total metabolite composition (the metabolome) of an organism is analyzed. By characterizing the metabolome of an organism at different developmental stages, or following exposure to different conditions, global shifts in its metabolism can be followed. This approach complements studies in which changes in the transcriptome and proteome are monitored. By combining global metabolomic analysis with transcriptomics and proteomics it is possible, for the first time, to gain a holistic view of the complex interactions between genes and metabolites. New metabolomic technologies will lead to a greater understanding of metabolism than was possible using previous profiling approaches limited to particular classes of compounds. In addition, metabolomics offers a powerful technology to characterise genes of unknown function, whereby the expression of the gene is manipulated by mutagenesis or genetic engineering and its identity inferred from the resulting metabolic changes. In particular, studies of plants and microbes are important in identifying molecular determinants of food quality, safety and nutrition, and opportunities for the exploitation of novel metabolic products. Plant Molecular Biology (PMB), Biochemistry of Metabolic Regulation in Plants (BOMRIP), Resource Allocation and Stress in Plants (RASP) and Prokaryotic Responses to Environmental Stress (PRES) are the research sub-disciplines for the generation of data. Over 40,000 biochemical compounds have already been identified in plants. Nevertheless, 10,000s and perhaps even over 100,000 compounds have still to be discovered. These compounds belong to a wide variety of metabolic classes that are all, in one way or another, inter-related. This multitude of chemical components, with many different carbon skeletons, functional groups and physiochemical properties makes plants an invaluable source of pharmaceuticals, health-promoting compounds, flavour & fragrance compounds, protectants, biocides, fine chemicals, toxins, etc.

Internet database sites: (Exemplary)

http://www.genome.ad.jp/kegg/metabolism.html

http://www.expasy.ch/cgi-bin/show_thumbnails.pl

http://www.expasy.ch/cgi-bin/show_thumbnails.pl?2

Glycomics

Glycomics is the study of interaction between proteins and carbohydrates (glycans). Maccabim-based Glycominds pioneered novel glycomics technologies to analyze interactions between complex carbohydrates and proteins, which will help identify many new drug targets as well as dramatically improve early stage drug development. Now, with the new GlycoChip, the first biochip microarray of glycans, one can analyze unprecedented numbers of glycan-protein interactions. We can see numerous applications of the GlycoChip in the pharmaceutical and biopharmaceutical industry, including identifying novel carbohydrate-binding proteins and potential inhibitors of glycan-protein interactions, as well as analysis of immunogenicity in serum samples for antibody-glycan binding. That is an important step in drug identification because more than 90 percent of therapeutic proteins are bound with glycans critical to the protein function. Combined with Glycominds' Glycomics Database, the world's largest comprehensive database of glycans, includes over 35,000 glycans from various biological sources describing more than 7,000 unique structures, 1,400 proteins and 1,550 interactions.

Internet database sites: (Examplery)

http://www.meddb.info/index.php.en?cat=8

http://www.functionalglycomics.org/glycomics/molecule/jsp/carbohydrate/carbMoleculeHome.jsp

Research in modern biology is progressing at a much greater space than ever before. Accordingly, data is being poured into the databases in hourly or daily basis. The near future would see a tremendous amount of data in every discipline discussed above. Further, new databases are also being added to the old ones to enrich the resources.

Chapter **4**

Correspondence Analysis in Study of Genomic Sequence of *Labeo rohita*

A.K. Roy & Sushma R. Martha

Introduction

Nucleotide and protein sequences construct the core of Bioiformatics. These biological sequences form complex and very large datasets. Various Statistical techniques have been used for analysis and extraction of precise information form small as well as large biological datasets. Evolution and natural selection makes biological sequences redundant and statistically biased. Once the mechanisms involved in such biases have been characterized, their analysis and systematic investigation become valuable tools for use in predicting the properties of other biological sequences. This biased nature results in relative codon usage value among the synonymous codons or Relative Synonymous Codon Usage (RSCU) value. Codon usage variation within a single genome can be an important source of information about gene expression levels and events of horizontal gene transfer (Medigue *et al.*, 1991). The analysis of codon usage patterns can be traced back to when first molecular sequence databases were being colleted (Grantham *et al.*, 1981). In order to evaluate the codon and amino acid usage variation, multivariate analysis options like Correspondence Analysis, Principal Component Analysis, Cluster Analysis, Factor Analysis, Multidimensional Scaling and Eigenanalysis Ordination are available (Morrison *et al.*, 1994). It is well established that codon usage by its nature is multivariate. These methods provide a low dimensional point

Applied Bioinformatics, Statistics and Economics in Fisheries Research: pp. 29-47

New India Publishing Agency, New Delhi (India)

representation of genes, where the proximity of gene specific points indicates (CoA) a similar codon usage of associated genes. Correspondence Analysis is an exploratory graphical approach that requires contingency tables and generates results in the form of tables, scattered plots and graphs. CoA can identify the major sources of variation in the dataset. The output from a CoA can be used to evaluate other aspects of genes, such as base composition, expressivity, aromaticity and location on the genome (Thiouse *et al.* 1997). Survey of literature suggests that Correspondence Analysis has been used to study codon usage in various species like *Escherichia coli* (Medigue *et al.*, 1991; Holm, 1986), *Bacillus subtilis* (Moszer *et al.*, 1999), *Borrelia burgdorferi* (Mclnerney, 1998; Lafay *et al.*, 1999), *Chlamydia trachomatis* (Romero *et al.*, 2000), *Mycoplasma genitalium* (Mclnery, 1997), *Helicobacter pylori* (Lafay *et al.*, 2000), *Pseudomonas aeruginosa* (Gupta, 2001), *Saccharomyces cerevisiae, Drosophilla, Arabidopsis thaliana*, and *Homo sapiens*. It has also been used for predicting coding regions in prokaryotes and eukaryotes (Fichant and Gautier, 1987), for studying the evolution of repeated sequences in primates (Quentin, 1988) and in rodents (Quentin, 1989) for analyzing trends in amino acid composition in E. coli (Lobry and Gautier, 1994) and for detecting sequencing errors like frame shifts (Fichant and Quentin, 1995). Any such analysis on fish especially in case of Indian major carp *Labeo rohita* is limited (Roy *et al.*, 2007a, 2007b, 2007c).

Source of Sequence Data

The presence of just four base pair A, T, G and C in different combinations and quantity makes the blue print of all available living organisms. These permutations and combinations bring into picture the concept of evolution and origin of numerous species. Storage and analysis of these sequences forms a major part of scientific research to understand the existence and phenomena of life. National Center for Biotechnology Information (NCBI) maintains a Nucleotide Sequence Database (Genbank) where the scientists from different corners of the world are depositing the nucleotide sequences of various organisms. On search it is revealed that till date, 141 core nucleotide sequences of *Labeo rohita* are present in the database. It includes, 7 complete coding sequences, 7 partial coding sequences, 33 complete sequences, 33 partial sequences, 76 clone sequences with 8 other sequences. Out of these 141 sequences, 100 sequences were taken in the present study for demonstration of steps involved in correspondence analysis.

Saleint Features of Codon Usage and Related Software Program

The genetic codes may have originated from natural interaction between RNA molecules and amino acids and evolved from reallocation of codons at separate occasions. Each codon can be recognized by a subset of cell's transfer ribonucleic acid molecules (Oswa and Jukes 1989; Osawa *et al.* 1990). In general, codons can be grouped into 20 disjoint families, one family for each of the standard amino acids, with a 21st family for the translation termination signal. Each family in the universal genetic code contains between 1 and 6 codons. Presence of 64 codons for only 20 amino acids supports these facts. Where present, alternate codons are termed as synonymous. Use of alternative synonyms might have a role in the regulation of gene expression level but the choice among synonymous codons might not alter the primary structure of a protein. The concept of these synonymous condons brings into picture codon usage and a relative codon usage value. Codon usage values are described either in terms of the number of times the codon is observed i.e the frequency of occurrence of the codon in a gene or the relative synonymous codon usage value. RSCU values are a reflection of how often a particular codon is used relative to the expected number of times that codon would be used in the absence of codon usage bias (Mclnerny, 1998). It is the relative frequencies of synonymous codons to code for same amino acid. The Codon Usage Program of *Sequence Manipulation Suite* is one of many such programs that calculate the RSCU values of individual raw sequences.

Brief Outline of Correspondence Analysis

Correspondence analysis (Greenacre, 1984) is the most popular and appropriate multivariate analysis for contingency table data such as codon usage values. It is an exploratory graphical technique to describe the relationships between the row and column variables of a contingency table (a table where row and column play equivalent roles and can be exchanged) in low dimensional space. Multiple correspondence analyses quantify nominal (categorical) data by assigning numerical values to the cases (objects) and categories so that objects from one category to objects from other category are clearly distinguished. The measure of correspondence can be any indication of the similarity, affinity, confusion, association, or interaction between the row and column variables. Correspondence Analysis allows one to examine the relationship between two nominal variables graphically in a multidimensional space. It computes row and column scores and produces plots based on the scores. Categories those are similar to each other in the plots. In this way, it is easy to see which categories of a variable are similar to each other or which categories of the two variables are related. The interpretation of correspondence analysis in terms of distances depends on the normalization method used. The correspondence analysis procedure can be used to analyze the differences between categories of a variable or analyze the differences between variables. With the default normalization, it analyzes the differences between the row and column variables. The correspondence analysis algorithm is capable of many kinds of analyses like centering the rows and columns and using chi-square distances corresponds to standard correspondence analysis and using alternative centering options combined with Euclidean distances allows for an alternative representation of a matrix in a low dimensional space.

Before proceeding further it is necessary to clarify the concept of the following terminologies like normalization, row and column principal, symmetrical and principal normalization related to CoA for easy understanding and interpretation of results arriving out of CoA.

Normalization: Normalization is used to distribute the inertia over the row scores and column scores. Some aspects of the correspondence analysis solution, such as the singular values, the inertia per dimension and the contributions, do not change under the various normalizations. Correspondence analysis has several ways to spread the inertia. The three most common ways include spreading the inertia over the row scores only, spreading the inertia over the column scores only, or spreading the inertia symmetrically over both the row and the column scores.

Row Principal: In row principal normalization, the Euclidean distances between the row points approximate chi-square distance between the rows of the correspondence table. The row scores are the weighted average of the column scores. The column scores are standardized to have a weighted sum of squared distances to the centroid of 1. Since this method maximizes the distances between row categories, one should use row principal normalization if one is primarily interested in seeing how categories of the row variable differ from each other.

Column principal: On the other hand, one might want to approximate the chi-square distances between the columns of the correspondence table. In that case, the column scores should be the weighted average of the row scores. The row scores are standardized to have a weighted sum of squared distances to the centroid of 1. This method maximizes the distances between column categories and should be used if one is primarily concerned with how categories of the column variable differ from each other.

Symmetrical: One can also treat the rows and columns symmetrically. This normalization spreads inertia equally over the row and column scores. It may be noted that neither the distances between the row points nor the distances between the column points are approximations of the chi-square distances in this case. This method may be used if one is primarily interested in the differences or similarities between the two variables. Usually, this is the preferred method to make biplots.

Principal: A fourth option is called principal normalization, in which the inertia is spread twice in the solution once over the row scores and once over the column scores. One should use this method if one is interested in the distances between the row points and the distances between the column points separately but not in how the row and column points are related to each other. Biplots are not appropriate for this normalization option and are therefore not available if one has specified the principal normalization method.

Steps Involved in Carrying out the CoA with Genomic Data of *Labeo rohita*

At first nucleotide sequences bearing unique Accession Numbers are downloaded from GenBank. Then, the codon usage values are calculated by running the Codon Usage program of Sequence Manipulation Suite for individual gene sequences separately followed by storage of these relative synonymous codon usage values in rectangular matrices in which the columns represent some measurements of codon usage or amino acid usage and the rows represent individual genes. Generally four punctuation codons including three termination and one initiation codon along with tryptophan codon are removed while constructing the matrices as they are unique and hence there exists no concept of biasness regarding them. These matrices of data are then imported into the Statistical Package (SPSS 14.0) for carrying out further study through Correspondence Analysis. Step by step illustrated analysis of *Labeo rohita's* genomic sequences downloaded from Genbank is being presented for better understanding and future use.

A view of one of the 142 nucleotide sequences in fasta format of *Labeo rohita* downloaded from Genbank with Accession Number AF 134200 is presented in the Fig 1. It is a growth hormone precursor.

Relative Synonymous Codon Usage Data View

Sample result of the codon usage value of one nucleotide sequence of *Labeo rohita* that is a complete coding sequence (cds) is displayed (Table 1). It is the growth hormone precursor with Accession Number AF 134200. In the table the first column represents amino acid, second column shows the synonymous codons of the corresponding amino acids and the third, fourth and fifth columns display the frequency of individual codons over the whole sequence, frequency per thousand codons and the RSCU values of the synonymous codons respectively. Similarly RSCU values of each sequence are computed separately using the Codon Usage software.

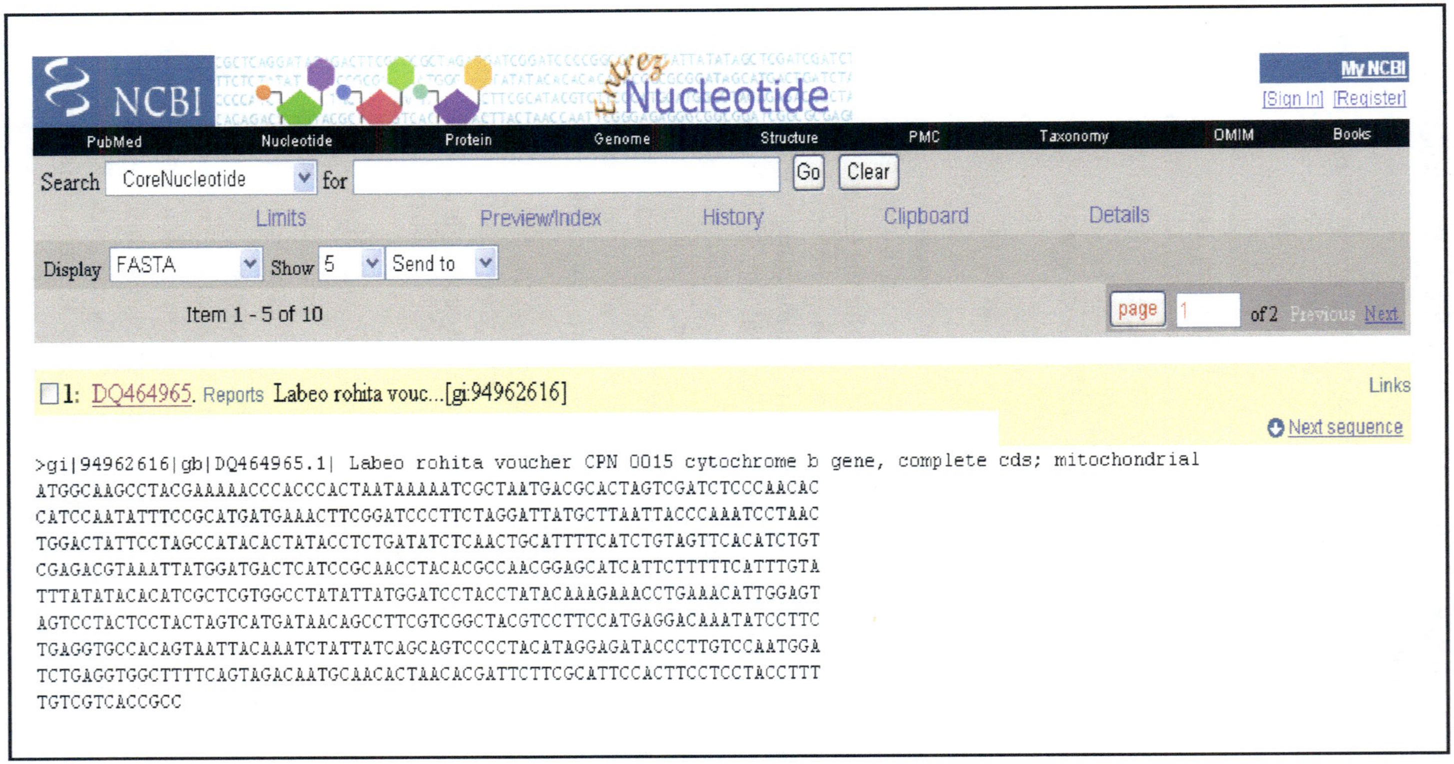

Fig 1. A view of nucleotide sequence data of *L. rohita*

Table 1. A Sample result of codon usage value of *Labeo rohita*

Results for 1162 residue sequence "gi|4836724|gb|AF134200.1 |AF134200 *Labeo rohita* growth hormoneprecursor, mRNA, complete cds" starting "GGCACGAGTT"

AmAcid	Codon	Number	/1000	Fraction ..
Ala	GCG	2.00	5.17	0.11
Ala	GCA	6.00	15.50	0.33
Ala	GCT	4.00	10.34	0.22
Ala	GCC	6.00	15.50	0.33
Cys	TGT	9.00	23.26	0.50
Cys	TGC	9.00	23.26	0.50

Construction of Matrix

After finding out the RSCU values of 100 sequences, these values are used to construct the matrix with codons as column variable and genes as row variable (Table 2).

SPSS Dialogue Box Showing CoA

After construction of the matrix, the matrix is being imported to the statistical package SPSS (14.0) to carry out Correspondence Analysis (Fig 2). Depending on the requirement of the experiment, before carrying out the analysis various parameters and variables with their data types and ranges are defined.

Table 2. A hypothetical matrix with codons as column variables and genes as row variables

Codons	Codan1	Codan2	Codan3	...	Codan62	Codan63	Codan64
Genes							
Gene 1	$RSCUg_1c_1$	$RSCUg_1c_2$	$RSCUg_1c_3$		$RSCUg_1c_{62}$	$RSCUg_1c_{63}$	$RSCUg_1c_{64}$
Gene 2	$RSCUg_2c_1$	$RSCUg_2c_2$	$RSCUg_2c_3$		$RSCUg_2c_{62}$	$RSCUg_2c_{63}$	$RSCUg_2c_{64}$
Gene 3	$RSCUg_3c_1$	$RSCUg_3c_2$	$RSCUg_3c_3$		$RSCUg_3c_{62}$	$RSCUg_3c_{63}$	$RSCUg_3c_{64}$
. . .							
Gene 98	$RSCUg_{98}c_1$	$RSCUg_{98}c_2$	$RSCUg_{98}c_3$		$RSCUg_{98}c_{62}$	$RSCUg_{98}c_{63}$	$RSCUg_{98}c_{64}$
Gene 99	$RSCUg_{99}c_1$	$RSCUg_{99}c_2$	$RSCUg_{99}c_3$		$RSCUg_{99}c_{62}$	$RSCUg_{99}c_{63}$	$RSCUg_{99}c_{64}$
Gene 100	$RSCUg_{100}c_1$	$RSCUg_{100}c_2$	$RSCUg_{100}c_3$		$RSCUg_{100}c_{62}$	$RSCUg_{100}c_{63}$	$RSCUg_{100}c_{64}$

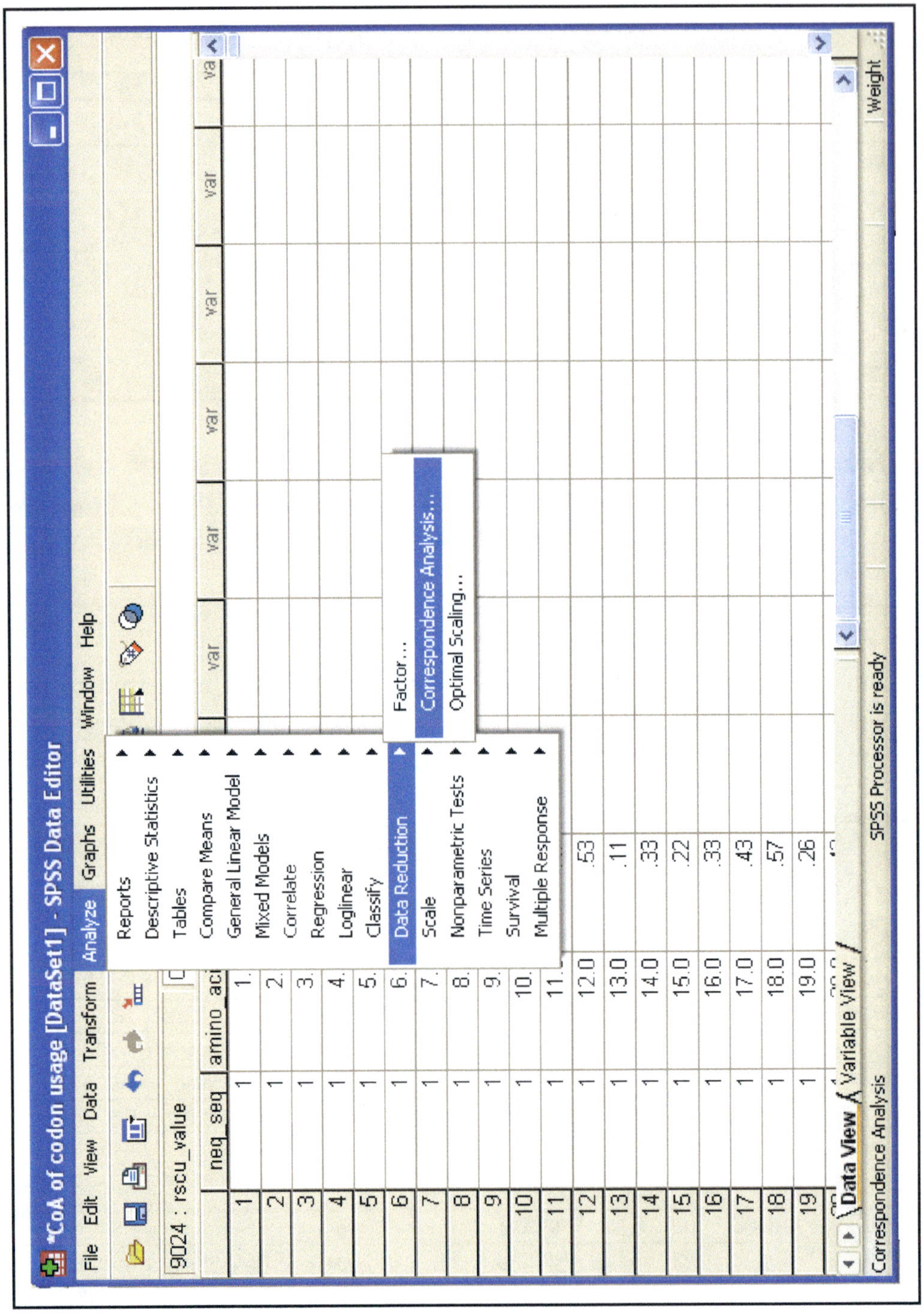

Fig 2. Dialogue Box for Correspondence Analysis of SPSS

Presentation and Interpretation of Output from SPSS

The SPSS provides output on the dimensionality, inertia of dimension, profiles and distances, row and column scores of dimensions after symmetrical normalization, mass and contribution of point of inertia of dimension. All these were taken into consideration and estimated along with confidence statistics of nucleic acid and codon scores for proper interpretation. All these output along with interpretation are presented below.

Correspondence Table

The correspondence table bellow is simply the cross tabulation of the row and column variables, including the row and column marginal totals, serving as input.

Table 3. Correspondence Table; weight cases by frequency variable (*L. rohita*)

Nucleic Acid	Amino Acid						
	GCG	GCA	GCT		TAG	TAA	Active Margin
1	.110	.330	.220		.300	.600	20.960
2	.000	.530	.130		.000	.000	21.000
3	.050	.470	.320		.320	.210	20.080
. . .							
99	.000	.330	.500		.500	.000	19.990
100	.000	.500	.250		.000	.000	19.980
Active Margin	12.900	34.540	30.190		18.570	31.820	1987.450

Profiles and Distances

Correspondence analysis takes into account marginal distribution as well as individual cell frequencies to determine the distance between categories. It computes row and column proportions for each cell based on marginal totals. The output of Nucleic Acid (row) and Amino Acid (column) are furnished in the table bellow.

The row profiles are the cell contents divided by their corresponding row total (.110/20.960=. 005 for the first cell) (Table 4 A). Likewise, the column profiles are the elements divided by the column marginal (.110/12.900= .009) (Table 4 B). This table also shows the row masses (row marginal as a percent of n). These are intermediate calculations on the way towards computing distances.

Table 4A. Nucleic Acid (row) profiles of *L.rohita*

Nucleic Acid	Amino Acid						
	GCG	GCA	GCT		TAG	TAA	Active Margin
1	.005	.016	.010		.014	.029	1.000
2	.000	.025	.006		.000	.000	1.000
3	.002	.023	.016		.016	.010	1.000
. . .							
99	.000	.017	.025		.025	.000	1.000
100	.000	.025	.013		.000	.000	1.000
Mass	.006	.017	.015		.009	.016	

Table 4B. Amino Acid (column) profiles of *L.rohita*

Nucleic Acid	Amino Acid						
	GCG	GCA	GCT		TAG	TAA	Mass
1	.009	.010	.007		.016	.019	.011
2	.000	.015	.004		.000	.000	.011
3	.004	.014	.011		.017	.007	.010
. . .							
99	.000	.010	.017		.027	.000	.010
100	.000	.014	.008		.000	.000	.010
Active Margin	1.000	1.000	1.000		1.000	1.000	

Dimensionality

Ideally a correspondence analysis solution represents the relationship between row and column variables in as few dimensions as possible. In the present case maximum number of dimensions is minimum of (100 or 64) - 1=63 (Table 3)

In Table 5 inertia per dimension is presented. In general two dimensional solutions is sufficient, as third dimension represents very in significant proportion of inertia. Singular values can be interpreted correlation between row and column scores. For each dimension, the singular values squared (eigen values) equals to inertia and this is another measure of importance of that dimension.

Table 5. Summery Table showing Inertia per Dimension of *L.rohita*

Dimension	Singular Value	Inertia	Chi Square	Sig.	Proportion of Inertia		Confidence Singular Value	
					Accounted for	Cumulative	Standard Deviation	Correlation
								2
1	.223	.050			.079	.079	.021	-.008
2	.209	.044			.070	.149	.021	
3	.190	.036			.058	.207		
. . .								
61	.005	.000			.000	1.000		
62	.001	.000			.000	1.000		
63	.001	.000			.000	1.000		
Total		.626	1243.264	1.000(a)	1.000	1.000		

a 6237 degrees of freedom

Mass is a measure that indicates the influence of an object based on its marginal frequency. Mass is affected by centroid, which is weighted mean row or column profile. Points with large mass pull the row centroid strongly together to its location and points with small mass pull the row centroid only slightly to its location. (Table 6)

Similarly column centroid is the mean of column profile. Points with large mass pulls the column centroid strongly to its location and points with small mass pulls the column centriod slightly to its location (Table 7). If one like to think of difference in terms of distance, then the greater the difference between the row profiles, greater the distance between points in a plot. Under column principal normalization, the Euclidean distance between column points in the full dimensional space equals to the chi square distances between columns of correspondence table. Under symmetrical normalization, these quantities are not equal. The total inertia defined as the weights are the masses. Rows with a small mass influence the total inertia, even when they are located close to the centroid. Similar phenomenon is applied to the columns.

Table 6. Row Scores (Symmetrical Normalization)

Nucleic Acid	Mass	Score in Dimension		Inertia	Contribution				
					Of Point to Inertia of Dimension		Of Dimension to Inertia of Point		
		1	2		1	2	1	2	Total
7	.010	1.043	-.178	.009	.047	.001	.270	.007	.278
18	.009	1.112	.147	.008	.050	.001	.312	.005	.318
19	.010	1.147	-.293	.007	.059	.004	.418	.025	.443
33	.010	-.321	-1.150	.008	.005	.064	.030	.357	.387
38	.008	.671	-1.366	.017	.016	.072	.046	.180	.226
54	.010	-1.277	-.321	.008	.070	.005	.414	.025	.439
60	.011	.242	1.115	.005	.003	.063	.025	.505	.530
68	.011	.238	1.080	.005	.003	.059	.026	.511	.537
. . .									
98	.010	-.042	-.595	.005	.000	.017	.001	.142	.143
99	.010	-.059	-.005	.004	.000	.000	.002	.000	.002
100	.010	.607	-.127	.007	.017	.001	.126	.005	.131
Active Total	1.000			.626	1.000	1.000			

a Symmetrical normalization

Contributions

Row and column points that contribute substantially to the inertia of a dimension are important to that dimension. The diagnostics that measure the contributions of points are an important aid in interpretation of CoA solution. Dominant points in the solution can easily be detected. For example Nucleic Acids 7, 18, 19, 54 etc. are dominant points in the first dimension contributing substantially to the inertia (Table 6). Among column points Amino Acids GAG, GAA, CAT, CAC, etc. contributes substantially in the first dimension (Table 7).

Table 7. Column Scores (Symmetrical Normalization)

Amino Acid	Mass	Score in Dimension		Inertia	Contribution				
					Of Point to Inertia of Dimension		Of Dimension to Inertia of Point		
		1	2		1	2	1	2	Total
GAC	.019	.688	1.122	.015	.040	.112	.134	.333	.467
GAG	.022	1.052	-.513	.014	.112	.028	.402	.090	.491
GAA	.023	-.964	.263	.012	.097	.008	.402	.028	.430
CAT	.022	-.715	-.654	.011	.050	.045	.233	.183	.416
CAC	.026	.634	.546	.010	.047	.037	.238	.166	.404
CAG	.029	.203	-.588	.008	.005	.048	.032	.254	.286
CAA	.020	-.416	.946	.011	.016	.086	.072	.351	.423
GTG	.018	.477	-.832	.009	.018	.060	.105	.300	.405
. . .									
TGA	.023	.531	-.119	.010	.029	.002	.150	.007	.157
TAG	.009	-.679	.243	.009	.019	.003	.102	.012	.115
TAA	.016	-.457	.138	.009	.015	.001	.081	.007	.088
Active Total	1.000			.626	1.000	1.000			

a Symmetrical normalization

Confidence Statistics

Assuring that the table to be analyzed is a frequency table and that the data are a random sample from unknown populations, the cell frequencies follow a multidimensional distribution.

From this, the standard deviations of dimensions and correlations of singular values, row scores and column scores can be computed (Table 8A and 8 B).

Row and Column points

The plot of two dimensional symmetrical normalization of Nucleic Acid shows that many points positionally 24,16,5,12,28,37,30,etc. are very close to the origin indicating that they differ little from average Nucleic Acid profile (row profile). These sequences are supposed to have more expression level then those of other ones located far away from origin (Fig 3 A).

Similarly examination of two dimensional symmetrical Normalization for Amino Acid of *Labeo rohita* reveals that only few points are close to the origin representing TTT, ACA, GCA, GCT, TTG, ATG, TGG, etc. (Fig 3 B).

Table 8 A. Confidence Row (Amino Acid) Points of *L. rohita*

Nucleic Acid	Standard Deviation in Dimension		Correlation
	1	2	1-2
1	.350	.473	.027
2	1.211	.943	.012
3	.423	.431	-.073
.			
.			
.			
98	.845	.625	-.097
99	.488	.561	-.034
100	.653	.976	.140

Table 8 B. Confidence column (Nucleic Acid) Points of *L. rohita*

Amino Acid	Standard Deviation in Dimension		Correlation
	1	2	1-2
GCG	1.418	1.685	.213
GCA	.658	.878	-.001
GCT	.692	.931	-.045
.			
.			
.			
TGA	.503	.801	.139
TAG	.744	1.118	.286
TAA	.552	.791	.201

Bi plot

Principal normalization cannot address the issue of relationship i.e. how are Amino Acids related to Nucleic Acids. To focus on how the variables are related to each other, symmetrical normalization is used because rather than spread of inertia twice as in case of principal normalization, symmetrical normalization divides the inertia equally over both the Amino Acid and Nucleic Acid representing rows and columns respectively. Distances between categories for a single variable cannot be interpreted, but distances between the categories for different variables are meaningful. This method is used to find out the differences or similarities between the two variables.

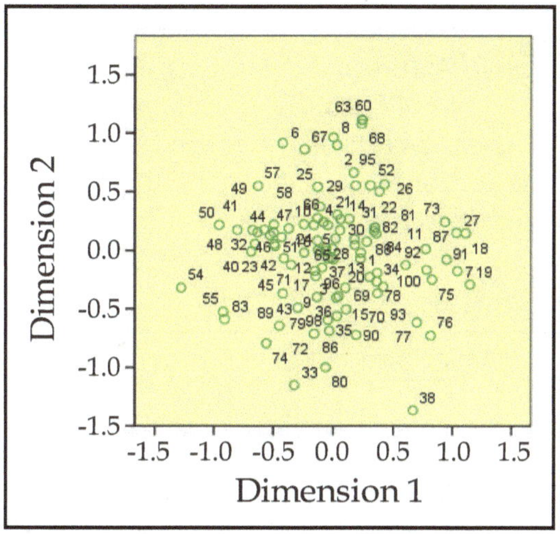

Fig 3 A. Plot of Symmetrical Normalization for Nucleic Acid of *L. rohita*

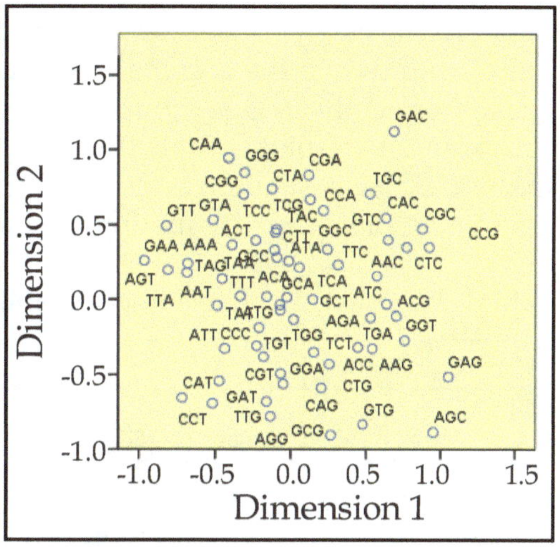

Fig 3 B. Plot of Symmetrical Normalization for Amino Acid of *L.rohita*

From the bi plot two distinct phenomena are identified. Firstly, it reveals the association between genes and codons. It is clear that the cluster of genes,

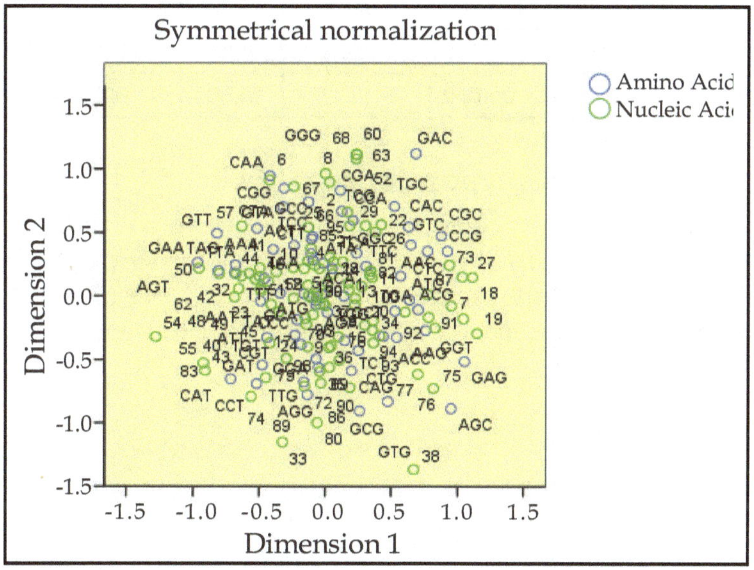

Fig 4. Bi plot of Nucleic Acid and Amino Acid of *L. rohita*

24, 16, 28, 5, 61, 37 and 24 are associated with codons ACA (Thr), GCA (Gly), TTG (Leu), GCT (Gly), TAT (Tyr), TTA (Leu) while the group of genes10, 66, 14, 30, 21, 30, 58 are having preference for GCC (Ala), ATA (Ile), CTT (Leu), TAC (Tyr), TTC (Ser), ACA (Thr). The codons GCT (Arg), GGA (Ala), GAT (Asp), TTG (Leu) occur frequently in the cluster having 36, 35, 98, 15 in their group while the codons AAA (Lys), TTA (Leu), AAT (Asn), AGT (Ser) occur frequently in the group of genes, 32, 44, 47, 46, 51, 48, 40, 45, 42, 41. Secondly, the bi plot reveals that the AT ending codons are more frequent in the nucleic acid sequences falling on the left side of the plot while the GC ending codons occur frequently on the right side of the plot.

Table 9. Permutated Correspondence Table According to Dimension 1

Nucleic Acid	Amino Acid						
	GAA	GTT	AGT		AGC	GAG	Active Margin
54	1.000	.500	.000		.000	.000	18.980
50	.000	.800	.630		.130	.000	17.010
55	.750	.000	.000		.250	.250	19.000
. . .							
7	.000	.000	.000		.250	1.000	19.000
18	.000	.080	.050		.050	.000	18.030
19	.000	.000	.080		.000	1.000	20.000
Active Margin	46.430	24.650	18.780		15.580	44.620	1987.450

Permutation of the Correspondence table

It is useful to order the categories of the rows and columns in increasing order according to the scores of the first dimension. Ordering can be achieved by taking the row and column scores as ordering variables. Table 9 shows the permutation of correspondence table along the first dimension.

Conclusion

Correspondence analysis requires contingency tables, which are counts of the joint occurrences of the rows and columns of a table. Correspondence analysis provides a simple visualization of distances by projecting the points from their original multidimensional space into low dimension, with genes with similar distances plotted as neighbors. The dimensionality of the analytical procedure using statistical package along with interpretations of Genomic sequence of *Labeo rohita* through CoA might help to understand and unearth the underlying phenomenon which in turn will improve the productivity of commercially important species like *Labeo rohita*, leading to the improvements of income, nutrition and food security of common people. The demonstration of correspondence analysis using sample data of *Labeo rohita* with the help of Statistical package will throw light on the possibility of in depth analysis of fish sequence data.

REFERENCES

Medigue, C., T. Rouxel, P. Vigier, A. Henaut and Danchin, A. (1991). Evidence for horizontal gene transfer in *Escherichia coli* speciation. *Journal of Molecular Biology*, 222: 851-856.

Holm, L. (1986). Codon usage and gene expression. *Nucleic Acid Res.*, 14: 3075- 3087.

Moszer, I., Rocha, E.P.C. and Danchin, A. (1999). Codon usage and lateral gene transfer in *Bacillus subtilis* genome. *Microbiology*, 141: 261-268.

Mclnerney, J.O. (1998). Replicational and transcriptional selection on codon usage in *Borrelia burgdorferi, Proc.Natl Acad.Sci.*USA, 95: 10698-10703.

Mclnerney, J.O. (1998). GCUA: General Codon Usage Analysis. *Bioinformatics*, 14: 372-373.

Lafay, B., Lioyd, A.T.,Mc Lean, M.J., Devine,K.M.,Sharp,P.M. and Wolfe,K.H. (1999). Proteome composition and codon usage in spirochaetes: species-specific and DNA strand-specific mutataional biases. *Nucleic Acids Res.*, 27: 1642-1649.

Romero, H., Zavala, A. and Musto, H. (2000). Codon usage in Chlamydia trachomatis is the result of strand-specific mutational biases and a complex pattern of selective forces. *Nucleic Acid Res.*, 28: 1084-2090.

Mclnery, J.O. (1997). Prokaryotic genome evolution as assessed by multivariate analysis of codon usage patterns. *Microbial Comp. Genomics*, 2: 1-10.

Lafay, B., Atherton, J.C. and Sharp, P.M. (2000). Absense of translationally selected synonymous codon usage bias in *Helicobacter pylori. Microbiology*, 146: 851-860.

Gupta, S.K. and T.C. (2001) Gene expressivity is the main factor in dictating the codon usage variation among the genes in *Pseudomonas aeruginosa*. Gene, 273: 63-70.

Fichant,G. and Gautier, C. (1987). Statistical methods for predicting protein coding regions in nucleic acids sequences.*Comput.Appl.Biosci.*, 3: 287-295.

Quentin, Y. (1988). The Alu family developed through successive waves of fixation closely connected with primate lineage history. *J. Mol.Evol.*, 3: 287-295.

Quentin, Y. (1989). Successive waves of fixation of BI variants in rodent lineage history. *J.Mol.Evol.*, 28: 299-305.

Lobry,J.R. and Gautier, C. (1994). Hydrophobicity, expressivity and aromaticity are the major trends of amino acid usage in 999 *Ecoli* chromosome-encoded genes. *Nucleic Acids Res.*, 23: 2900-2908.

Fichant, G.A. and Quentin,Y. (1995). A frameshift error detection algorithm for DNA sequencing projects. *Nucleic Acids Res.*, 23: 2900-2908.

Cancilla, M.R., A. J. Hillier and Davidson, B. E. (1995). *Lactococcus lactis* glyceral dehyde-3-phosphate dehydrogenase gene. Gap. Further evidence for strongly biased codon usage in glycolytic pathway genes. *Microbiology U.K.*, 141:1027-1036.

Freirepicos, M. A., Gonzalezsiso, N.I., Rodriguezbelmonte, E., Rodrigueztorres, A. M., Ramil, E. *et al.*, (1994). Codon usage in *kluyveromyces lactis* and in yeast cytochrome c-ending genes, *Gene* 139: 43-49.

Gharbia, S. E., Williams, J. C., Andrews, D. M.A. and Shab, H.N. (1995). Genomic clusters and codon usage in relation to gene-expression in oral gram-negative anaerobes. *Anaerobe*, 1: 239-262

Delorme, C., Godon, J.J., Ehrlich, S. D. and Renault, P. (1994). Mosaic structure of large regions of the *lactococcus lactis* subsp. cremoris chromosome. *Microbiology U.K.*, 140: 3053-3060.

Groisman, E. A., Saier, M.H.J. and Ouchman, H. (1992). Horizontal transfer of a Phosphatase gene as evidence for mosaic structure of the Salmorella. *EMBO Journal*, 11: 1309-1316.

Grantham,R., Gautier,C., Gouy,M., Jacobzone,M. and Mercier,R. (1981) Codon Catalog Usage is a genome strategy modulated for gene expressivity. *Nucleic Acids Res.*, 9: r43–r74.

Thioulouse, J., Chessel, D. and Dolédec, S. (1997). ADE-4: A multivariate analysis and graphical display software. *Stat. Comput.*, 7: 75–83.

Greenacre,M.J. (1984). Theory and Applications of Correspondence Analysis. Academic Press, London.

Ames, B., and Hartmann, P. (1963). The histidine operon. *Cold Spring Harbor Symposium Quantitative Biology*, 28: 349-356.

Osawa, S., Muto, A., Jukes, T. and Ohama, T. (1990). Evolutionary changes in the genetic code. *Proceedings of the Royal Society of London Series B*, 241: 19-28.

Osawa, S. and Jukes, T. H. (1989). Codon reassignment (codon capture) in evolution. *Journal of Molecular Evolution*, 29: 271-278.

Morrison, D.A., Ellis, J., and Johnson, A.M. (1994) An empirical comparison of distance matrix techniques for estimating codon usage divergence. *Journal of Molecular Evolution* 39: 533-536.

Roy, A. K. and Rath, S. N. (2007a). Use of Correspondence Analysis in Gene Expression Study *of Labeo rohita*. 12th National Conference of Gwalior Academy of Mathematics and Sciences (GAMS) and All India Symposium on Computational Biology at Gwalior, (Abs. No. CB -20)

Roy, A. K, Martha, S. R and Rath, S. N. (2007b). Correspondence Analysis of Codon Usage in *Labeo rohita*. International Symposium on Chromosome to Genomes (c2g) at CCMB, Hyderabad, held during July 3-5, 2007 (Abs. No. 9)

Roy, A. K, and Martha, S. R. (2007c). Correspondence Analysis in Study of Genomic Sequence of *Labeo rohita*. 8th Asian Fisheries Forum (8aff) held at Cochi during 20-30 November, 2007 (Abs No. GMO-008)

Chapter **5**

Statistical Design and Analysis of Microarray Data of Carps

A.K. Roy, Chinmayee Sahoo & Sushma R. Martha

Introduction

Recent advances in bioinformatics such as microarray analysis are bringing about a revolution in our understanding of the molecular mechanisms underlying normal and dysfunctional biological processes. With the help of microarray technologies, biologists are currently capable of observing the abundance of transcripts from tens of thousands genes in biological samples, enabling the exploration of the dynamics of transcription and interaction between genes on a genome-wide scale. With the accumulation of gene expression dataset, the challenging task of all microarray experiments is how to extract meaningful and trustworthy information out of thousands of genes that do not contribute in the designed experiments. To achieve this goal, many rigorous mathematical tools and computational software were introduced to the field, like statistical techniques for data normalization, clustering algorithms, class prediction methods, ANOVA, and gene-gene interaction studies. Microarray technology has been used widely in various areas of agriculture and allied field. Application of statistics in microarray is limited particularly in experiments involving fishes. In this chapter it is attempted to present various statistical methods applied to accessory microarray experiments with illustrated example of fish microarray data.

Applied Bioinformatics, Statistics and Economics in Fisheries Research: pp. 49-72

New India Publishing Agency, New Delhi (India)

Various Applications of Microarray Experiments

Microarray in Agriculture and allied fields

It is widely excepted that microarray are used in monitoring the expression level of genes in comparison between two conditions. Expression profiling can be used to determine the function and regulation (up-, down or unchanged) of particular genes during a particular state. In agriculture microarray have been used to identify genes that are involved in the ripening of vegetables like Microarray have been used to map the cellular, regional, or tissue-specific localization of genes and their respectively encoded proteins. Pharmacological studies use microarray as means of discerning the mechanism of action of therapeutic agents and as a corollary to develop new drug targets. One of the most exciting areas of application is the diagnosis of clinically relevant diseases. The ability to identify cancer cells based on gene expression represents a novel methodology that has real benefits. Microarray have been used as a shortcut to characterize the genes within an organism and determine whether those genes are expressed in a similar way to a reference organism. Comparative genomic hybridization (CGH) is another important application where gene copy numbers are compared between two samples and genomic DNA rather than RNA transcripts are labeled. CGH can detect gene amplification or deletion events underlying tumor genesis. It can also be used to detect genomic rearrangement or abnormal events with serious health consequences in humans such as trisomies. SNP microarray are designed to detect the presence of single nucleotide differences between genomic samples. Titling arrays interrogate the transcript abundance at 30 base pair resolution across an area of the genome. These arrays hold the promise of identifying not only novel transcripts, but also their intron-exon boundaries.

Microarray in Fisheries

Most of the Microarray studies have been carried out in aquaculture is on zebrafish, medaka and fugu (Hasiao *et al.,* 2005). The zebrafish is a powerful system for understanding the vertebrate genome, allowing the combination of genetic, molecular, and embryological analysis (Clark *et al.,* 2001). The zebrafish (Danio rerio) provides an opportunity to investigate the molecular mechanisms underlying these interactions through genetic and chemical screens that take advantage of its transparency during larval and juvenile stages. Comparisons of gene expression profiles between heart and fin regeneration revealed a set of regeneration core molecules as well as tissue-specific factors. Zebrafish heart regeneration is associated with sequentially up regulated wound healing genes and growth factors and suggest that PDGF signaling is required (Lien *et al.,* 2006). Study on zebrafish microarray experiments resulted in an up-regulation of vitellogenin up to 850-fold. In addition, several other genes were identified to be up regulated by estrogens (Alberti *et al.,* 2005). Using zebrafish DNA microarrays, several new targets of Wnt/ beta -catenin signaling was identified (Weidinger *et al.,* 2005). Several large-scale ESTs sequencing projects have been initiated worldwide in both model fish (zebrafish, medaka and fugu) and cultured fish (carp, salmon and catfish) (Hsiao, *et al.,* 2005).

Using microarray technology, few studies studies are carried out in other species also. Preservation of fish egg quality after ovulation-control protocols is a major issue for the development of specific biotechnological processes (e.g. nuclear transfer) (Bonnet *et al.,* 2007). Identity and expression patterns of those genes support the theory comparing ovulation to an inflammatory-like reaction (Bobe *et al.,* 2006). Survey of literature suggests that in the field of eco-toxicology study, ecological risk assessments have been done on fathead minnow *Pimephales promelas* (Larkin *et al.,* 2007). Brevetoxins produced by *K. brevis* lead to massive fish mortalities and contamination of shellfish stocks, and they pose risks to human health. Microarray study used to investigate and understand the correlation between gene expression of *K. brevis* and the physiological and ecological nature of the organism is reported (McLean and Crawford, 2006). There is the potential of correlating gene expression profile with rank position within dominance hierarchies, thus identifying targets for candidate gene approaches (Sneddon et al., 2005). In vaccinated Japanese flounder among the 1187 analyzed genes, 42 genes were up regulated during the course of infection either in vaccinated or non-vaccinated fish. These genes included immune-related genes, such as MMP-9, MMP-13, CXC chemokine, CD20 receptor and hepcidin. By determining how gene expression profiles change toxicant dose will improve the utility of arrays in identifying biomarkers and modes of toxic action (Hook *et al.,* 2006). Study on Antibacterial responses of Atlantic salmon indicated that there were both temporal differences and tissue differences in the transcriptional response to bacterial exposure, potentiality of relevance to the establishment of protection (Martin *et al.,* 2006). Use of Human cDNA microarray for identification of differentially expressed genes in Atlantic Salmon Liver during *Aeromonas salmonicida* infection demonstrates the feasibility of using human microarray to facilitate the discovery of differentially expressed genes in Atlantic salmon, for which no homologous microarray are available (Tsoi *et al.,* 2003). In all cases the microarray data can be statistically confirmed by using different tests depending on data, and can be clustered into groups based on similarity in the gene expression patterns by using a cluster program. Thus microarray can be confirmed as a powerful tool for investigating the evolutionary processes of adaptive divergence among natural populations (Derome *et al.,* 2006). The simplicity and the ability to re-use the sensor array without loss of sensitivity makes this a promising procedure for further development, including deployment on moored instruments (Anderson *et al.,* 2006).

Design of Experiments

It is reported that the design of scientific experiments is an art of balancing considerations *viz.* skill, cost, equipment, and accuracy. For making a series of experiments and to compare the results, it may be ensured that the designs and hybridization conditions are similar. There are several variables in the experimental design that are common to most studies. The experimentalist is likely to know the most about the scientific question and therefore have the greatest input into the experimental design. There are many factors influencing the successful outcome of a microarray study, and attention to as many as possible will greatly aid in the final outcome. The consideration of the strengths and weaknesses of each variable on both the biological and statistical level would help the investigator for a better design of his microarray study.

Controls

Control is common in all designed experiment to compare the outcome with treatment. The treatment can be chemical (drugs, toxins), biological (viruses, microbes, transgenes) or environmental (stress, irradiation) in nature. The individual treatments can be given in a time series or at different doses. In all these cases, the control should be matched as closely as possible genetically. Tissue are matched by taking different regions from the same tissue, one that is healthy to serve as the control, and one that carries the disease as the treatment sample. Controls for cell based experiments generally consist of identical cultures without the physiological, physical, or chemical treatments. Because of all these potential variables that could affect the microarray results, according to expert views it is desirable to create a design with more than one control within the study. Number of microarray required depends on the goals of the study, the resources, and the reliability of the technology, specifically how accurate the chips are, and how often a hybridization fails. Six samples per condition allows meaningful permutation tests, which can give more accurate, and less conservative, estimates of p-values and false discovery rates.

Replicates

Replicates means repetition of the same experiment under identical conditions. Replicates provide a measure of the experimental variation, such as in RNA isolation, labeling efficiency, or in chip quality. It is necessary to include replicates in design of experiment, otherwise the inherent experimental variation can not be established. The question of how many replicates to do depends on how small the differences are desired to be detected, and the noise level in your system. Different systems have different noise levels, and the only way to estimate the noise is to do three or four replicate hybridizations. To do meaningful clustering requires at least 20 samples, and generally many more. Including a large number of both controls and treatments without replicates is another design strategy that has been used.

Pooling

Sometimes the amount of sample from any one individual sample is insufficient for hybridization and in that case, pooling is a practical necessity. In theory, if the variation of a gene among different individuals is approximately normally distributed, then pooling of independent samples would result in reduction of variance.

Choosing a Technology

It is established that for expression arrays, there are two principle platforms viz. spotted arrays and *in situ* synthesized arrays such as Affymetrix. A control reaction is performed and hybridized and then the treatment reaction is performed on a different chip. The results are then compared after having been normalized by independent spiked standards.

Gene Clustering and Classification

It is often important to identify genes and their related expression levels that predict the experimental group they are derived from (gene classification) than to find groups of genes that act in a particular way after treatment (gene clustering). For classification studies, large numbers of samples for each treatment groups are needed in order to robustly find a set of genes that predicts their original treatment group. The large number of samples is also necessary to validate the predicted classification based on training sets derived from the data.

Generation of Microarray Images

There are a variety of microarray platforms that have been developed to accomplish and the basic idea for each is a glass slide or membrane that is spotted or "arrayed" with DNA fragments or oligonucleotides that represent specific gene coding regions. Purified RNA is then fluorescently or radioactively labeled and hybridized to the slide/membrane. In some cases, hybridization is done simultaneously with reference RNA to facilitate comparison of data across multiple experiments. After thorough washing, the raw data is obtained by laser scanning or autoradiographic imaging. The final step of the laboratory process is to produce an image of the surface of the hybridized array. Fluorescently labeled microarrays can then be "read" with commercially available scanners scanning confocal microscopes with lasers exciting at wavelengths specifically for Cy3 and Cy5. The amount of signal emitted is directly in proportion to the amount of dye at the spot on the microarray and these values are obtained and qualitated on the scanner. This results in the image that is typical microarray picture.

Preprocessing

Preprocessing of microarray data comprises analytical or transformational procedures that is directed at resolving the systematic error and bias which is considered very important.

Missing values: There are often many missing values in that text formed microarray data because of deviation in the intensity value. This is because of absence of expression of gene or Background intensity higher than the spot intensity. There are many options for the treatment of missing values as they may be replaced with estimated values in a process called imputation or can be deleted from the further analyses.

Checking the background Information: Background and foreground intensities together form the spot intensity. If the spot intensities are dependent on the background intensities, it is possible either not to apply any background correction to the data or discard the deviating observations from further analyses. The background corrected intensity values are calculated spotwise by subtracting the background intensity from the spot (foreground) intensity.

Calculation of expression change: Calculation of expression change is very essential. There are three commonly used measures of expression change. Intensity ratio is the raw expression value, and log ratio and fold change are transformationally derived from it.

(i) *Intensity ratio:* After background correction, expression change is calculated. The simplest approach is to divide the intensity of a gene in the sample by the intensity level of the same gene in the control. Intensity ratio can be calculated from background corrected or uncorrected data. The formula for two-color data is:

$$\text{Intensity ratio} = \frac{\text{Cy3'}}{\text{Cy5'}}$$

For affymetrix data, substitute Cy3′ and Cy5′ with the appropriate intensitiesfrom the sample and control chips. This intensity ratio is one for an unchanged expression, less than one for down regulated genes and larger than one for up-regulated genes. The problem with intensity ratio is that its distribution is highly asymmetric or skewed. Up-regulated genes can take any values from one to infinity (1 to ¥) whereas all down regulated genes are squeezed between zero and one (0 to 1). Such distributions are not very useful for statistical testing.

(ii) *Log ratio:* To stabilize variance and to make the distribution more symmetric log-transformation can be applied. The most commonly used log-transformation is 2-based (*log* 2), The formula for the calculation of *log*2-transformation is:

$$\text{Log ratio} = \text{log2 (Intensity ratio)} = \text{log2}\left(\frac{\text{Cy3'}}{\text{Cy5'}}\right)$$

After the log-transformation, unchanged expression is zero, and both upregulated and down-regulated genes can take values from zero to infinity. Log ratio has some properties compared with other measures of expression. It makes skewed distributions more symmetrical, so that the picture of variation becomes more realistic. In addition, normalization procedures become additive. The following intensity ratio results for two replicates (A and B) after normalization:

$$\text{Gene A} = \frac{120}{60} = 2.0$$

and

$$\text{Gene B} = \frac{30}{60} = 0.5$$

The mean of these replicates is 1.25 instead of 1, which would have been expected. If the 2-based logarithmic transformation is applied, the log ratios are:

$$\text{Gene A} = \log 2\left(\frac{120}{60}\right) = 1.0$$

and

$$\text{Gene B} = \log 2\left(\frac{30}{60}\right) = -1.0$$

The mean of these log ratios is 0, which corresponds to the mean intensity ratio of 1. It is used because of its additive property. One drawback of the log-transformation is that it introduces systematic errors in the lower end of the expression value distribution.

Outliers and Filtering bad & uninteresting data: Outliers in chip experiments can occur at several levels. There can be entire chips, which deviate from all the other replicates. Or there can be an individual gene, which deviates from the other replicates of the same gene. Outliers should consist mainly of quantification errors. In practise, it is often not very easy to distinguish quantification errors from true data, particularly if there are no replicate measurements. If the expression ratio is very low (quantification errors) or very high (spot intensity saturation), the result can be assumed to be an artifact, and should be removed. Most of the actual outliers should be removed at the filtering step (those that have too low intensity values). This is often equivalent to a filtering, where observations with too low or high intensity values are excluded from further analyses.

Skewness and normality: From the statistical point of view skewness and normality should be checked after normalization. The distribution of the data (log-transformed intensity ratio) can be compared before and after normalization.Normality means how well your data fits to the normal distribution. This should be checked, because most of the statistical procedures assume that the data is normally distributed. For checking the skewness, comparison of the mean and median is complementary to this graphical method.

Linearity: After skewness and normality, checking of linearity is a must. Linearity means that in the scatter plot of channel 1 (red colour) versus channel 2 (green colour), the relationship between the channels is linear. It is often more informative to produce a scatter plot of the log-transformed intensities, because then the lowest intensities are better represented in the plot. In this kind of a plot, the data points fit a straight line, if the data is linear.Another way to test linearity is to plot the log ratio versus the intensity of one channel. This should be done for both channels independently.

Normalization

Normalization is the process of removing many sources of systematic variation in microarray experiments that affect the measured gene expression levels. Sources of systematic bias or variation may be due to dye effect, scanner malfunction, uneven hybridization, error in printing tip, plate and reporter effects, batch effect and array design and other experimenter issues. There are three broad groups of normalization techniques, each of which can contain multiple methods.

(i) *Normalization :* It can be defined as the process of transforming a statistic so that it will approximate normal distribution, or be more normal-like. Usually a log-transformation works quite well as a normalization method for microarray data. The simple ratio flattens all the under expressed genes between 1 and 0. Log transformation removes this bias. Log-transformed data gives a more realistic sense of variation, and makes the variation of intensities and ratios of intensities more independent of absolute magnitudes. Log-transformation also stabilizes the variance of high intensity spots.

(ii) *Standardization:* is the process of expanding or contracting the distribution of a statistic so that the experimental values can be compared with those from another experiment. In statistical sense standardization means converting the intensity ratios to Z-scores which are distributed as a standard normal distribution.

(iii) *Centralization:* It is a process of moving a distribution so that it is centered over the expected mean (balancing the two channels). For the logtransformed intensity ratios, an intensity dependent centralization (*e.g.*, lowess) might help to correct the dye bias.

Steps in Performing Normalization

Choice of the method: The selection of the best normalization method depends also very much on how the results turn. Therefore, after inspecting the quality of your data in general it is also advisable to try several different methods and see which one gives the most realible results for your experiment. In any case, inside one experiment the same normalization scheme should be applied uniformly to all chips.

Basic ideas: The basic idea behind all the normalization methods is that the expected mean intensity ratio between the two channels (two-color data) or two chips (one-color data) is one, because only about 10–20% of all the genes in a cell are expressed at any one time.

Control genes: Suitable controls are housekeeping genes and spiked controls.

When using spiked controls, a known labeled DNA target is added to the labeled cDNA sample which is RNA or DNA from another species than the studied one that do not cross-hybridize, but behave similarly to sample RNA or DNA. The expected intensity ratio of the spiked controls is one. If the observed ratio deviates from expected, the observed ratio is used for normalization.

Linearity of data matters: There are several mathematical procedures and different algorithms for normalization. Before deciding on the method, the linearity of data should be checked. If the data is linear median centering can be applied. I the data is non-linear, lowess smoothing or any other local method should rather be applied.

Statistical Analysis of Microarray Data

DNA microarray technology has revolutionized biological sciences in terms of the amount of data obtained from a single experiment. This wealth of array data cannot be interpreted with a simple glance. Before a researcher can identify the genes of any given regulatory circuit their expression levels must be measured with accuracy and confidence. However, because of the presence of experimental and biological variances inherent in high dimensional DNA microarray experiments, this is not a simple task. It is necessary to apply statistical methods to address these fundamental problems. The statistical analyses are performed on normalized expression values to accelerate the research on the DNA microarray data.

The Purposes of Statistical Tests

Microarray studies often aim to identify genes that are differentially regulated across different classes of samples; examples are: finding the genes affected by a treatment, or finding marker genes that discriminate diseased from healthy subjects. Statisticians have developed many new procedures that seem more useful than conventional tests for microarray data. Some studies also aim to identify groups of genes that act together, or to uncover molecular similarities among subsets of samples. The section on Exploratory Analysis describes appropriate techniques for these studies. Some investigators confirm results found by exploratory analysis by using formal statistical methods, or do cluster analysis on differentially expressed genes. Such procedure can be circular in logic; better procedure is to identify differences between groups defined on biological grounds. Microarray data is often used as a guide to further, more precise studies of gene expression by qt-PCR or other methods. Then the goal of the statistical analysis is heuristic: to provide the experimenter with an ordered list of good candidate genes to follow up. Sometimes the experimenter plans to publish microarray results as evidence for changes in gene abundance; in this case it is important to state the correct degree of evidence: the 'p-value'. Many microarray papers still present p-values, as if each gene had been tested in isolation, even though many genes were actually tested in parallel; these p-values are wrong in the context of testing thousands of genes. Perhaps a better way to specify the confidence of microarray results is the 'false discovery rate'.

Exploratory Statistical Analysis

Now a days DNA microarray technology has revolutionized biological sciences in terms of the amount of data obtained from a single experiment. This wealth of array data cannot be interpreted with a simple glance. Before a researcher can identify the genes of any given regulatory circuit their expression levels must be measured with accuracy and confidence. However, because of the presence of experimental and biological variances inherent in high dimensional DNA microarray experiments, this is not a simple task. It is necessary to apply statistical methods to cope with these fundamental problems. It is established that exploratory analysis aims to find patterns in the data that aren't predicted by the experimenter's current knowledge or pre-conceptions. Some typical goals are to identify groups of genes expression patterns across samples are closely related or to find unknown subgroups among samples. First step in all analyses is to identify outliers among samples that appear suspiciously far from others in their group.

Finding Differentially Expressed Genes

Good quality data can be further filtered so that only the genes that show some changes in the expression during the experiment are preserved in the dataset. Often the differentially expressed or otherwise interesting genes are stored as simple lists of gene names. Most typically gene lists are created by driving the data through a certain filter, but they can also include lists of genes, which have been produced by clustering analyses or other more complex methods. Sometimes it is sufficient to get information about genes that are either under or over expressed during the experiment.

Filtering by absolute expression change: If the log-transformed data is used for analysis, the over expressed genes have an expression above zero and under-expressed genes have an expression below zero. However, often the experimental errors are of the order of two, and for finding the over and under expressed genes the cutoffs of -1 and 1 are used. All the genes that fulfill the filter criteria are saved in a new gene list.

Statistical single chip methods: In the absence of replicates, there are still a few statistical methods that can be used, if we want to have statistical significance values for expression changes. However, using such methods carries the risk that the most unstable probes or mRNAs are identified as differentially expressed.

Noise envelopes: It is a method where the standard deviation is used as a cut-off for differentially expressed genes. The variance in the reported gene expression level is a function of the expression level so that the variance is higher in the lower expression values and lower when the expression is high. Thus, the identification of the differentially expressed genes calls for a statistical model of the variance or noise. The simplest model could be constructed on the basis of the standard deviation of expression values. Such a model can be constructed relatively easily using statistical programs. The method relies on a segmental (sliding window) calculation of standard deviation.

Confidence: Significance can only be assessed, if replicate measures of the same gene were performed during the experiment. The chips (or expression of invidual genes on them) can be compared by the t-test (two conditions) or ANOVA (more than two conditions).

When some treatments have replicates: The standard deviation is determined for every gene, and the expression change is compared with the standard deviation. The more the change exceeds the standard deviation between replicates, the more significant the gene is. Note that this method does not allow for the calculation of p-values.

When all the treatments have replicates: two-sample t-test: The two-sample t-test looks at the mean and variance of the two distributions (say, control and treatment chip log ratios), and calculates the probability that they were sampled from the same distribution. The t-test can be applied successfully in situations, where both the control and treatment have been repeated. After calculating the t-test p-values for the replicated genes, the ones with the lowest p-value can be saved into a new gene list and used in further analyses, for example cluster analysis.

When all the treatments have replicates: one-sample t-test: Up and down regulated genes can be more effectively found, if the chips are replicated. In such cases, the one-sample t-test for the deviation from zero (expected mean) can be performed. Finding the differentially expressed genes is even easier, if the Volcano plot is produced. This plot of log ratio versus statistical significance shows also which is the lowest expression change we can be statistically confident in. In our example, 2-fold down-regulated genes have a *p*-value of 0.01. In contrast, 1.4-fold upregulated genes have a similar *p*-value. This indicates that new chip technology coupled with an appropriate number of replicates can produce reliable data even for very small expression changes.

Comparison of Two Groups of Samples: The simplest and most common experimental set-up is to compare two groups: for example, Treatment vs. Control. The issues arising in simple comparisons arise also in more complex settings; it is easier to explain these in the simpler context. The long-time standard test statistic for comparing two groups is the t-statistic:

$$t = (x_{i,1} - x_{i,2}) \; / \; s_i,$$

where $x_{i,1}$ is the mean value of gene i in group 1, $x_{i,2}$ is the mean in group 2, and s_i is the (non-pooled) within-groups standard error (SE) for gene i.

Permutation Tests : Permutation testing is an approach that is widely applicable and copes with distributions that are far from Normal; this approach is particularly useful for microarray studies because it can be easily adapted to estimate significance levels for many genes in parallel. Latest software packages, like SAM, implement permutation testing in a menu-driven interface. The meaning of a p-value from a permutation procedure differs from the meaning of a model-based p-value. The model-based p-value is the probability of the test statistic, assuming that the gene levels in both the treatment and control groups follow the model. A permutation-based p-value tells how rare that test statistic is, among all the random partitions of the actual samples into pseudo-treatment and pseudo-control groups.

Comparison between Several Groups – Analysis of Variance: In order to test for a statistical significance between means, we need to compare or analyze variances. The purpose of Analysis of Variance (ANOVA) is to test for significant differences between the means of several groups. Unlike the t-test, ANOVA does not specify which of the groups are significantly different but only that there exist significant differences. Many current microarray studies compare more than two groups. Sometimes the question is to determine differences among three or more cell lines, or strains of experimental animal. Another common design compares the effect of a particular treatment (often a ligand for a receptor), on cell lines (or animals) with wild-type and mutant versions of the receptor. Usually the experimenter wants to know which genes are actively regulated during treatment in both cell lines, or wants some criterion for selecting those that are differentially regulated among groups.

These questions belong in the tradition of analysis of variance. An empherical example by ANOVA carried out with microarray data on brain expression of *C. carpio* under exposure for varied degrees of temperature is presented in Table 1.

Table 1. Analysis of Variance on brain expression genes of *C.carpio* obtained using SPSS

	Cluster		Error		F	Sig.
	Mean Square	**df**	**Mean Square**	**df**		
B30D23A forward	1348.117	4	3.580	7601	376.593	.000
B30D23A reverse	4252.562	4	.785	7601	5419.244	.000
B30D23D forward	266.151	4	2.956	7601	90.039	.000
B30D23D reverse	852.639	4	2.632	7601	323.904	.000
B30D23E forward	2198.571	4	2.287	7601	961.384	.000
B30D23E reverse	5932.312	4	1.249	7601	4748.196	.000
B30D23F forward	1682.369	4	2.979	7601	564.750	.000
B30D23F reverse	509.221	4	3.504	7601	145.327	.000
B30D2C forward	253.990	4	1.682	7601	150.967	.000
B30D2C reverse	63.920	4	1.821	7601	35.109	.000
B30D2B forward	301.365	4	1.637	7601	184.152	.000
B30D2B reverse	386.911	4	2.520	7601	153.553	.000
B30D2E forward	293.469	4	1.711	7601	171.482	.000
B30D2E reverse	174.594	4	2.192	7601	79.637	.000

Source : Roy *et al.* 2007

Cluster Analysis of Microarray Data

Principle : The basic concepts of clustering are to try to identify and group together similarly expressed genes and then try to correlate the observations to biology. The idea is that co-regulated and functionally related genes are grouped into clusters. Clustering provides the framework for this analysis. Basic principles of clustering are as follows:

1. Clustering organizes the data into a small number of homogeneous groups. The methods are used to find similar expression motifs irrespective of the expression level. Therefore, both low and high expression level genes can end up in the same cluster if the expression profiles are correlated by shape.
2. Clustering methods can be grouped as supervised and unsupervised. Supervised methods assign some predefined classes to a data set whereas in unsupervised methods no prior assumptions are applied.

Types : Different types of clustering involves Hierarchical clustering, K-means, self-organizing maps (SOMs), principal component analysis (PCA), and K-nearest neighbor (KNN) have been commonly used. There are also other methods, such as Correspondence Analysis, Fuzzy K-Means Clustering Multidimensional scaling (MDS), minimum description length (MDL), gene shaving (GS), decision trees, and support vector machines (SVMs).

1. Hierarchical clustering

The hierarchical clustering can be represented as a tree, or a dendrogram. Branch lengths represent the degree of similarity between the genes. Hierarchical clustering consists of two separate phases. The hierarchical clustering is performed as follows:

(i) Each gene is assigned to a cluster of its own.

(ii) The closest pair of clusters is found and merged into a single cluster.

(iii) The distances (similarities) between the new clusters are computed and each of the old clusters using either single, average or complete linkage method.

(iv) Steps 2 and 3 are repeated until all genes are clustered.

In single linkage clustering, the distance between one cluster and another is considered to be equal to the shortest distance from any member of one cluster to any member of the other cluster. In complete linkage clustering, the distance between one cluster and another cluster is considered to be equal to the longest distance from any member of one cluster to any member of the other cluster. In average linkage, the distance between one cluster and another cluster is considered to be equal to the average distance from any member of one cluster to any member of the other cluster. An example of hierarchical clutering carried out with microarray data of *C. carpio* is presented in figure 1 for visualization.

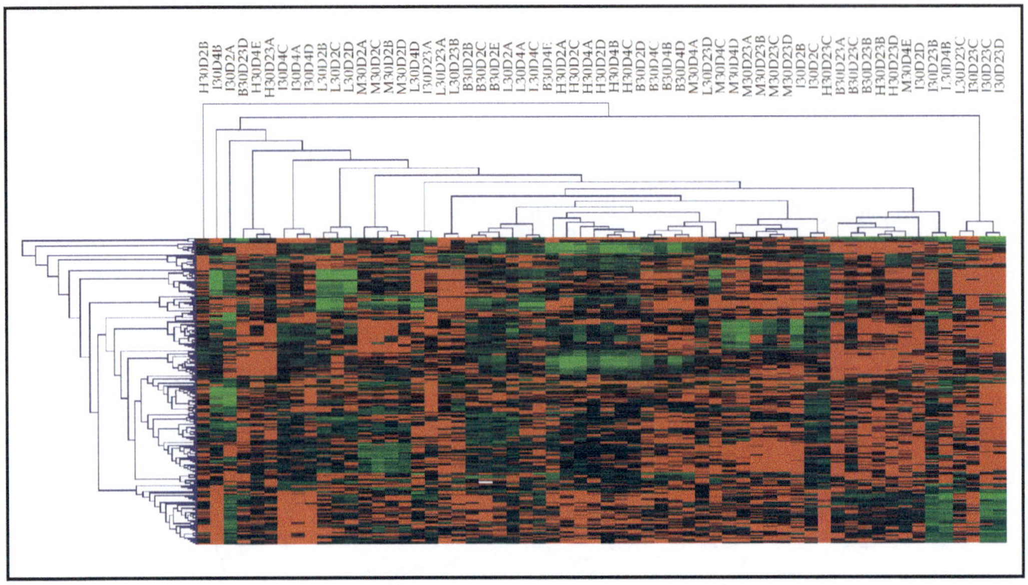

Fig. 1. An Example of Hierarchical clustering of *C. carpio* obtained using MeV software

2. Self-organizing map

Kohonen's self-organizing map (SOM) is a neural net that uses unsupervised learning. SOM tries to learn to map similar input vectors (gene expression profiles) to similar regions of the output array of nodes. The method maps the multidimensional distances of the feature space to two-dimensional distances in the output map. In SOMs, the number of clusters has to be predetermined. The dimensions of the two-dimensional grid or array give the value. One has to experiment with the actual number of clusters and find such array dimensions that there is a minimum number of poorly fitting genes in the clusters. The actual number of clusters is also difficult to assign, but the square root of the number of genes is a good initial estimate.

3. K-means clustering

K-means is a least-squares partitioning method for which the number of groups, K, has to be provided. The algorithm computes cluster centroids and assigns each object to the nearest centroid. However, it is also possible to estimate K from the data, taking the approach of a mixture density estimation problem.

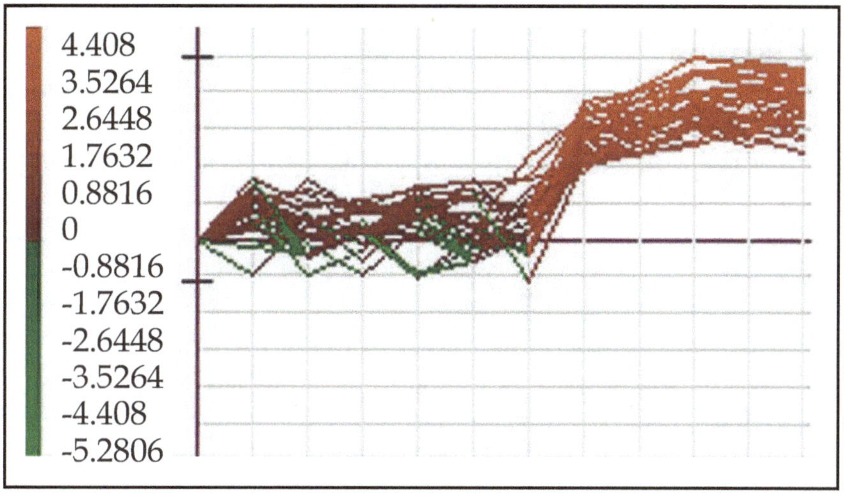

Fig. 2. An example of self-organizing map (Source; Hovatta *et al.* 2005)

(i) The genes are arbitrarily divided into K centroids. The reference vector, *i.e.* location of each centroid, is calculated.

(ii) Each gene is examined and assigned to one of the clusters depending on the minimum distance.

(iii) The centroid's position is recalculated.

(iv) Steps 2 and 3 are repeated until all the genes are grouped into the final required number of clusters.

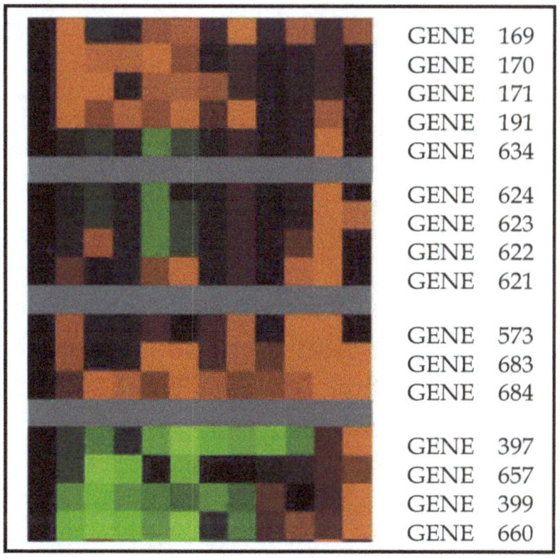

Fig. 3. An example of k-means clustering with four clusters
Source: Hovatta *et al.*2005

K-means partitioning is a so-called NP-hard problem, thus there is no guarantee that the absolute minimum of the objective function has been reached. Therefore, it is good practise to repeat the analysis several times using randomly selected initial group centroids, and check whether these analyses produce comparable results. An example of K-means expression graph of *C. carpio* generated by using MeV software is presented in figure 4.

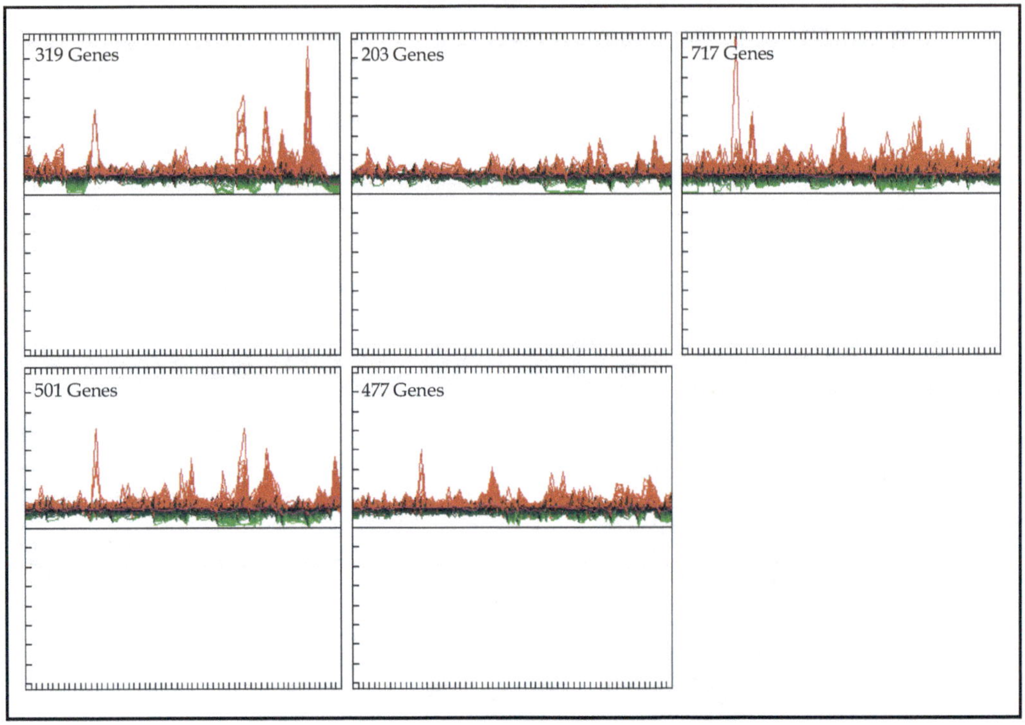

Fig. 4. K-means Expression Graph of *C. carpio* by MeV software
Source : Roy et al., 2007

4. KNN classification - a simple supervised method

K-nearest neighbor (KNN) is a simple classification method used for finding a set of genes that differentiates two or more groups of samples (here, chips). The idea is to build a classifier, *i.e.*, find the set of genes that differentiates the groups of interest. For optimal performance we would need a training set and a test set. Training set is used for building the classifier, and the test set is used for the validation of the classifier. If no test set is available, the classifier can be validated using a leave-one-out cross-validation, but the performance is likely to be inferior. The KNN method has three phases, neighborhood analysis, class prediction, and validation.

(i) In *neighborhood analysis*, every gene is represented by an expression vector that consists of the gene's expression in every tumour sample. The tumour classes are represented by an idealized expression vector, where one group is marked with 0, and the other group with 1. A correlation between the gene's expression vector, and the idealized expression vector is calculated using some correlation measure, and the result is verified using permutation testing. Only the genes with the best correlation measures are selected for the class prediction phase.

(ii) In the *class prediction* phase, the group is predicted for every sample based on votes. The vote is based on the similarity (correlation) of the gene's expression to the mean expression of the groups which later summed to obtain total votes. The sample is assigned to the group with a higher number of votes.

(iii) In the last phase the performance of the classifier is tested or *validated*. Validation is needed to rule out the effect of sampling error on the construction of the classifier. The most reliable way to test the classifier is to collect a new group of samples, and test how well the classifier performs on them. If the performance is bad, a more reliable classifier might be built, and the whole process starts again. If the performance is good, the classifier might be used. If no test set is available, cross-validation can be used. In cross-validation, one by one, the samples are left out of the training set, the classifier is built, and its preformance is tested on the sample that was left out. This leave-one-out procedure goes through every sample of the set, and stops when all the samples have been left out once. The genes that perform consistently for every cross-validation sample are then selected for the final classifier.

5. Principal component analysis

Objectives of principal component analysis (PCA) are to discover or to reduce the dimensionality of the data set and to identify new meaningful underlying variables. PCA transforms a number of correlated variables into a smaller number of uncorrelated variables called principal components. The basic idea in PCA is to find the components that explain the maximum amount of variance possible by n linearly transformed components. The first principal component accounts for as much of the variability in the data as possible, and each succeeding component accounts for as much of the remaining variability as possible. PCA can be also applied when other information in addition to the actual expression levels is available. This is also applied to SOM and K-means methods.

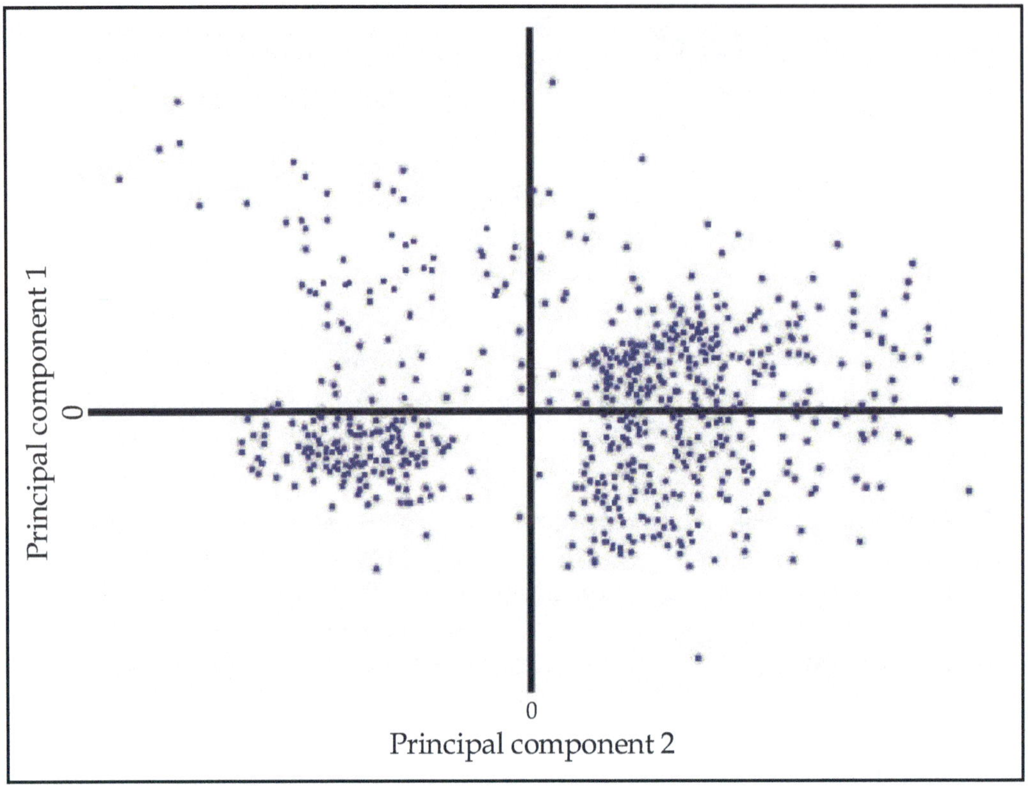

Fig. 5. An example of principal component analysis. The two most significant principal components have been selected as the axes of the plot (Source; Hovatta *et al.*, 2005)

6. Correspondence Analysis

Correspondence analysis is an explorative method to study associations between variables. (Fellenberg *et al.* 2001, Culhane *et al.* 2002) Like principal components, it displays a low-dimensional projection of the data. But, in this case, both genes and samples can be projected onto the same space, revealing associations between them. Correspondence analysis requires an expression matrix with no missing values. Genes that lie close to one another on the plot tend to have similar profiles, regardless of their absolute value. The same is true for samples. If some genes and samples lie close to one another on the plot, then these genes are likely to have a high expression in the nearby samples relative to other samples that are far away on the plot. On the other hand, if a set of genes are on the opposite side of the plot from a set of samples relative to the origin, then the expression of that set of genes is likely to be depressed in those samples relative to samples that might be positioned close to those genes. The farther the points are from the origin, the stronger the association between genes and samples. Correspondence analysis works by decomposing a matrix of chi-squared values derived from the rows and columns of the expression matrix. The first two or three axes are the most informative in showing associations among genes and experiments. The amount of information explained by a given axis is quantified by its inertia, which may be thought of as the proportion of the total chi-squared value of the matrix explained by that axis. The inertia values are provided under the corresponding node under the main COA analysis node.

7. Fuzzy K-Means Clustering

The fuzzy k-means clustering (Fig. 6) is done with based on following equation.

$$J(K,m) = \sum_{k=1}^{K} \sum_{i=1}^{N} (u_{ki})^m \, d^2 (x_i, c_k)$$

K and N are the number of clusters and genes in the data sets, m is a parameter which relate to 'fuzziness' of resulting clusters, uki is the degree of membership of gene x_i in cluster *k*, $d^2 (x_i; c_k)$ is the distance from gene x_i to centroid c_k. The parameters in this equation are the cluster centroid vector ck and the components of the membership vectors u_{ki}. These unknown parameters can be optimized by La- grange method. Calculated u_{ki} shows the belonging ratio to a cluster k and centroid c_k shows the representative gene expression profile of a cluster k. In this study, a parameter m was set to 2.0 and the number of clusters was set to 5. For the number of the clusters in the other clustering method, we selected the same number as that of clusters using Fuzzy k-means clustering in order to compare the clustering results.

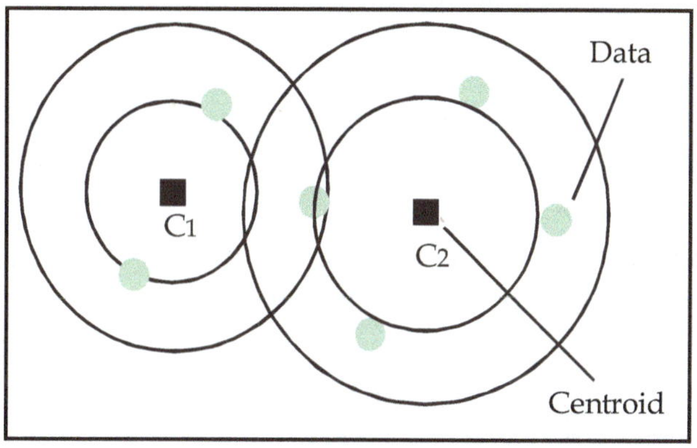

Fig. 6. Fuzzy k-means clustering (Source Arima *et al.*)

Multidimensional Scaling

Multidimensional scaling (MDS) is an approach to multivariate analysis aimed at producing a spartial or geometrical representation of complex data. It is used to produce set of points that represent the objects in some low-dimensional Euclidean space when the measure of similarity or dissimilarity between a set of objects is known. Given an n x n dissimilarity matrix D=(dij) for n objects (e.g. genes or samples), multidimensional scaling (MDS) tries to find n points in Euclidean space with a similar distance structure D′ = (d′ij). Multi-dimensional scaling (MDS) makes the most useful graphical displays. Classical MDS is identical to PCA for most data sets; however if you fix a dimension a modern iterative algorithm can place the points in an arrangement that is more representative of true distances, than are the same number of principle components.

Gene Regulatory Networks

Gene expression data permits us to go beyond the traditional line of research and to study finer structures of molecular pathways exposing causal regulation relations between genes. It gives more intricate view of molecular interactions offering further possibilities for medical interventions. Inferring regulatory networks from expression data, a process called reverse-engineering, is not a computationally simple problem because an enormous amount of time is needed even when a trivial approach is applied.

Visualization of Microarray Gene Expression Data

After preprocessing has been performed on the data set, CEBS provides the users with exploratory plots. Exploratory plots are an intuitive way to obtain gene expression data and to judge the comparability of the data sets that are selected from single or multiple experiments. The visual presentation of data can serve as part of an explorative analysis to help users determine the general trends in the data. To construct the above exploratory plots in a convenient and intuitive way, a series of color codes are used to represent samples from different experiments.

1. The *Box-Whisker Plot* portrays the essential data distribution characteristics, such as maximum, minimum, median, lower quartile, upper quartile, and outliers, for each of the arrays the user selected for analysis.
2. The *Dendrogram* allows the users to view the relative similarities between arrays by clustering the arrays with similar patterns and displaying the results in a hierarchy. The distances used for clustering are 1-pearson

correlations between expression values (in log scale) from arrays.< The left axis shows the distances between arrays or average distances between clusters of arrays.

3. *The Multi-dimensional Scaling* projects the relative distances between arrays to two-dimensional space with a multi-dimensional scaling algorithm and portrays the relative similarities between arrays, i.e. arrays with expression similarities cluster close to one another.

Conclusion

Well designed experiments with a clearly defined objective and with data stored in sufficient detail in standard formats should provide a valuable resource for discovering many different aspects of gene interactions, metablic pathways and regulatory effects. By combining these data sets with other types of genomic and proteomic data, the computational biology community can build models of major biological processes.

REFERENCES

Hsiao, CD; Chen, YC; Lin, WD; Chuang, NY; Shieh, YE; Wang, YC; Ho, JM; Hwang, PP. (2005). Construction of Integrated Database for Fish Expressed Sequence Tags. *Journal of the Fisheries Society of Taiwan*. 32:1, p. 11.

Clark, MD; Hennig, S; Herwig, R; Clifton, SW; Marra, MA; Lehrach, H; Johnson, SL. (2001). An Oligonucleotide Fingerprint Normalized and Expressed Sequence Tag Characterized Zebrafish cDNA Library. *Genome Research*. 11:9, pp. 1594-1602.

Lien, Ching-Ling; Schebesta, Michael; Makino, Shinji; Weber, Gerhard J; Keating, Mark T. (2006). Gene Expression Analysis of Zebrafish Heart Regeneration. *PLoS Biol*. 4(8):[np].

Alberti, Martin; Kausch, Ulf; Haindl, Stefanie; Seifert, Martin. (2005). Gene Expression Analysis for Exposure to Estrogenic Substances. *Acta Hydrochimica et Hydrobiologica*. 33:1, pp. 38-44.

Weidinger, G; Thorpe, CJ; Wuennenberg-Stapleton, K; Ngai, J; Moon, RT*.(2005). The Sp1-Related Transcription Factors sp5 and sp5-like Act Downstream of Wnt/ beta - Catenin Signaling in Mesoderm and Neuroectoderm Patterning *Current Biology*. 15:6, pp. 489-500. 29

Bonnet, E; Fostier, A; Bobe, J*. (2007) Microarray-based analysis of fish egg quality after natural or controlled ovulation. *BMC Genomics*. 8(55): 17.

Bobe, J; Montfort, J; Nguyen, T; Fostier, A. (2006). Identification of new participants in the rainbow trout (Oncorhynchus mykiss) oocyte maturation and ovulation processes using cDNA microarrays. *Reproductive Biology and Endocrinology*. 4, 39, 16 pp.

Larkin, Patrick; Villeneuve, Daniel L; Knoebl, Iris; Miracle, Ann L; Carter, Barbara J; Liu, Li; Denslow, Nancy D; Ankley, Gerald T. (2007) Development and Validation of a 2,000-Gene Microarray for the Fathead Minnow (Pimephales Promelas). *Environ. Toxicol. Chem.* 26(7):1497-1506.

McLean, TI; Crawford, DL. (2006). Towards a Genetic Understanding of Karenia brevis: Using Microarrays to Discover Gene Expression Patterns as They Relate to K. brevis Ecology EOS, Transactions, American Geophysical Union [*EOS Trans. Am. Geophys. Union*]. 87(36): suppl., [np]. suppl.

Sneddon, LU; Margareto, J; Cossins, AR. (2005). The Use of Transcriptomics to Address Questions in Behaviour: Production of a Suppression Subtractive Hybridisation Library from Dominance Hierarchies of Rainbow Trout. Physiological and Biochemical Zoology [*Physiol. Biochem. Zool.*]. Vol. 78, no. 5, pp. 695-705.

Hook, SE; Skillman, AD; Small, JA; Schultz, IR. (2006). Dose-response relationships in gene expression profiles in rainbow trout, Onchorhyncus mykiss, exposed to ethynylestradiol. *Mar. Environ. Res.* Vol. 62, no. suppl. 1, pp. S151-S155.

Martin, SAM; Blaney, SC; Houlihan, DF; Secombes, CJ. (2006). Transcriptome response following administration of a live bacterial vaccine in Atlantic salmon (Salmo salar) Molecular Immunology [*Mol. Immunol.*]. Vol. 43, no. 11, pp. 1900-1911.

Tsoi, SCM; Cale, JM; Bird, IM; Ewart, V; Brown, LL; Douglas, S. (2003). Use of Human cDNA Microarrays for Identification of Differentially Expressed Genes in Atlantic Salmon Liver During Aeromonas salmonicida Infection. *Marine Biotechnology [Mar. Biotechnol.*]. Vol. 5, no. 6, pp. 545-554. Dec 2003.

Derome, N; Duchesne, P; Bernatchez, L*. (2006). Parallelism in gene transcription among sympatric lake whitefish (Coregonus clupeaformis Mitchill) ecotypes *Mol. Ecol.*. Vol. 15, no. 5, pp. 1239-1249.

Anderson, DM; Kulis, D; Erdner, D; Ahn, S; Walt, D. (2006). Fibre optic microarrays for the detection and enumeration of harmful algal bloom species. *Afr. J. Mar. Sci.* 28(2): 231-235.

Roy, A.K., Sahu, C. and Martha, S. R. (2007). Cluster analysis of gene expression data of *C. carpio*. International Conference in Applied Bio-engineering held at Chennai, 5-7 Dec, 2007 organized by Sathyabhama University.

Hovatta, I., Kimppa, K., Lohmussola, A., Parsenen, T. (2005). DNA Microarray Data Analysis, 2nd ed. Pleasel Oy Helsinlai, 2005.

Fellenberg, K., Hauser, N.C., Neutzner, A., Houeisel, J.D., and vingren, M. (2001). Correspondence analysis applied to microarray data, PNAS, 98:10781-10786.

Chapter **6**

Recent Biotechnological Approaches in Fish Disease Diagnosis

P.K. Sahoo

Introduction

The United Nations' Food and Agricultural Organisation reported that global aquaculture is increasing by 11 percent per year and is the world's fastest growing food-producing sector. More than 38 million people are employed in aquaculture and associated industries worldwide, and 131 aquatic species are currently being commercially cultured. In 2002, Asia led the world in the aquaculture production, with about 12.4 million tonnes, while Europe and America together produced 1.2 million tonnes.

Among the Asian countries, India ranks second in aquaculture and third in capture fisheries. Freshwater aquaculture in India is mainly based on Indian major carps (*Labeo rohita*, *Catla catla* and *Cirrhinus mrigala*) with an annual production of 1.8 million tonnes, and exotic carps, which together contribute to over 87% of the total aquaculture production of 2.2 million tonnes. Disease problems represent important constraint to the aquaculture industry. The need to control endemic diseases imposes severe year-on-year costs on producers. The advances made in the development of vaccines for diseases of Indian major carps are very meager.

Applied Bioinformatics, Statistics and Economics in Fisheries Research: pp. 73-92

New India Publishing Agency, New Delhi (India)

1. Disease as a major constraint

The incidence and emergence of new infectious diseases has almost paralleled the growth of aquaculture industry. The occurrences of diseases are mostly due to poor husbandry, intensive culture, and lack of health management practices, expansion of ornamental fish trade, unexpected negative interaction between cultured and wild fish populations, poor biosecurity measures, introduction of sick fish to a healthy zone, slow awareness of countries to emerging diseases, climate change, human-mediated movement of aquaculture commodities and misunderstanding or misuse of specific pathogen free (SPF) stocks (eg. shrimps).

2. Socio-economic impacts

The movement of fish, eggs and genetic material from country to country or region to region has resulted in the introduction of new diseases. The economic impact of fish disease is difficult to determine, but it may be as high as 10 to 15 percent of the total value of fish production worldwide (Klesius et al., 2007). Certain diseases may destroy the entire production chain and result in the destruction of healthy fish in affected area to control the epizootic form spreading to other regions or countries. ADB/NACA (1991) reported that the minimum, conservatively estimated loss of farm production in 1990 to fish diseases such as epizootic ulcerative syndrome (EUS) of fish, penaeid shrimp diseases and a variety of other diseases causing losses in finfish pond culture and marine cage culture in 15 developing Asian countries was US$1.36 million. At the global level, combined estimated losses in production value due to shrimp diseases from 11 countries from 1987 to 1994 were about $3.01 billion (Bondad-Reantaso and Subasinghe, 2005).

If we consider one state of India i.e., Andhra Pradesh, that produces more than 40,000 tonnes of Indian major carps valued at more than 600 million rupees a year suffers a loss of about 40 million rupees due to diseases. Similarly, an economic loss of 9450 million rupees due to viral diseases in shrimp farming in 1995-96 has been recorded (Shankar *et al.,* 2000). Even one bacterial pathogen, *Aeromonas hydrophila* has been recorded to cause mortality of 30-70% in rearing carp ponds (Shankar *et al.,* 2000). Thus, to sustain aquaculture industry, appropriate health management measures with development of rapid diagnostics and vaccines are necessary.

3. Disease diagnostic system

The rapid detection of pathogens in infected fish, both clinically and sub-clinically, is desirable for effective health management in aquaculuture. If pathogens can also be detected and identified in the environment, for example, during the course of culture, before restocking or between harvesting or before disease outbreaks, then this can be extremely useful to act as early warning signals. Development of sensitive, specific, simple, rapid and cost-effective diagnostics are essential to detect various stages of infections, serotyping of pathogens and epidemiological studies. There is a need to detect infectious agents in genetic material or live fish or fish products with a rapid diagnostic tool (at least 10 copies of infectious agent). Traditional methods of disease diagnosis or detection of pathogens rely upon culture and biochemical identification that may take few days to weeks with poor sensitivity and specificity. Recently, development of antibody-based and nucleic acid based detection systems have come up that can be completed in few hours to days. These monoclonal antibody based detection and PCR based detection systems have revolutionized the whole world in disease diagnosis. Further modification and introduction of real time PCR and bacteriophage-based detection along with engineered antibodies are recent approaches that have to be taken up in aquaculture. This chapter aims to provide the recent developments in fish disease diagnosis worldwide and also highlighted some of the important tools that are coming up in the human or animal sector for disease diagnosis.

4. Latest developments in the detection system

4.1. Traditional methods

Traditional bacteriology, virology, histology, parasitology and mycology are appropriate for detection of common and easily culturable pathogens; however, these are quite expensive, time-consuming and might not lead to definite diagnosis in many cases. Particularly in cases of slow growing or fastidious organisms which take long time to grow in culture media poses a great problem due to overgrowth of contaminants many times. Therefore, combination of methods have been developed that are based on immunological and molecular techniques.

4.2. Immunlogical methods

Immunological mthods, such as fluorescent antibody technique (FAT), indirect FAT (IFAT), immunohistochemistry (IHC), ELISA, dot blot or western blot enable the rapid and specific detection of pathogens without a need for isolation of pathogens. The development of monoclonal antibody technique has made the above techniques more easy, sensitive and specific. However, each method has its own advantages and disadvantages. The techniques like FAT and IFAT, which can be completed within two hours and are very sensitive; but requires fluorescent microscope and experienced personnel to diagnose by avoiding background staining. Nevertheless, these are being in use in diagnosis of viral diseases of fish such as infectious salmon anaemia virus (ISAV) (Adams and Thompson, 2006). Similarly, IHC that requires only light microscope and diagnosis from formalin-fixed tissues followed by routine histology with incubation in pathogen-specific antibody. But it takes a long time of 2-3 days and it is less sensitive than FAT. However, it helps in visualizing surrounding cells histological changes associated with the disease. Similarly, the ELISA, a powerful technique used to detect either antigen or antibody that takes few hours to complete and requires a reader for quantitation to process large number of samples at a time, is also not free from limitations particularly with reference to detection of sub-clinical cases. ELISA is used for sero-surveillance, health monitoring for a particular pathogen, broodstock screening or post-vaccination antibody titration. Some of the most common ELISA variations used in diagnostics are: dot-ELISA, direct ELISA, indirect ELISA, TAS-ELISA, competitive ELISA, sandwich ELISA, biotin-avidin ELISA.

4.2.1. Monoclonal antibodies

The importance of antibodies in disease diagnosis, serotyping, antigen characterization, identification of microbes, epidemiology and vaccine development is well known. The limitations of conventional antibody techniques viz., cross reaction, quantity limitation, unwanted background reactions, inability to discriminate antigen at epitope level led to the development and popularization of monoclonal antibodies (Shankar et al., 2000).

After the initial development of Kohler and Mielstein in 1975, monoclonal antibody (MAbs) has made tremendous impact in developing disease diagnosis tools. MAbs production involves antibody producing lymphoid cells from immunized mice. These cells are then immortalized by hybridization with an established myeloma cell line followed by separation of individual hybridoma clones. The clones obtained are grown in tissue culture to produce sufficient quantities of MAbs.

Monoclonal antibodies are used to improve or develop ELISA methods or immunofluorescent antibody tests that are used to screen infected fish. The signal molecule incorporated onto the monoclonal antibody can be an enzyme, gold particles, fluorescent dye or bioluminescent compounds.

Depending on the kind of monoclonal antibody signal molecule, the antigen-antibody complex can be detected by

- photometric apparatus through a substrate chemical reaction
- an electron microscope

- a bioluminescent apparatus through light emissions, or
- a light microscope through visible fluorescent reaction.

MAbs are currently available for many of fish bacterial, viral and fungal pathogens world-wide. However, there is only handful of MAbs available in India. To name, these are against *Aeromonas hydrophila* infection of fish, *Aphanomyces invadans* associated with EUS lesions of fish and white spot virus infection of shrimps.

4.2.2. Engineering antibodies for clinical applications

Molecular engineering has contributed a lot to the clinical success of antibodies in recent years. It has been possible to modify the molecular size, pharmacokinetics, immunogenicity, binding affinity, specificity and effector function of antibodies. In addition fusion proteins of antibodies with various proteins and peptides yielded targeted biological modifiers, toxins and imaging agents (Jain *et al.*, 2007). The development in this sector is yet to take up in aquaculture. The below mentioned process briefly says about the process of production of engineered antibodies (Fig. 1).

Cloning and sequencing of target genes/gene of interest

(eg. viral coat protein; bacterial surface protein or toxin gene; parasite cilia protein)

↓ ligation

Expression vector

↓ transformation

Expression systems (*E. coli*, yeast or insect cell lines)

↓ expression, purification and scale up

Purified recombinant antigen

↓

Immunoassays (ELISA/FAT)

Fig. 1. Production of recombinant antigens for antibody based detection systems

4.2.3. Multiplex testing systems

It is a testing system that uses Luminex technology to detect a variety of pathogens or immune response molecules (viz., lysozyme, cytokines, IgM) simultaneously. It offers a new dimension in fish health management and epitope mapping for vaccine development.

4.3. Molecular methods

4.3.1. Nucleic acid probes

Nucleic acid probes are segments of DNA or RNA that have been labeled with enzymes, antigenic substrates and chemiluminescent moieties or radioisotopes. They can bind to nucleic acid with high specificity. Probes are 20 to 1000 bases long and can be directed both against DNA or RNA as target. Oligonucleotide probes (< 50 bases long) have the advantage of hybridizing more rapidly to target molecules and can be chemically synthesized. Now a days the advent of non-radioactive labels such as biotin and digoxigenin has made the technique wide in its use for many bacterial and viral disease diagnosis with higher sensitivity.

4.3.2. Polymerase Chain Reaction (PCR)

The advent of PCR in the mid-1980s revolutionized molecular biology methods of detecting pathogens. The basic application of PCR is the amplification of small amount of template DNA or RNA into large quantities in few hours. This is the most senisitive and specific method, which is used to detect pathogens both from the environment as well as fish samples. Besides disease diagnosis, this is also used for epidemiological studies for strain differentiation, and in particular, the differentiation of closely related strains. rRNA genes (5S, 16S and 23S rRNA) are very often used as the target DNA for detecting pathogens because they contain both highly conserved region and variable species-specific regions (Adams and Thompson, 2006). Some of the most common PCR variations used in diagnostics are: nested PCR, semi-nested PCR, multiplex PCR, touch down PCR, random amplified polymorphic DNA (RAPD), reverse transcriptase PCR (RT-PCR), reverse cross blot PCR (rcb-PCR), restriction fragment length polymorphism-PCR (RFLP-PCR), repetitive extragenic palindromic PCR (rep-PCR), ERIC-PCR (enterobacterial repetitive intergenic consensus PCR), immunocapture PCR, loop-mediated isothermal amplification PCR (LAMP-PCR), real time PCR; and RT-PCR enzyme hybridization assay, which offers the advantages of detecting live pathogens and a large sample throughput. Besides, in situ hybridization and in situ hybridization-PCR are techniques used to visualize DNA or RNA sequences within cells for localization of infections, portal of entry studies, map gene sequences to chromosomes and identify sites of gene expression.

4.3.3. Realtime PCR

Real-time reverse-transcriptase (RT) PCR quantitates the initial amount of the template most specifically, sensitively and reproducibly, and is a preferable alternative to other forms of quantitative RT-PCR that detect the amount of final amplified product at the end-point. Real-time PCR monitors the fluorescence emitted during the reaction as an indicator of amplicon production during each PCR cycle (ie, in real time) as opposed to the endpoint detection. The real-time progress of the reaction can be viewed in some systems. Real-time PCR does not detect the size of the amplicon and thus does not allow the differentiation between DNA and cDNA amplification, however, it is not influenced by non-specific amplification unless SYBR Green is used. Real-time PCR quantitation eliminates post-PCR processing of PCR products (which is necessary in competitive RT-PCR). This helps to increase throughput and reduce the chances of carryover contamination. In comparison to conventional RT-PCR, real-time PCR also offers a much wider dynamic range of up to 10^7 fold (compared to 1000-fold in conventional RT-PCR). Dynamic range of any assay determines how much target concentration can vary and still be quantified. A wide dynamic range means that a wide range of ratios of target and normaliser can be assayed with equal sensitivity and specificity. It follows that the broader the dynamic range, the more accurate the quantitation.

The real-time PCR system is based on the detection and quantitation of a fluorescent reporter. This signal increases in direct proportion to the amount of PCR product in a reaction. By recording the amount of fluorescence emission at each cycle, it is possible to monitor the PCR reaction during exponential phase where the first significant increase in the amount of PCR product correlates to the initial amount of target template. The higher the starting copy number of the nucleic acid target, the sooner a significant increase in fluorescence is observed. A significant increase in fluorescence above the baseline value measured during the 3-15 cycles indicates the detection of accumulated PCR product.

Real-time PCR based diagnosis of fish diseases has taken up in a large scale for many of the diseases such as infectious pancreatic necrosis virus, infectious haematopoietic necrosis virus and infectious salmon anaemia virus infections of fish and white spot syndrome virus and yellow head virus infections of shrimp.

4.3.4. Nucleic acid based sequence amplification (NASBA)

NASBA is a robust amplification technology that is highly appropriate for diagnosis of RNA viruses as well as other pathogens. This has already been developed for fish nodavirus and Taura syndrome virus of shrimps. NASBA is a single-step, enzyme-based RNA amplification reaction employing 3 enzymes: AMV-RT, ribonuclease H and T7 RNA polymerase. Two target specific primers, including one with a 5′ extension containing the promoter sequence for T7 RNA polymerase is to be designed to amplify the RNA target in exponential kinetics through double-stranded cDNA intermediates. The amplicons are mainly single stranded RNA, with a polarity opposite to that of the target. It is extremely sensitive, rapid and specific (Deiman *et al.*, 2002).

4.3.5. Microarray technology

Molecular biology research evolves through the development of the technologies used for carrying them out. It is not possible to do research on a large number of genes using traditional methods. DNA microarray is one such technology which enables the researchers to investigate and address issues which were once thought to be non traceable. One can analyze the expression of many genes in a single reaction quickly and in an efficient manner. DNA microarray technology has empowered the scientific community to understand the fundamental aspects underlining the growth and development of life as well as to explore the genetic causes of anomalies occurring in the functioning of the human body.

A typical microarray experiment involves the hybridization of an mRNA molecule to the DNA template from which it is originated. Many DNA samples are used to construct an array. The amount of mRNA bound to each site on the array indicates the expression level of the various genes. This number may run in thousands. All the data is collected and a profile is generated for gene expression in the cell.

An array is an orderly arrangement of samples where matching of known and unknown DNA samples is done based on base pairing rules. An array experiment makes use of common assay systems such as microplates or standard blotting membranes. The sample spot sizes are typically less than 200 microns in diameter usually contain thousands of spots. Thousands of spotted samples known as probes (with known identity) are immobilized on a solid support (a microscope glass slides or silicon chips or nylon membrane). The spots can be DNA, cDNA, or oligonucleotides. These are used to determine complementary binding of the unknown sequences thus allowing parallel analysis for gene expression and gene discovery. An experiment with a single DNA chip can provide information on thousands of genes simultaneously. An orderly arrangement of the probes on the support is important as the location of each spot on the array is used for the identification of a gene.

DNA microarray technology helps researchers learn more about different diseases such as heart diseases, mental illness, infectious disease and especially the study of cancer. Until recently, different types of cancer have been classified on the basis of the organs in which the tumors develop. Now, with the evolution of microarray technology, it will be possible for the researchers to further classify the types of cancer on the basis of the patterns of gene activity in the tumor cells. This will tremendously help the pharmaceutical community to develop more effective drugs as the treatment strategies will be targeted directly to the specific type of cancer. This technique is now coming up in aquaculture sector and in next few years its use in couple of laboratories may be seen.

4.4. Bacteriophage used in bacterial detection

This is latest development in diagnostic science where bacteriophages are used in diagnosis. This concept is yet to pick up for aquatic microbes. Bacteriophage are viruses that infect bacteria. Phage are either virulent (lytic) or temperate and use the host bacterium as a factory for their own replication. In fact, it is difficult to over-emphasize the huge impact that phage research has had on molecular biology. For example, phage research was instrumental in defining central dogma; the identification of DNA as genetic material, gene mapping, discovery of mRNA as an intermediate for the production of proteins besides its use in molecular biology (transducing phage, restriction enzymes, phage promoter based expression systems, random mutagenesis, phage based cloning vectors and genomic libraries and phage display techniques) (Petty *et al.,* 2007).

Although the use of phage in the synthesis of nanostructures is still in development, their use in bacterial detection is more advanced. Many methods for detecting specific bacteria are available, but the low-cost and ready production of large numbers of phage, added to their specificity for a target bacterial species, makes them ideal for bacterial detection. The major advantage of phage based detection is detection of live bacteria which reduces false positive cases that are being detected by PCR system. Classically, bacterial pathogens were identified by phage typing and by enrichment on selective growth media, in addition to biochemical tests, which is a lengthy procedure. Recently, recombinant phage based detection system has come up. Upon infection of their hosts, these engineered phage deliver reporter genes, such as the *luxAB* genes from *Vibrio harveyi*. Replication of the viral genome results in production of many copies of the reporter gene, and subsequent expression of these reporter genes ensures the amplification of the initial phage-bacterium interaction into a signal that can be readily detected - bioluminescence in the case of *luxAB* (Petty *et al.,* 2007). Otherwise, phage infection of the pathogenic bacterium results in release of many phage particles that can be detected by a second non-pathogenic sensitive bacterial strains. This technique has been used successfully to detect *Mycobacterium tuberculosis*. Besides, phage based detection system has been used to detect bacteria like *Pseudomonas aeruginosa, Staphylococcus aureus, E. coli, Salmonella typhimurium* and *Listeria monocytogenes.*

5. Bioinformatics in fish disease diagnosis

Bioinformatics and computational biology involve the use of techniques including applied mathematics, informatics, statistics, computer science, artificial intelligence, chemistry, and biochemistry to solve biological problems usually at the molecular level. Research in computational biology often overlaps with systems biology. Major research efforts in the field include sequence alignment, gene finding, genome assembly, protein structure alignment, protein structure prediction, prediction of gene expression and protein-protein interactions, and the modeling of evolution. Now-a-days number of softwares are pouring in to simplify the molecular tools. Few important websites and their brief utility for fish biotechnology research including disease diagnosis and related areas are collected from the web and described hereunder.

A. Primer design webservers for PCR, mutagenesis and RNAi

- AntiSense Design - Design antisense primers at IDT
- AutoPrime - designs primers that are specific for expressed sequences (mRNA)
- CODEHOP - COnsensus-DEgenerate Hybrid Oligonucleotide Primers
- ELXR - Exon Locator and eXtractor for Resequencing - Design exon amplification primers for a complete gene
- ELXRdb - Exon Locator and eXtractor for Resequencing database - Retrieve precalculated primers for a complete gene
- Exonprimer - Design primers for the amplification of exons with intronic primers
- GAP (Genome- wide Automated Primer finder servers) - Design thousands pairs of primers for amplifying a set of open reading frames over an entire genome
- Genefisher - Interactive PCR primer design
- MEDUSA - A tool for automatic selection and visual assessment of PCR primer pairs (Karolinska)
- Methprimer - Design primers for methylation PCR

- MPrime - an interface which allows the efficient detection of multiple primers or oligonucleotides for genic regions in either the human, mouse, rat, zebrafish, or fruit fly genomes
- NetPrimer - Primer design service of Premier Biosoft
- Oligodb - a web-based system for interactive design of oligo DNA for transcription profiling (hybridization) of human genes. The oligodb system uses the human DNA-transcripts of ENSEMBL
- Osprey - Oligonucleotide Design Software for Sequencing and Gene Expression
- PCR suite - a collection of programs to search overlapping primers, genomic primers for exon amplification, SNP- and cDNA flanking primers
- Pira-PCR - PCR Primer design for restriction analysis of single nucleotide polymorphisms
- PRIDE - The less automated webversion of PRIDE (a.o. 50-70 mer oligo design)
- Primaclade - a web-based application that accepts a multiple species nucleotide alignment file as input and identifies a set of PCR primers that will bind across the alignment
- Primer3 - a common used software for designing primers
- mPrimer3 - modified Primer3
- PrimerQuest - Primer design at IDT
- Primer Generator - Automated generator of primers for site-directed mutagenesis
- PrimerX - Automated design of primers for site-directed mutagenesis
- ProbeWiz Server - The CBS ProbeWiz WWW server predicts optimal PCR primer pairs for generation of probes for cDNA arrays
- PUNS - Primer-UniGene Selectivity Testing - compares primer sequences against the both the genome and transcriptome to assess the potential for multiple amplicons
- RNAi Design - Design primers for RNAi at IDT
- ROSO - Software to design optimized oligonucleotide probes (size over 25 nucleotides) for microarrays

- Sirna - Target accessibility prediction and RNA duplex thermodynamics for rational siRNA design
- SNPbox - a modular software package that automates the design of PCR primers for large-scale amplification and sequencing projects in a standardized manner resulting in high quality PCR amplicons with a low failure rate
- Soligo - Target accessibility prediction and rational design of antisense oligonucleotides and nucleic acid probes
- SOP3 - Selection of Oligonucleotide Primers for PCR and Pyrosequencing

B. PCR Primer design software for local installation

Freely available

- Amplify - a freeware Macintosh program for simulating and testing polymerase chain reactions (PCRs)
- AmplifX - Software to test, manage and design your primers for Macintosh and Windows
- Fast PCR - PCR primer design, DNA and protein tools, repeats and own database searches
- MutaPrimer - Designs primers for Stratagene's QuikChange site directed mutagenesis kits
- OligoPicker - OligoPicker picks specific oligos by skipping regions with contiguous bases common in other sequences. In addition, oligo specificity is double-checked by NCBI BLAST. Sequence regions similar to non-coding RNAs are avoided because total RNA is often used for array hybridization. Low-complexity regions are also filtered out to maintain oligo specificity. Oligos and sequence regions that may form secondary structures are discarded since both the probes and the sequence target sites should be easily accessible for hybridization
- PRIMEGENS- PRIMEGENS (PRIMEr Design Using GEN Specific Fragments) is a computer program to select gene-specific fragments and then design primer pairs using Primer3 for PCR amplifications
- PrimerD - The primerD program implements a novel algorithm for the design of unique degenerate primer pairs
- Primer3 - a common used software for designing primers for microarray construction

- mPrimer3 - modified Primer3
- ProMide - ProMide is a collection of command-line tools for Probe selection and Microarray Design
- STACKdb - The STACKdb, Sequence Tag Alignment and Consensus Knowledgebase, is generated by processing EST and mRNA sequences obtained from Genbank through a pipeline consisting of masking, clustering, alignment and variation analysis steps. The STACKdb database is created using tools called "stackPACK". Also available through Electric Genetics.

Not freely available or commercial packages

- AlleleID - For real time PCR based pathogen detection and bacterial identification. TaqMan probe design supported
- Beacon Designer - Real time PCR primer and probe design for single tube and multiplex PCR assays
- OligoChecker - An oligo database program which quickly checks which oligos available in a lab can be used on a given template (Shareware).
- PRIDE and GenomePRIDE - (a.o. 50-70 mer oligo design)
- Visual OMP - multiplex primer and probe design optimized to reduce cross-hybridization between oligos and targets, an integrated folding engine for visualizing target and oligo structures, thermodynamics modeling, and built-in BLAST and ClustalW. Product of DNA Software, Inc.

C. Microarray primer design webserver

- ROSO - Software to design optimized oligonucleotide probes (size over 25 nucleotides) for microarrays
- MEDIANTE - Freely accessible database of human and mouse RNG/MRC oligonucleotide probes for microarrays

D. Microarray primer design software for local installation

Freely available

- OligoArray2 - a free Java program that computes gene specific oligonucleotides for genome-scale oligonucleotide microarray construction
- OligoWiz Site - Download the OligoWiz Java client to access the CBS OligoWiz WWW server and predict optimal oligonucleotides for generation of spotted arrays.

- Probepicker - Featurama's Open Source Probepicker 0.7 for custom designed oligonucleotide microarrays
- Primer3 - a common used software for designing primers for microarray construction.

Not freely available or commercial packages

- Array Designer 2 - Design hundreds of primers for DNA or oligonucleotide microarrays. Product of Premier Biosoft
- PRIDE and GenomePRIDE - (a.o. 50-70 mer oligo design)
- Sarani - Sarani Gold (Genome Oligo Designer) is a software for automatic large-scale design of optimal oligonucleotide probes for microarray experiments. Thousands of gene sequences can be analyzed together and best available oligonucleotide probes with uniform thermodynamic properties and minimal similarity to non-specific genes can be selected. Product of Strand Genomics.

D. In-silico PCR

- Genome tester - tests 1) whether PCR primers have excessive number of binding sites on template sequence and 2) how many PCR products would be amplified from the template DNA and where are they located.
- UCSC in-silico PCR - In-silico PCR on human genomic DNA at UCSC
- In-silico experiments with complete genomes - In-silico experiments (including PCR) on bacterial and lower eukaryotic genomes.

E. Primer property calculators

- AutoDimer Rapid screen of previously selected multiplex PCR primers for primer-dimer and hairpin interactions in short DNA oligomers (< 30 nucleotides)
- Oligonucleotide Properties Calculator
- OligoAnalyzer 3.0 at IDT

F. Primer Selection Tools

OSP: Oligo Selection Program (OSP) is a computer program developed to aid in selecting oligonucleotide primers for DNA sequencing and for the polymerase chain reaction. OSP allows the user to specify (or use default) constraints for primer and amplified product lengths, %G+C contents, and (absolute or relative) melting temperatures; for primer 3' nucleotides; and for the maximum allowable primer-self, primer-primer, and primer-product annealing propensities. Candidate primer sequences are screened against a user-supplied data set of other sequences (e.g. repetitive element or vector sequences) to help minimize the possibility of non-specific priming. Primers meeting all constraints are ranked and displayed in order of increasing overall 'score', which is a user-definable weighted sum of the above parameter values. These programs are developed by LaDeana Hillier and Phil Green at Washington University Medical School, St. Louis.

Primer: PRIMER is a computer program for automatically selecting PCR primers. PRIMER tests oligos for annealing temperature, complementarity to genomic repeat sequences (e.g. Alu), ability to form primer-dimer, and other criteria. Importantly, PRIMER's annealing temperature calculation is based on thermodynamic parameters (with the base stacking term dominant), and is far more accurate than calculations based on GC/AT ratios. PRIMER was developed and produced by Steve Lincoln, Mark Daly, and Eric Lander of MIT center for Genome Research and Whitehead Institute for Biomedical Research.

G. Comprehensive Genomic Sequence Analysis Packages/Suites

GCG : Commercial software package, based on the Wisconsin Package, developed by Genetics Computer Group, Inc. Databases available to be used with GCG are: GenBank, EMBL, PIR-Protein, SWISS-PROT, PROSITE, Restriction Enzymes (REBASE).

GDE: The Genetic Data Environment is part of a growing set of programs for manipulating and analyzing "genetic" data. It differs in design from other analysis programs in that it is intended to be an expandable and customizable system, while still being easy to use. GDE provides an X Windows based GUI to a large number of publicly available programs for sequence analysis. The software used SUN's XVIEW, efforts are underway to port the software to MOTIF.

H. Genetic Mapping Tools/Utilities

CRI-MAP : The main purpose of CRI-MAP is to allow rapid, largely automated construction of multilocus linkage maps (and facilitate the attendant tasks of assessing support relative to alternative locus orders, generating LOD tables, and detecting data errors). Crimap is developed by Phil Green of Dept. of Genetics, Washington University School of Medicine.

QUICKMAP : QUICKMAP is a compact database and navigation tool for the physical map of the human genome. The backbone of the map is the

Genethon 1993 genetic map. This includes 2067 STS genetically mapped to a chromosomal assignation and a genetic position. QUICKMAP was developed at GENETHON, France by P.Rigault and E. Poullier.

SIGMA: SIGMA (System for Integrated Genome Map Assembly) is a software tool for creating, maintaining and editing integrated genome maps. SIGMA is produced by the Human Genome Information Resources as Los Alamos National Laboratory and is freely available on request (although all rights are reserved).

I. Sequence Similarity, DB Search Tools

Blast Output Browser: Blast Output Browser, *BoB*, is an an X window application to browse through blast output results from either NCBI Blast server, DNA workbench server or the BLITZ email server in an environment where a lot of BLAST processing is done (for example in cDNA projects). One or two blast output results can be browsed simultaneously. While browsing through results, you can immediately view alignments, blast results file, and also the actual database entry. BoB program is developed by John Powell and Rao Parasa of CIT at NIH.

CLUSTAL W: Clustal W is a general purpose multiple alignment program for DNA or proteins.

Multiple alignments are carried out in 3 stages:

All pairs of sequences are aligned separately (pairwise alignments) in order to calculate a distance matrix giving the divergence of each pair of sequences;

A guide tree (like a phylogenetic tree) is constructed from the distance matrix;

The sequences are progressively aligned according to the branch order in the guide tree.

ClustalW program is a major update and rewrite of clustalV program, distributed with GDE (Genetic Data Environment) package. ClustalW is produced by Julie D. Thompson, Toby Gibson of European Molecular Biology Laboratory, Germany and Desmond Higgins of European Bioinformatics Institute, Cambridge, UK.

Fasta: This program package contains sequence database search programs:

- *Library search programs:* FASTA, TFASTA, FASTX, TFASTX, SSEARCH
- *Local homology programs:* LFASTA, PLFASTA, PCLFASTA, LALIGN, PLALIGN, FLALIGN

- *Statistical significance:* PRDF, RELATE, PRSS
- *Global alignment:* ALIGN, ALIGN0
- Utility programs: BESTSCOR, FROMGB

In addition, included are several programs for protein sequence analysis, including a Kyte-Doolittle hydropathicity plotting program (GREASE, TGREASE), and a secondary structure prediction package (GARNIER).

Local BLAST: BLAST (Basic Local Alignment Search Tool) is the heuristic search algorithm employed by the programs blastp, blastn, blastx, tblastn, tblastx and blast3.

The programs are used for the following purposes:

- lblastp - to compare an amino acid query sequence vs. a protein sequence database.
- lblastn - to compare a nucleotide query sequence vs. a nucleotide sequence database.
- lblastx - to compare a nucleotide query sequence translated in all reading frames vs. a protein sequence database.
- ltblastn - to compare a protein query sequence vs. a nucleotide sequence database dynamically translated in all reading frames.
- ltblastx - to compare a nucleotide query sequence translated in all reading frames vs. a nucleotide sequence database translated in all reading frames.
- lblast3 - protein database search for three-way alignments, using the BLAST pairwise search algorithm.

Databases to be searched by local blast programs must first be processed either by the program *setdb* for protein sequence databases (re: blastp and blastx) or the program *pressdb* for nucleotide sequence databases (re: blastn, tblastn and tblastx). The input database format is FASTA/Pearson.

Point accepted mutation (PAM) matrices of various generations can be produced automatically with the *pam* program. The output can be saved in a file whose name is then specified in the M=filename option of a blastp, blastx, or tblastn query. Also various BLOSUM matrices are supplied.

J. Network Entrez

Entrez is a molecular sequence retrival system devloped at NCBI, a division of the National Library of Medicine (NLM). Entrez porvides an integrated approach for gaining access to nucleotide and protein sequence information, to the MEDLINE citations in which the sequences were published, and to a sequence-associated subset of MEDLINE. The sequence records are derived from a variety of database sources, including GenBank, EMBL, DDJB, PIR, SWISS-PROT, PRF and PDB. Network Entrez is a client program, which requires a network server to access the databases.

K. Sequence Analysis Tools

Signal Scan: Signal scan is a program developed to facilitate the analysis of DNA sequences for known eukaryotic signals. It incorporates the data from Dr. David Ghosh's Transcriptional Factor Database. What SIGNAL SCAN does is to find homologies of published signal sequences in your sequence, most of these being transcriptional elements. It cannot, at this time, predict what it finds has any meaning. The interpretation of those results are up to you. Most signal elements found probably will not have any meaning, as the elements are in the wrong milieu, wrong cell type, or wrong organism. Consequently, there will be many more erroneous signals found by SIGNAL SCAN than significant ones. Sigscan program is developed and produced by Dan S. Prestridge of University of Minnesota at St. Paul.

xgrail: xgrail is a client-server implementation of a group of analysis tools for sequence exploration and gene discovery. This package allows the user to find protein coding regions in anonymous DNA sequences, to assemble gene models, translate part or all of these models, and search these translations against various databases. It also provides information about GC content, functional site identification (splice junctions, poly A sites, Pol II promoters, and CpG Islands) and the location of a variety of repetitive DNA sequences. xgrail is produced by M.B.Shah, X.Guan, J.R.Einstein, S.Matis, Y.Xu, R.J.Mural and E.C.Uberbacher of informatics group at Oak Ridge National Laboratories.

L. Molecular biology software that are available freely

- AiO - DNA/Protein Programs and Oligo Management
- Biocatalog
- Genamics SoftwareSeek
- GeneDoc
- BioEdit
- EnzymeX
- Free Molecular Visualization Software
- Protein Explorer
- Biotechnology Software

- Free Software
- Genotator
- BOXIT - Database for Molecular Biology Laboratories
- JaMBW - Java Molecular Biology Directory
- MOLMOL - MOLecule analysis & MOLecule display
- PaperWorks Paper Organizer
- PC Software for Molecular Modeling
- pDRAW32 - DNA Analysis Software
- PHYLIP Software for Phylogenetics
- PhyloDraw - A Phylogenetics Tree Drawing System
- RasMol - Molecular Visualization Freeware
- SEQtools
- SeqVerter - Sequence Format Converter
- The Sanger Institute - Software
- TIGR Software
- Software for Computational Molecular Biology
- Stuffit Expander
- Sequence Analysis Tools for Mac OS X
- TreeView Phylogeny Software
- TREE-PUZZLE
- VMD - Visual Molecular Dynamics
- WU-BLAST
- Affordable Biocomputing for Everyone
- Apollo User Guide
- Apollo Genome Annotation Curation Tool
- JustBio - Tools
- Uniblast
- ClustalX
- Jmol - Molecular Viewer
- Tools at PSU Bioinformatics Lab
- SWEET - Saccharide Structure Utility
- MatchMiner
- Alfresco
- ANTHEPROT
- Biolinux Software

- SeqVISTA
- Software for Kinetics
- ACGT
- GeneMine
- DNATools
- Librarian - PDF Article Organizer for Bioscientists
- Genetics Analysis Software
- MEGA - Molecular Evolutionary Genetics Analysis
- DARWIN - Data Analysis and Retrieval With Indexed Nucleotides
- Artemis - DNA Sequence Viewer

REFERENCES

Jain, M., Kamal, N. and Batra, S.K. (2007). Engineering antibodies for clinical applications. *Trends in Biotechnology*, 25 (7): 307-316.

Klesius, P.H., Evans, J.J., Shoemaker, C.A. and Lim, C. (2007). A US perspective on selected biotechnological advancements in fish health. Part II: Genetic stock improvement, biosecurity tools and alternative protein sources in fish diets. *Aquaculture Health International*, May, 18-20.

Shankar, K.M., Mohan, C.V., Anil, T.M. and Vidya, R. (2000). Monoclonal antibodies in fish and shellfish health management in India. Naga, *The ICLARM Quarterly*, 23(4): 10-12.

Petty, N.K., Evans, T.J., Fineran, P.C. and Salmond, G.P.C. (2007). Biotechnological exploitation of bacteriophage research. *Trends in Biotechnology*, 25(1): 7-15.

Bondad-Reantaso, M.G. and Subasinghe, R.P. (2005). Aquatic animal diseases and their economic impact: A global perspective. *Aquaculture Health International*, May: 4-5.

Adams, A. and Thompson, K.D. (2006). Biotechnology offers revolution to fish health management. *Trends in Biotechnology*, 24(5): 201-205.

Deiman, B., van Aarle, P. and Sillekens, P. (2002). Characteristics and applications of nucleic acid sequence-based amplification (NASBA). *Molecular Biotechnology*, 20:163-179.

Chapter 7

Present Trends in Aquaculture Probiotics

S. Mohanty & N.K. Maiti

Introduction

Fish production in aquaculture has increased dramatically over the last 20 years with an increment of 9-10% per year over the last decade. Microbial disease problems are prevalent in aquaculture as in any other intensive animal or plant production sector. Vaccines are used as disease prevention and can be very effective, however, they do not work at all developmental stages of fish or in all species. Therefore, antibiotics play a major role in control of fish bacterial diseases, although serious concerns have been raised with respect to development of antibiotic resistance and its transfer to human pathogenic bacteria. Also, new EU regulations call for specific license for antibiotic use in aquaculture and this is too costly even for fish farmers.

The urgent need for development and application of non-antibiotic disease control measures has led to research in probiotic organisms. Probiotics are "live microorganisms which when administered in adequate amount, confer a health benefit on the host (FAO/WHO 2001)" and the principle is most widely known from the belief that ingestion of high numbers of lactic acid bacteria (fermented foods) can control human pathogenic organisms in the intestine. Fish pathogenic agents, however, can invade via skin and gills and probiotic organisms may therefore be as efficient if applied to the outer environment of the fish.

Applied Bioinformatics, Statistics and Economics in Fisheries Research: pp. 93-104

New India Publishing Agency, New Delhi (India)

What is a Probiotic ?

The word "probiotic" was introduced by Parker (1974). According to his original definition, probiotics are 'Organisms and substances which contribute to intestinal microbial balance'. Fuller (1989) revised the definition as 'A live microbial feed supplement which beneficially affects the host animal by improving its intestinal microbial balance'. This revised definition has put forward the importance of live cells as the essential component of a potential probiotic and it clears the confusion created by the use of the term '*substances*'. However, " an effect on intestinal microbial balance" has been defined and demonstrated only in a few cases. This was noted by Tannock (1997), and he proposed the definition "living microbial cells administered as dietary supplements with the aim of improving health". The concept of microfloral manipulation was first appreciated by Metchnikoff (1907) who examined the consumption of yoghurt and found an effect on the longevity of Bulgarian peasants. A beneficial effect by application of certain beneficial bacteria in human, pig, cattle and poultry nutrition has been well documented. On the other hand, the use of such probiotics in aquaculture is a relatively new concept. With interest in treatments with friendly bacterial candidates increasing rapidly also in aquaculture. Among the still limited information available concerning probiotics in aquaculture, several reports deal with enhancement of growth of oyster larva with addition of certain bacterial mixture. Interest is also increasing in the search for microorganisms that may contribute to a more beneficial gut flora in fish, because it has become clear that the microflora in the gut plays an important role with respect to the well-being and health of fish. In spite of the limited information available on the effect of beneficial bacteria on the growth and health of fish, some results indicate the applicability of probiotics in intensive farming as well. According to Austin et al., (1995) the bacteria *Vibrio alginolyticus* was found to reduce diseases in Atlantic salmon (*Salmo salar*) caused by infection of common pathogenic strains (*Aeromonas salmonicida, Vibrio anguillarum*) and thus to have a probiotic effect. Olsson *et al.,* (1992) investigated bacterial strains associated with intestinal and skin mucus of adult marine turbot (*Scophthalmus maximus*) and dab (*Limanda limanda*), and found them to suppress the growth of the fish pathogen *Vibrio anguillarum*. A number of preparations of bacterial mixture are now-a-days commercially available and have been introduced to fish farming as feed additives,or are incorporated in live fish feed in order to increase the performance. As an example, a commercial preparation of the bacteria *Streptococcus faecium* incorporated in the feed improved growth and feed efficiency of Israeli carp (Noh et al., 1994; Bogut et al., 1998). Recently, Moriarty (1998) proposed that bacterial mixtures distributed by water as "water additives" instead of "feed additivee" may also have a beneficial effect on aquaculture production. Queiroz and Boyd (1998) reported that a commercially prepared bacterial mixture of *Bacillus* spp. mixed into the rearing water increased survival and production of channel catfish (*Ictalurus punctatus*).

Mechanisms of Actions of Probiotics

Although many publications about probiotics in aquaculture have emerged during the last decade, the approach was generally empirical and the arguments with regard to the mode of action were often circumstantial. The exact modes of action of the probiotics were rarely completely elucidated. In human and agriculture application, probiotic research has enjoyed much more attention through history and several modes of action have been supported by unambiguous experimental data. It is clear that the experience obtained with terrestrial animals has been used in aquaculture, especially with regard to the use of lactic acid bacteria. Also, the suppression of plant root pathogens by fluorescent *Pseudomonas* spp. has been examined in detail. Although each of these disciplines has its specific characteristics, several of the microbial-ecological mechanisms involved are similar, allowing some transfer of experimental evidence.

Considering the possible probiotic effect in vivo, one has to make a distinction between the intrinsic ability of the strain to positively influence the host and its ability to reach and maintain itself in the location where the effect is to be exerted. For instance, the production of siderophores or inhibitory compounds in sufficient amounts and even under the conditions prevailing in the gut is of no relevance if the strain is not ingested by the host. This is important, since it has been demonstrated that, both selective ingestion and digestion of microbes by the bivalve *Mytilus edulis*. Similarly, if the candidate probiotic is not capable of efficient proliferation in the gut after being ingested, it is improbable that it will exert strong effects unless it is added regularly through the diet. Hence, the possible modes of action require implicitely that the candidate probiotics be able to reach the location where their probiotic effect is required. Those modes are as follows: production of inhibitory compounds; competition for chemicals or available energy; competition for adhesion sites; enhancement of the immune response; improvement of water quality; interaction with phytoplankton; source of macro- and micronutrients; and enzymatic contribution to digestion.

Production of Inhibitory Compounds

Microbial populations may release chemical substances that have a bactericidal or bacteriostatic effect on other microbial populations, which can alter interpopulation relationships by influencing the outcome of competition for chemicals or available energy. The presence of bacteria producing inhibitory substances in the intestine of the host, on its surface, or in its culture medium is thought to constitute a barrier against the proliferation of (opportunistic) pathogens.

In general, the antibacterial effect of bacteria is due to the following factors, either singly or in combination: production of antibiotics, bacteriocins, siderophores (which are dealt with in a later section), lysozymes, proteases, and/or hydrogen peroxide and the alteration of pH values by the production of organic acids.

Lactic acid bacteria are known to produce compounds such as bacteriocins that inhibit the growth of other microorganisms. There are many reports of the inhibitory activity of lactic acid bacteria mediated by bacteriocins, mostly but not exclusively against gram-positive bacteria. However, almost all the pathogens involved in aquaculture are gram negative. The possible involvement of lactic acid bacteria as probiotics in aquaculture is discussed by Ringø and Gatesoupe (1998). Since lactic acid bacteria normally account for only a small part of the gut microbiota of fish and since they are generally considered to be nonpathogenic (with some exceptions), it can be questioned if the inhibition of closely related species by bacteriocins produced by lactic acid bacteria can effectively contribute to the health status of the higher organism.

Compounds other than bacteriocins and antibiotics have been suggested to play a role in the amensalism that may occur between bacterial species. Nair et al. (1985) showed that a large proportion of marine bacteria produced bacteriolytic enzymes against *V. parahaemolyticus*. Furthermore, Imada et al. (1985) isolated and characterized *Alteromonas* sp. strain B-10-31, isolated from nearshore seawater of Japan, which produces an alkaline protease inhibitor called monastatin. In an in vitro assay, the purified and concentrated monastatin showed inhibitory activity against a protease from *Aeromonas hydrophila* and a thiol protease from *V. anguillarum*, both pathogenic to fish.

Many authors assign the inhibitory effects detected in in vitro antagonism tests to bacteriocins or antibiotics without looking for any other causes. It has been argued that observed growth inhibition can, in many cases, be accounted for by primary metabolites or simply by a decrease of the pH.

Many studies have demonstrated the presence of bacterial strains showing in vitro inhibition toward pathogens known to occur in aquaculture. This shows that the ability to inhibit other bacteria is not uncommon for bacteria found in aquaculture environments. However, it has not been demonstrated that production of such inhibitory compounds occurs under in vivo conditions, and the ecological relevance of the production of inhibitory compounds towards other bacteria is still unclear.

The in vitro production of inhibitory compounds towards known pathogens for the considered species has often been used in the selection of putative probiotic strains. At this stage, however, the association between amensalistic activity and in vivo probiotic activity is very weak and circumstantial. Typically, a correlation is made between the in vitro ability of the probiotics to inhibit pathogens and the in vivo protection of the cultured aquatic species, but in none of the studies published so far has it been shown unequivocally that the production of inhibitory compounds is the cause of the observed in vivo probiotic activity of the strains. Hence, future research in this field is required. Similar research has already been performed in the field of plant disease suppression, where the production of inhibitory compounds by some fluorescent *Pseudomonas* spp. is now recognized as an important factor in the disease suppression ability of these strains.

Competition for Chemicals or Available Energy

Competition for chemicals or available energy may determine how different microbial populations coexist in the same ecosystem. On theoretical grounds, it seems likely that competition for nutrients will operate in the mammalian gut, but the evidence for its occurrence in humans and terrestrial animals is not good. Competition for nutrients can theoretically play an important role in the composition of the microbiota of the intestinal tract or ambient environment of cultured aquatic species, but to date there have been no comprehensive studies on this subject. Hence, successful application of the principles of competition to natural situations is not easy and remains a major task for microbial ecologists.

The microbial ecosystem in aquaculture environments is generally dominated by heterotrophs competing for organic substrates as both carbon and energy sources. Specific knowledge of the factors governing the composition of the microbiota in aquaculture systems is required to manipulate it. This knowledge, however, is generally not available, and one must therefore rely on an empirical approach. Nevertheless, Rico-Mora *et al.* (1998) selected a bacterial strain for its active growth in organic-poor substrates and inoculated it into a diatom culture, where it prevented the establishment of an introduced *V. alginolyticus* strain. Since the inoculated strain had no in vitro inhibitory effect on *V. alginolyticus*, it was suggested that the strain was able to outcompete *V. alginolyticus* due to its ability to utilize the exudates of the diatom.

Verschuere *et al.* (2000) selected several strains with a positive effect on the survival and growth of *Artemia* juveniles. The in vitro antagonism tests and filtrate experiments showed that no extracellular inhibitory compounds were involved in the protective action of these strains against *V. proteolytics* CW8T2 but that living cells were required to protect *Artemia* against the pathogen. It was suggested that the selected bacteria exerted their protective action by competing with the pathogen for chemicals and available energy.

Competition for Iron

Virtually all microorganisms require iron for growth. Siderophores are low-molecular-weight (<1,500), ferric ion-specific chelating agents which can dissolve precipitated iron and make it available for microbial growth. The ecological significance of siderophores resides in their capacity to scavenge an essential nutrient from the environment and deprive competitors of it. Successful bacterial pathogens are able to compete successfully for iron in the highly iron-stressed environment of the tissues and body fluids of the host. The ecological significance of siderophores in soils as important tools for iron acquisition by microorganisms and plants and their involvement in suppression of plant root pathogens have been established.

The requirement for iron is high for many pathogens, including *V. anguillarum*. In a challenge test with this bacterium, salmon mortality increased linearly with dietary iron content. Observations indicate the importance and the biological role of ferric iron in the establishment of the microbiota associated with cultured aquatic species.

Harmless bacteria which can produce siderophores could be used as probiotics to compete with pathogens whose pathogenicity is known to be due to siderophore production and competition for iron or to outcompete all kind of organisms requiring ferric iron from solution. The possible effectiveness of siderophore-producing probiotics can be illustrated by the study of Gatesoupe (1997), in which the addition of the bacterial siderophore deferoxamine to live food (rotifers) increased the resistance of turbot larvae challenged with the pathogenic *Vibrio* strain P. The addition of a siderophore-producing *Vibrio* strain E protected the turbot larvae slightly more.

Pybus *et al.* (1994) tested thirty strains of *V. anguillarum* as potential probiotics against the salmon pathogen *V. ordalii* by the deferred-antagonism test. Only one strain (*V. anguillarum* VL4335) inhibited strains of *V. ordalii* in vitro, and this effect was blocked when iron salts were added to the medium, indicating that the growth inhibition was linked to iron deficiency. Using the chrome-azurol sulfate assay to measure siderophore production, *V. anguillarum* VL4335 yielded significantly higher values than other *V. anguillarum* strains.

Smith and Davey (1993) showed that the fluorescent pseudomonad F19/3 is capable of inhibiting the growth of *Aeromonas salmonicida* in culture media and that this inhibition is due to competition for free iron. Strain F19/3 was also capable of excluding *A. salmonicida* from Atlantic salmon (*Salmo salar*) presmolts with stress-inducible infections.

Gram *et al.* (1999) examined the applicability of the siderophore-producing *Pseudomonas fluorescens* AH2 isolated from frozen freshwater fish. This latter strain is inhibitory to several gram-positive and gram-negative bacteria, particularly when iron availability is limited. In vitro tests revealed that the growth of *V. anguillarum* was inhibited by the filter-sterilized supernatans from iron-limited cultures of *P. fluorescens* AH2 but not from iron-replete cultures. During coculture, *P. fluorescens* AH2 inhibited the growth of *V. anguillarum* independently of the iron concentration, when the initial level of the siderophore producer was 100 to 1000 times greater than the level of the fish pathogen. The in vitro tests were confirmed by an in vivo antagonism test, in which the mortality of rainbow trout juveniles due to *V. anguillarum* infection was decreased by 46% when the culture was treated with *P. fluorescens* AH2.

Siderophore production by fluorescent pseudomonads in soil is known to be influenced by a great variety of factors. Therefore, detection of in vitro production of siderophores does not necessarily mean that they are produced in significant amounts in vivo in order to have a significant biological control effect. Similar to the production of inhibitory compounds, the evidence for the participation of competition for chemicals or available energy and, more specifically, of free iron or siderophores in the mode of action of probiotics is still circumstantial.

Competition for Adhesion Sites

One possible mechanism for preventing colonization by pathogens is competition for adhesion sites on gut or other tissue surfaces. It is known that the ability to adhere to enteric mucus and wall surfaces is necessary for bacteria to become established in fish intestines. Since bacterial adhesion to tissue surface is important during the initial stages of pathogenic infection, competition for adhesion receptors with pathogens might be the first probiotic effect. In human medicine, adherent strains are key candidates for probiotic therapy.

Adhesion can be nonspecific, based on physicochemical factors, or specific, involving adhesin molecules on the surface of adherent bacteria and receptor molecules on epithelial cells. Inhibition of the adhesion of pathogens to human or other mammal cells has already been demonstrated in vitro by several authors. Competitive exclusion resulting from presumtive colonization has been shown for the cecal walls of chickens, which kept their effect after the ceca were washed four times in buffered saline. As far as is known, similar research approaches have not yet been followed with aquatic species, and competition for adhesion sites as a mode of probiotic action is still hypothetical.

Adhesion capacity and growth on or in intestinal or external mucus has been demonstrated in vitro for fish pathogens like *V. anguillarum* and *A. hydrophila* and for candidate probiotics such as *Carnobacterium* strain K1 and several isolates inhibitory to *V. anguillarum*. In one of these studies, the aim was to measure the in vitro capacity of the strains to adhere to and grow in turbot intestinal mucus in order to investigate their potential to colonize farmed turbot as a means of protecting the host from infection by *V. anguillarum*. The intestinal isolates generally adhered much better to a film of turbot intestinal mucus, skin mucus, and bovine serum albumin than did *V. anguillarum*, indicating that they could compete effectively with the pathogen for adhesion sites on the mucosal intestinal surface.

Adhesion of probiotics to the gut wall or other tissues does not necessarily imply competition for adhesion sites as the (only) mode of action for the probiotic effect. It is conceivable that bacteria are able to colonize, for example, the intestinal gut wall of a fish and exert their protective action against a pathogen by excreting inhibitory compounds.

Enhancement of the Immune Response

Immunostimulants are chemical compounds that activate the immune systems of animals and render them more resistant to infections by viruses, bacteria, fungi, and parasites. Fish larvae, shrimps, and other invertebrates have immune systems that are less well developed than adult fish and are dependent primarily on nonspecific immune responses for their resistance to infection.

Observations obtained in experiments with warm-blooded animals indicate that probiotic (lactic acid) bacteria administered orally may induce increased resistance to enteric infections. There are many reports that bacterial compounds act as an immunostimulant in fish and shrimp, but only specific cell compounds or nonliving cells were used in these studies. It has also been suggested that ingestion of bacteria and subsequent endocytosis in cod and herring larvae are involved in stimulation of the developing immune system. However, at present it is not clear whether bacteria administrered as probiotics could have a beneficial effect on the immune response of cultured aquatic species, but such a mode of action cannot a priori be excluded.

Improvement of Water Quality

In several studies, water quality has been recorded during the addition of the probiotics, especially *Bacillus* spp. The rationale is that gram-positive *Bacillus* spp. are generally more efficient in converting organic matter back to CO_2 than are gram-negative bacteria, which would convert a greater percentage of organic carbon to bacterial biomass or slime. It is reasoned that by maintaining higher levels of these gram-positive bacteria in the production pond, farmers can minimize the buildup of dissolved and particulate organic carbon during the culture cycle while promoting more stable phytoplankton blooms through the increased production of CO_2 . However, several studies utilizing one or more bacterial species such as *Bacillus, Nitrobacter, Pseudomonas, Enterobacter, Cellulomonas*, and *Rhodopseudomonas* spp. in the culture of shrimps or channel catfish could not confirm this hypothesis. This shows that the published evidence for water quality improvement is poor, except with regard to nitrification.

The start-up of biofilters by transferring medium from an existing filter is a common practice in aquaculture. However, the use of nitrifying cultures as inocula to speed up the establishment of nitrification is not so common. A lot of bacterial cultures containing nitrifying bacteria to control the ammonia level in culture water are available commercially and are aimed especially at aquarium hobbyists. Nitrifiers are responsible for the oxidation of ammonia to nitrite and subsequently to nitrate. Seeding of biological filters with nitrifying bacteria is effective in reducing the activation time of new biofilters.

Nitrifying cultures could also be added to the ponds or the tanks when an incidental increase of ammonia or nitrite levels is observed. Besides ammonia, nitrite toxicity is a common problem in fish culture.

Interaction with Phytoplankton

Recent reports demonstrate that many bacterial strains may have a significant algicidal effect on many species of microalgae, particularly of red tide plankton. Bacteria antagonistic toward algae would be undesirable in larval rearing where unicellular algae are added (e.g., the green-water technique) but would be advantageous when undesired algal species develop in the culture pond.

Claims for bacterial amendments include a decrease of the proportion of cyanobacteria. One such bacterial suspension consisted of *Bacillus, Nitrobacter, Pseudomonas, Enterobacter, Cellulomonas,* and *Rhodopseudomonas* spp. and was applied to channel catfish ponds. Positive effects of bacteria on cultured microalgae have also been observed. It is conceivable that bacteria can indirectly influence the health or the zootechnical performance of the cultured aquatic animals through their effect on the microalgae used as food or in the green-water technique. When probiotic bacteria are selected to be used in a culture environment comprising algae, their possible interaction with these unicellular algae must be taken into consideration when the mode of action is being investigated.

Safety of Strains

A very important issue is the safety of these strains to the target species, the environment and humans. Strains should be tested for their sensitivity to antibiotics. For some strains, tools are available that will allow the screening of specific toxins, or genes associated with the toxin production, but this is not the case with less common species. Hence, probiotic products should not include *Vibrio* (risk of transfer of virulent genes) or coliforms. Within Bacillus and their related genera *Paenibacillus* and *Brevibacillus*, the *B.anthracis* group is pathogenic for animals. The species within this group (*B.anthracis, B.cereus, B.mycoides* and *B.thuringiensis*) must not be present in any probiotics and, furthermore, *Bacillus* probiotics in other generally safe group, like the *B.subtilis* group must be checked to ensure they do not carry any genes for toxins from the anthrax group. The strains should also be evaluated in challenge tests and rearing trials in order to confirm their lack of toxicity to the host or other animals found in the production facilities, at doses 10 or 100x higher than the recommended doses. For some strains, safety is based on historical use; they have been used in human food for long period of time and thus considered as safe. However, their safety to aquaculture organisms should still be investigated. Once the possible probiotic effect has been demonstrated in vivo, the ability of the strain to remain active in the aquaculture environment, to perform the expected function and to maintain itself should be demonstrated. The ability of *Bacillus* species to colonise in the intestinal tract of shrimp, rather than exist as transient flora during feeding, is now the subject of investigation.

Conclusion

The potential for probiotic use in aquaculture is immense because farmers must avoid prophylactic use of antibiotics, however, bacterial and fungal pathogens will always be present. The use of safe, natural bacteria (probiotics) to modify the microbial communities in aquatic animals and their environment and thus control the pathogen is, therefore, a sustainable alternative to the use of antimicrobial chemicals. However, one should not consider them as miracle products solving health and related problems.

REFERENCES

Austin, B., Stuckey, L.F., Robertson, P.A.W., Effendi, I., Griffith, D.R.W.1995. A probiotic strain of *Vibrio alginolyticus* effective in reducing diseases caused by *Aeromonas salmonicida, Vibrio anguillarum* and *Vibrio ordalii*. *Journal of Fish Dis.*, 18:93-96

Bogut,I., Milakovic, Z., Bukvic, Z., Brkic, S., Zimmer,R.1998. Influence of probiotic (*Streptococcus faecium* M74) on growth and content of intestinal microflora in carp (*Cyprinus carpio*). *J. Anim.Sci.* 43:231-235

Burr, G., D. M. Gatlin III., and S. Ricke. 2005. Microbial ecology of the gastrointestinal tract and the potential application of probiotics and prebiotics in finfish aquaculture. *J. of the World Aquacult. Soc.*, 36:425-436

Fuller, R., 1992.Problems and prospects.In: Fuller, R(Ed), Probiotics.The Scientific Basis.Chapman & Hall, London, UK, pp.377-386

Gatesoupe, F. J. 1997.Siderophore production and probiotic effect of Vibrio sp. associated with turbot larvae, *Scophthalmus maximus.Aquat.Living Resour*, 10:239-246

Gram, L., J. Melchiorsen, B. Spanggaard, I. Huber and T. Nielsen. 1999. Inhibition of Vibrio anguillarum by Pseudomonas fluorescens strain AH_2-a possible probiotic treatment of fish.*Appl.Environ.Microbiol.*, 65:969-973

Huber, I., B. Spanggaard, K.Appel, L. Rossen,T. Nielsenand and L.Gram. 2004.Phylogenetic analysis and insitu identification of the intestinal microbial community of rainbow trout (*Onchorhynchus mykiss*,Walbaum).*J.Appl. Microbiol.*, 96:117-132

Imada, C., U.Simidu and T.Naga.1985.Isolation and characterisation of marine bacteria producing alkaline protease inhibitor.*Bull.Jpn.Soc.Sci.Fish*, 51:799-803

Imada, C., M.Maeda and T. Naga. 1985. Purification and characterisation of the protease inhibitor "monastatin" from a marine Alteromonas species with reference to inhibition of the protease produced by a bacterium pathogenic to fish. *Can. J.Microbiol*, 31:1089-1094

Irianto, A. and B. Austin. 2002. Probiotics in aquaculture. *J. Fish. Dis.*, 25:633-642

Metchnikoff, E. 1907. Prolongation of Life.Heinemann, London

Moriarty, d.J.W., 1998. Control of luminous *Vibrio* species in penaeid aquaculture ponds.*Aquaculture*, 164:351-358

Niir, S., K. Tsukamoto and U. Shimidu.1985. Distribution of bacteriolytic bacteria in the coastak marine environments of Japan. *Bull. Jpn. Soc. Sci. Fish*, 51:1469-1473

Noh, S.H., Han,K., Won, T.H., Choi, Y.J.,1994.Effect of antibiotics, enzyme, yeast culture and probiotics on growth performance of Israeli carp.*Korean Journal of Animal Science*, 36:480-486

Olsson, J.C., Westerdahl, A.,Conway, P.L.,Kjelleberg, S., 1992.Intestinal Colonization Potential of Turbot (*Scopthalmus maximus*) and Dab(*Limanda limanda*) Associated Bacteria with inhibitory effect against *Vibrio anguillarum.Applied and Environmental Microbiology*, 58(2):551-556

Parker, R.B., 1974. Probiotics, the other half of the antibiotics story.*Animal Nutrition and Health*, 29:4-8

Patterson, J.A. and K. M. Burkholder. 2003.Application of prebiotics and probiotics in poultry production.*Poultry Science*, 82:627-631

Pybus, V., M.W. Loutit, I. L.Lamont and J. R.Tagg.1994.Growth inhibition of the salmon pathogen *Vibrio ordalii* by a siderophore produced by *Vibrio anguillarum* strain VL4335.*J.Fish Dis.*, 17:311-324

Queiroz, J.F., Boyd,C.E.,1998. effects of a bacterial inoculum in channel catfish ponds. *J. World Aquacult.Soc.*, 29(1):67-73

Rico-Mora, R., d.Voltolina and J.A.Villaescusa-Celaya.1998.Biological control of *Vibrio alginolyticus* in *Skeletonima costatum* (Bacillariophyceae) cultures.*Aquacult.Eng.*, 19:1-6

Ringo, E., and F. J. Gatesoupe.1998 Lactic acid bacteria in fish: a review. *Aquaculture* 160:177-203

Smith, P., and S. Davey.1993. Evidence for the competitive exclusion of *Aeromonas salmonicida* from fish with with stress inducible furunculosis by a fluorescent pseudomonad.*J.Fish Dis.*, 16:521-524

Tannock, G.W. 1997. Modification of the normal microbiota by diet, stress, antimicrobial agents and probiotics. In:Mackie,R.I., with B.A.Issacson, R.E.(Eds.),Gastrointestinal Microbiology,Vol.2, Gastrointestinal Microbes and host interactions. Chapman and Hall Microbiology Series, International Thomson Publishing, New York,pp.434-465

Verschuere, L., G.Rombaut, G.Huys, J.Dhont, P.Sorgeloos and W. Verstraete. 2000. Protection of Artemia ahainst the pathogenic effect of *Vibrio proteolyticus* CW8T2 by selected bacterial strains. *Appl.Environ. Microbiol.*, 66:1139-1146.

Chapter **8**

Infection and Immunity of Brood Carps

P. Swain, P.K. Nanda & S. Dash

Introduction

Brood stock management is an essential component of fish farming since the success of aquaculture industry mainly depends on quality of brood stock in order to ensure the production of top quality offspring. Their health condition is also important from culture point of view, which in turn increases the fish production. Unfortunately, in India, less attention is given to the brood stock management during non-breeding seasons. The pond aquaculture is also becoming over utilized due to high stocking beyond the carrying capacity, addition of huge quantities of chemicals, drugs, fertilizers and supplementary feeds etc. Physical injury due to rough handling, temperature fluctuations, extremes in oxygen levels can also work in a cumulative way and produce stress and make the fish susceptible to variety of diseases. Environmental deterioration, which occurs due to dumping of pesticides and chemicals alongwith metabolic waste products, left out feed materials and organic load of the pond bottom regularly come in contact with certain vital organs and tissues of fish. All such factors can cause deterioration of fish health and magnify the risk of outbreak of diseases. Disease outbreaks are one of the stumbling blocks in the development of aquaculture industry affecting both the economic development and socio-economic revenue

Applied Bioinformatics, Statistics and Economics in Fisheries Research: pp. 105-119

New India Publishing Agency, New Delhi (India)

from this sector. Today, there is no availability of exact figure on loss due to diseases in fish farming. The phenomenon is more common in carp Industry, and mortalities due to disease outbreaks cause major management problems and economic loss every year.

Diseases in aquaculture are not only caused due to a single event but are the end result of a series of linked events involving the interactions between the host, the environment and the presence of a pathogen. Pathogens may include viruses, bacteria, parasites and fungi; either a single species or a mixture of different pathogens may cause diseases. It is very essential to identify these infectious diseases in fish even in the initial stage and control the same before spreading.

Bacterial Diseases

Bacterial diseases are frequently encountered in all stages of fish causing heavy mortality and production loss. These microorganisms become pathogenic when fishes are immunocompromised by some form of stressor. Bacteria, that are significant pathogens of carps are predominantly Gram -ve types. These pathogens come in contact with their potential host either being carried over by water or by direct contact or by their presence in the food. Depending on the virulence and immunological status of the host, the symptoms of the disease are expressed in various forms. Some of the common bacterial diseases encountered in carps are given in Table 1.

To know the bacterial disease status of the fish, study was conducted in three major aquaculture zones of India viz. coastal areas of Orissa (Zone I), 925 sq. km. area of Kolleru Lake of Andhra Pradesh (Zone II), freshwater and wastewater aquaculture areas of West Bengal (Zone III). Information regarding different culture practices and disease status, collected from progressive fish farmers and state fisheries department officials from the year 2003-2005 alongwith the disease incidence data from 1999 to 2003 obtained from fish health management division of CIFA, were analysed. Bacterial diseases such as dropsy, columnaris, ulcer, septicaemia, fin rot/tail rot, bacterial gill disease and diseases due to mixed bacterial infections were recorded in all the three-aquaculture zones. The percentage of incidence of these diseases in all the three zones over the period is depicted in Table 2.

Table 1. Common bacterial diseases reported in carps

Disease	Causative agent (s)	Stage of fish	Organ (s) affected	Symptoms	Causes	Control measures
Bacterial gill disease	*Flavobacterium branchiophilum*	Grow out and adult	Gills	Pale coloured necrotic gill tissues, sluggish movement, loss of appetite, asphyxiation, surfacing, mortality up to 50%	Overcrowding, nutritional deficiency, decreased oxygen level, increased ammonia	Short bath for 3-10 minutes in 3% Nacl, alkyl benzalkonium chloride 2 ppm (1 hr)
Columnaris disease	*F. columnare Flexibacter/ Cytophaga columnaris*	All stages	Head, fins and gills	Greyish white discolouration / patches, skin lesions, 20-30 % mortality	Fluctuation in temperature, poor water quality, overcrowding	Benzalkonium chloride, antibiotics @ 80 mg / kg fish/day, $CuSo_4$ 0.5 - 1.0 ppm, $KMnO_4$ 2-3 ppm
Infectious dropsy	*A. hydrophila P. fluorescens*	Adult	Skin, abdomen and viscera	Loosening of scales, ulcers, exophthalmia, abdominal distensions, 100 % mortality	High temp. overcrowding environmental stress	Pond treatment with $KMnO_4$ (2-3 ppm) or acriflavin (3-10 ppm), oxytetracycline with feed @70-80 mg/kg fish/day
Edwardseillosis	*E. tarda*	Juveniles	Kidney, spleen, skin and muscle	Anemia, cutaneous lesions, abscesses in muscle, necrosis of the hepatic and renal tissues, 100 % mortality in young ones	High stocking, poor water quality, high temp.	Oxytetracycline with feed @50 mg/kg fish/day up to 10 days, need to improve water quality
Eye diseases	*A. liquifaciens S. aureus*	Adult	Eyes, optic nerves and brain	Opaque eyes, necrosis and mass mortality	Poor water quality, high temp.	Disinfection with 0.1 ppm $KMnO_4$ followed by liming 300 ppm.

Contd...

Haemorrhagic septicaemia	*A.hydrophila*	Adult	Whole body	Septicaemia, haemoharrgic and bleeding spots on skin, abdominal wall and visceral organs; 100 % mortality	Fluctuation in temperature, poor water quality, overcrowding, unhygienic conditions, prior external parasitic infections	Acriflavin 3-10 ppm, antibiotic therapy
Fin rot / Tail rot	*A.hydrophila* *P.fluorescens*	Mostly grow out and Juveniles	Caudal fin and tail	Putrefecation, whitening of and degeneration of caudal fin and tail	High stocking, poor water quality, high temp.	Benzalkonium chloride (1-2ppm) for 1 hr., treatment with antibiotics, Disinfect wound with 1 % Silver nitrate
Red sore/ Red pest	*A.hydrophila* *P.fluorescens*	All stages	Body surface	Red patches, spots on body surface may lead to ulceration in later stage, exophthalmia, rupture of minor blood vessels	Sudden changes in environment, nutritional deficiency	Disinfection with $KMnO_4$, liming (300 ppm)

Table 2. Incidence of different bacterial diseases in Indian major carps in different aquaculture zones of India

Diseases	Zone-I	Zone-II	Zone-III
Dropsy	12.76 ± 2.21	9.4 ± 1.36	12.18 ± 2.52
Columnaris	15.4 6 ± 1.52	12.1 ± 1.48	13.80 ± 1.56
Ulcer	12.85 ± 2.13	12.8 ± 2.58	11.9 ± 1.81
Septicaemia	8.6 ± 0.86	7.75 ± 1.83	8.14 ± 1.45
Fin rot/ Tail rot	13.85 ± 2.17	11.28 ± 2.07	10.7 ± 2.02
Mixed bacterial infections	7.79 ± 4.51	6.9 ± 0.97	6.0 ± 1.29
Bacterial gill disease	16.9 ± 4.22	30.2 ± 5.46	16.14 ± 2.24
Others	11.79 ± 4.5	9.57 ± 2.62	21.14 ± 4.32

Fungal Diseases

Fungal spores are ubiquitous in all freshwater systems. Fungal infections of fish by Oomycetes, commonly known as water moulds, are wide spread in freshwater ecosystems affecting cultured fishes. Usually fungal diseases occur by secondary infections, poor farm management practices, infestation by parasites etc. But there are several reports of these occurring as primary infectious agents of fish. Major groups of fungus causing fish disease belong to *Achyla, Aphanomyces, Branchiomyces, Dictyuchus, Fusarium, Leptomities, Pythium* and *Saprolegnia*. Among these, *Saprolegnia and Branchiomyces* are more specific for carps. But saprolegnia is considered as dreaded disease since it affects fish at all stages and even fish eggs. *Saprolegnia* sp. can affect a wide range of fish species, including carp especially the brood stock, during the post-spawning period. Some of the common fungus associated with carp diseases is presented in Table 3.

Table 3. Common fungal diseases reported in cultivable carps

Disease	Causative agent(s)	Organ (s) affected	Symptoms	Causes	Treatment/ Preventive measures
Saprolegniasis (Cotton wool)	*Saprolegnia parasitica* (fish moulds)	Fin, gills, mouth, eye, muscle	White to brown cotton like patches, skin lesions, lesions of internal organs, fungal hyphae penetrating into dermis and muscle, fish becomes weak and lethargic	Poor handling, high stocking, decreased water level, netting injury	Bath in malachite green (1-2ppm) for 1 hr., Formalin 1:500 for 15 min.,3% salt treatment (10 min.), KMnO4 10 ppm for 90 min.
Branchiomycosis (gill rot)	*Branchio-myces sanguinis*	Gills	Gill discolouration, blotchiness of the gills, gill necrosis, hyphae in blood capillaries, asphyxia, imbalance, gasping at the surface of water	Poor handling, high stocking, decreased water level, netting injury.	Malachite green (0.1 ppm), Chlorine 20 ppm, Copper sulphate (0.5 – 1.0 ppm), strict hygiene

Protozoan Diseases

Protozoan diseases are among the most significant of all parasitic diseases. Large-scale mortalities of fry and fingerling of carp species are common due to such infections. Most of the organ systems are infected by these parasites, which can cause mortality in fish. Protozoans become pathogenic to their host and can cause clinical disease when the fish becomes immunosuppressed or otherwise debilitated, perhaps due to adverse environmental conditions. Ciliates, *Mastigophora* and *Myxosporadia* are three main phyla responsible for various protozoan diseases of fish. Some of the common protozoan diseases found/ isolated frequently in carp are given in Table 4.

Helminth Diseases

Worm diseases of fishes are caused by the groups monogenia, digenia, and cestode. Although many of them do not cause much harm to carp species, these parasites in association with other parasite groups cause serious damage to the host fishes resulting in growth retardation and mortality. Among monogenetic trematodes, *Gyrodactylus* and *Dactylogyrus* sp. are most

Table 4. Common protozoan diseases recorded in carps

Disease	Causative agent(s)	Organ (s) affected	Symptoms	Causes	Treatment
Ichthyophthiriosis (White spot disease)	*Ichthyophthirius multifiilis*	Skin, fin, gills	Cysts of white spots, respiratory stress, restlessness, excess mucus secretion, lifting and separation of gill epithelium	Low oxygen level, rise in water temperature	Prolonged bath (4 times alternate days) with 0.1 ppm malachite green and 25 ppm formalin, 2-3% formalin 200 ppm (daily one hr.), 2-3% NaCl (daily one hr.).
Trichodiniasis	*Trichodina* sp.	Skin, gills	Pale colour gills, excess mucus secretion, mild hyperplasia, darkening and patches on skin	High stocking, poor water quality	Chelated copper compounds, 25 ppm formalin, 2-3% NaCl (5-10 min).
Costiosis	*Ichthyobodo necator (Costia necatrix)*	Gills, skin	Excess mucus secretion, blue grey film over skin and gills, hyperplasia of gill epithelium, erratic movement, flashing	Poor water quality	Bath in 3% NaCl or 1:2500 formalin for 10 minutes
Myxosporodiosis	*Myxosporidian* sp.	Gills, scale, skin, Intestinal tissue	Cysts on body and internal organs, excess mucus secretion, weakness, emaciation, necrosis and destruction of target tissue	Low chloride content (<400) ppm,, Temp. fluctuations in winter	2 % NaCl, Segregation of age groups, chlorinated lime, drying of ponds.

important. *Gyrodactylus* infects the skin whereas *Dactylogyrus* affects only the gills. Some of the common helminth diseases in carps are listed in Table 5.

Table 5. Common helminth diseases recorded in carps

Disease	Causative agents	Organ (s) affected	Symptoms	Treatment
Dactylogyr osis	*Dactylogyr-us sp.* (gill fluke)	Gills, opercula	Gill colour fades, excess mucus secretion, dropping of scales, restlessness, gathers near inflow of water and gasp for air, appear anaemic	Bath in Dipterex (0.25 - 0.5 ppm) Chlorophus (0.25 ppm), $KMnO_4$ (3-5 ppm)
Gyrodactyl osis	*Gyrodactyl-us cyprini* (skin fluke)	Skin	Dark blue or opacity of skin, flashing, gather at water inflow and gasp for air	Dip in Trichlorphon 1 % for 2-3 minutes, formalin-25 ppm, $KMnO_4$ – 5ppm, Potassium dichromate–20 ppm
Diplostomi asis (Black spot disease)	*Diplostom-um pigmentata*	Skin, eye	Black pigments on skin nodules, opacity of lens, exophthalmia, blindness	Praziquantel 300 mg/ kg fish with feed for one week, dip treatment in picric acid (3: 100,000) for 1 hr.
Ligulosis	*Ligula intestinalis*	Intestine	Abdominal distension, rupture of wall, compression of visceral organ, retarded growth, decreased feeding	Prevent contamination of water by infected bird faeces, application of praziquantel
Sanguinico liasis	*Sanguinico -la inermis*	Heart, gills, liver, spleen	Damaged and pale gills, bulging opercula, exophthalmia, lethargy, emaciation, spiral movement	Disease free stocks, application of $CuSo_4$

Crustacean Diseases

Crustacean parasites are frequently found to infest fishes. These are mainly ectoparasites and 3 groups viz. Branchiura (*Argulus* sp.), Copepoda (*Lernaea* sp., *Ergasilus* sp.) and Isopoda (*Isopod* sp.) are responsible for fish diseases. Among these, two crustacean parasites *Lernaea* sp (anchor worm) and *Argulus* sp (fish louse) are most widespread and commonly found parasitizing major carp species and sometimes causing large-scale damage in nursery, rearing and stocking ponds. However, low infestations of crustacean parasites do not cause serious damage to fish except causing irritation and localized ulceration. Some of the common crustacean diseases in fish are cited in Table 6.

Table 6. Common crustacean diseases recorded in carps

Disease	Causative agent	Organ (s) affected	Symptoms	Treatment
Argulosis	*Argulus foliaceus* *A. engalensis*	Skin, fin	Loss of appetite, erratic swimming, restlessness, skin loss, haemorrhagic and ulcerative lesions over body	Bath in Trichlorophon (0.2 ppm) for 24 hr, $KMnO_4$ (10 ppm for 30 min) Removal by forcep.
Ergasilosis	*Ergasilus sieboldi*, *Murgulicola* sp.	Gills, operculum	Necrosis of respiratory, epithelial cells and tissues, excess mucus secretion, surfacing, hyperplasia, retarded growth	Treatment for one week with 0.5 ppm copper sulphate + 0.2 ppm ferric sulphate Sanitary control, drying of pond, application of lime.
Lernaeasis	*Lernaea cyprinacea* *L. bengalensis*	Muscle, tissue, scale	Ulceration and sloughing off. Erratic movement, emaciation, scale loss, necrosis of tissues	Treatment with 250 ppm Formalin (30 min.) for three weeks, Chlorine 1 ppm (3 days)

Viral Diseases

All known viruses are infective agents. They are often host-specific and in some instances they have high specificity for certain tissues. As viral diseases of fish are geographically limited, disease free areas consider them as exotic diseases. Although many viruses have been isolated from the carp family, fish diseases due to viruses have not been reported in India as far as carp species is concerned. Some of them produce high mortality in young fish and little or no loss in adults, which may become carriers. Some viral diseases in carps reported from other countries are presented in Table 7.

Table 7. Common viral diseases recorded in carps

Disease	Agent	Sps mostly affected	Organ(s) affected	Clinical signs	Treatment
Spring viraemia of carp	Rhabdovirus carpio (birnavirus)	Common carp, Koi carp, Grass carp	Skin, gills	Fish gathering at outflows, dark colouration, loss of balance, inflammation, haemorrhagic of skin and gill	Bath immunization, vaccination, application of antibiotics to prevent secondary infection, variation in water temperature
Grass carp haemorrhag ic disease	Reo virus Piconovirus	Grass Carp	Internal organs, skin	Ventral haemorrhagic inflammation, bleeding in scale base	Bath vaccination of eggs, bath immunization of fingerlings with attenuated virus vaccine.
Carp pox	Cyprinid herpesvirus-1	Koi carp	Body surface	White plaque like nodular lesions, epidermal hyperplasia	Surgical removal of pox or culling
Koi herpes virus	Cyprinid herpesvirus-3	Common carp, Koi carp, gold fish	Intestine, gills, spleen	Mottled appearance, respiratory distress, red and white patches, sunken eyes, blisters on skin, lethargic, gill necrosis	Not to import fish from infested areas

Immunity of Fish

Immunity of fish like all other vertebrates plays major role in protection against pathogens. It can be either non-specific, which is an innate defence mechanism, rendering the host's resistant to infection, or specific, which is an acquired immunity by the host's response following an infection or immunization. Since many fish have short life span and live in aquatic environments, it slows the development of specific immune response. Fish, therefore, rely more on non-specific defense mechanisms as compared to mammals to encounter pathogenic organisms. The non-specific immune assays include lysozyme, myeloperoxidase, phagocytic assay, bacterial agglutination, haemagglutination and haemolysin activity etc. and specific parameter include antibody level against a particular pathogen. Majority of teleostic fishes including Indian major carps secrete antibody molecule of IgM type and it occurs in monomeric, dimeric and tetrameric forms depending on the requirement and site. For example, serum IgM is tetrameric whereas fish secrete monomeric form of IgM that is transferred to offspring through the egg yolk. These immune parameters vary round the year due to many reasons including physico-chemical parameters, particularly water temperature. Assays of these responses can provide useful measures of immunosupression induced by environmental contamination, stress and system of culture. It has been observed that some of the non-specific immune parameters of tropical fishes significantly drop during winter season when water temperature falls below 20^0 C making the fish more prone to microbial infections. Some of these non-specific parameters and their seasonal variations in Indian major carp, rohu (*L. rohita*) has been given in Table 8.

Table 8. Seasonal variations of different non-specific immune parameters of brood Indian major carp, *Labeo rohita*.

Non-specific immune parameters	Units	Summer (March-May)	Rainy (July-Sept)	Winter (Nov-Jan)
Lysozyme activity	mg/ml	11.59 ± 1.30	12.93 ± 1.66	7.26 ± 0.87
Myeloperoxidase activity	OD	0.97 ± 0.16	1.31 ± 0.19	0.54 ± 0.11
Bacterial agglutination	Log_2	3.40 ± 0.60	8.70 ± 1.70	4.00 ± 0.89
Haemagglutination titre	Log_2	5.20 ± 0.80	3.20 ± 0.61	3.80 ± 0.53
Haemolysin titre	Log_2	3.60 ± 0.78	2.80 ± 0.61	4.00 ± 0.73

Source: Swain *et al.* (2007)

Immunity of Young Fish

The immunity of a species depends on the development of lymphoid organs, which only starts after birth in all animals. Similarly in fish, the ability to mount a specific immune response against a pathogen starts only on / after 3 weeks of age before which they either depend on maternally derived immunity or specific immunity. Since fish at their early stages of development are prone to several infectious diseases, significant loss due to larval mortalities in various fishes has been reported throughout the world. To check this mortality, it is desirable to increase the immune status of these larvae. For this, Immunity of mother plays an important role in protection of offspring against any infections as immunity is transferred to the offspring through eggs in many fish species to combat infectious agents in early phases of life. It is important to know more about these defence mechanisms and if the health status of the mother might affect the survival rate of larvae. Such considerations are important for both farmed and wild fish populations.

Vaccination and Immunomodulation

A number of approaches including sanitary prophylaxis, disinfections, and chemotherapy with particular emphasis on the use of antibiotic and in recent times, vaccinating against specific diseases have been applied in an attempt to solve this problem. However, there are practical difficulties and undesirable consequences associated with some, if not with all these approaches. For instance, prophylaxis based on sanitary isolation is difficult to achieve due to the presence of other fish species, invertebrates and water itself. Antibiotic therapy is undesirable, as it has the potential for enhanced microbial resistance and the accumulation of residues in tissues of the fish. Vaccines up regulate the specific immunity by triggering the production of specific antibodies towards one particular pathogen. The first documented attempt to develop a vaccine for disease prevention in fish dates back to the furunculosis vaccination trials. Although, vaccination has become a useful prophylaetic measure for infectious diseases of fish and several vaccines are already commercially available, the development of vaccines is often difficult against the unknown pathogens causing diseases to fishes. Therefore, the immediate control of all the fish diseases using only vaccines is impossible. In this context, much attention is deviated towards improving the innate immune system of fish because the defence system of fishes are based on non-specific defence mechanism rather than relying on the more complex specific immune system that have developed in higher mammals. Vaccination, although highly effective, but in some instances, it is time consuming, labour intensive, costly and protection is often pathogen specific. Thus, an alternative approach is to boost or stimulate the innate immune system of cultured fish.

Immunostimulants are increasingly viewed as an integral part of disease prevention and management in modern aquaculture. Besides, aquaculture may benefit from the use of such immunostimulants during developmental phases of the organisms when they are particularly susceptible to infectious agents. Immunostimulants can be given to them to activate non-specific defence mechanisms or they may be administered with a vaccine to activate non-specific defence mechanisms and heightening a specific immune response as well. The compounds termed as immunostimulants include bacteria and bacterial products, complex carbohydrates, nutritional factors, animal extracts, plant extracts, cytokines, lectins and synthetic drugs. A large number of putative immunostimulants have also been investigated in fish that include killed pathogens and their products, fungal ß-glucans, chitin, unidentified animal extracts, synthetic chemicals like levamisole, plant molecules, nutrients, RNA and host/animal molecules. Although lot of reports on development of immunostimulants in fish is pouring in daily from abroad, still its application has a long way to go, particularly in India.

Conclusion

One of the major issues in fish production through aquaculture is loss associated with disease. Improper and faulty management practices followed in fish culture system are often stressful to fish. Under stress conditions, fish suppresses the immune responses and subsequently suffer from diseases. There are two strategies for disease management viz. prevention of disease through prophylactic measures and treatment and control of disease outbreaks using the disease therapeutics. But preventive measures have always big advantage over curative practices as there is a common saying that an ounce of prevention is worth a pound of cure. Apart from good farm management practices, good water quality, stocking ponds in proper ratio and quantity, providing nutritionally balanced diet, conducting effective disinfection practice in hatcheries, practicing crop holiday, using carrier free and disease free broods for breeding programme are some of the measures to be followed. The aquaculturist should strive to decrease the stress causing factors and take appropriate measures to prevent the entry of pathogenic organisms by strictly adhering to the fish health-monitoring programme. Traditionally, synthetic chemicals and antibiotics have been used to prevent or treat fish diseases and have achieved at least partially success. But use of these expensive chemotherapeutants and antibiotics for controlling diseases has been widely criticized for their negative impact resulting in reduced consumer preference for fish treated with antibiotics. Hence, to prevent infections in brood fish and to handle the disease outbreaks, proper quarantine protocols should be followed. Serious thoughts should be given to develop rapid diagnostic methods, use of probiotics and biocontrol agents, application of immunostimulants, development of vaccines etc. Last but not the least, creating environmental awareness, proper surveillance and monitoring, carefully planning and timely application of suitable prophylactic / curative measures can surely prevent the occurrence of many diseases in brood fishes of carp.

Important Websites Dealing with Fish Diseases

www.oie.int
www.fao.org
www.enaca.org
www.efishbuisness.com
www.frs-scotland.gov.uk
www.defra.gov.uk
www.countryside.wales.gov.uk/fisheries
www.scotland.gov.uk
www.dardani.gov.uk
www.fishdoc.co.uk
www.gov.au.westfish
www.fws.gov/aah

Table 9. Journals dealing with fish diseases

Name of Journal	Publisher
Fish and Shellfish Immunology	Elsevier
Veterinary Immunology and Immunopathology	Elsevier
Comparative Immunology, Microbiology and Infectious Diseases	Pergamon
Journal of Fish Diseases	Blackwell
Aquaculture	Elsevier
Aquaculture Research	Blackwell
Diseases of Aquatic Organisms	Inter Research
Asian Fisheries Science	Asian Fisheries Society
Annual Review of Fish Diseases	Pergamon

REFERENCES

Anderson, D. P. 1992. Immunostimulants, adjuvants and vaccine carriers in fish; applications to aquaculture. *Annual Review of Fish Diseases*. 2: 281-307.

Anderson, D. P. 1996. Environmental factors in fish health: immunological aspects. In The fish Immune System. Organism, Pathogen and Environment. (Iwama, G. and Nakanishi, T. (eds). *Academic press*, London, pp: 289-310.

Avtalion, R. R. 1981. Environmental control of the immune response in fish. In: CRC Critical Reviews of the Environmental Control. *CRC Press*, pp: 163-188 .

Dehadrai, P.V. 2006. Indian fisheries resources and production. In: Handbook of Fisheries and Aquaculture. Published by Directorate of Information and Publications of Agriculture, ICAR, New Delhi – 110 012. pp: 1-35.

Du Pasquire, L. 2001. The immune system of invertebrates and vertebrates. *Comparative Biochemistry and Physiology*, 129 (B): 1-15.

Fletcher, T. C. 1975. Defense mechanisms in fish. In : Biochemical and Biophysical Perspectives in Marine Biology, (Malins, D. C. and Sargent, J. R. eds. *Academic Press*, London, pp: 189-192.

Peddie, S. and Secombes, C. J. 2005. An overview of fish immunostimulant research. *Fish Veterinary Journal*, 8: 1-31.

Press, C. .M. 1998. Immunology of fishes. In: Handbook of Vertebrate Immunology (Pastoret, P. P., Griebel, P., Bazin, H. and Govaerts, A. (eds.) *Academic press*, London, pp: 3-62.

Raa, J., Rorstad, G., Engstad, R and Robertson, B., 1992. The use of immunostimulants to increase resistance of aquatic organisms to microbial infections. *In: Diseases in Asian Aquaculture.* Schariff, M., (Subasinghe, R. P. and Arthur, J. R. eds). Fish health section, Asian Fisheries Society, Manila, Phillippines, 26-29, pp: 39-50.

Rijekers, G. T. 1982. Non lymphoid defence mechanism in fish. *Developmental and Comparative Immunology*, 6: 1-13.

Sakai, M. 1999. Current research status of fish immunostimulants. *Aquaculture*, 172: 63-92.

Swain, P., Dash, S., Sahoo, P.K., Routray, P., Sahoo, S.K., Gupta, S.D., Meher, P.K. and Sarangi, N., 2007. Non-specific immune parameters of brood Indian major carp, *Labeo rohita* and their seasonal variations. *Fish and shellfish Immunology*, 22: 38-43.

Zapta, A. G., Varas, A. and Torroba, M. 1992. Seasonal variation in the immune system of lower vertebrates. *Immunology Today*, 13 : 142-147 (Vol. 1 & 2).

Chapter **9**

Induced Breeding Experiments of an Indian Major Carp *Labeo rohita*- A Cause Effect Study

S.D. Gupta & S.C. Rath

Introduction

Indian major carps viz. *Catla catla, Labeo rohita* and *Cirrhinus mrigala* are the fast growing, consumer preferred and the stay component species in Indian pond aquaculture system. These carps breed spontaneously in natural water viz. river and bundh type tanks where fluviatile condition prevails (Hora, 1945; David 1959; Quasim and Qayyam, 1962; Natarajan, 1971). Above three species under culture condition attain sexual maturity but do not spawn in confined water (Chaudhuri, 1960; Chaudhuri and Singh, 1984). Khan and Jhingran (1975) in their review described several factors viz. temperature, run off water, photoperiod, turbidity and water depth that are playing important roles in initiating natural spawning in 'bundhs' and rivers. They further described that no single factor can be considered responsible for spawning separately. The act spawning involves a chain of interrelated conditions as prerequisite to spawn. Monsoon, flood forms the spawning ground and initiates natural spawning. Although fish bred in its natural environment, it was difficult to understand why some species do not breed in confined water where as other species in the same environment breed somoto.

The initial attempt to induce breed IMC (mrigal) was made during 1937. Khan (1938) was able to induce ovulation in Cirrhinus mrigala by

Applied Bioinformatics, Statistics and Economics in Fisheries Research: pp. 121-129

New India Publishing Agency, New Delhi (India)

administration of mammalian pituitary extract but eggs were not fertilized. No remarkable success on induced breeding of Indian major carps was made till 1957. Indian major carps were induced to breed for the first time on 10th July, 1957 by the injection of aqueous carp pituitary extract at Govt. Fish Farm, Angul of Orissa state by the scientific team from pond culture substation, Cuttack (Chaudhuri and Alikunhi 1957). Eversince 1957, the induced breeding of carps have been tried by many permutation and combination, addition and alteration of inducing agents.

Over the years the techniques of induced breeding of fish was simplified and standardised. Such work is repeated and reviewed by several workers time and again (Alikunhi et al., 1960; Chaudhuri, 1960; Das and Khan, 1962; Chondar, 1970; Khan and Jhingran, 1975; Moitra and. Sarkar, 1975, Chaudhuri and Singh, 1984; and Gopakumar et al., 2000). The wide spread adoption of hypophysation technique for induced breeding of confined water culturable species opened the opportunities for collection and marketing of carp pituitary glands which in turn became more commercial and sometimes the quality of the gland, concerned with gonadotropin hormone was not dependable for successful induced breeding.

Subsequently alternatives had been searched out for carp pituitary gland. Pituitary extract of marine cat fishes, *Trachysurus thalassinus* and *Trachysurus jella* were found effective in induced breeding of major carps (Verghese et al., 1975; Verghese and Rao 1976). Bhowmick et al. (1979) opined that pituitary gland extract of *Arius* sp. (marine castfish) and *Mystus* sp. (brackish water cat fish) do not help for induced breeding of Indian major carps

Use of human chorionic gonadotropin (hCG); fish and mammalian gonadotropin (GtH); gonadotropin releasing hormone (GnRH) and GnRH analogues are the land mark attempts for induced breeding of fish. Ibrahim and Chaudhuri (1966) suggested that the aqueous solution of carp pituitary extract can be preserved in glycerin and ampouled in small airtight glass bottles for use for several years. Human chronic gonadotropin, a polypeptide hormone, similar with LH and FSH in function has been used in induced breeding of carps in India with varied degrees of success. hCG can be collected in crude form from the pregnant women urine. For some years in history of induced breeding of fish, the role of hCG was much discussed and debated. It was opined that hCG could help for gonadal maturation and induced breeding. Investigators have also reported the negative result of hCG on induced breeding (Das and Singh, 1990). Human chorionic gonadotropin was tried successfully in the induced breeding of carps in India (Bhowmick, 1979; Chondar, 1984; Somashekarappa *et al.*, 1990). Partially purified fish gonadotropin was administered and carps were bred by Chaudhuri et al. (1977) and Sinha (1971). Gupta et al. (1988) studied the oocyte maturation and spawning of Indian major carps by administration of winstrol and progesterone. Das and Rath (1999) have studied the efficacy of some synthetic hormones alone and in combination with dopamine antagonist.

Induced breeding by administration of pituitary extract popularly known as hypophysation. To standardize the technology of induced breeding and to produce adequate quantity of seed of *Labeo rohita* (Ham.) several breeding experiments have been conducted over the years. Further it was attempted to reveal the effect of different inducing agents and factors based on 421 experiments that were conducted from July, 1970 to August, 1982 with carp pituitary extract (CPE), noncarp pituitary (NPE) and gonadal concerned hormone (GCH) as inducing agents. Again the inducing agents have been administered in different combination and in different protocols. Experiments have been conducted within the temperature range of 27.5 to 35°C. Brood body weight (wt.) ranges from 0.3 - 2.7 kg (Male) and 0.4 - 3.5 kg (Female). Spawning fecundity varies from 0.03 lakh eggs/kg to 4.18 lakhs/kg body wt. of the female. Fertilization rate ranges from 0 to 95 percent and spawn recovery ranges from 0 to 2.83 lakhs/kg body wt. of female (Gupta *et al.*, 2004).

Inducing Agents and Spawning Responses

Twenty seven types of inducing agents have been used in 421 breeding experiments on *L. rohita*. These inducing agents are broadly classified as carps pituitary extract, noncarp pituitary extract and marine fish pituitary extracts. Pituitary extracts alone, with isolated pituitary hormone and in combination with salmon pituitary powder etc. and their spawning response is presented on Table-1. Again carp pituitary extract in aqueous medium for immediate use and in glycerine medium for instant use. Glycerine medium extracts have been tried after 0 year, 1 year, and 2 years, 3 years, 4 years and 5 years intervals

Temperature and Breeding Response

Water temperature plays a vital role in carp breeding. In the present analysis of non-breeding response of *Labeo rohita* about 73.5% of breeding failure is attributed to the water temp prove than 32°C. Only 26.5% of the non-responded instances found in the temperature range less than 31.5°C.

Fertilization Efficiency

Spawn production depends upon the rate of fertilization of the ovulated eggs as a factor. Fertilization efficacy less than 50% is considered as poor fertilization instances (PFI).

Table 1: Spawning response in *Labeo rohita* (Ham.) with different inducing hormones

	Inducing Agents	Spawning Percentage		Remarks On Negative (-ve) Spawning
		(+ ve)	(- ve)	
	Acetone preserved carp pituitary in aqueous extract (ACPAE)	79.7	20.3	Inadequate diet and improper gonadal maturation
CPE	Carp pituitary aqueous extract (CPAE)	88.9	11.9	High temp., unripe gonads, incorrect doses
	Carp pituitary glycerin extract (CPGE)	57.2	42.8	High temp. unripe gonads, and some other unrevealed factors.
GCH	Pituitary extracts and other hormone combination (PEOMC)	52.4	47.6	Loss of potency in more than two years stock of pituitary adverse weather condition, and improper gonadal maturation
NPE	Noncarp pituitary extract (NPE)	39.6	60.4	Pituitary extract other than freshwater catfishes, carps and salmon and single dose of salmon pituitary powder.

Spawn Recovery

In the present study the fertilized eggs are incubated in both out door hapa system (OHS) and in indoor hapa system (IHS). If spawn recovery is less than 70% out of fertilized eggs is taken as poor recovery. The OHS showed 80.5% of poor recovery and IHS showed 33.3% poor recovery.

Partial Spawning Response and Reduced Spawn Recovery

As the induced carp breeding is an eco-physiological phenomenon. Some times it resulted in partial spawning. In the present study it is observed that 76% of partial spawning instances showed < 50% (0 - 49%) spawn recovery.

Flop Spawning

Six per cent of the total breeding response was flop. Eggs were not fertilized. This may be due to unprime gametes (sperm or ovum).

Statistical Analysis

Data on breeding experiments (421 sets) of *L. rohita* conducted during 1970-1982 were computerized and analyzed statistically to study the breeding performance of these fishes in relation to biotic and abiotic factors. Basic statistics of brooder wt. average temperature, quality of eggs and total dose is furnished in table 2.

Table 2. Basic statistics of various parameters of *L. rohita* breeding experiments.

	Mean	**Std. Dev.**
Female (Kg)	1.67	2.26
Average Temp.(°C)	30.65	1.27
Quantity of eggs (Lakhs)	2.82	1.79
Total Dose (mg/Kg)	15.78	4.96

Source : Gupta *et al.*, 2004

Pituitary extract and breeding response : The treatments used in breeding were divided into following broad heads. Carp Pituitary Extract (CPE) - H_1: Non-carp Pituitary Extract (NCPE) – H_2 and Gonadal Concern Hormone (GCH) - H_3. The cross tabulation of qualitative characters (treatment and response) were reported as follows showing the number of occurrences under each category along with percentages (Table 3).

Table 3. Breeding response of *L. rohita* under various hormone treatments

Treatments	**Response**		
	Positive response (%)	**Lack of response (%)**	**Total**
H_1 (CPE)	274 (82.30)	59 (17.7)	333
H_2 (NCPE)	12 (28.6)	30 (71.4)	42
H_3 (GCH)	22 (47.8)	24 (52.2)	46
Total	**308**	**113**	**421**

(Values in parenthesis indicate percentage) *Source: Gupta et. al.2001*

Test of Independence of Treatment and Breeding Response

Two-way classification of treatment and breeding response gave comparative performances of various treatments over response and X^2 (chi-square) test of the data reveals that (X^2 tab = 5.99<X^2 cal = 48.45) breeding response depends on the treatments.

Hormones and Spawn Recovery

Reports on Analysis of Variance (ANOVA) carried out to study the effect of hormone on egg release did not demonstrate any significant difference. However, maximum average spawn recovery was observed in case of H_1 (3.11) followed by H_2 (3.04) and H_3 (3.0 lakhs). ANOVA performed to study the effect of hormone on spawn recovery did not reveal any significant difference. However, apparent highest spawn recovery was observed in case of H_3 (1.53 lakhs) followed by H_2 (1.22 lakhs) and H_1 (1.09 lakhs). ANOVA revealed significant difference in percentage fertilization due to hormone at 5% level. Highest percentage fertilization was demonstrated by H_3 (56.01). (Gupta *et al.* 2001).

Brood Weight, Dose, Temperature, Quantity of Eggs Released and Spawn Recovery

In order to quantify the possible effect of different independent variables like weight of female, doses of hormone, temperature, quantity of eggs released and percentage fertilization on the dependent variable like spawn recovery a multivariate approach of exploratory data analysis has been reported (Gupta et al. 2001) The stepwise regression analysis for each of the hormone groups (*viz.* H_1, H_2 and H_3) with independently variables like weight of female brood, dose of hormone, temperature, quantity of eggs released and percentage fertilization has been reported (Gupta *et al.*, 2001) separately. In case of Carp Pituitary Extract (H_1) independent variables like female weight, temperature and quantity of eggs together explain 46.74% of the total variability in spawn recovery (R^2 = 0.4674; P<0.01; n=212). Similarly in case of non-carp pituitary extract (H_2) quantity of eggs alone explains 85.91% of total variability in spawn recovery (R^2 = 0.8591%; P<0.01; n = 11). Further for gonadal hormone (H_3) quantity of eggs demonstrated 21.33% of total variability (R^2=0.2133; P<0.05; n = 16). Interrelationship among variables are furnished in table 4.

Significant positive correlation exists between spawn recovery and each of the variables like quantity of eggs and female weight. Negative correlation exists between spawn recovery and temperature (Table 4).

Model Building

By applying backward eilimination procedure of stepwise regression retained only dependent variables like female weight, average temperature, quantity of eggs in the equation after eliminating variables for contributing significantly (Table 5).

Table 4. Correlation of matrix of various factor of *L. rohita* breeding experiment.

	Spawn recovery	Female	Avg. Temp	Quantity Eggs.	Total Dose
Spawn recovery	1.000	-	-	-	-
Female wt.	.313	1.000	-	-	-
Avg. Temp.	-.183	.053	1.000	-	-
Quantity of Eggs	.651	.318	-.024	1.000	-
TOT Dose	-.090	.056	.081	-.097	1.000

Source: Gupta *et al.* 2001

Table 5. Variables retained in the equation after elimination of insignificant variables

Variables	B	SEB	Beta	. T	SigT
Female wt.	.053744	.019786	.129709	2.716	.0070
Avg. Temp.	-.129677	.033394	-.175833	-3.883	.000
Quantity of Eggs	.031901	.02455	.605728	12.699	.000
Constant	3.861106	1.027701	-	3.757	.0002

$\hat{Y}$ *(spawn recovery) = 3.8611 + .0537 (female weight) - .1297 (temperature) + .0319 (quantity of eggs)*
($R^2 = 0.4674$; $P < 0.001$)

Validation of Model

The fitted equation is capable of explaining 46.74% of total variability of the spawn recovery by the identified independent variables like body weight of female brood, average temperature and quantity of eggs. All these are biologically agreeable (Matsui, 1957; Quasim and Qayyum, 1963; Khan and Jhingran, 1975). The findings of the study will be helpful for further research work in the breeding of carps.

REFERENCES

Alikunhi, K.H; M.A. Vijayalakshmanan and K.H. Ibrahim, 1960. Preliminary observation on the spawning of Indian major carp, induced by injection of pituitary hormones. *Indian J.Fish*, 7(1) 1-19.

Bhowmick, R. M. 1979. Observations on the use of human chorionic gonadotropin prepared in the laboratory, in inducing spawning in major carp *Labeo rohita* (Ham.). Tech. Paper 4, *Sym. on Inland Aquaculture (Abst.)* Barrackpore, India.

Chaudhuri, H. 1960. Experiments on induced spawning of Indian major carps with pituitary injection. *Indian J. Fish*, 7 (1): 20 - 49.

Chaudhuri, H. and K.H. Alikunhi 1957. Observations on the spawning of Indian carps by hormone injection, *Curr. Sci.*,26(12) 381-382.

Chaudhuri, H. and S.B. Singh 1984. *Induced breeding of carps*, Publication of Indian Council of Agricultural Research, New Delhi, p. 76.

Chaudhuri, H., R.M. Bhowmick, G.V.Kowtal, R.K. Jana and S.D. Gupta. 1977a. Efficacy of partially purified salmon gonadotrophin (SG-G-100) in inducing spawning in Indian major carps. *J. Inland Fish*. 9,162 – 164.

Chondar, S. L. 1970. Handbook on breeding of Indian major carps by pituitary hormone injection. Satish Book Enterprise, Agra, pp. 100.

Chondar, S.L 1984. Induced spawning of Indian major carps on commercial scale by crude HCG combined with fish P.G. In. 12th and 13th state conference of the state Junior Fishery Officer Association West Bengal Dec 8-9.

Das, P.C. and R.K. Rath. 1999. Effect of exogenous hormones in combination with dopamine antagonists in breeding of Indian major carps. *J. Aqua*. 7, 1-10.

Das, S. M. and H. A. Khan 1962. The pituitary and pisciculture in India with an account of the pituitary of some Indian Fishes and a review of technique and literature on the subject Icthyology(1and 2) 43-58.

David, A.1959. Observations on certain spawning movement of the Gangetic carps and carp seed resources in India. *Indian Geogr, J.*, 27 (3and 4): 1-17.

Gopakumar, K; B. N. Singh, and V. R. Chitranshi 2000. Fifty years of Fisheries Research in India. Indian Council of Agricultural Research, New Delhi, 270 p.

Gupta, S.D., A. K. Roy, S.C. Rath and P. K. Satpathy (2004). Database of induced breeding experiments in an Indian Major Carp *Labeo rohita* (Ham.): In *Bioinformatic and Statistics in Fisheries Research* Vol.I (Eds. A.K. Roy, A. Antony and M. Rout): pp.117-118.

Gupta, S.D., H.A.Khan, M.S. Tantia and R.C. Das 1988. Effects of winstrol and progesterone in vivo oocyte maturation and spawning in *Labeo rohita*. Proc. of Carp Seed Production Technology, 2 - 4 Sep. Eluru, 24 - 28.

Hora, S.L. 1945. Analysis of the factors influencing the spawning of Indian major carps. *Proc. Nat. Inst. Sci. India* (B), 11 : 303 –12.

Ibrahim, K. H. 1961. Some observation on the carp fry resources of the river Godavari near Rajahmundry, Andhra Pradesh. *Indian J. Fish* 8(2) 403-412.

Khan, H. 1938. Ovulation in fish (effect of administration of anterior lobe of pituitary gland). *Curr. Sci.* 7(50) 233 –234.

Khan, H.A.and V.G Jhingran, 1975. Synopsis of biological data on rohu, *Labeo rohita* (Ham.) 1822, FAO Fish Synops (III) 100pp.

Moitra, S.K. and S.K. Sarkar. 1975. A new method for induced breeding by hypophysation of some major carps in dry bundhs of Bankura, West Bengal. *Zool. Sec.* Calcutta 28, 41 - 50.

Natarajan, V 1971. Biology and fishery of *Labeo calbasu* (Hamilton) in Bhabanisagar reservoir, Madras. *J. Fish.* 6, 14-56.

Quasim, S. Z. and Qayyam, A. 1962. Spawning frequencies & breeding seasons of some freshwater fishes with special reference to those occuring in the plains of northern India *J. Fisheries* 8(1) 24-43.

Sinha, V.R.P. 1971. Induced spawninmg of carp with fractionated fish pituitary extract. *J. Fish Biol.* 3, 263-272.

Somashekarappa, B; H.N. Chandrashekaraiah and M.C. Nandeesha 1990. Advancement of maturity in Indian major carp (*Catla catla*) through improved diet and human chorionic gonadotropin administration. *Proc. of workshop on carp seed production technology*, (2): 29-33.

Verghese , T.J. and G.P.S. Rao 1976. On the use of marine catfish pituitary gland for induced spawning of silver carp and catla. *Curr. Sci.*, 45: 302.

Verghese, T.K; G.P.S. Rao; K.V. Devraj and B. Chandrasekhar 1975. Preliminary observation on the use of marine catfish pituitary gland. *Curr. Sci.* 44 (31), 72-75.

Chapter **10**

Freshwater Prawn Aquaculture Status and Scope

Bindu R. Pillai

Introduction

Majority of the commercially important freshwater prawns belongs to the genus *Macrobrachium* of family Palaemonidae. Among the commercially important freshwater prawns of India the most important cultivable species at present is the giant freshwater prawn *Macrobrachium rosenbergii* due to its superior cultivable attributes such as fast growth rate, large size, wide range of salinity tolerance, hardiness and compatibly to grow with carps. The species enjoys good consumer preference and domestic as well as export market demand. Freshwater prawns are cultured in freshwater or low saline brackish water (salinity <10 ppt) and can be cultured either alone (monoculture) or in combination with fishes such as carps, tilapia and milkfish (polyculture). Freshwater prawn culture can be carried out in earthen ponds, cement cisterns, in pens or in cages. However, most of the operations are being carried out in earthen ponds.

Global Status of Freshwater Prawn Farming

Freshwater prawns have become a major contributor to global aquaculture both in terms of quantity and value (New and Valenti, 2000). Though there are several large species of freshwater prawns that can be cultured, most of the present production is from only two species *M. rosenbergii*

Applied Bioinformatics, Statistics and Economics in Fisheries Research: pp. 131-140

New India Publishing Agency, New Delhi (India)

and *M. nipponense.* While *M. rosenbergii* is cultured in many countries, *M. nipponense* is cultured only in China. The first recorded global data for farmed giant prawn showed a production of 5246 t in 1984. By 1989 the production showed a three-fold increase to 17,608 t (FAO, 2000) with a value of about $75 million. From 1990 to 1995 production did not show significant increase averaging 21,013 t/yr. In 1992, Asia contributed more than 92% of farmed production of freshwater prawns. North, central and south America contributed 7.7% (2200-2400 t), Africa contributed 0.2% and Europe reported no production. The rapid global expansion in freshwater prawn farming since 1995 was mainly due to the huge production from China and rapid take off of farming in India and Bangladesh (New, 2005). The average annual expansion rate of freshwater prawn farming between 1992 and 2001 was over 29%. Between 1999 and 2001 the global expansion rate was more rapid (48%). However, 83% of the expansion during the decade occurred due to the phenomenal increase in the production in China. Other than China, the rapid expansion of freshwater aquaculture occurred in Bangladesh, Thailand and India. The global production of freshwater prawn *M. rosenbergii* stood at 1,94,159 t in 2004 with a value of over US$810million (FAO, 2006). The production of *M. nipponense* in 2004 was 2,13,078 t with a value of US$724 million. Thus the total global production of freshwater prawns in 2004 was nearly 4,07,237 t with a value of over US$1.6 billion. This figure excludes the production of *M. malcolmsonii* reported from India (245 t) and other freshwater prawns belong to family palaemonidae which are reported under the FAO statistical category 'freshwater crustaceans nei' (28,422 t).

China is the top producer with a production of 98,413 t in 2004. China started reporting their production in 1996. From 1996 till 2001 there was rapid increase in the production in China; highest production was achieved in 2001 (1,28,338 t). In 2002 the production fell to 1, 13,743 t and 2003 saw further decline in production to 87,252 t, in 2004 however, there was an increase in production (98,413 t). However, the production in China still far exceeds the production from other countries. India ranks second with 38,720 t production. Thailand is the third largest producer with 28,500 t production in 2004. Other major producers are Bangladesh (17,123) and Taiwan (10,039 t). Asia continues to dominate in the production contributing to more than 99% of the production.

Aquaculture Production of *M. rosenbergii* in India

The farmed production of this prawn in India has shown phenomenal increase since mid nineties, the production increasing from less than 500 t in 1997 to more than 38,720 t in 2004 (FAO, 2006). In 2005-06 the total production was 43,395 t and 43,433 ha area was under scampi culture

(MPEDA, 2006). The average production per ha in 2005 was 985 kg/ha with the productivity ranging from 65 kg in the reservoirs of Maharashtra to 1250 kg/ha in the highly productive aquafarms of Nellore, 62% of the production comes from Andhra Pradesh and 90% of which comes from Nellore district (Bojan *et al.*, 2003). Export of scampi increased from 2678 t worth Rs.123 crore in 1999-2000 to 10,379 t worth Rs. 447 crore in 2002-03. Japan is a major consumer of scampi and there is a steady increase in the global demand for the product.

Table 1: Aquaculture production of *M. rosenbergii* by major producers, 1997-2004.

(tonnes)

S. No	Country	1997	1998	1999	2000	2001	2002	2003	2004
1	China	42851	61868	79055	97420	128338	113743	87252	98413
2	India	1500	3900	7000	16600	24230	30500	35870	38720
3	Thailand	7856	4764	8494	9917	13310	15393	35102	28500
4	Bangladesh	4534	5751	5394	5504	9471	9559	10200	17123
5	Taiwan	7554	8165	7223	8149	6859	7026	10045	10039
6	Vietnam	4070	2918	2544	3513	4933	5552	5961	-
7	Brazil	446	279	227	450	450	450	450	363
8	Malaysia	143	281	653	1338	752	535	627	317
9	Indonesia	-	-	-	-	-	400	246	290
10	Dominican Rp.	27	87	16	82	62	78	78	97
11	Puerto Rico	2	-	7	12	7	8	104	40

Source : FAO, 2006

Table 2: State-wise details of Scampi culture 2004-05

State	Area Developed (ha)	Area under culture (ha)	Production (MT)
Andhra Pradesh	32,431	28,096	34,541
West Bengal	4,647	4395	3,193
Orissa	3,188	3188	470
Kerala	911	911	268
Tamil Nadu	428	192	144
Karnataka	206	165	44
Gujarat	4	4	3
Maharashtra	143	43	60
Total	**41,958**	**36,994**	**38,723**

Source: MPEDA, 2006

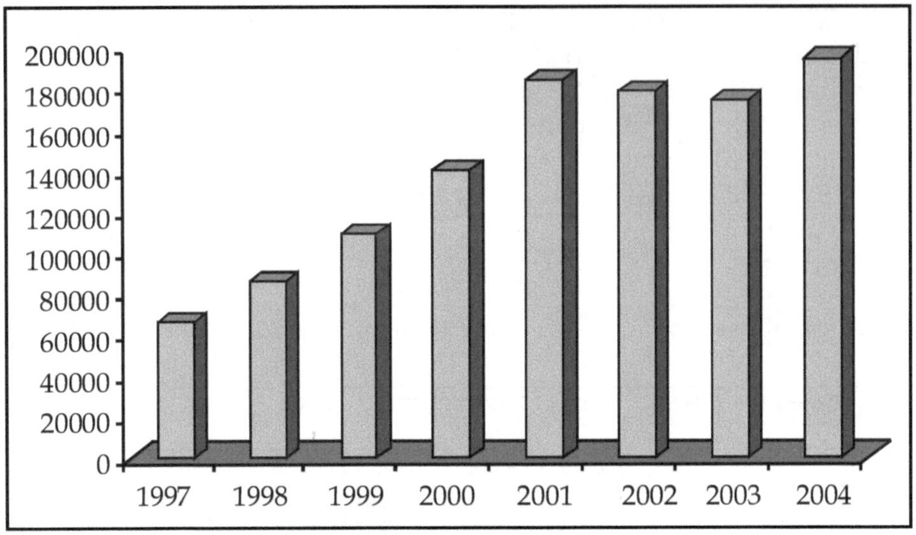

Fig. 1. Global Aquaculture Production of *Macrobrachium rosenbergii*

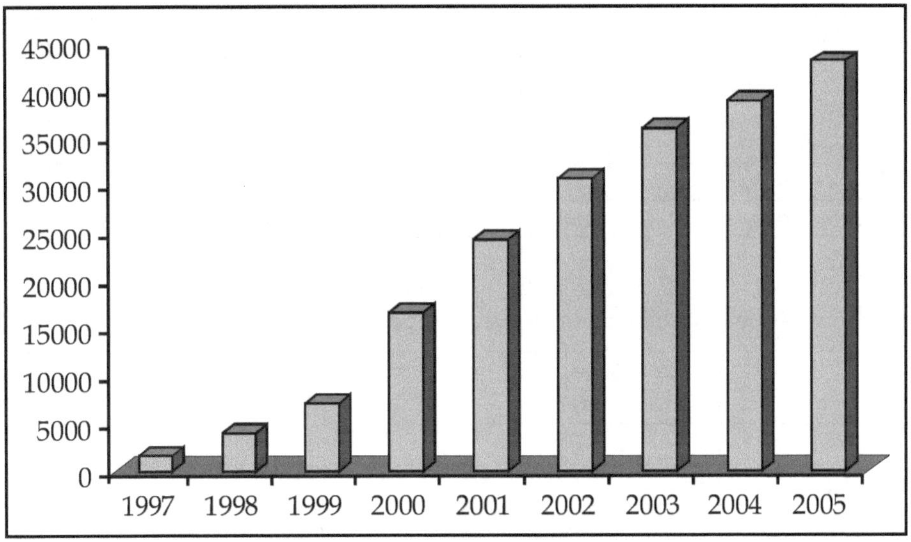

Fig. 2. Aquaculture Production of *M. rosenbergii* in India

Scope of Freshwater Prawn Farming: India has vast natural resources in the form of 2.03 million ha of reservoirs, 2.38 million ha of ponds and tanks, 0.798 million ha of bheels and oxbow lakes that can be made use for freshwater prawn culture. In addition, some of the 1.44 million ha of available brackish water area in the country can also be made available for freshwater prawn culture. Already nearly 0.9 million hectare is under fish culture in the country.

With little concerted efforts freshwater prawn can be easily incorporated into this system, as prawns are compatible with major carps thus helping the farmer to generate additional income. Availability of standard breeding and culture techniques will help in expanding the culture operations. Utilization of these vast resources needs extension of developed freshwater prawn culture technologies throughout the country. For successful extension, some basic information on biology, breeding and seed production, hatchery and culture technology are important which are furnished below.

Distribution and habits: This species is widely distributed in India and is found in most of the rivers in both east and west coast of the country. It lives in lower reaches of rivers and migrates to the estuaries for completion of its larval cycle in brackish water. It is slow moving and sluggish by nature and hides under shades and shelters in the shallow areas of rivers, canals, lakes and ponds during day time and is very active during night time

Food and feeding habits: The species is a bottom feeder and omnivore and accepts a variety of food items ranging from grains, worms, flesh pieces of molluscs, crustaceans and fish etc. It locates its food mostly by touch with feelers. The first pair of chelipeds is the chief organs of food capture and is assisted by the second pair.

Molting and Growth: Growth is not continuous because of the hard exoskeleton covering the body and appendages. Visible growth takes place only at the time or immediately after molting. Frequency of molting depends on the age, quality and quantity of food consumed and environmental parameters. Growth rate is rapid and is characterized by sexually dimorphic patterns and highly variable rates. Both the males and females grow at similar rates until the females begin to divert much energy intake into ovarian development and less into growth.

Reproduction: The sexes are separate in *M. rosenbergii* and the male have proportionately larger head than female. The females are small in size with smaller head and broad abdominal space to serve as brood chamber for incubating eggs. The females attain first maturity at about 150 mm in total length (avg. wt. 40 g) in 4-7 months (Rao, 1991). The males also attain maturity in this period. The sexually mature female completes the pre-spawning molt before mating. Breeding season differs in different river systems. Mating and incubation takes place in freshwater and brackish water as the species especially prawns with ripe ovary and berried ones are migratory in habit in natural systems. Although hatching can take place in freshwater, larval survival and growth takes place only in brackish water environment. After metamorphosis the post-larvae migrate back to freshwater region in the river and grow into adult.

Hatchery Technology: The first successful rearing of all the larval stages of *M. rosenbergii* in captivity (Ling, 1969) generated interest in many countries in controlled breeding and seed production. Central Institute of Freshwater Aquaculture, a premier research institute involved in developing and standardizing technologies for increasing the freshwater aquaculture production in the country has developed a commercially viable hatchery technology for the production of good quality freshwater prawn seed after more than 10 years of research (Rao and Tripathy, 1993; Pillai and Rao, 1999). The hatchery technology of giant freshwater prawn, *M. rosenbergii* developed by CIFA includes a)Generation of quality broodstock in captivity b) A two-phase, clear water technique of larval rearing for production of prawn seed.

Generation of quality broodstock in captivity: Broodstock constitutes an essential component in the hatchery phase. Freshwater ponds are ideally suited for the development of brood stock to have an uninterrupted supply of berried prawns to the hatchery. To develop captive broodstock advanced juveniles of *M. rosenbergii* (>5g) are stocked @ 1-2/m^2 at a male female ratio of 1:4 in well-prepared earthen ponds. The prawns are fed with specially formulated broodstock diet (Crude protein - 38%, Lipid - 8%) twice daily @ 10% of the biomass for the first two months and subsequently at rates ranging from 2-5% of the biomass. The ponds are monitored and managed on a regular basis to maintain optimum water quality. The prawns attain maturity in 2-3 months and females bearing advanced embryos are collected from these ponds regularly for the seed production purpose. For the operation of hatchery the berried prawns (bearing grey embryos) are collected at regular intervals from the brood stock ponds using a cast net.

Semi-closed, two phase clear water larval rearing technology : CIFA has developed and standardized a two-phase clear water technology for larval rearing of *M. rosenbergii* suitable for non-coastal hatcheries. This technology can be suitably modified to suite other locations also. The various steps involved in the technique is described below.

Healthy mother prawns (bearing grey eggs on their pleopods; >50g) are selected from the brood stock pond/tank and disinfected with 0.3 ppm copper sulphate or 30 ppm formalin for 30 min. Mother prawns are then stocked @100-150g/m^2 (2-3 nos of ~ 50g female) in brackish water (salinity - 5ppt) and reared till hatching. Tanks are checked daily for appearance of larvae. Twenty four hours after the larvae appear in the media the spent female is removed from the tank and released back to the brood stock pond. The newly hatched larvae are planktonic and require brackish water (10-12 ppt) for survival and growth. The salinity of the larval rearing medium is increased to 12 ppt and the rearing is continued in the same tank or the larvae are estimated and stocked in another tank for first phase rearing. In the first phase the larvae (Stage I or Zoea I) are stocked in conical tanks at a high density (200-400 larvae/l). About 50% of the medium is usually exchanged every other day with fresh medium of identical salinity. The larvae are reared for about 10-12 days in this phase. In the second phase, the advanced larvae are stocked in large cement tanks with a greater surface area at the rate of 50-80 per litre and reared till metamorphosis.

For the first two days after hatching the larvae need not be fed as they depend on their internal yolk reserves for nutrition. From third day onwards the larvae need to be fed. The freshly hatched *Artemia* nauplii are given as live food to the prawn larvae, 4-5 times per day in the early stages (Stages II to V or VI) and later, once during late evening in combination with wet larval feed which is usually given during day time. The brine shrimp nauplii are fed to the prawn larvae at the rate of 5 to 50 nauplii per larva per day. Wet larval feed (egg custard, minced fish/mollusc flesh; protein level> 50%) is fed @50-200 μg/larva/day depending on the larval stage. The wet feed is given during daytime. Feeding commence in the early morning and continue till late afternoon at an interval of one hour with constant supervision. The larval rearing tanks are cleaned daily by siphoning off excess food particles and metabolic waste from the bottom of the tank.

Water quality in a hatchery deteriorates due to accumulation of metabolites and decomposition of unutilised feed. Water exchange is done once every alternate day after checking pH and ammonia levels. Daily monitoring of temperature, salinity, pH and dissolved oxygen levels is done to maintain the water quality at optimum levels. Larvae normally take one to four days to go through one stage. Healthy larvae swim at the water surface up to stage V and feed actively and unhealthy larvae accumulate at the tank bottom. The giant prawn larval metamorphosis is non-synchronous in nature. The appearance of first post-larva is usually observed 20 days after hatching, normally between 22 and 32 days (at 28-32°C) and 90% larvae metamorphose within next 10 days.

The post-larva is characterised by a radical change in behaviour and appearance. For the first time they resemble miniature adult prawns and swim freely and crawl or cling to the tank surface. The post-larval production normally ranges between 35-40 per litre and the cycle lasts for 35-40 days. The post larvae are gradually acclimatized to freshwater. After acclimatization to freshwater the post-larvae (seed) are weaned from larval diet and fed on formulated diet @100% of the bio mass daily. After a week or fortnight the post-larvae are suitable for stocking in grow-out ponds.

Grow-out Culture Technology

CIFA has developed and standardized an environmentally friendly and sustainable culture technology for this prawn. Grow-out technology developed at CIFA aims at a sustainable production of 1-1.5 tonnes/ha/crop. The technology involves an initial nursery phase for a period of 45-60 days followed by a grow-out phase of six months.

Nursery rearing: In the nursery phase hatchery produced post larvae are reared in well-prepared earthen nurseries (0.02-0.1 ha) @ 20-50/m^2 for 45-60 days. Aeration is provided from an aerator for ~8 h/day. Post larvae (8-10 mg) are fed with pellet diet (crude protein- 35%; lipid-8%) in crumble form @ 100% of the biomass during the first fortnight. The feed is broadcasted in the pond twice daily in the morning and in the late evenings. In nursery ponds approximately 10% of the pond surface is covered with floating weeds with dense root system such as *Eichhornia* Sp. to improve the survival rate of post-larvae. The weeds are kept inside a PVC frame or bamboo frame to avoid their spreading in the pond. A fortnight after stocking sampling of post larvae is done to observe the growth using cast net or fry net. The feed quantity is increased as the post larvae grow in size. During nursery rearing water quality parameters such as pH, dissolved oxygen, transparency and depth are checked once every week and are maintained in optimum ranges. Loss of water due to seepage and evaporation is compensated by water addition at least once every fortnight. At the end of the nursery rearing the juveniles (>1.0g) are collected by dewatering the pond.

Grow-out phase: The grow-out pond is prepared prior to stocking in the same way as done for nursery pond. Hideouts in the form of earthen or other pipes, bamboo pieces etc are provided in the pond (1000-1500 nos./ha). The nursed juveniles (~1 g) harvested from the nurseries are stocked in well prepared earthen production ponds (0.1-0.4 ha) @ 3-5/m^2. The preferred stocking density in grow-out pond is 40,000/ha. The prawns are fed daily with formulated pellet diet (2-3 mm size) @ 10% of the biomass initially and then reduced to 3% of the biomass at the end of the culture period. Daily monitoring of critical water quality parameters such as dissolved oxygen, pH, and temperature is essential to prevent any loss of stock due to poor water quality. Phytoplankton bloom and decaying waste material are the main reasons for dissolved oxygen depletion usually seen in prawn ponds. Regular monthly sampling with cast nets or small mesh seine nets is done to assess growth of the prawns. After four months prawns that achieved marketable size (>40g) are removed by using a seine net of suitable mesh size. Selective harvesting is continued once every 3-4 weeks for another 3-4 months and finally the pond is harvested by complete draining. Prior to harvesting arrangements for marketing the prawns should be made. Prawns after harvesting is cleaned in freshwater and is packed in ice layer by layer to avoid loss of quality.

Conclusion

Freshwater prawn farming has many advantages over marine shrimp farming. Unlike marine shrimp farming the culture sites of freshwater prawns are located far inland and thus does not competes for coastal resources nor harms coastal ecosystems. It can be more easily integrated with other agricultural activities. Freshwater prawns are suitable candidates for inclusion in fish culture. By incorporating freshwater prawns fish farmers will get an opportunity to increase their production and profit with little extra investment. They have good demand in the international market. Most of the freshwater prawn production comes from low intensity culture and hence there is no risk of 'auto pollution'. However, there are some constraints for the development of freshwater prawn farming. The major problem during freshwater prawn culture is the size heterogeneity in the harvested crop, which demands additional effort to market them. Tail yield of freshwater prawns (40-50%) is less than that of marine shrimp (60%). Freshwater prawns require more care in processing than marine shrimp. There are some reports of reduced growth rate in grow-out phase from some countries including India which has been attributed to 'inbreeding depression' thus it is essential to take up selective breeding programme of this species to improve the growth rate and other desirable characters.

Finally it can be said that though there are some minor concerns freshwater prawn farming has a bright future especially in tropical and subtropical areas of the globe. More concerted efforts are needed to transfer the developed technologies to the end users and to solve the existing problems.

REFERENCES

Bojan, J., V. B. Bhat, P. N. Vinod, 2003. Role of freshwater prawn farming in coastal aquaculture diversification. Marine Products Export Development Authority Newsletter, 8:19-25.

FAO, 2000. Aquaculture Production Statistics 1988-1998, FAO Fisheries Circular 815, FAO, Rome, Italy.

FAO, 2006. Yearbook of Fisheries Statistics - Aquaculture Production Vol. 98/2 Food and Agricultural Organization, Rome, Italy, 89.

Ling, S. W., 1969. Methods of rearing and culturing *Macrobrachium rosenbergii* (DE MAN) *FAO UN Fish. Rep.* 57(3), 607-619.

New, M. B. & W. C. Valenti, 2000. Freshwater prawn culture: The farming of *Macrobrachium rosenbergii*, Blackwell Science, Oxford, UK.

New, M. B. 2005, Freshwater prawn farming: global status, recent research and a glance at the future. *Aquaculture Research* 36:210-230.

Pillai, B. R. and K. J. Rao, 1999. Effect of stocking density on the post-larval production of *Macrobrachium rosenbergii* in a two-phase larval rearing system. *J. Aqua.*, 5: 71-76.

Rao, K. J and S. D. Tripathy, 1993. A manual on giant freshwater prawn hatchery. Manual Series- 2, Central Institute of Freshwater Aquaculture, pp.50.

Rao, K. J., 1991. Reproductive biology of the giant freshwater prawn *Macrobrachium rosenbergii* (de Man) from Lake Kolleru (Andhra Pradesh). *Indian J. Anim. Sci.*, 61(7): 780-787.

Chapter 11

Artificial Propagation and Cryoconservation of Fishes

P. Routray

Introduction

Global diversity in domestic animals and aquatic organisms is considered to be under threat worldwide. Worldwide 11 % birds, 25 % mammals and 34 % of fish species are threatened (Lovell-Badge, 2001; Vemuganti and Balasubramanian, 2002) and aquatic ecosystem is becoming more vulnerable to disasters due to run off from agriculture and industrial sources, oil spills or sudden environmental changes leading to total elimination of stocks from different ecosystems. The multifarious human activities in open waters have threatened the fish diversity all round the world. Fish worldwide are in crisis. The effective population sizes in fish hatcheries are getting smaller and smaller that causes loss to farmers worldwide. The World Resources Institute has reported that nearly 70 percent of the world's marine fish stocks are over fished or are being fished at their biological limit. Apart from fish, across all live stock species there is a shrinking pool of genetic diversity. For instance, the effective population size for all dairy cattle breeds is less than 60 animals. To preserve the gene pool of unique populations by protecting them from extinction, loss of genetic variability (inbreeding) and dilution due to inter breeding with unrelated populations (hybridization) gene banks or germplasm repository can play a vital role. Germplasm repositories in fisheries could be used for: 1) restoration of threatened species, 2) brood stock

Applied Bioinformatics, Statistics and Economics in Fisheries Research: pp. 141-149

New India Publishing Agency, New Delhi (India)

improvement through genetic modification of a targeted population, and 4) information and technology development including gene pool database management.

Fish population germplasm repository can be of two types: live fish banks which are maintained at multiple locations to provide a continuous source of progeny, or tissue gene banks based on gamete, zygote or somatic tissues maintained in cryogenic condition for later retrieval. In fishes, mostly sperm cells are used as source with greater accuracy in producing results (Routray, 2003; Routray *et. al.*, 2003). Gene banks may be used for decades for restoration of a species and for genetic manipulation studies. During the prolonged and regular use of gene banks, a huge amount of information may be generated and must be stored in an appropriate form in a user friendly manner. The only way all this information can be managed effectively for gene bank managers and other users is with a well designed computerized data base.

Prerequisites for Establishing a Gene Bank

Before establishing a gene bank a thorough thinking must be there about the type of gene bank to be established. For instance if a live gene bank to be established then the following things needs to be ensured: 1) A suitable land with perennial water source; 2) high embankment of ponds or tanks to avoid inter mixing of populations; 3) the species must be acclimatized to any particular agro climatic conditions. There are limitations of live gene banks as these are to be maintained at multiple locations, problems of mass mortality due to disease, unforeseen calamities like heavy rain fall or flood etc. The tissue gene bank has the advantages of overcoming the above problems and easy to manage.

Role of Germplasm Repositories

The germplasm repositories across different countries generally aimed at addressing the characterization and conservation of animal genetic resources. The following roles are ideal for germplasm repositories:

i. Understanding the population structures and demographic changes in population size, location and number of producers raising specific breeds;
ii. Acquiring germplasm (semen, embryos, ova and DNA) for storage in the repositories;
iii. Evaluation, development and implementation of cryopreservation protocols for long term germplasm cryoconservation;
iv. Development of an information database and tracking system that links genetic preservation (*ex-situ* and *in-situ*) and population management;

v. Act as a nodal centre for easy accessibility to genetic material for different regions on species restoration during disasters.

State of the Art in Cryopreservation Technology

Cryopreservation allows virtually indefinite storage of biological material without deterioration over a time scale of at least several years (Mazur, 1985), but probably much longer. Many important mile stones in cryobiological research could be achieved in the second half of the previous century. Much progress resulted from empirical studies. In later years, progress was also strongly stimulated by the development of fundamental theoretical cryobiology. The principle of glass transition is well applied in different cryopreservation protocols which involve the cooling of biological materials at a range of cooling rates that are fast enough to prevent slow cooling damage but are slow enough to allow sufficient dehydration of the cells to prevent intracellular ice crystal formation. The preservation of fish semen and blastomeres are a success story for germplasm conservation through cryopreservation. More than 200 fish species with external fertilization have been tested for sperm cryopreservation. The present state of art for many species of fish seems to be adequate for the purpose of gene banking and germplasm conservation.

During cryopreservation of spermatozoa of fish, many times common mistakes do occur in coding the visotubes or cryovials that leads to several experimental errors and some times loss of important materials. At liquid nitrogen temperature also several marker inks gets erased due to the extreme low temperature. After thawing the general protocol of toxic cryoprotectant removal is often overlooked at many hatcheries that also lead to very poor fertilization.

Oocytes

In the last decades, considerable progress has been made with cryopreservation of oocytes in a great number of animal species (Woelders *et al.*, 2003). Viable oocytes also have been recovered after freezing and thawing in some species. Post-thaw oocyte maturation, fertilization and embryo development have been reported in a number of species. Production of offspring from frozen thawed oocytes is still less efficient than that from cryopreserved embryos. Cryopreservation of fish oocytes is not yet successful (Blesbois *et al.*, 2003) due to large size, low permeability, sensitivity to chilling injury, high lipid content and the polar organization (vegetal and animal pole).

Embryos and Embryonic Cells

In spite of a few claims (Wang *et al,* 1987: Chen *et al.*, 1988: Zhang *et al.*, 1989; Zhang *et al.*, 1997), there is no reproducible cryopreservation protocols for fish embryo largely because of some limitations as in the case of fish oocytes i.e., large size, the high lipid content and polar organization in the fertilized eggs and the early embryos. It has been reported that the yolk syncytial layer (YSL), a vitelline membrane formed by a single cell that surrounds the yolk and not the yolk or chorion embryos blocked the permeation of CPAs into the yolk atleast in Zebra fish (Zhang and Rawson, 1996b; Hagedon *et al.*, 1996; 1997b). An understanding of the physiology and removal of the permeability barrier could be a key to the future successful cryopreservation of fish embryos.

However, cryopreservation of isolated embryonic cells in fish is an option. Post-thaw survival of blastomeres was achieved in rainbow trout, carp and medeka (Blesobois *et al.*, 2003). Embryonic cells can be used to produce chimeric embryos provided that the gonads of the recipient embryos are implanted and populated with primordial germ cells (PGCs) from the donor embryo. Such chimeric embryo can be utilized to produce progeny of the donor genotype (Nakagawa *et al.*, 2002).

Somatic Cells

Cryopreservation of somatic cells was demonstrated for a number of cell types. Simple procedure of adding a suitable CPA to the suspended cell in culture medium followed by placing tubes at –80 ^{0}C in a mechanical freezer proved to be effective till today. Use of controlled rate freezer was reported in few cases.

Germ Cell Transplantation

Transplantation of ovarian tissues and germ cells (eg. primordial germ cells, PGCs or spermatogonial stem cells, SSCs) are emerging technologies with potential for future conservation programme. Development of transgenic strains in a species of commercial importance and conservation needs time, labor and high financial supports, besides its unpredictable result. Germ cell transplantation is better option to overcome those difficulties. The introduction of foreign gene in to the PGCs and selection of transformants facilitate the *in vitro* selection of transformants. The method is advantageous because of easy selection procedure of transformants, which can be performed in a petri-dish instead of rearing large number of fishes to identify the desired transgenic candidate and development of an individual fish using chimera technique is more convenient. Successful germ cell transplantation has been achieved in fish with fascinating results of allogenic transplantation in rainbow trout (Takeuchi *et al.*, 2003) leading to production of live fry with donor-derived phenotype from gametes of PGC-recipients.

Management of Gene Bank Information

A huge amount of information is required to manage the broad range of activities that are essential for an operating gene bank such as population sampling procedures, sample identification, tissue processing, cataloging, information retrieval, quality control and sample disposition. Gene pool preservation requires that a sufficient number of individuals be collected to effectively sample the total available gene pool and to achieve the gene and genotype frequencies which are representative of the source population. The actual number of individuals necessary to achieve this goal is ambiguous and debatable, but the criteria most commonly used is the smaller of a predetermined percentage of the total population (10-100%) or an upper limit number of individuals (100, 400 or more) that sampling theory indicates is sufficient (Kincaid, 1983, Ryman,1991). Once milt is collected, a sample identification system is applied to ensure positive identification of each sample during all handling procedures (species, date, time, processing, storage, monitoring and retrieval operations). A quality control program is needed to ensure that high viability is maintained. A milt replacement program is also needed to ensure that the deteriorated milt is replaced or removed for breeding programs.

Gene bank information is primarily of two types: 1) Gene bank operations information to catalog and manage the stored tissues, and 2) gene bank user information to characterize the source population and determine the most effective applications for stored tissues. The database must include the information needed to accomplish the following functions:

1. Sample and population identification procedures.
2. Collection procedures.
3. Sample processing procedures.
4. Sample storage and retrieval operations.
5. Quality monitoring.
6. Inventory maintenance.
7. Information retrieval.
8. Milt replacement program.
9. Milt discards procedures.
10. User application programmes.

The decision to preserve a specific population gene pool requires the collection of baseline information to determine population uniqueness and the traits which define this uniqueness. Baseline information includes population name, location, range and distribution, genetic characterization, breeding history, life history, habitat preferences and any specific trait pertaining to the species. Species with no cryogenic preservation history should be studied to determine a species specific cryogenic processing and storage procedures before gene banking operations can begin.

The information regarding gene bank users should be based on the population characterization data, stored milt quality and quantity data and the effective population size represented in the preserved milt. Information on population origin, breeding history, source habitat parameters, life history traits, genetic analysis and performance characteristics in culture and natural habitats should be clearly brought out. This information would help the users to match the preserved stock with the existing ones and change the management accordingly.

Gene Bank Information Database

A database is a collection of data arranged in a systematic order for quick search and retrieval operation in subsets. The first step involving the designing of database is crucial and the essential requirements are: development of data structure based on the identification of specific data elements. Essential information includes data needed by the gene bank managers to facilitate decisions on management and research applications where cryopreserved milt will be used.

The initial data element or trait list is developed by answering in complete details a series of questionnaires as shown in Table-1. The questions are of two types; 1) for the gene bank operations and 2) the additional information needed to assist user groups. During the development of initial trait list, representatives of all the potential users including Government officials should be involved.

Development of the information database management should be a priority when establishing a gene bank. The gene bank information management data base is essential because it is the management system that will catalog, summarize, query and retrieve information from the data set required to establish and operate gene bank. Fisheries mangers and hatchery owners using cryopreserved milt will rely on the data base for accurate information on source population, milt quality, milt availability etc. India should ensure that the genetic diversity is preserved as the agricultural industry

Table 1. A sample questionnaire for initial trait list development for a new gene bank information.

Type:I For Gene bank operations and management		
S No	**Information description**	**Example traits or characteristics**
1	Population sample identification system	Population identifier, sample identifier, ampule or straw identifier etc.
2	Population sampling plan	Population size, number of collection sites, nuber of individuals collected per site, total numbers etc.
3	Milt at field	Date spawned, sample and fish identification, capture method, spawning method,, milt holding method, milt quality measurement, etc.
4	Milt transportation system?	Cooling and freezing methods, shipping temperature, transit time, transportation mode, etc.
5	Source population characteristics	Origin, breeding history, life history, reproductive traits, behavioral traits, habitat preference etc.
6	Source population genetic characteristics	Analysis method, allele frequencies, heterozygosity, distinctive characteristics etc.
7	Source fishery characteristics	Location, water body type, water quality, food source, predator species etc.
8	Gene bank processing procedure	Milt quality, additives, diluents, dilution rate, freezing procedure, storage unit type, etc.
9	Quality control and monitoring	Milt testing schedule, test sample collection procedure, fertility rate, etc.
10	Milt replacement procedures	Program identification, date removed, replacement criteria
11	Brood improvement programme	Prog. Identification, date removed, milt quality, number progeny produced, brood stock improvement results
Type:II For Gene bank user information (addition to type-I)		
1	Source population characteristics?	Additional traits and characteristics of special interest to milt user: origin, source, habitat, life history, cultural performance etc.
2	Source population status	Threatened or endangered, effective population size, disease and health status, etc.
3	Source population genetic characteristics	Source population, allele frequencies, heterozygosity, effective population size, genetic analysis etc.
4	Milt availability	Gene bank population milt inventory, gene bank milt replacement programme, restoration and enhancement etc.

appears to be headed towards a few lines of proven performers. In India, the fisheries sector has one bureau (National Bureau of Fish Genetic Resources at Lucknow) for germplasm inventory, evaluation and gene banking of freshwater fishes but for marine species there is as such no organization. At Central Institute of Freshwater Aquaculture, Bhubaneswar protocols for cryopreservation of fish semen, blastomeres and other cells of freshwater fishes has been developed (Gupta *et. al.*, 1995, Routray, 2003, Routray *et. al.*, 2006) but maintenance of germplasm of many species are not being looked after as Scientists are working on the upgradation of genetic stock by different methods. In Indian fisheries, there is no full fledged fish gene bank in operation, so it is essential to establish a gene bank to save and propagate our rich fish biodiversity in these sub-continent and other places also.

Emerging reproductive technologies

Transplantation of ovarian tissue and germ cells (e.g., primordial germ cells (PGCs) or spermatogonial stem cells (SSCs) are emerging technologies with potential for future use in conservation programs. Successful transplantation of embryonic cells has been reported in some fish species. Production of xenogenic transplantation of PGCs for production of surrogate fish has been reported. Similarly androgenic restoration of fish has been also successful.

REFERENCES

Blesbois, E., and Labbé, C., 2003. Main improvements in semen and embryo cryopreservation for fish and fowl. In: Workshop on Cryopreservation of Animal Genetic Resources in Europe, Paris 2003. (Ed. D. Planchenault) pp 55-65.

Chen, X.H., X.S.Zhang, H.Y. Zhu and K.M. Yao 1988. Experimental study of freezing damage factors and feature analysis of microscopic images of fish embryos. *Cryobiology* 25:564.

Gupta, S. D., Rath, S. C., and Dasgupta, S. (1995)Fertilization efficiency of cryopreserved carp spermatozoa over four years at different time intervals after thawing. *Geobios,* 22, 208-211.

Hagedorn, H., E.W. Hsu, F.W. Klainhans, and D.E. Wildt (19967b). New approaches for studying the permeability of fish embryo towards successful cryopreservation. *Cryobiology* 34:583.

Kincaid, H. L. (1983). Inbreeding in fish populations used for aquaculture. *Aquaculture,* 33:215-227

Lovell-Badge, R., 2001. The future of stem cell research, *Nature,* 414, 88.

Mazur, P. (1985). Basic concepts in freezing cells. *Proceedings of the 1st International Conference on deep freezing of boar sample.* Uppasala, 1985 (L. A. Johnson and K. Larsson, eds), pp. 91-111.

Routray, P. (2003). Evaluation of cryopreserved milt of a selected carp, *Labeo calbasu* for fertilization and seed production, Ph. D. Thesis, Sambalpur University, Sambalpur, Orissa.

Routray, P., Gupta, S. D., and Behera, M. K. (2003). Cryogenics in fish hatchery technology. *Fishing Chimes*, 23:7-9.

Routray P., Choudhary A. K., Dash S. N., Verma D. K., Dash C., Swain P., Jena J. K., Gupta S. D. & Sarangi N. (2006). Cryopreservation of dead fish spermatozoa several hours after death of Indian major carp, *Labeo rohita* and its successful utilization in fish production. Aquaculture doi: 10. 1016/*J. Aquaculture*, 261:1204-1211.

Ryman, N. (1991). Connservation genetics considerations in fisheries management. *Journal of Fish Biology, Supplement A* 39:211-224.

Vemuganti, G. S., Balasubramanian, D., (2002). Heralding the dawn of cultured adult stem cell transplantation. *Indian J Biotechnol* 1-39.

Wang, D.R., T.Hua and H. Chang 1987. Feasibility of vitrification of fish embryos. *Cryobiology* 24:583.

Woelders, H., Matthijs, A., Zuidberg, C.A. and A.E. Chaveiro. (2005). Cryopreservation of boar semen: equilibrium freezing in the cryomicroscope and in straws, *Theriogenology* 63: 383-395.

Zhang, X.S., L.Zhao, T.C.Hua and H.Y.Zhu. (1989). A study on the cryopreservation of common carp (*Cyprinus carpio*) embryos. *Cryo-Letters* 101: 271-278.

Zhang,K.J., Y.D.Lou, Y.J. Zhang and P. Wu. 1997. Studies on cryopreservation of three fishs' embryos. *J Fish. China* 21: 366-372.

Zhang,T., and D.M. Rawson (1996b). Permeability of the vitelline membrane of zebrafish (*Brachydanio rario*) embryos to methanol and propan-1,2-diol. *Cryo-Letters* 17: 273-280.

Chapter **12**

Methods of Assessment of Microbial Population and their Role in Aquatic Ecosystem

N.K. Maiti, R.N. Mandal & S. Mohanty

The term 'Ecology' deals with the study of organisms from all sources and their interaction with surrounding environments. The aim of ecology is, therefore, to understand and also to explain the relationships of all organisms to their environment. In this respect, the duty of the ecologist is to study a particular environment, estimate the population of each species, recognize the communities, and determine their activities and their interactions with both related species, the rest of the community, and their environment. Macroecological studies always initiate by defining the species composition and then proceed to draw certain conclusions on the role of these species in the ecosystem. On the other hand microbial, especially bacterial classical taxonomy (based on isolation techniques and determination of morphology and in vitro biochemical characteristics of isolates) for determination of each species of aquatic bacteria is almost impossible, except in a few cases. Consequently, the microbial ecologist has to depend upon certain methods that can measure the biomass or activity of a very large conglomerate of coexisting microbial taxa in an aquatic environment. Probably, this has caused the very slow development of microbial ecology concepts and application of a general ecological theory to the microbial ecology.

Applied Bioinformatics, Statistics and Economics in Fisheries Research: pp. 151-164

New India Publishing Agency, New Delhi (India)

Aquatic microbal ecology has developed rapidly since the middle 1970s when this science was recognized as a separate discipline and when different biochemical methods were applied in a consistent way (Overbeck, 1974; King and White, 1977). Many methods are now available for estimating bacterial biomass (Floodgate, 1980), growth rates (Moriarty, 1986), bacterial production (Fuhman and Azam, 1982; Simon and Azam, 1989), and the activities of microorganisms comprising the biomass (Meyer-Reil, 1984, Chrost, 1989). The concept of the ecology of aquatic microorganisms has changed significantly in recent years. In some respects, the change is sufficiently dramatic. In this connection, heterotrophic microorganisms play a significant role in the cycling of organic and inorganic matter in waters. Thus, a large amount of the primary production is not consumed directly by herbivorous consumers. It is then passed through detritus organic matter via bacterial production to phagotrophic microorganisms, thereby creating the concept of the "microbial loop" (Azam *et al.,* 1983).

In spite of microbial ecological advancement, many of the fundamental questions of the same are still unanswered. Microbial ecology still describes the effects of microbial activity rather than its causes. Advance molecular biology says that DNA makes the microorganism and contains the signals that affect gene expression, resulting in RNA synthesis and enzyme production, and that leads to understanding of molecular mechanisms of metabolic regulation. This has the utmost importance in assessing microbial community structure and function.

The complex of interactions of aquatic microorganisms with abiotic factors of the environment, mechanisms of metabolic regulation and adaptation, as well as mutual relationships between the microorganisms and higher biota, lead finally to the establishment of a particular microbial community. Therefore, the relationship between the structure of the community and its function necessarily depends upon the information about microbial community in general and bacterial population in particular. However, classical microbiological methods (based on isolation and then identification of isolates) have been found to be limited in studies of bacterial population dynamics in natural waters as because a very small fraction of the native aquatic bacteria can be cultivated and thus isolated on laboratory media. Some advances in methods of molecular biology and immunology can be now applied to aquatic microbial ecology.

Methods for Identification of Microorganisms

One of the easiest methods for identifying bacteria directly in water samples, and thereby estimating their natural abundance, is an electrophoretic analysis of low-molecular weight RNA (SS rRNA and tRNA) that is considered to be the most useful. Analysis of bacterial RNA profiles provides results for only the main members of the native aquatic community. This approach is one of the first attempts that allow microbial ecologists to determine directly, i.e., without cultivation, the major groups of native aquatic bacteria. The use of radiolabeled nucleotide precursors in combination with the RNA profile analysis may even provide insights into the physiological activities of single taxa. By this method, analysis of stable RNA could be a major step towards a unifying theory of microbiology and ecology that may permit a broad assessment of the structure and function of natural bacterial communities.

Adaptation of Microbial Community

Microbial communities, however, do not remain constant. Besides environmental reasons, structural and functional dynamic changes may also reflect genetic interactions between microorganisms, enabling them to colonize new habitats, by means of genetic change. Among the different potential mechanisms of gene transfer, there is some evidence that transduction of viral nucleic acids, transformation, and conjugation of chromosomal and/or plasmid DNA may be important processes in genetic adaptation of microorganisms to aquatic environments. It is now well documented that many bacterial bear plasmids and plasmid-mediated gene transfers were observed decades ago, though the role of bacterial plasmids in aquatic environments is still at the initial state.

Role of Microbes on Transformation of Organic Matter

It is essential to know the habitat status of aquatic ecosystem in relation to microbial role. Because organic matter in aquatic environment is considered to be a major substratum and microbial activities are necessary for its transformation. There are two principal sources of organic matter in aquatic environments, living (biomass) and nonliving (detritus) (Odum and ruz, 1963; Wetzel, 1983). It has long been noted that nonliving organic matter plays an important role in the structure and function of aquatic ecosystems (Krogh and Lange, 1932; Birge and Juday, 1934; Ohle, 1934; Odum, 1963; Saunders, 1977). The detritus food chains were established along parallel lines with the classical phytoplankton-zooplankton grazing food chains. Wetzel et al. (1972) have developed a detailed scheme and model for the detritus food chain and have emphasized its fundamental differences from the classical grazing food chain. The nonpredatory loss of organic matter was found to be essential for understanding the whole aquatic ecosystem. Detrital organic matter tends to accumulate in aquatic environments until a quasi equilibrium is attained. In this respect many environmental processes, such as decomposition, utilization, aggregation, and sedimentation are involved to regulate this equilibrium.

Dissolved Organic Matter (DOM) in pond ecosystem

The unutilized pond is characterized by high input of allochthonous organic matter. On the other hand, the autochthonous production is mainly mediated by macrophytes, periphyton, and phytoplankton. In pond, macrophytes and periphyton communities have a remarkable complexity in their structure and function. The large surface area of macrophytes and suspended particles in pond are colonized by eipphytic communities composed of various taxa of microorganisms (algae, bacteria, fungi, protozoa) and fauna (snails, Diptera, Oligochaeta). They altogether generate a very active biotic complex (periphyton), which is highly efficient for DOM production and utilization. The productivity of emergent and submergent plants and their epiphytic microbial community may exceed that from open water by 5 to 10 times (Wetzel and Haugh, 1973; Lieczynska 1976). A significant portion of photosynthetically fixed organic carbon is released extracellularly and directly enters the DOM pool (Wetzel and Manny, 1972; Hough and Wetzel, 1975; Sondergaard, 1981). The amount of released dissolved organic matter (RDOM) is variable and depends on environmental conditions and species composition. The majority of RDOM excreted by macrophytes is composed of low-molecular-weight compounds (among them: glucose, xylose, sucrose, glycin), which are rapidly utilized by microheterotrophs and transformed into molecules of higher molecular weight (Wetzel and Manny, 1972; Sondergaard, 1981). The amount of RDOM may vary between 1 and 40% of the total organic carbon generated in photosynthesis and averages about 10% (Sondergaard, 1981; Wetzel, 1983).

Role of Autotrophs and Heteroptrophs

Autotrophic and heterotrophic interactions between macrophytic, epiphytic, and epipelic communities are very important in autochthonous and allochthonous detritus processing in ponds. Microbial degradation and transformation of both dissolved and particulate fractions of detritus are the key processes of organic and inorganic matter cycling. Prokaryotic and eukaryotic microorganisms utilize different substrates at different vegetation periods of water plants. Fungal biomass is higher at the early growth stage, whereas bacterial biomass increases during the decay stage of hydrophytes. Those microorganisms having the capacity for the production of different hydrolytic enzymes have an advantage in detritus processing and substrate utilization (Sinsabaugh and Linkins, 1988).

Role of Phytoplankton to Release Dissolved Organic Matter

Several studies indicate that Released Dissolved Organic Matter (RDOM) released by phytoplankton is an important carbon and energy source for heterotrophic bacteria. Interestingly, these bacterial populations are capable of rapid response to RDOM flux pulses in pelagic waters and algal cultures (Nalewajko, 1977; Chrost, 1978a, 1978b, 1983; Chrost and Bczeska, 1978; Herbst and Overbeck, 1978; Meffert and Overbeck, 1979; Cole *et al.*, 1982; Coveney, 1982; Brock and Clyne, 1984; Herbst, 1984; Riemann and Sondergaard, 1984, 1986; Overbeck, 1985). There have been a number of important estimates of RDOM in relation to bacterial secondary production. It is evident from algal RDOM that bacterial secondary production in natural waters requires from 10% to 100% of the carbon (Larsson and Hagstrom, 1982; Lancelot and Billen, 1984; Riemann and Sondergaard, 1984; Bell and Kuparinen, 1984). It is, however, difficult to estimate the difference between RDOM used to bacterial metabolism and quality and quantity of RDOM released by phytoplankton population. Usually, natural bacterial populations can utilize from 30 to 90% of RDOM during photosynthetic production and release. Gross rates of RDOM exudation varied from 9 (sunrise) to 40 μg Cl^{-1} h^{-1} (noon) and followed the total photosynthetic organic matter (PhOM) production rates. The percentage of release of RDOM, however, was found to be the highest at the sunset (32%) and sunrise (26%), while the daytime constituted approximately 22% of the total produced photosynthetic carbon. The bacterial utilization rates of RDOM were found significantly higher in the afternoon and evening (12 to 17 μg Cl^{-1} h^{-1}) than in the morning and forenoon (2.5 and 10 μg organic Cl^{-1} h^{-1}, respectively). Further, the percentage of utilized RDOM by bacteria also increased markedly from 30% (forenoon) to 64 to 90% (late afternoon to evening). The above results showed diel temporal succession of the photosynthetic organic matter production by phytoplankton and heterotrophic RDOM utilization by bacteria.

Factors relating to RDOM utilization

Bacterial production and algal exudation of RDOM decreased with depth of water body, when the same increased in the surface (0 to 2 m) water of water body. The rates of bacterial utilization of RDOM correlated with phytoplankton primary production and release rates. Bacteria used RDOM at the rates from 2 to 19 μg organic Cl^{-1} h^{-1}, i.e, 40 to 86% of released organic carbon was utilized within the epilimnion. Bacterial cell numbers and production rates were also the highest in the epilimnion (4.6 to 8.7 x 10^9 cells 1^{-1} and 3.1 to 3.6 μg organic Cl^{-1} h^{-1}, respectively) and displayed a similar distribution pattern to that of PhOM production and RDOM. The estimates of RDOM utilization indicated that from 70 to 75% and 25 to 60% of the used RDOM was respired and incorporated, respectively, into bacterial cells in the water column.

The studies on RDOM utilization by natural bacterioplankton populations showed that organic compounds with molecular weight less than 1,000 daltons were preferentially used, and the percentage of RDOM utilization was lower with increasing molecular weight of RDOM (Chrost and Faust, 1983; Chrost, 1986). The composition of algal RDOM remaining after a 24-hour incubation period of water samples with natural bacterial populations changed significantly, and high-molecular-weight compounds predominated.

The patterns observed demonstrate an interesting scenario for growth and utilization of RDOM by natural populations of bacteria in relation to the possible changes in the dissolved organic matter released by phytoplankton. Rapid bacterial utilization of algal RDOM prevents these compounds from increasing above low concentrations in natural waters and suggests the presence of bacterial assemblages adapted for the rapid uptake and metabolism of released compounds. It is possible that this is due to specific algal-bacterial associations that develop as a result of bacterial species selection within the algal phycosphere (Bell and Mitchell, 1972; Bell, 1980). The utilization of low molecular weight compounds of algal RDOM and/or the subsequent release of high-molecular weight molecules by bacteria (Nalewajko and Lean, 1972) may be one of the factors responsible for the predominance of high-molecular-weight DOM in natural waters.

Approaches to Bacterial Population Dynamics

Studies on the structure of phyto and zooplankton communities have been numerous in aquatic ecology. They have contributed considerably to our knowledge of species composition, genetic population structure, and environmental factors regulating local, spatial, and temporal variability. However, our knowledge of the structure of the indigenous bacterial flora that participates in the flux of matter in aquatic ecosystems is scarce. This may be partly because identification of bacterial species is very laborious and often still uncertain. Today, due to the development of modern molecular (nucleic acid homology, RNA sequencing, RNA probing) and immunological techniques (genus- and species-specific fluorescent antibodies, monoclonal antibodies), solutions to these problems may be anticipated.

But on the other hand, studies on microbial community structure are not without conceptual difficulties. For bacteria, there is no clear-cut, generally accepted definition of a species, as there is for eukaryotic organisms. Moreover, for most species, no clear relationship between ecological role and taxonomy can be defined. The variability among strains that are assigned to the same species is too high. Therefore, the structure of the bacterial flora is not sufficiently described by a listing of the species composition. Rather, an ecologically relevant characterization of the morphological, physiological, biochemical, and genetic features of the members is required.

If we consider each bacterial cell as a unique combination of characters, then every strain occupies a certain position within a mutlidimensional space. The strains are scattered within this space, forming clusters around certain combinations. For taxonomic purposes, the strains within a cluster are assigned to the same species, whereas others are regarded as "outliers" (Gyllenberg, 1984). Thus, each species represents a set of character combinations, which cause a large genotypic and phenotypic variability within populations.

Numerical Taxonomy to Analyze Bacterial Population

Several numerical taxonomic methods for determination of the relationships between bacterial species are available (Sneath and Sokal, 1973). The current techniques in numerical taxonomy (Goodfellow *et al.*, 1985), and the computer programs available (Sackin, 1987) have been reviewed recently. Usually, all these methods use equal weighing of each character (Krambeck and Witzel, 1983) and can be used with phenotypic as well as genotypic properties. Two main lines of grouping procedures are used, hierarchical and nonhierarchical.

Cluster analysis is the common hierarchical grouping. Various similarity coefficients (Austin and Colwell, 1977) can be calculated between each pair of strains, which are then arranged into groups by the use of cluster algorithms (Anderberg, 1973). These techniques have been used successfully for a numerical taxonomic characterization of bacterial populations from many aquatic habitats, including lakes, fish ponds, solar salterns, marine, and estuarine habitats (reviewed by Goodfellow and Dickinson, 1985).

Cluster analysis has been used to show seasonal variation in the bacterial community in Plußsee (Witzel et al, 1982a). Typical populations succeeded from spring to summer and autumn. The growth of corresponding specific bacterial populations was induced by seasonal successions of phytoplankton populations that provided different organic substrates. In winter only the essential core remained to provide the inocula for the next seasons. The populations were not distinguishable only by one or a few significant characters, but there were distinct combinations of morphological and functional properties that characterized the population dynamics (see Witzel *et al.,* 1982a). Temperature played a minor role, if the appropriate substrates were available.

Immunological Approaches

Antibodies produced against almost any organic substance are highly specific. Due to this specificity, antibodies react with the target molecule and form an antigen-antibody complex and is used for the detection and identification of organisms. The location of the antigens can be observed in the microscope when the antibodies have been labeled with gold or fluorescent dyes. These features provide the potential for a wide range of ecological applications to identify and quantify specific microorganisms in the microbial community.

Double Staining Techniques

General staining technique can be combined with the fluorescent antibody technique to obtain information on a specific part of the microbial community. Brayton and Colwell (1987) combined the nalidixic acid technique of Kogure *et al.* (1979) with fluorescent antibody labeling to distinguish between viable and dead cells of *Escherichia coli* and *Vibrio cholerae* in water samples. With this technique it was possible to identify individual cells by the fluorescent antibodies and recognize their viability by elongation after substrate addition, while cell division was blocked by nalidixic acid.

For recognition of a specific group of microorganisms within the local bacterial community, double staining techniques have been successful. Muyzer *et al.* (1987) enumerated *Thiobacillus ferrooxidans* with FITC-conjugated antibodies, whereas total bacteria had been stained in the same sample with ethidium bromide. The combination of ethidium bromide and FITC staining allows the simultaneous viewing of both stains in the same image field.

Genetic and Chemical Approaches

Over last few years there have been rapid progress in molecular genetic studies, particularly in genetic engineering, nucleotide sequencing, and design of nucleic acid probes. Many of these are becoming useful tools for the study of natural bacterial populations.

Identification of Bacteria by Molecular Marker

Chemcial markers are used for the classification of microorganisms. These include DNA base composition, DNA/DNA and DNA/RNA reassociation studies (Schleifer and tackebrandt, 1983), 16S ribosomal RNA sequencing

(Stackebrandt *et al.*, 1985), 5S ribosomal RNA patterns, one- or two-dimensional electrophoresis of proteins (Kersters and DeLey, 1980).

Hybridization studies have been successfully applied for identification of species and their phylogenetic relationships. The main disadvantage of this technique is that DNA-DNA hybridization is relatively unaffected by point mutations or by insertions and deletions of short DNA segments .

Sequence similarities of DNA in different organisms could be ued to determine individual relatedness. However, for studies with hundreds of strains of the same species, sequence analyses are too laborious and impractical. It is now possible to sequence DNA molecules directly from water samples by combining sequencing technique with PCR. This would allow the determination of genetic information about organisms that cannot be cultivated by present techniques.

Protein profiling determined by one- or two-dimensional polyacrylamide gel electrophoresis of cell extracts have been used to study the taxonomic relationships within several bacterial taxa (Costas et al., 1987). Electrophoresis of the total protein of microbial cells provides information on the number and physiochemical properties of the gene products. Electropherograms of cellular proteins can be used as fingerprints of a particular organism and can thus be helpful for identification and classification.

RNA Chemotaxonomy

Chemotaxonomy is the use of chemical characteristics of an organism for its taxonomic classification and identification (Goodfellow and Minnikin, 1985).

RNA chemotaxonomy is the use of chemical characteristics of stable RNA (rRNA and tRNA) for taxonomic classification and identification of microorganisms. Biochemical characterization of the RNA molecule due to its molecular structure and function provides taxonomically valuable information for average isolates obtained from aquatic environments. The low-molecular-weight RNAs (5S ribosomal RNA and transfer RNA) fulfill this requirement because, despite the fact that the 5S rRNA is a ribosomal cistron, the main emphasis for the chemotaxonomic use of LMW RNAs is on the use of the total tRNA pool. High-resolution gel electrophoresis is a very powerful and versatile approach for characterizing LMW RNAs. The major advantages of gel electrophoresis are that a variety of different detection techniques are applicable and many samples can be run in parallel. Furthermore, this method allows the determination of a second dimension of biochemical information.

The method developed for the analysis of LMW RNA, generally applicable to any bacterial biomass sample. There are several steps involved in the protocol that are of general importance for any analysis of nucleic acid from bacterial biomass. Enough material usually 5 to 10 mg of dry weight of bacterial mass required for isolation of low-molecular-weight RNAs. Equal amount of extraction buffer pH 5.1, 1% sodium dodecyl sulfate (SDS) in 50 mM sodium acetate and 10 mM EDTA) and buffer saturated phenol are added to the pellet and extracted for 10 min at 60°C. After the extraction, cell debris and phenol are separated by centrifugation, and the supernant is reextracted and purified with a second phenol extraction followed by a chloroform extraction. The final supernatant is precipitated overnight with ethanol precipitation mixture (ethanol, 2 M sodium acetate and 1 M magnesium chloride, at a ratio of 100:10:1, v/v). The ethanol-precipitated total RNA can be kept safe from ribonuclease degradation at –20°C or below.

Total RNA was separated by denaturing high power polyacrylamide gel electrophoresis. Optimal arcylamide concentration for the separation of the whole range of LMW RNAs was14%. Higher acrylamide concentrations (e.g., 16%) result in sharper bands but lack resolution in the tRNA region and lower concentrations (e.g., 12%) suffer from band broadening.. The vacuum dried ethanol precipitate was resuspended in 4 µl of a loading solution (300 mg ml^{-1} sucrose, 460 mg ml^{-1} urea, 10 µl ml^{-1} 20% SDS, 1 mg ml^{-1} bromophenol blue, 1 mg ml^{-1} xylene cyanol, all substances dissolved in TBE buffer). After pre-electrophoresis at 40 W for 1 to 2 hours, 10 to 15 µg of total RNA sample is loaded on the gel and run at a constant power of 40 W (28 V cm^{-1}) at 60°C for 1 hour to separate high-molecular-weight RNA (23 S and 16S rRNA) from LMW RNA. For the next 1.5 hours, electrophoresis was run at 80 W (42 V cm^{-1}). For the last half hour, power was increased to 160 W (65 V cm^{-1}) until xylene cyanol reached the bottom of the gel. This stepwise increase in power obtained the best results in terms of resolution, reproducibility, and band sharpness.For the detection of nontradioactive RNA, silver staining is the most sensitive technique. The dried and silver-stained gel can be directly photographed.

REFERENCES

Anderberg, M.R. 1973. *Chuster Analysis for Applications*. Academic Press, London.

Austin, B. and Colwell, R.R. 1977. Evaluation of some coefficients for use in numerical taxonomy of microorganisms. *International Journal of Systematic Bacteriology* 27:204-210.

Azam, F., Fenchel, T., Field, J.G., Gray, J.S., Meyer-Reil, L.A., and Thingstad, F. 1983. The ecological role of water-column microbes in the sea. *Marine Ecology Progress Series* 10:257-263.

Bell, R.T. and Kupariran, J., 1984. Assesing phytoplankton and bacterioplankton production during early spring in Lake Erken, Sweden. *Applied and Environmental Microbiology* 48: 1221-1231.

Bell, W.H. 1980. Microbial ecology of algal extra cellular products. The specificity of algal-bacterial interactions. *Biological Bulletin* 159:456-487.

Bell, W.H. and Mitchell, R. 1972. Chemotactic and growth responses of marine bacteria to algal extracellular products. *Biological Bulletin* 143:265-277.

Birge, E.A. and Juday, C. 1934. Particulate and dissolved organic matter in inland lakes. *Ecology and Oceanography* 4:440-474.

Brayton, P.R. and Colwell, R.R. 1987. Fluorescent antibody staining method for enumeration of viable environmental *Vibrio cholerae* O1. *Journal of Microbiological Methods* 6:309-314.

Brock, T.D. and Clyne, J. 1984. Significance of algal excretory products for growth of epilimnetic bacteria. *Applied and Environmental Microbiology* 47:731-734.

Chrost R.J. 1978b. Extracellular release in Chlorella vulgaris culture and the role of bacteria accompanying-algae in this process. *Acta Microbiologica Polonica* 27:55-62.

Chrost, R.J. 1978a. The estimation of extracellular release and heterotrophic activity of aquatic bacteria. *Acta icribiologica Polonica* 27:139-146.

Chrost, R.J. 1983. Plankton photosynthesis, extracellular release and bacterial utilization of released dissolved organic carbon (RDOC) in lakes of different trophy. *Acta Microbiologica Polonica* 32:275-287.

Chrost, R.J. 1986. Algal-bacterial metabolic coupling in the carbon and phosphorus cycle in lakes. pp.360-366 in Megusar, F., and Gantar, M. (Editors), *Perspectives in Microbial Ecology*. Slovene Society for Microbiology, Ljubljana.

Chrost, R.J. and Brzeska, D. 1978. Extracellular release of organic products and growth of bacteria in *Anabaena cylindrica* (blue-green algal) culture. *Acta Microbiologica Polonica* 27:287-295.

Chrost, R.J. and Faust, M.A. 1983. Organic carbon release by phytoplankton: its composition and utilization by bacterioplankton. *Journal of Plankton Research* 5:477-493.

Chrost, R.J., 1989. Characterization and significance of ß-glucosidase activity in lake water. *Limnology and Oceanography* 34:660-672.

Cole, J.J., Likens, G.E. and Strayer, D.L. 1982. Photosynthetically produced dissolved organic carbon: An important carbon source for planktonic bacteria. *Limnology and Oceanography* 27:1080-1090.

Costas, M., Leach, R.H., and Mitchelmore, D.L. 1987. Numerical analysis of PAGE protein patterns and the taxonomic relationship within the "Mycoplasma mycoides Cluster". *Journal of General Microbiology* 133:3319-3329.

Coveney, M.F., 1982. Bacterial uptake of photosynthetic carbon from fresh water phytoplankton. *Oikos* 38:8-20.

Floodgate, G.D. 1980. the assessment of marine microbial biomass and activity. Pp.217-252 in Colwell, R.R. and Foster, J. (editors), aquatic Microbial Ecology. A Maryland sea Grant Publication, College Park.

Fuhrman, J.A. and Azam, F. 1982. Thymidine incorporation as a measure of heterotrophic bacterioplankton production in marine surface waters : Evaluation and field results. *Marine Biology* 66:109-120.

Goodfellow, M. and Dickinson, C.H. 1985. Delineation and description of microbial populations using numerical methods. pp.165-225. in Goodfellow M., Jones D., and Priest, F.G. (editors), *Computer-Assisted Bacterial Systematics*. Academic Press, London.

Goodfellow, M. and Minnikin, D.E. 1985. *Chemical Methods in Bacterial Systematics*. Academic Press, London.

Gyllenberg, H. 1984. automated identification of bacteria: An overview and examples. pp.329-339 in Bergan, T. (editor), *Methods in Microbiology*, vol. 16, Academic Press, London.

Herbst, V. 1984. Physiologische Untersuchungen zur Kopplung der Stoffwechel von Oscillatoria redekei van Goor and Begleitbakterien. Archiv fur Hydrobiologie, Supplement 69:525-594.

Herbst, V. and Overbeck, J. 1978. Metabolic coupling between the alga Oscillatoria redekei and accompanying bacteria. *Naturwissenschaften* 65:598-599.

Hough, R.A. and Wetzel, R.G. 1975. The release of dissolved organic carbon from submerged aquatic macrophytes diel, seasonal and community relationship. Internationale Verinigung fur Theoretische and Angewandte Limnologie, Verhandlungen 19:939-948.

Kersters, K. and DeLey, J. 1980. Classification and identification of bacteria by electrophoresis of their proteins. pp.273-297 in Goodfellow, M. and Board, R.G. (editors), *Microbiological Classification and Identification*, Academic Press, London.

King, J.D. and White, D.C. 1977. Muramic acid as a measure of microbial biomass in estuarine and marine samples. *Applied and Environmental Microbiology* 33:777-783.

Kogure, K., Simidu, U., and Taga, N. 1979. A tentative direct microscopic method for counting living marine bacteria. *Canadian Journal of Microbiology* 25:415-420.

Krambeck, H.J. and Witzel, K.P. 1983. Classification of aquatic bacterial strains: An example of numerical taxonomy in limnology. *EDV in Medizin and Biologie* 14:45-49.

Krogh, A.A. and Lange, E. 1932. Quantitative Untersuchungen uber Plankton, Kolloide and geloste organische and anorganische Substanzen in dem Furesee. *Internationale Revue der Gesamten Hydrobiologie* 26:20-53.

Lancelot, C. and Billen, G. 1984. Activity of heterotrophic bacteria and its coupling to primary production during the spring phytoplankton bloom in the southern bight of the North Sea. *Limnology and Oceanography* 29:721-730.

Larsson, U. and Hagstrom, A. 1982. Fractionated phytoplankton primary production, exudate release, and bacterial production in a Baltic eutrophication gradient. *Marine Biology* *:57-70.

Meffert, M.E. and Overbeck, J. 1979. Regulation of bacterial growth by algal release products. *Archiv fur Hydrobiologie* 87:118-121.

Meyer-Reil, L.A. 1984. Bacterial biomass andheterotrophic activity in sediments and overlying waters. Pp.523-546 in Hobbie, J.E. and Williams, P.J.LeB. (editors), Heterotrophic Activity In The Sea. Plenum Press, New York.

Moriarty, D.J.W. 1986. Measurements of bacterial growth rates in aquatic systems from rates of nucleic acid synthesis. *Advances in Microbial Ecology* 9:245-292.

Muyzer, G., de Bruyn, A.C., Schmedding, D.J.M., Bos, P., Westbroek, P., and Kuenen, G.J. 1987. A combined immunofluorescence DNA-fluorescence staining technique for enumeration of Thiobacillus ferroxidans in a population of acidophilic bacteria. *Applied and Environmental Microbiology* 53:660-664.

Nalewajko, C. 1977. Extracellular release in freshwater algae and bacteria: extracellular products of algae as a source of carbon for heterotrophs. pp.589-626 in Cairns, J. (editor). *Aquatic Microbial Communities*. Garland, New York.

Nalewajko, C. and Lean, D.R.S. 1972. Growth and excretion in planktonic algae and bacteria. *Journal of Phycology* 8:361-366.

Odum, E.P. 1963. Primary and secondary energy flow in relation to ecosystem structure. *Proceedings of 16th International Congress of Zoology* 4:336-338.

Odum, E.P. and de la Cruz, A.A. 1963. Detritus as a major component of ecosystems. *Bulletin of American Institute of Sciences* 13:39-40.

Ohle, W. 1934. Uber organische Stoffe in Binnenseen. *Internationale Vereinigung fur Theoretische and Angewande Limnologie, Vehandlungen* 6:249-262.

Overbeck, J. 1984. Microbiology and biochemistry. Mitteilungen der Internationalen Vereinigung fur Theoretische and angewande *Limnologie* 20:198-228.

Overbeck, J. 1985. Stoffwechselkopplung zvschen Phytoplankton and heterotrophen Gewasserbakterien-ein Schlussel zum Verstadnis von Stoffkreislaufen in Gewasser. *Rhenisch-Westfalische Akademic der Wissenschaften Vortage* 337:7-34.

Riemann, B. and Sondergaard, 1986. Carbon Dynamics in Eutrophic, Temperate Lakes. Elsevier, Amsterdam.

Riemann, B. and Sondergaard, M. 1984. Bacterial growth in relation to phytoplankton primary production and extracellular release. pp.233-248 in Hobbie, J.E. and LeB. Williams, P.J. (editors), *Heterotrophic Activity in the Sea*. Plenum, New York.

Sackin, M.J. 1987. Computer programs for classification and identification. pp. 459-494 in Colwell, R.R. and Grigorova, R. (editors), *Methods in Microbiology*, vol. 19. Academic Press, London.

Saunders, G.W. 1977. Carbon flow in the aquatic systems. pp.417-440 in Cairn, J. Jr (editor), *Aquatic Microbial Communities*. Garland, New York.

Schleifer, K.H. and Stackebrandt, E. 1983. Molecular systematics of prokaryotes. *Annual Review of Microbiology* 37:143-187.

Selander, R.K., Caugant, D.A, Ochman, H., Musser, J.M., Gilmour, M.N. and Whittam, T.S. 1986. Methods of multilocus enzyme electrophoresis for bacterial population genetics and systematics. *Applied and Environmental Microbiology* 51:873-884.

Simon, M. and Azam, F. 1989. Protein content and protein synthesis rates of planktonic marine bacteria. *Marine Ecology Progress Series* 51:201-213.

Sneath, P.H.A. and Sokal, R.R. 1973. *Numerical Taxonomy. The Principles and Practice of Numerical Classification*, Freeman, San rancisco.

Sondergaard, M. 1981. Kinetics of extracellular release of [°C]-labelled organic carbon by submerged macrophytes. *Oikos* 36:331-347.

Stackebrandt, E., Ludwig, W. and Fox, G.E. 1985. 16S ribosomal RNA oligonucleotide cataloguing. pp.75-107 in Gottschalk, G. (editor), *Methods in Microbiology*, vol. 18. Academic Press, London.

Wetzel, R.G. 1983. *Limnology*, 2nd ed. Saunders College, Philadelphia.

Wetzel, R.G. and Hough, R.A. 1973. Productivity and role of aquatic macrophytes in lakes: An assessment. *Polskie Archiwum Hydrobiologii* 20:9-19.

Wetzel, R.G. and Manny, B.A. 1972. Secretion of dissolved organic carbon and nitrogen by aquatic macrophytes. *Internationale Vereinigung fur Theoretische and Angewande Limnologie, Verhandlungen* 18:162-170.

Wetzel, R.G., Rich, P.H., Miller, M.C. and Allen, H.L. 1972. Metabolism of dissolved and particulate detrital carbon in a temperate hard-water lake. *Memorie Dell'Instituto Italiano di Idrobiologia* 29:185-243.

Witzel, K.P., Overbeck, H.J. and Moaledj, K. 1982a. Microbial communities in lake Pluâsee-an analysis with numerical taxonomy of isolates. *Archiv fur Hydrobiologie* 94:38-52.

Chapter **13**

Present Scenario of Ornamental Fish Farming and Trade

Saroj K. Swain

Introduction

Ornamental fish keeping in aquarium has emerged as the second most popular hobby next to photography. As these fishes usually are kept in aquarium, they are popularly known as 'Aquarium Fishes'. These beautiful fishes form an important commercial component of aquaculture, providing for aesthetic requirements and upkeep of the environment. At present the market for ornamental fish in the world for public aquaria is less than 1% and over 99% of the market is still confined to hobbyist. In India the hobby of ornamental fish keeping is nearly 70 years old. The brilliant, flamboyant color and exotic appearance of fish appeal almost for every one, children and aged alike. According to psychiatrists, placing aquaria with ornamental fishes in the patient's house could treat certain type of mentally disordered problems. The relatively minimum requirement of space or attention compared to other pet animals is the reason of growing interest in keeping aquaria all over.

Global Trade of Ornamental Fishes

The trade of ornamental fish has been increasing since the 1980's. It has been estimated to the tune of US $ 4.5 billion in 1995 and striding further, with an annual growth rate of about 14 per cent per year (FAO, 1999). According to FAO, 2000 the world trade of ornamental fish is estimated to

Applied Bioinformatics, Statistics and Economics in Fisheries Research: pp. 165-184

New India Publishing Agency, New Delhi (India)

US $ 9 billion, out of which the total retail trade is at about US $ 3 billion (Live animals for aquarium).

The E.U. market

The European Union is currently the largest trade block for ornamental fish, where over 3.2 million houses keeping aquaria. The import statistics says, it imported US$ 121.1 million worth of ornamental fish in 2002, representing 51% of world production. Popular species are neon tetra, cardinal tetra, guppy, platy, swordtail, siamese fighter, angel, corydoras, rasbora, gourami, and loach. The top importing countries in the year 2002 were Germany (20.1% of imports), UK (17.2%), France (17.2%), Belgium (8.4%), Italy (8.5%) and the Netherlands (8.2%). The recent statistics reveals that, aquarium keeping is most popular in Germany, which is the largest market among the European Union (EU) countries. The major suppliers to EU markets are Singapore, The Czech Republic, Israel, Japan, Malaysia, Indonesia, China (HK), Sri Lanka, Thailand, USA and Brazil.

The American market

It has been estimated that 1.2 million aquarists in USA with more than 10% (9.4 million houses) keep aquarium fish in their houses and 40% of these houses have more than one aquarium. The US market now the largest single market accounting for US$ 40 million worth of imports during 2002. The important freshwater species are under import, mainly two species i.e. guppy and neon tetra, accounting for 40% of the market. Other important species are molly, platy, swordtail, discus, angel, African cichlid and zebra danio. The major suppliers to this market are Singapore, Thailand, Indonesia, Philippines, Malaysia and Brazil. About 80% of the import to USA originates from Asia and the rest from South America, which contributes over 80% of bred freshwater varieties and 15% of wild caught fishes.

The Japanese market

Japan is the second largest single market constituting around 10% (1.2 million houses) keep aquarium. In 2002 Japan imported ornamental fish of worth US$ 25.6 million. 28% of Japanese market comprises only guppy varieties. The other demanded species are tetra, tiger barb, harlequin rasbora, discus, angel, fighters, gourami, platy, swordtail and corydoras. The top suppliers were Singapore, Brazil, Indonesia, Malaysia, USA and China (Hong Kong).

The Asian market

Based on the Singapore Primary Production Department's announcement, during the year 1998, about 73 millions Singapore dollars worth of ornamental fish have been exported and 60% of this is contributed by Malaysia breeders. Singapore, the world's largest exporter of ornamental fish with more than 30 per cent of the global supply, has already produced numerous lucrative new varieties through selective breeding. At present, 60% of the total export trade of aquarium fishes is from Asian countries.

The total world exports of ornamental fish were valued at US$ 189.5 million in 2002 with Singapore 21.9%, Malaysia 9%, Czech Republic 7.1%, Indonesia 6.7%, China-HK 5%, Japan and USA 4% each and Sri Lanka, Peru and Philippines 3% as the main exporters. The total imports for the same year 2002 were US$ 234. 2 million with the main importing countries being USA 16.0%, Japan 10.9%, Germany 10.4%, UK 10.1%, France 8.9%, Singapore 4.8%, Italy 4.4%, Belgium and Netherlands 4.3%, China 4.1%, and Canada 2.8%. (Singh, 2005). Since 1982 the value of world exports has been increasing markedly from US $ 44.5 million to almost US $ 204.8 million during 1996. Export again dropped to US $ 159.2 million in 1998 and this was due to the financial crisis in the world during that period. Exports then rose to US$ 189.5 million in 2002.

The trend in world Import of ornamental fishes has increased markedly from US $ 50 million in 1982 to a peak of around US $ 330 million in 1994-96. Imports then fall to US$ 262 million in 1998 because of World over financial crisis during that period. The total import stood at 234.2 million in 2002. The main importers are the United States (16.9%), Japan (10.9%), Europe, particularly Germany (10.4%), The United Kingdom (10.11%) and France (8.8%). Other important destinations were Singapore 4.8%, Italy 4.4%, Belgium 4.3.0%, Netherlands 4.3%, China 4.1% and Canada 2.8% (Singh, 2005).

It is said that, 90% of freshwater species are farmed, against 10% collected from the wild. For marine species almost 99% are collected from the wild and just 1% is reared. With regard to the value of freshwater species represent about 90% against 10% for marine species in international trade. The general trend of export activity is gradually sifting from the traditional markets of Japan and USA to the new markets in Europe. Export activity also sifting from developed countries to developing countries.

Species are in the trade : The species that dominate the market are all freshwater origin, particularly from the family, *Poeciliidae, Cyprinidae, Characidae* and *Cichlidae* However, the marine species are gaining more popularity particularly from the family, the *Pomacentridae, Acanthuridae, Chaetodontidae, Balistidae, Labridae, Pomacanthidae, Syngnathidae* and the invertebrates. Among the freshwater candidates, the dominating species are guppy, platy, swordtail, molly among livebearers and among egg layers, angel, neon tetra, gold fish, zebra danio and discus. The important marine species are clownfish, damsel, chromis, beaked coralfish, butterfly fish, sailfish, wrasse, marine angel, scorpion fish, goby, trigger fish and sea horse. The marine aquarium with live coral reefs is the trend for the 21st century.

Common freshwater Indigenous ornamental fishes: Indian water possesses a rich biodiversity of about 2118 species, out of which more than 600 species exist in freshwater (Baseer *et al.,* 2001). The country further possesses a rich diversity of colourful ornamental fish, with over 100 varieties of indigenous species, in addition to a similar number of exotic species that are bred in captivity. Further, the vastness of the country's natural resources comprising 2.36 million ha ponds and tanks, 1.3 million ha of beels, jheels and derelict waters, in addition to 0.12 million canals, 2.05 million ha of reservoirs and 75,000 km of river stretch (FAO, 2000) in comparison to those of potential countries like Singapore, Sri Lanka, Malaysia, Indonesia and many African countries offers great scope for exploitation and commercial production. With regard to important indigenous freshwater ornamental fishes, their captive commercial breeding has not been tried in large scale. The major shares of country's export are captive collection from North Eastern States, which comprise as much as 85% of total trade. Rest 15% is contributed by the exotic varieties, mainly from Southern India. The domestic market with regard to indigenous fishes is limited due to preference of many colourful exotic varieties. The most demanded species are loaches, barbs, eels, catfishes and gobbies. It has been estimated that, among the attractive Indian fishes, about 28 species of loaches, 40 catfishes, 7 eels, 20 barbs, 6 danios, 3 rasboras, 13 gobbies, 5 puffers, 5 gouramis, 11 snake heads and 9 species of cichlids are available in different water bodies of the country. In order of availability of these species in different regions of the country it is estimated that, about 33% are available in North Eastern region, 24% in Southern region, 23% in Eastern region, 6% in Western region, 3% in Northern region, 2% in Central region and 10% through out India. So far commercial breeding of these species has not yet been initiated in India. Breeding and larval rearing of few indigenous species like rosy barb, flying barb, rainbow Gourami (Swain and Das, 2001) have been successfully bred and reared at CIFA. Recently few species like Indian rosy barb (*Puntius conchonius*), Black line rasbora (*Parluciosoma daniconius*), Flying barb (*Esomus danricus),* Rainbow gourami (*Colisa fasciata*), Zebra fish (*Brachidanio rerio*), Leopard danio (*Brachidanio frankei*), Sophore barb (*Puntius sophore*), Ticto barb (*Puntius ticto*) and Melon barb (*Puntius fasciatus)* have been bred and reared successfully. Many other indigenous species are under trial at CIFA ornamental fish breeding and culture unit.

Table 1. Important indigenous ornamental fish species and their distribution

Common name	Scientific name	Distribution
Banded loach	*Shistura beavani*	North Bengal
Black border tail orange cap	*Osteochilichthys nashii*	Western Ghats of Karnataka
Black knife fish	*Notopterus notopterus*	All over India
Black line rasbora	*Parluciosoma labiosa*	Nasik, Maharastra
Black spot barb	*Puntius filamentosus*	Karnataka
Blue dotted hill trout	*Barilius bakeri*	Western Ghat, Kerala
Burjors brilliance	*Chela dadiburjori*	Kerala, Goa
Burmese puffer	*Chelonodon nigroviridis*	Gangetic delta, Orissa
Butter catfish	*Ompok bimaculatus*	Through out India
Cheetah catfish	*Laguvia shawi*	Himalaya, Darjeeling, North Bengal
Clown catfish	*Gagata cenia*	North and North East India
Copper catfish	*Glyptothorax telchitta*	Northern India, Orissa
Day's panchax	*Aplocheilius dayi*	Kerala
Devil catfish	*Chaca chaca*	Bengal, Assam, Orissa, Bihar
Dwarf chameleon	*Badis badis badis*	North east, Orissa, Bengal
Dwarf gourami	*Colisa lalia*	Throughout India
Emerald puffer	*Tetradon cutcutia*	North-Eastern India
Filament barb	*Puntius filamentosus assimillis*	Kerala
Finger fish	*Monodactylus argenteus*	Indo West Pacific
Flying barb	*Esomus barbatus*	Tamilnadu, Karnataka
Four spot barb	*Puntius arulius tambraparnei*	Rivers of Tamilnadu
Giant danio	*Danio aequipinnatus*	North Eastern India
Giant glass fish	*Parambasis thomassi*	Western Ghat
Giant mouth catfish	*Hara filamentosa*	Dibru River Assam
Goalpara loach	*Lepidocephalus goalparensis*	Goalpara Assam, Orissa

Contd...

Table 1. Contd...

Golden banded loach	*Botia dario*	Assam, Bengal, Bihar, Orissa
Grizzled loach	*Shistura sikmaiensis*	Manipur Valley
Gulper catfish	*Ompak pabda*	North east India
Hi fin barb	*Oreichthys cosuatis*	Eastern India
Hi fin glass fish	*Pseudambassis ranga*	Throughout India
Honey gourami	*Colisa sota*	North East India
Indian coolie loach	*Pangio pangia*	North-East Bengal, Eastern Madhya Pradesh
Indian rosy barb	*Puntius conchonius*	North India
Indian tiger shark	*Pangasius pangasius*	Large Rivers of India
Jaguar loach	*Somileptes gongota*	North Bengal, Assam
Jewel panchax	*Aplocheilius parvus*	Karnataka
Leaf fish	*Nandus nandus*	Throughout India
Leopard loach	*Acanthocobitis botia*	North Eastern India
Long nosed needle fish	*Xenentodon cancila*	Most places in India
Long snout barb	*Puntius dorsalis*	Krishna River System
Long snouted loach	*Nemachilichthys ruppelli*	Karnataka
Malabar danio	*Danio malabaricus*	Western Ghats
Malini's barb	*Puntius mahecola*	Annamalai Hills
Many banded loach	*Schistura multifaciatus*	Eastern Himalayas
Moustache danio	*Danio dangila*	Bihar, Assam
Neon hatchet	*Chela cachius*	Assam
Noble gourami	*Ctenops nobilis*	Eastern India
Olivaceous loach	*Schistura devdevi*	Teesta river drainage
Panther loach	*Lepidocephalus gunthea*	Northern and Eastern India
Pencil gold labeo	*Labeo nandina*	Assam
Polka dotted loach	*Schistura corica*	North Bengal

Contd...

Table 1. Contd...

Puma loach	*Acanthocobitis rubidipinnis*	Upper Assam
Red chameleon fish	*Badis burmanicus*	Upper Assam
Red line torpedo barb	*Puntius denisonii*	Kerala
Red panchax Red scat	*Aplocheilius panchax* *Scatophagus argus*	Orissa, North & Eastern India Hooghly estuary, Chilka
Red Tailed eel	*Macrognathus jacobbi*	North Bengal, Orissa
Red tailed silver shark	*Gonoproktopterus curmuca*	Western Ghat
Rhinoceros algae eater	*Garra bicornuta*	Thunga river, Karnataka
Ring loach	*Shistura denisoni dayi*	Bihar
Scaly barb	*Gonoproktopterus lithopidos*	South Karnataka
Sidewinder loach	*Aborichthys bijulensis*	Garo hills, Meghalaya
Slender rasbora	*Parluciosoma daniconius*	Throughout India
Spotted scat	*Swcatophagus argus argus*	Estuaries, Chilka, Orissa
Striped gourami	*Colisa fasciata*	Eastern India
Striped loach	*Botia striata*	Tunga river, Kolhapur, Maharastra
Stripped glass catfish	*Pseudeutropius atherinoides*	Through out India
Stripped panchax	*Aplocheilius lineatus*	Orissa, Western and Southern India
Tail spot loach	*Lepidocephalus annandalei*	Upper Assam
Birdi loach	*Botia birdi*	Punjab, North India
Turquoise danio	*Danio devario*	Eastern India
Twin banded loach	*Botia rostrata*	Assam
Tyre Track eel	*Mastacembelus armatus*	Bengal, Orissa, Bihar
Velvet catfish	*Rita pavimentatus*	Andhra Pradesh
Victory loach	*Shistura scaturigina*	Eastern sub Himalayas
Yellow sunfish	*Pristolepis marginata*	Kerala
Zebra danio	*Brachydanio rerio*	All over India except NEH
Zodiac loach	*Mesonoemacheilus triangularis*	Western Ghats, Kerala

Source : Swain, 2004

Exotic ornamental fishes : The aquarium fishes are mainly grouped into two categories *viz.*, egg-layers (oviparous) and live-bearers (ovo-viviparous), though some are categorized under egg layers with no care (non-guarders), egg-layers with care (guarders), egg burriers, mouth-incubators, nest-builders and egg-carriers. Important groups of egg-layers are barbs, rasboras, goldfish, tetras, danios, bettas and gouramis and the major livebearers are guppies, platies, mollies and swordtails. Barbs are the most important group and most species of the group are known to have originated from India, *viz.*, rosy barb (*Puntius conchonius*), melon barb (*P.fasciatus fasciatus),* aruli barb (*P. arulius*). *etc.* Among the exotic barbs, tiger barb (*P. tetrazona*) and its aquarium developed strains like green tiger barb and red tiger barbs are fascinating in modern days. The major species of the group danios include zebra danio (*Brachidanio rerio*), pearl danio (*B. albalineatus),* giant danio (*Danio aequipinnatus*), turquoise danio (*D. devario*), and malabar danio (*D. malabaricus*). Zebra danio is one of the typical examples of ornamental fish species of Indian origin, which can be bred and reared easily. Among rasboras, the slender rasbora (*Rasbora daniconius*), glowlight rasbora (*R. pauciperforata*), harlequin rasbora (*R. heteromorpha*) and scissors tail (*R. trilineata*) are important ones. The gold fish, *Carassius auratus,* is the most commonly available fish, preferred by the hobbyists because of its attractive colouration ranging from pure gold to red, orange, black and albino. Further, the various types of morphological characteristics like, divided caudal and anal fins of varying sizes, egg shaped body due to shortened vertebrae, enlargement and protrusion of eyes, presence of pearl-like glittering scales, thickening of skins on the top of the head, absence of dorsal fins, transparent and dazzling colours, obtained through cross breeding or selective breeding, make the species more attractive. Some of the common varieties of gold fish available are comet, lion head, oranda (a modification of lion head), fringe tail, veil tail, fantail, shubunkin, telescopic eye, etc. These fishes grow up to 20 cm in length, but start breeding when they are only 6 cm long. Many aquarium bred strains of ornamental koi, *Cyprinus carpio* (var. koi) can be bred easily by available hormone injections. Its origin is from East Asia but strains like orenji ogon, kin matsuba, kohaku, and shiro-bekoo are being cultured worldwide.

The tetras are small fishes of 3-8 cm long; majority of which has originated from South America. The most common species of the group are black widow tetra (*Gymnocorymbus ternetzi*), serpae tetra (*Hyphessobrycon sarpae*), lemon tetra (*H. pulchripinnis*), flame tetra (*H. flammeus*), neon tetra (*Paracheriodon innesi*), cardinal tetra (*P. axelrodi*), black neon tetra (*Hyphessobrycon herbertaxelrodi*) and pretty tetra (*Hemigrammus pulcher*). The species, *betta splendens* commonly called as Siamese fighting fish occurs in varied colours like green, red, blue, albino and sometimes with a combination of two or three shades. The attractive colour and hardiness of the species are the characters of the species for its wide adoption by the hobbyists. The males are brightly coloured with beautifully spread-over fins. They show aggressive behaviour, only when other males are present. Angel fish, *Pterophyllum altum* and *P. scalare* are important candidate species widely preferred for aquaria, with different developed varieties such as black, veil tail, gold and black marble, platinum, gold pearl, pink ghost and albino. The filamentous lower fins and their compressed body shape with their elegant movements are pleasant to look at. Among gouramies, threespot gourami (*Trichogaster trichopterus*), pearl gourami, (*T. leen*), moon light gourami (*T. microlepis*), snakeskin gourami (*T. pectorails*), dwarf gourami (*Colisa lalia*), honey gourami (*C. sota*), striped gourami (*C. fasciata),* noble gourami (*Ctenops nobilis*) and kissing gourami (*Holostoma temmincki*) are the important species. The bettas and gouramies are most popular among nest-builders, characterised by possession of accessory respiratory organs, thus making the species hardier. Among catfishes the exotic variety of sucker mouth catfish (*Hypostomus plecostomus*) is very popular. being originated from South and Central America it is available in every aquarist. Catfishes like butterfly catfish (*Hara hara*), dwarf anchor (*H. jerdoni*), cheetah catfish (*Laguvia shawi*), copper catfish (*Glyptothorax telchitta*) are beautifully coloured catfish available in Indian waters.

Livebearers are the second groups of ornamental fishes, giving birth to young ones and produce only a few numbers of offspring in comparison to egg layers. Breeding of most of the livebearers is relatively easy. Development of young ones takes place inside the female and they released after about four weeks. The species of livebearers include guppy (*Poecilia reticulata*), black molly (*Poecilia sphenops*), swordtail (*Xiphophorus helleri*) and platy (*X. maculatus*). The number of babies produced by a livebearer is normally 50-60 only, though some larger swordtails and guppies may produce as many as 100 nos. If the livebearers are fed properly with natural feed and supplemented with better artificial feed, the mother produces more than 100 young ones. Soon after the baby comes out from the mother, they have to be separated out to reduce mortality caused by predation. (Swain and Bandyopadhyay, 2002).

Table 2 : Important exotic ornamental fish traded in India

Category	Common name	Scientific Name	Family	Remarks
Live-bearers Swordtail				
1.	Red velvet swordtail	*Xiphophorus helleri*	Poecillidae	PE
2.	Black swordtail	*Xiphophorus helleri*	Poecillidae	PE
3.	Marvel swordtail	*Xiphophorus helleri*	Poecillidae	PE
4.	High-fin tangerine swordtail	*Xiphophorus helleri*	Poecillidae	PD
5.	High-fin lyre tail	*Xiphophorus helleri*	Poecillidae	PD
6.	pineapple swordtail	*Xiphophorus helleri*	Poecillidae	PE
7.	Tuxedo swordtail	*Xiphophorus helleri*	Poecillidae	PE
8.	Tangerine swordtail	*Xiphophorus helleri*	Poecillidae	PE
9.	Orange pineapple swordtail	*Xiphophorus helleri*	Poecillidae	PE
10.	Green swordtail	*Xiphophorus helleri*	Poecillidae	PE
Molly				
11.	Black molly	*Poecilia spenops*	Poecillidae	PE
12.	Spotted molly	*Poecilia spenops*	Poecillidae	PE
13.	Chocolate molly	*Poecilia spenops*	Poecillidae	PE
14.	Silver molly	*Poecilia spenops*	Poecillidae	PE
15.	Albino molly	*Poecilia spenops*	Poecillidae	PE
16.	Albino lyre-tail molly	*Poecilia spenops*	Poecillidae	PE
17.	Sail fin molly	*Poecilia latipinna*	Poecillidae	PE
18.	Veil-tail molly	*Poecilia latipinna*	Poecillidae	PE
19.	Lyre-tail molly	*Poecilia latipinna*	Poecillidae	PE
Guppy				
20.	Neon guppy	*Poecilia reticulata*	Poecillidae	PE
21.	Cobra guppy	*Poecilia reticulata*	Poecillidae	PE
22.	Butterfly guppy	*Poecilia reticulata*	Poecillidae	PE
23.	Mixed guppy	*Poecilia reticulata*	Poecillidae	PE
24.	Golden guppy	*Poecilia reticulata*	Poecillidae	PE
25.	Black guppy	*Poecilia reticulata*	Poecillidae	PE
26.	Swordtail guppy	*Poecilia reticulata*	Poecillidae	PE
Platy				
27.	Variatus platy	*Xiphophorus varietus* hybrid	Poecillidae	PE
28.	Maculatus platy	*X. maculatus* hybrid	Poecillidae	PE

Contd...

Table 2. Contd...

29.	Red wagtail platy	*X. maculatus* hybrid	Poecillidae	PE
30.	Gold crescent platy	*X. maculatus* hybrid	Poecillidae	PE
31.	Blue platy	*X. maculatus* hybrid	Poecillidae	PE
32.	Bleeding heart platy	*X. maculatus* hybrid	Poecillidae	PE
33.	Red tuxedo platy	*X. maculatus* hybrid	Poecillidae	PE
Egg-layers Gold fish				
34.	Shubunkin gold	*Carrassius auratus*	Cyprinidae	PE
35.	Red cap gold	*Carrassius auratus*	Cyprinidae	PE
36.	Red cap oranda	*Carrassius auratus*	Cyprinidae	PE
37.	Black moor	*Carassius auratus*	Cyprinidae	PE
38.	Pearl scale gold	*Carassius auratus*	Cyprinidae	PE
39.	Calico gold	*Carassius auratus*	Cyprinidae	PE
40.	Peacock tail gold	*Carassius auratus*	Cyprinidae	PE
41.	Bubble-eye	*Carassius auratus*	Cyprinidae	PD
Shark				
42.	Red-tailed Black shark	*Epalzeorhyncus bicolor*	Cyprinidae	PD
43.	Tiger shark	*Pangassius species*	Cyprinidae	PD
44.	Silver shark	*Balantiocheilius melanopterus*	Cyprinidae	PD
Gourami				
45.	Kissing gourami	*Heleostoma temminckii*	Anabantidae	PE
46.	Golden gourami	*Trichogaster trichopterus*	Anabantidae	PE
47.	Siamese fighting Fish	*PEtta splendens*	Anbantidae	PE
48.	Chinese paradise fish	*Macropodus opercularis*	Anabantidae	PD
49.	Pearl gourami	*Trichogaster leeri*	Anabantidae	PD
50.	Blue (three-spot) gourami	*T.trichopterus sumatranus*	Anbantidae	PE
51.	Moonlight gourami	*Trichogaster microlepis*	Anbantidae	PE

Contd...

Table 2. Contd...

Cichlids **Angel and Discus**				
52.	Pearl scale angel	*Pterophyllum scalare*	Cichlidae	PD
53.	Vail tail angel	*P. scalare*	Cichlidae	PD
54.	Ghost angel	*P. scalare*	Cichlidae	PD
55.	Black angel	*P. scalare*	Cichlidae	PD
56.	Marble angel	*P. scalare*	Cichlidae	PD
57.	Marble pearl scale angel	*P. scalare*	Cichlidae	PD
58.	Discus (pompadour fish)	*Symphysodon discus*	Cichlidae	PD
Severum				
59.	Albino severum	*Cichlasoma severum*	Cichlidae	PE
60.	Convict cichlid	*C. nigrofasciatum*	Cichlidae	PE
61.	Fire-mouth cichlid	*Cichlasoma meeki*	Cichlidae	PE
62.	Ramerez's dwarf cichlid	*Papiliochromis ramirezi*	Cichlidae	PE
63.	Albino oscar	*Astronotus ocellatus*	Cichlidae	PD
64.	Tiger oscar	*A. ocellatus*	Cichlidae	PD
65.	Red oscar	*A. ocellatus*	Cichlidae	PD
Catfish				
66.	Sucker mouth catfish (Crocodile fish)	*Plecostomus spp.*	Loricariidae	PD
67.	Albino sucker mouth cat	*Plecostomus spp.*	Callichthydae	PD
68.	Leopard catfish		Callichthydae	PD
69.	Bronze catfish	*Corhdoris melanistius*	Callichthydae	PD
Tetra				
70.	Congo tetra	*Micralestes interuptus*	Characidae	PD
71.	Diamond tetra	*Moenkheusia pittieri*	Characidae	PD

Contd...

Table 2. Contd...

72.	Bleeding heart tetra	*Hyphessobrycon erythrostigma*	Characidae	PD
73.	Black-widow tetra	*Gymnocorymbus ternetzi*	Characidae	PD
74.	Neon tetra	*Paracheirodon innesi*	Characidae	PD
75.	Cardinal tetra	*Cheirodon axelrodi*	Characidae	PD
76.	Buenos Aires tetra	*Cheirodon axelrodi*	Characidae	PD
77.	Albino Buenos Aires	*Cheirodon axelrodi*	Characidae	PD
78.	Golden widow tetra	*Cheirodon axelrodi*	Characidae	PD
79.	Serpae tetra	*Hyphessorbrycon serape*	Characidae	PD
80.	Boehlke's penguin	*Thayeria boehlkei*	Characidae	PD
Barbs				
81.	Rosy barb	*Puntius conchonius*	Cyprinidae	PE
82.	Tiger barb (Sumatra barb)	*Puntius tetrazona*	Cyprinidae	PE
83.	Albina Tiger barb	*Puntius tetrazona*	Cyprinidae	PE
84.	Green tiger barb	*Puntius tetrazona*	Cyprinidae	PE
Koi carp				
85.	Two color koi (Asia Koi)	*Cyrpina carpio* var. koi	Cyprinidae	PE
86.	Nishiki koi (3 color koi)	*Cyrpina carpio* var. koi	Cyprinidae	PE
87.	Manila koi (mongrel koi)	*Cyrpina carpio* var. koi	Cyprinidae	PE
Knife fish				
88	Black ghost knife fish	*Apteranotus albifrons*	Gymnotidae	PD
89	Glass knife fish	*Eigenmannia virescens*	Gymnotidae	PD

(PE- Propagation easy; PD-Propagation difficult)

With regard to exotic ornamental fishes the species acclimatised in captivity range from cheap guppies to costly neon tetra, discuses etc., which are bred by the aqua-culturists and available with retailers at every city and town. But for commercial production, the infrastructures needs to develop are a number of cement tanks of various sizes, aerators, overhead tank, bore well, electric/diesel pump, a hatchery shed and other necessary items. For the same one may require Rs 80,000 to few lakhs of rupees depending on the production capacity.

Table 3. Breeding season (months) of ornamental fishes inhabiting in tropical water conditions

Variety	Jan.	Feb	Mar	Apr	May	Jun	Jul.	Aug.	Sept.	Oct.	Nov.	Dec
Goldfish	■	■	■				■	■	■			■
Koi	■	■				■	■	■				■
Manilacarp	■	■				■	■	■				■
Molly			▒	■	■	■	■	■	■	■		
Platy			■	■	■	■	■	■	■	■	■	▒
Sword tail					■	■	■	■	■	■		
Guppy	▒	▒	■	■	■	■	■	■	■	■	■	
Gourami				■	■	■	■		■	■		
Angel			■	■	■				■			
Tiger shark				■	■	■	■	■	■	■		
Fire-mouth cichlid			■	■	■	■						
widow tetra				■	■	■	■	■	■			
Neon tetra	■	■				■	■	■	■			
Serpaetetra						■	■	■	■			
Danio's						■	■	■	■			
Rosy barb			■	■	■	■	■	■	■			
Siamese fighter				■	■	■	■	■	■	■		
Tiger barb			■	■	■	■	■	■	■			

■ Breeding period (months) ▒ Breeding uncertain

Ornamental fish farming practice in India : Ornamental fish breeding and culture is concentrated in and around the metro cities of India *viz.* Mumbai, Kolkata and Chennai. Recently in states like Andhra Pradesh, Orissa, Kerala, Assam and Chhatisgard, people have started producing freshwater ornamental fishes in commercial scale. This could be possible due to training and guidance of various Central Institutes, Universities and State Fisheries Departments. Kolkata is the first city and so to say a gateway of popular aquarium fishes imported from different countries from where they are distributed to other places. With regard to breeding of ornamental fishes in Kolkata, it is mostly focused on livebearers, gold fishes and gouramies. There is an ornamental fish village called Udairampur in Amtala, West Bengal, where over 200 families are producing ornamental fish in their back yard units. Those units consisting of 2-6 cement tanks of different size, utilized for breeding of livebearers, gouramies, fighter and gold fishes. The farm produce are sold in a weekly market at Galib Street near Hathibagan, Kolkata. The produce sometimes is sold to the whole sellers situated in cities from where the fishes are distributed to different parts of the state. The secret of good production of ornamental fishes may be due to easy abundance of live foods *viz., Tubifex* and zooplankton. According to NABARD, there are about 500 units in the districts of Howrah, Hoogly, South 24 Paraganas and North 24 Paraganas, engaged for ornamental fish production. There are farms existing in Kolathur area in Chennai, produces livebearers, koi carps, cichlids etc in large scale. In this area, people collect plenty of bloodworms from the nature and feed the fishes. There are some farms where different varieties of gold fishes are produced and exported to foreign countries. A farm at Tuticorin produces a wide range of cichlids for export and domestic markets. They collect bloodworms from Madurai city and transported through train for feeding their fishes. In Mumbai there are many breeders who breed only rare and expensive fishes like Angels, Discus and Tetras. Due to non-availability of many natural live foods, breeders mostly depend on brine shrimps (*Artemia* nauplii) and often-prepared artificial feeds for their fishes. Recently it has come to the notice that Mumbai breeders are able to breed silver dollar, emperor tetra, rummy nose, neon tetra and flowerhorns besides a variety of discus strains. Besides the metro cities, people recently have come forward to start small-scale units at various places like Bhubaneswar, Cuttack and other nearby places in Orissa state. The breeders mostly depend upon the metro cities like Kolkata, Chennai, and Mumbai for procuring ornamental fishes and accessories. Among the three cities since Kolkata is nearer to Orissa and for easy transportation through rail takes 6-8 hours, the retail out lets depend on it. Presently some traders and farmers are collecting ornamental fishes and accessories from Chennai by air or rail which takes 2-24 hours respectively. By seeing the demand and enthusiasm among the farmers of Andhra Pradesh and Orissa there is every possibility of mushrooming of small scale units in coming five years in both the states.

Prospects for commercial production and export: Wide ranges of ornamental fishes of International demand are available in Indian waters. Apart from this, exotic species that are acclimatized in various farms are of high economic importance. Several species of Indian barbs, catfishes, danios, rasboras, and anabantids are very popular in International trade. At present time the ornamental fishes are very much popular-as model animals for biotechnological research. (Swain *et al.* 2007). In India, we are blessed with the best climate for production of such ornamental fishes in mass scale. In this regard ICAR has launched mega seed project in 2006 for ornamental fish and for which many research ornganisations like CIFA, CIFE, CMFRI, CIFRI and SAU'S are involved in production of quality seed.

The export of ornamental fishes from the country at present is mainly confined to freshwater varieties and limited to the fishes from the Northeastern states (85%) and a few bred varieties of exotic species (15%). The export of ornamental fishes is limited to about INR. 80 lakhs during (1995-96) that of course has shown marked increase over the years from about INR 15 lakhs during 1990-91. As per the MPEDA statistics in recent years (2000-2001) India exported a worth about INR 228 lakhs of ornamental fishes to Japan (24.1%), Singapore (20.2%), USA (19.7%), China (10.5%), Germany (6.1%), UK (4.4%), Taiwan (6.4%), Thailand (2.6%), Hong Kong Special Administrative Region (2.6%), Sri Lanka 1.3%), Netherlands (1.2%), France (0.8%), Bangladesh (0.8%), Belgium (0.7%), Malaysia (0.7%), Nepal (0.6%), Switzerland (0.2%), Finland (0.2%) and Maldives (0.1%). According to an estimate presently there are 350 fulltime and more than 2000 part time fish breeders are available in the country (Nair, 2001).in the year 2003-04 country exported INR 30 million (US$0.7 million) worth of ornamental fish. Though the above figure is quite negligible in world trade, it involves 150 species of fish. The unit price varies from US$ 0.10 to US$10.00/ fish. Besides the exotic species Kolkata is the gateway for export of indigenous fishes. During the year 1998-99 the export through West Bengal State is worth INR 142.56 Lakh. Most of the wild caught popular fishes from North Eastern States are being exported to Europe and other countries through Kolkata airport. A few exporters are also having farms where they acclimatize the wild caught fishes before shipment. Kolkata is the largest exit point followed by Chennai and Mumbai.(Swain, 2005, 2006).

Table 4 : Export of Ornamental Fish from India.

YEAR	Rupees (Lakh)
1992-93	35.42
93-94	55.32
94-95	65.03
95-96	81.67
96-97	108.34
97-98	79.42
98-99	158.23
00-01	228.00
2003-04	Over 300

Source : MPEDA, Cochin

Table 5 : Main Ornamental Fish Suppliers of Asia during the year 2002

S. No	Country	%
1	Singapore	37.0
2	Malaysia	16.0
3	Indonesia	11.0
4	China	10.0
5	Japan	7.0
6	Sri lanka	5.0
7	Thailand	5.0
8	Taiwan	2.0
9	India	1.0

Source: Infofish International, Number 3/2005, May/June, pp.12

Upliftment of rural livelihood: Ornamental fish keeping not only gives aesthetic pleasure but also a money earning technology for rural farmers, farm-women, and unemployed youth. The farmers of the rural area produce ornamental fish and sell their fish to the retailers situated at cities and get back the appropriate price. The space requirement for establishment of a unit is comparatively less than a carp or prawn farm. The investment is also less than the other aquaculture species farming and the techniques are also easy to adopt. There are about 3000 units locates at different parts of India. The cost of the back yard units established at different places ranges from Rs 10,000 to 50,000. In commercial scale the farmer has spent Rs. 80,000-5 lakh for their units. The total income from the unit varies from Rs. 500 to 2000/ month for backtyard units, whereas for commercial units the return is Rs 5000-25,000/ month. In some villages rural women are also involved for production system and earning their livelihood. The women self help groups are also involved in such farming for their livelihood. The cultivable species among egg layers are a variety of gold fishes (Oranda, Shubunkin, fantail, Comet, Black moor, Redcap etc.), Gouramies like blue, golden, pearl scale and Indigenous Gouramies like *colisa* sp., Koi carps (manila carps, two to three colour koi), Barbs like rosy barbs, tiger barbs, among Danios *zebra danio* and few indigenous species, cat fishes like tiger sharks (*Pangassius* species), different types of angels (*Pterophyllum* species) etc. are very common among the farmers. Among livebearers many varieties of molly, platy, guppy and swordtail are being cultured by the beginners. Farmers generally collect their brood stock from retailers and known traders.They feed their fishes mostly prepared food made from Groundnut cake, Wheat bran, Soybean meal, Rice polish, Fish meal in combination or individually. The backyard farmer feeds their fishes the carbohydrate wastes removed from their kitchen and live food like Tubifex, Zooplankton collectyed from the nature.The exact commercial farming in purely scientific way, is yet to start by the producers of India. However, there is a tremendous scope for this sector to grow in rural areas with easy marketing network.

Constraints in ornamental fish farming : (i) Insufficient number of international airports for easy export to international markets, (ii) Lack of trained manpower., (iii) Lack of knowledge for catching live wild stocks, (iv) Non-existent of major production centers with organised marketing system in different parts of the country, (v) Non-availability of modern equipments like filtration system, aerators, pumps etc., (vi) Lack of Institutional support for training and dissemination of package of practice/ technologies for breeding and culture of ornamental fish to the rural farmers/entrepreneurs in different regions of the country, (vii) difficulties in Institutional finance for poor farmers intend to take up ornamental fish farming, (viii)Lack of awareness among the trade about the potential for damage that of exotic carnivorous ornamental fishes represent and no such regulations to dictate the conditions of import.

Suggestions for improvement : (i) Popularization of the enterprise through training programmes, workshops, seminars, symposia and exhibitions, (ii) Interested farmers /entrepreneurs should be given freehand long-term loans from various agencies like banks to construct their farms and to develop their infrastructure, (iii) Survey of statewise different indigenous species and stock assessment of natural resources are highly essential, (iv) In many places indigenous ornamental fishes having more commercial value are used for food due to ignorance. So the extension officials and various non-government organisations should create general awareness about their commercial value as ornamental fish, (v) Since the indigenous fishes have more export value and fetch more prices, judicious harvest from different potential places should be done and handed over to organised co-operative societies, or fish farmer federations, so that the harvester gets better price. The same may be applicable for the exotic fish producers, (vi) Strong regulation for importing and selling of exotic carnivorous fishes like Piranha and Arowana should be followed, which may damage the aqua- biodiversity, (vii) Public aquaria should be developed in every state capital for tourist point of view through out the country, (viii) In every potential state, export license should be given to the interested persons, so that they can export farm produced ornamental fishes, (ix) Genetic studies with regard to indigenous ornamental fishes are highly necessary to study genetic characterization, gene mapping, and selective breeding, (x) Greater emphasis on research thrust should be given through

International collaboration, (xi) Considering the high potentials of ornamental fish culture for domestic markets as well as exports, it is necessary to provide a greater research and development support to the sector, (xii) Live food culture are to be encouraged viz. earthworms, tubifex, zooplanktons (xiii) Development of repacking and acclimatizing centers should be constructed near International airports.

Fish Health and Medications

Since the success of this trade depends on best colouration, healthy fish showing good swimming behaviour, the farmers take care of their fishes but due to management faults sometimes diseases are occur. They are mostly the tropical diseases like fin and tail rot due to bacterial contamination, white spot disease due to myxosporidians, Mouth fungus, ulcerations etc. Since these fishes are not consumed by human being, the farmers use locally available chemicals for cure of such diseases. They are Potassium permanganate, Methyline blue, Acryflavin, Tincture iodine etc. for different diseases. Some use human or animal grade antibiotics in water. Recently a medicine (CIFACURE) has been developed for curing certain diseases and commercialised is in progress. Works on Koi Herpes virus are going on to identify the occurrence of such virus anywhere in the country through funding from Department of Biotechnology.

Conclusion

Seeing the potentiality of ornamental fish culture, emphasis may be given on systematic cataloguing of commercially important aquarium fishes, bringing new ornamental fish species for diversification, detailed studies on breeding, biology and behaviour aspects, nutrition and feed formulations, disease diagnosis and comprehensive health management, fabrication of aquarium associated equipments through education and training.

REFERENCES

Basheer, V. S., A.Gopalakrishnan, K. K. Lal and A. G. Ponniah, 2001. Needs for diversifying research on new aquaculture species, *NBFGR-NATP Publication No. 3*, Paper No. 23.

Food and Agriculture Organization (FAO, 1999). Ornamental aquatic life: What's FAO got to do with it? *FAO News & Highlights* (1 September).

FAO year book, 2000, Vol.90/2

Globefish. Research Programme, Volume 67, FAO, 2001

Nair, K. S., 2001. Status of Indian ornamental fish trade. Captive breeding for aquaculture and fish germplasm conservation, *NBFGR-NATP Publication No. 3*. Paper No. 61.

Singh, T., 2005. Emerging trends in world ornamental fish trade, Infofish international, number, 3, May/June, pp.16.

Swain, S.K and B. K. Das, 2001. Captive mass breeding of dwarf Gourami *Colisa lalia* in the cement cisterns, *NBFGR-NATP Publication No. 3*, Paper No. 10.

Swain S.K. and P. Bandopadhyay, 2002. Commercial Live bearing Ornamental fish breeding: A beginners Delight, *Fishing Chimes*, 22:3, 56-60pp.

Swain, S.K., 2004. Training manual on Ornamental fish breeding and culture for entrepreneurship development. CIFA (TTC) publication, 84pp.

Swain S.K. and R.K. Jana, 2005. Freshwater ornamental fish farming in India: present status, prospects and constraints, *BIOVED*, 16(1,2),139-146 pp.

Swain, S.K. 2005. Breeding of ornamental fish species and entrepreneurship development. In training cum workshop on ornamental fish culture and trade, *edited by Das, S. K.*, College of Fisheries, AAU, Raha, Nagaon, Assam,32-38pp.

Swain, S.K., Kuldeep Kumar and G. S. Saha, 2006. Training manual on Ornamental fish farming for entrepreneurship development. Sponsored by NFDB, Hyderabad, Organized by CIFA, Kausalyaganga, CIFA publication. 102 pp.

Swain. S.K. (2006) Ornamental fish breeding and culture, In : *Hand Book of Fisheries and Aquaculture*, ICAR, New Delhi, DIPA publ, 354-377 pp.

Swain S.K., D. Mallik, S. Mishra, B. Sarkar and P. Routray, 2007. Ornamental fish-as model animals for biotechnological research. *In : Environmental Biotechnology* (Ed. Mishra and Juwarkar), APH Publ. Corp. Delhi, pp. 293-328.

Chapter **14**

Spatial Information Technology in Aquaculture and Inland Fisheries

M. N. Reddy, G.P. Reddy & B.S. Sontakki

Introduction

Activities to promote and monitor the growth of aquaculture inherently have a spatial component, because of the variations among biophysical and socio-economic characteristics from location to location. Biophysical characteristics may include water quality, soil type and climate. Socio-economic characteristics in aquaculture development may include administrative regulations, competing resource uses, market conditions, infrastructure support, and availability of technical expertise. Geographical Information Systems (GIS) provides the spatial information needed for decision-makers who evaluate such biophysical and socioeconomic characteristics as part of aquaculture planning efforts more effectively and timely (Kapetsky and Travaglia, 1995). GIS is potentially a powerful tool for assisting this class of decision-makers, and is already being effectively used for such purposes in some places (Carswell, 1998; Arnold *et al.,* 2000) where advanced capabilities in terms of infrastructure and trained personnel exists. Nath *et al.* (2000) discussed various applications of GIS in aquaculture. In general spatial information technologies are being effectively used for sustainable management of aquaculture and inland fisheries. Some potential applications are as follows:

Applied Bioinformatics, Statistics and Economics in Fisheries Research: pp. 185-195

New India Publishing Agency, New Delhi (India)

- Spatial distribution and Resource inventory of water bodies including small water bodies.
- Site selection for targeted fish species
- Trade-offs among alternative uses of natural resources
- Environmental impact assessment
- Sustainable fish farming development through selection of suitable sites for fish ponds etc.

Some of these applications are briefly discussed in the following sections.

Spatial Information Technology

In recent years spatial information technologies of GIS(Geographical Information Systems), GPS(Global Poisoning system) and Remote Sensing(RS) have made major advances in management of natural resources for their sustainability. GIS is a spatial data management and information analysis system that is able to capture, synthesize, generate, retrieve, analyze, and display spatial information. A GIS can also be considered as a spatial Database Management System which allows users to store, retrieve and manipulate data, integrated with a series of routines which allow sophisticated spatial analysis and display (Burrough, 1986). GIS have been developed rapidly in the past decade and begun to be used for aquaculture and fisheries development studies. A major benefit of a GIS is that it can collect, store and present information at different spatial scales. GIS and Remote Sensing (RS) are increasingly being used as natural resource management tools including fisheries and aquatic science. Satellite remote sensing technologies are providing high quality in terms of high resolution digital data about the objects on the earth surface. This enabled to monitor spatial and temporal dynamics of natural resources and the impact of various technologies over space and time. GIS, combined with other analytical tools and models, can allow for improved spatial monitoring, analyses, and eventually better and more effective and sustainable management of natural resources. GPS is a precision location navigation system and records geographic coordinates along with altitude of a given location on the earth surface. Geographical identities are essential inputs required for spatial data management applications using GIS software.

GIS have been developed rapidly in the past one and half decade and begun to be used for aquaculture and fisheries development studies. The GIS would also likely benefit from integration of its output with economic analysis and marketing tools which can allow a more comprehensive examination of the costs as well as benefits associated with aquaculture and agriculture. The GIS should also be expanded to address land use options besides aquaculture and agriculture (e.g. urban development, forestry, and livestock rearing) if it is to serve as a comprehensive tool for sustainable resource use planning.

Potential Applications in Aquaculture and Inland fisheries

Inventory of Water Bodies for Fisheries Management and Aquaculture Development

Spatial information technologies proved to be efficient and effective tools for natural resources inventory and mapping. Some of the questions to be answered in the study area are.

- How many water bodies are there?
- Where are they?
- What is their size distribution?
- What amount of water do they maintain during the dry season?
- Are their present uses compatible with fisheries and aquaculture?
- How productive are they biologically?

If answers can be found to these questions it would be possible to devise a plan for fisheries and aquaculture development. Satellite RS could be a rapid, cost-effective means of providing this required information on spatial distribution of water bodies. Fig.1 shows the spatial distribution of water bodies of West Godavari district in Andhra Pradesh derived using IRS Satellite digital data in GIS.

The below maps are of great use in developing optimum spatially balanced sampling design for selecting primary data through sample survey. Because of the availability of Good quality of satellite RS data in terms of spatial resolution, it is technically feasible for the detection of small water bodies. Visual analyses would be less costly than computer processing, but at the same time would limit the scope and utility of results.

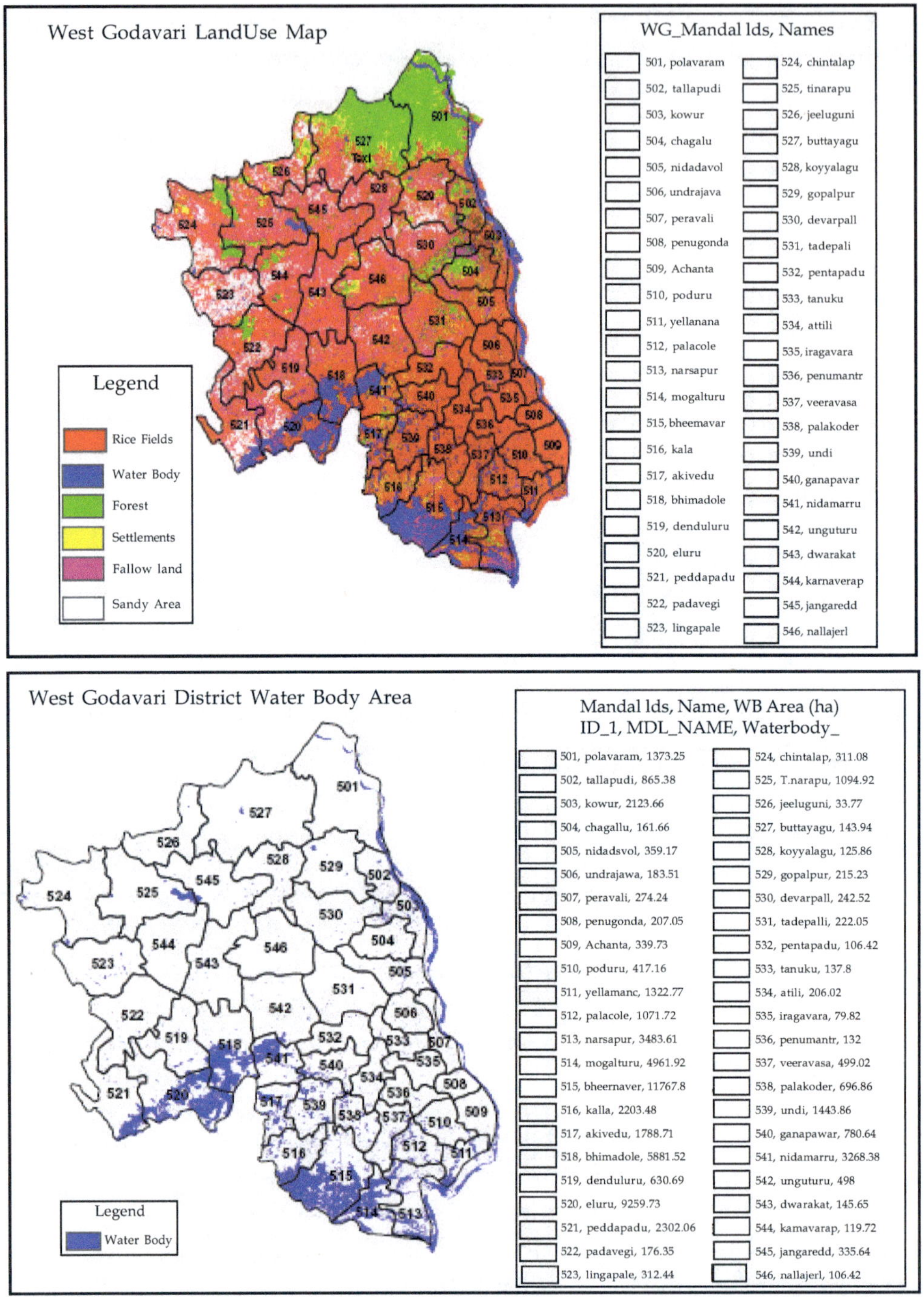

Fig. 1. Spatial Distribution of Land Use and Water Bodies in West Godavari district of Andhra Pradesh

Fig. 1 shows only the distribution of water bodies in different mandals, and the map does not show the spatial distribution of small Aqua cultural ponds. In this regard the Google Earth application software (http://earth.google.com) could provide spatial distribution of small water bodies (Fig.2).

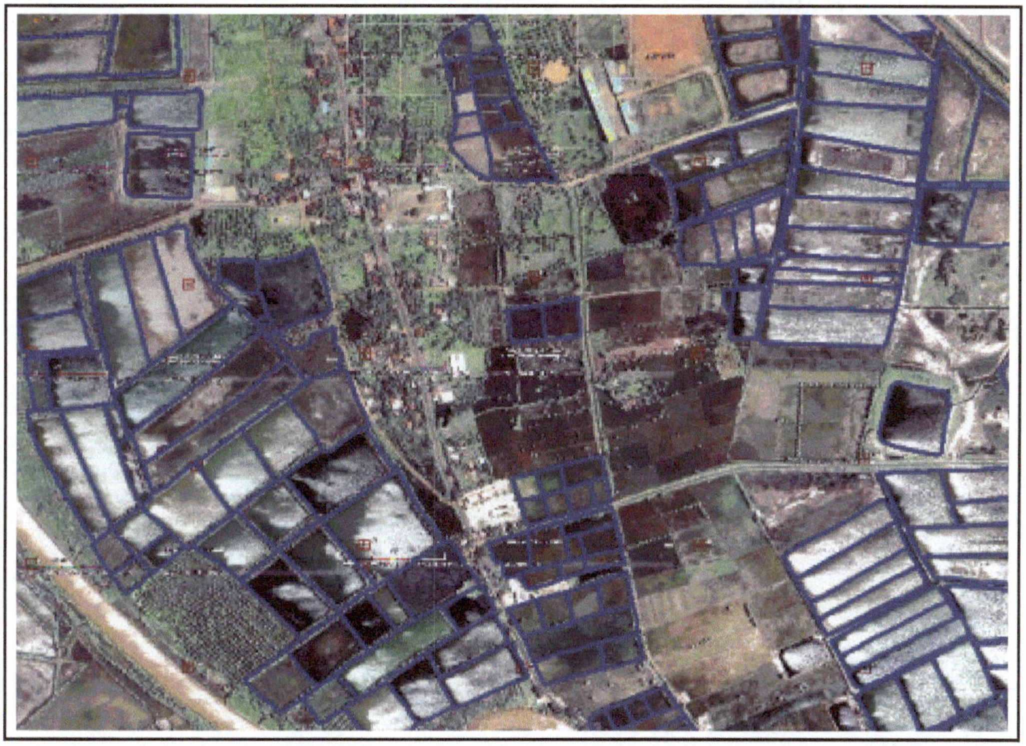

Fig. 2. Spatial Distribution of Aquacultural Ponds (http://earth.google.com)

Land Suitability Analysis for Aquaculture

We come across many successful applications of GIS to assess land suitability for agriculture and aquaculture at different scales say, district, state and national levels. The analysis basically involves overlaying various theme maps in GIS and assessing land suitability by imposing certain constraints on land use depending on the purpose. Some of the themes to be overlaid in GIS are as follows:

- Water resources (rivers, streams, lakes, dams etc.)
- Land use type (agriculture, irrigation, shrimp farming, forestry etc.)
- Environmental characteristics (temperature, elevation, soils etc.)
- Protected areas (Mangroves, conservation zones etc.)

- Pollution sources (location of industries, sugar factories, domestic waste etc.)
- Urban development (Cities, towns and villages)

For example suitability maps for agriculture and aquaculture for a state in Mexico are shown in Fig.3, which are derived by overlaying various thematic maps in GIS Aguilar-Manjarrez and Ross, 1992, 1996).

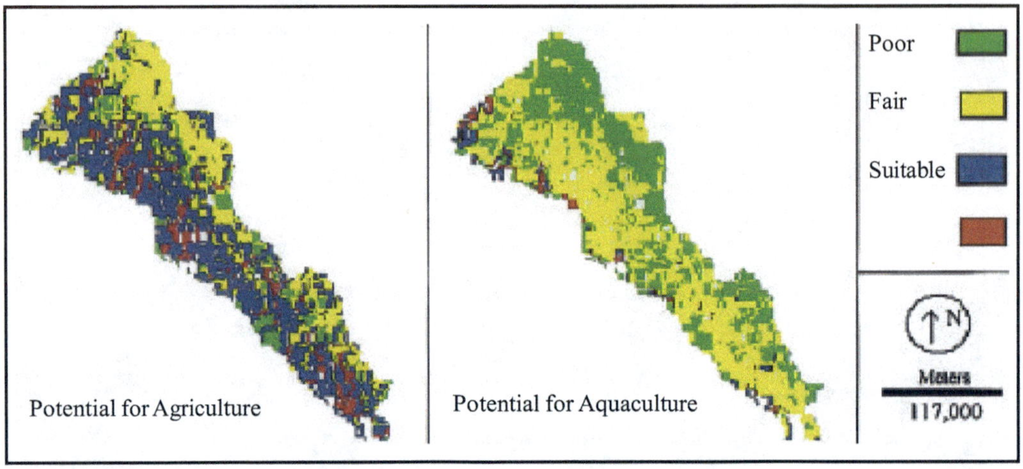

Fig. 3. Land suitability maps for Agriculture and Aquaculture (*Source*:Nath *et al.* 2000)

Lake size dynamics for inland fisheries development

Studies showing lake size fluctuations over time could be vital for the successful implementation of inland fisheries development. Remote sensing images derived in GIS programme are found to be very effective to monitor changes in lake size or shape over time. For example in a study by Nuriddinov, (1989), remote sensing imagery derived in GIS had enabled the determination of the direction of the lake's growth and to forecast the position of its shoreline. Analysis of the RS imagery in GIS provided a temporal picture of the development of the lake.

In this study, analysis of the space imagery for 1975 and 1976 clearly revealed areas that showed water logging of these areas. And likewise, analysis of space imagery in 1980 and 1985 again confirmed that areas appearing to be waterlogged always appeared to be flooded on later images. Since this flooding is likely to continue, i.e. because there are no plans to halt the irrigation processes, then it is now possible to draw accurate maps which look at the future incremental growth of the Lake. The modeling of this process would clearly be an example of a useful study which could be greatly and efficiently expedited if performed as a GIS exercise.

Identifying areas for Fish Pond development

GIS can be used to locate and map inventory areas which are suitable for fish pond development. There are many ways in which it could be improved. Some of the location criteria would include spatial information on the following themes:

i. Ground water depth.
ii. Water quality.
iii. Soil chemical characteristics - as they affect water productivity and quality.
iv. Distance from feed suppliers and fish processing plants - as they affect transport costs.
v. Spatial development cost variations, e.g. cost penalties for siting within the 100year floodplain area.
vi. A slope map could have been incorporated to eliminate the subjectivity in assessing relative ratings

The study by Kapetsky *et al.* (1988) shows that an effective GIS for aquaculture planning need not be complex or expensive. Once the databases have been captured, updating is easy and the data can be manipulated to achieve various simulations, e.g. what effect would raising level of bank heights have on diminution of flood prone areas?

Impact Assessment on Mangroves using GIS and PRA tools

An information revolution brought on by rapidly advancing digital data processing technologies is now influencing rural development in isolated regions of the world. GIS provide a potentially useful tool for spatial analysis of natural resource management in developing nations. However, this evolving technology must blend with current social changes. Development philosophies worldwide now stress the need to incorporate the knowledge and perspectives of local stakeholders. Participatory Rural Appraisal (PRA) technique and GIS tools are suggested to study the impact of fish culture at local mangrove resources. Data Collection using PRA tools (Fig.4):

- Semi-structured interviews with key informants
- Time-line and change analysis
- Village and resource mapping
- Analysis of Secondary data from topographical maps and remote sensing imageries
- SWOT analysis of technologies & interventions of government and non
- government agencies.

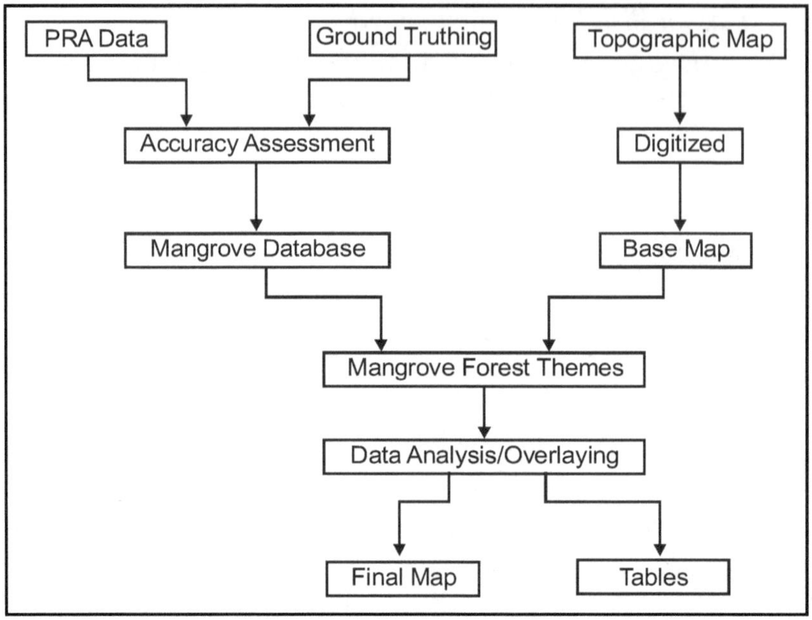

Fig. 4. GIS and PRA methodology (*Source*: Agdalle & Boromthanarat, 2003)

Data to be generated using the PRA workshops:

1. Local mangrove resource characteristic
2. Government and community use and livelihood
3. Issues, problems and potential of the mangroves resources
4. Land cover and land use changes

GIS is the tool for integrating spatial and non-spatial data on the following factors collected through PRA and through secondary sources:

- environmental
- economic
- legal and
- social issues

GIS themes to be generated are:

1. The map of Mangrove Forests attribute over time
2. Community identified potential uses of Mangroves and their changes

3. Map of Navigation Route
4. Map of Mangrove Buffer zone, and

PRA integrated with GIS is a useful tool to draw the local knowledge of the people about the resource by allowing participation of stake holders.

To assesses the impact of shrimp farming on environmental parameters

The science of impact assessment has developed rapidly in the last few years. Despite significance advances, methods of impact assessment are required to be fine-tuned to site-specific nature of agricultural research. Impact assessment attempts to estimate the effects that research has had in the past or that it may have in future. An impact assessment may look at whether farmers accept or reject new technology, or it may focus on increase in yield and production that can be attributed to new technology. It can also estimate changes in employment, nutritional status, pollution, erosion, or rural - urban migration.

Shrimp farming and Aqua cultural activities are mostly carried out in mangroves and coastal regions all over the world. The intensive shrimp farming has significant impact on mangroves and marine environment as well. It has been realized that the polluted water discharge from shrimp farms causes deterioration of water quality of adjacent water bodies like mangroves and coastal zone. Costal zones are places of ending changes. Effective inventorying and monitoring of these changes are required for the understanding and managing this environment (Tripathi *et al.* 2003).

The Environmental Impact Assessment (EIA) is systematic, well documented and multidisciplinary procedure, where we identify, describe and assess the direct and indirect effects of a project or an activity on different environmental factors such as soil, water, air, climate, landscape, cultural heritage, flora, fauna, human beings, and also interaction among these factors. The main tasks involved in EIA: a) to manage huge and diverse nature of database (spatial, non-spatial, dispute, continuous and temporal) and b) to make final interpretation based on these data, which by most of the conventional methods is very subjective GIS can be a potential tool for EIA. GIS acts as an integrated framework for generation, storage and display of thematic information.

There are many conventional methods to find out the impact of all concentrations of the ecosystem. These conventional methods only give us the concentration of pollutant at a particular point or so but can not show the dispersion pattern or extent of pollution at the same time it is much harder to detect the pollution that has had great length of time to disperse and hence cover large area, or pollution that does not emanate from a point source (Clark, 1993). Remote sensing provides better tool to serve all these purposes to give synoptic coverage of water bodies. The greatest advantage of remote sensing technology is that it provides augmented view of spatial and temporal variability of water quality and land use status which can not be obtained by any conventional measuring techniques.

REFERENCES

Agdalipe L and S.Boromthanarat (2003) www.gisdevelopment.net

Aguilar-Manjarrez, J., 1992. Construction of a GIS for Tobasco State Mexico. Establishment of technical and social decision models for aquaculture development. MSc Thesis, University of Stirling, UK.

Aguilar-Manjarrez, J., 1996. Development and evaluation of GIS-based models for planning and management of coastal aquaculture: a case study in Sinaloa, Me´xico. PhD Thesis. Institute of Aquaculture, University of Stirling, Scotland, UK.

Arnold, W.S., Norris, H.A., White, M.W., 2000. Integrated resource management using geographical information system technology: shellfish aquaculture in Florida, USA. Aquacultural Eng., in press.

Burrough, P.A., McDonnell, 1998. Principles of Geographic Information Systems, 2nd ed. Oxford University Press, New York.Clark C.D., 1993, Satellite and remote sensing of marine pollution. *International Journal of Remote Sensing* 14:16: 2985-3004.

Carswell,B., 1998. BCAS: an information system for aquaculture and marine resource planning. Ministry of Agriculture, Fisheries and Food, British Colombia. Unpublished Report.

Clark C.D., 1993. Satellite and remote sensing of marine pollution. *International Journal of Remote Sensing* 14:16 pp 2985-3004

Kapetsky, J.M., Hill, J.M., Worthy, D.L., 1988. A geographical information system for catfish farming development. *Aquaculture* 68, 311–320.

Kapetsky, J. M., Hill, J. M., Worthy, D. L., Evans, D.L., 1990a. Assessing potential for aquaculture development with a geographic information system. *J. World Aqua. Soc.* 21, 241–249.

Nath, S. Bolte J. P., Ross G. L. and J. Aguilar-Manjarrez. (2000). Applications of geographical information systems (GIS) for spatial decision support in aquaculture. *Aquacultural Engineering* 23 pp. 233-278.

Nuriddinov, O. S., 1989. Use of Remote Sensing in the study of the changing shoreline of Sarykamsh Lake. *Mapping sciences and Remote sensing* Vol 26. No1 pp 74-77

Tripathi, N. K., Annachchatre,A and A.A.Patil. (2003). www.gisdevelopment.net

Chapter 15

Role of GIS and Remote Sensing in Inland Fisheries Development

Sanjeev Kumar Sahu

Introduction

India is an agrarian society where 70% of the population depends on agriculture, animal husbandry and fisheries. Fishery is an important sector, its planning and development largely depends on a sound and reliable data and information on resource, production, marketing and distribution. Generating reliable information basically needs identification, conceptualization, classification and sound scientific techniques for collection, compilation and analysis of data. Geographical Information System emerged as an efficient tool to access, organize, update and analyse the information and to utilize it in an optimal way.

A proper computer based information system is the need of day to organize such complex data in order to draw reliable conclusions which may help in drawing action programmes. This multidimensional nature of information on topography, physico-chemical parameters, socio-economics and fish production data can only be handled and processed by Geographical Information System (GIS).

GIS is a generic term implying the use of computers to create and display digital maps. The attribute data which describe the various features presented in maps, may relate to physical, chemical, biological, environmental,

Applied Bioinformatics, Statistics and Economics in Fisheries Research: pp. 197-216

New India Publishing Agency, New Delhi (India)

technological, social, economic or other earth's surface properties. GIS allows mapping, modeling, querying, analyzing, and displaying large quantities of such diverse data, all held together within a single database. Its power and ability to integrate quantities of information about the environment and wide repertoire of tools provided to explore the diverse data.

GIS produces and reads maps. Its major advantage is that it permits identifying spatial relationship between different map features. It can create map in different scale, projection and colours. But it is not a map-making tool. It is a primarily analytical tool that provides new ways of looking at, linking and analyzing data by projecting tabular data into maps and integrating data from different diverse resources.

Principle of Remote Sensing

Remote sensing is a multi disciplinary science of deriving information about an object from measurement made at a distance far from the object without coming in contact with it. Remote sensing refers to the identification of earth features by detecting the characteristics electromagnetic radiation that is reflected / emitted by earth surface. Every object reflects / scatters a portion of the electromagnetic radiation depending upon its physical properties in addition; objects also emit electromagnetic radiation depending upon their temperature and emmissivity. Reflectance / emittance pattern at different wavelengths for each object is different; this enables identification and discrimination of objects.

Presently, the commercial availability of remotely sensed satellite data of Indian as well as foreign satellites is easily available. These remotely sensed data have high spatial, spectral, radiometric and temporal resolution, eases the monitoring and surveying of the water resources. The spatial resolution of Indian satellite varies from 2.5 m to 350 m, spectral resolution varies from 1 band to 8 bands and temporal resolution varies from 2 days to 25 days.

Remote sensing is not a new concept. Visual perception of human eye is a best example of remote sensing. The human eye (sensor) captures light (visible radiation) from apples (objects), placed at a distance (remote). Different objects return different amount and kind of energy in different band (range) of the EM spectrum incident upon it. This unique property depends on the property of material (structural, chemical, and physical), surface roughness, angle of incidence, intensity and wavelength of radiant energy.

The remote sensing is a science, which includes a combination of various disciplines such as optics, spectroscopy, photography, computer, electronics, telecommunication, satellite launching etc. All these technologies are integrated to act as one complete system in itself, known as remote sensing.

There are number of stages in a Remote Sensing System, working as links in complete, and each of them is important for successful operation.

Stages in Remote Sensing

1. Origin of electromagnetic energy (Sun, transmitter carried by the sensor).
2. Transmission of energy from the source to the surface of earth and its interaction with the intervening atmosphere
3. Interaction of energy with the earth surface (reflection/ absorption/ transmission) or self-emission.
4. Transmission of the reflected / emitted energy to the remote sensor placed on a suitable platform.
5. Detection of energy by the sensor converting into photographic image or electrical output.
6. Transmission / recording of the sensor output.
7. Pre-Processing of the data for generation of the data product.
8. Data processing and interpretation.

Thus the remote sensing system consists of a sensor to collect the radiation and a platform which can be satellite, rocket, aircraft and balloon on which sensor can be mounted. The information received by the sensor is suitably manipulated and transported back to earth may be telemeter or brought back through films, magnetic through tape etc. The data are reformatted and processed on the ground to produce photographs, computer compatible magnetic tape (CCT), CD ROM, DVD etc.

Need for GIS

Development of an information system should be based on various aspects which are environmentally friendly, economically feasible and socially acceptable. For planning accurate, reliable and timely information which supports rules for management such as variety of maps, large amount of spatial data, selectively extract information relevant to a planning task is required. This type of need can be fulfilled by developing an information system based on geographical data.

Definitions of GIS

Geographic : Data referenced by some coordinate system to locations on the surface of the earth.

Information : Provides answers to queries about the geographic world.

System : This is the environment which allows data to be managed and questions to be posed.

A GIS is basically a computerized information system like any other database, but with an important difference that all information in this are linked to a geographic (spatial) reference (latitude/longitude or other spatial coordinate system)

GIS = G + IS

Geographic reference + Information system

Data of spatial coordinates on the surface of the earth (MAP)-location data	Database of attribute data corresponding to spatial location and procedures to provide information for decision making

GIS = Information System with geographically referenced data

There are no clear-cut definitions for GIS, different people defined GIS according to capability and purposes for which it is applied. Few of them are:

Clarke defined it as a computer system for the capture, storage, retrieval and display of spatial data, within a particular organization.

Burrough defined it as a powerful set of tools for collecting, storing, retrieving as well as transforming and displaying spatial data from the real word.

ESRI defined this as an organized collection of computer hardware, software, geographic data and personal designed to efficiently capture, store, update, manipulate, analyze and display geographically referenced information.

Ducker defined GIS as a special case of information system where the database consists of observations on spatially distributed features, activities or events, which are definable in space as points, lines or areas. A GIS manipulates data about these points, lines or areas to retrieve data for adhoc queries and analysis.

The United State Geological Survey (USGS) defined - "*A GIS as a computer hardware and software system designed to collect, manage, analyze and display geographically (spatially) referenced data.*" This definition is fairly comprehensive and is suitable for agricultural application of GIS.

From the above discussion it may be derived that GIS is a system which utilizes spatial databases to provide answers to queries of geographical nature using set of tools for efficient information generation.

It is a database system in which most of the data is spatially indexed upon which a set of procedures operated in order to answer queries about spatial entities in the database.

It can be defined as an Information system that handles spatially referenced data using a powerful set of tools for collecting, storing, retrieving, transforming and displaying - thus answering queries about spatial entities.

Basic questions that can be investigated using GIS

Location	-	Where is it................?
Condition	-	How is it................?
Trend	-	What has changed ....?
Routing	-	What is the best way ....?
Modelling	-	What if..........?

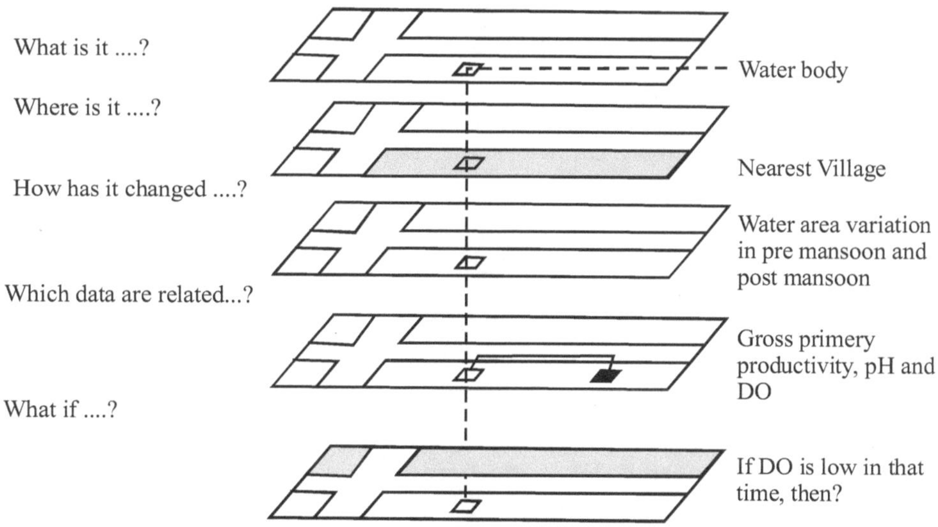

Required Functions of GIS

Functional components of GIS

Data Input	- Capture, Digitize, Edit, Transform.
Structure and store	- Data Model - Vector, Raster and Hybrid
Mode of storage	- RLE and Quadtree
Analysis	- Simple retrieval, Reclassify
Combination	- Overlay and Boolean operations
Neighborhood Network	- Buffer and Proximity
Display /Present	- Maps, Graphs, Tables, Reports
Model	- Query, Spatial (Three-dimensional)

Two kinds of databases

Specific characteristics of a location - Slope, soil type, rainfall etc.

Attribute - Description / non-locational data about the features Statistics or written text, tables etc.

Attributes at a location can be monitored through time - Temporal

Acquisition of data for GIS

GIS can not work without data. The more is the data more is the versatility and greater potential functionality of GIS. There are two type of data *viz.* primary data and secondary data. Primary data collection consists of acquiring various types of original data by a number of persons or direct methods and techniques. Secondary data represents available data or information, which may be availed in different formats.

Data acquisition and input

First Step - Developing database for GIS

- ❒ Acquire data
- ❒ Input to GIS

Data

- ❒ Spatial
- ❒ Non-spatial

e.g. Simple object in space - Location of a water well specifying any latitude/longitude (spatial). For certain application - wide range of additional information required e.g. depth, volume of water, quantity of water etc (attribute or non-spatial)

Mode of data Input

- Keyboard entry
- Manual locating device- digitization
- Automated devices - e.g. Scanning
- Conversion directly from other digital sources
- Digitization operation-Establish relation between digitiser and map Converts all digitizer x,y coordinates to map co-ordinates
- Modes : point mode-Stream mode

Scanners

- Video camera
- Electromechanical scanner

Data representation

Two ways of representing the information

Vector model : Mosaic of interconnecting lines and points which represent the location and boundaries of entities

Grid Model : Representing an entity by its presence or absence within a specific grid cell.

Collection of Primary Data

Data acquired by primary collection techniques could be in many forms like photographic numeric, digitally encoded etc.. Some of the important methods employed for primary data collection includes the following

Manual methods of data collection using no equipments

- Direct mapping or survey sketching
- Interviewing
- Questionnaires and counts
- Form filling

Methods of data collection using simple, non electronic equipments

- Photography
- Site surveying and measurement
- Meteorological data gathering
- Trawl surveys

Methods of data collection using simple electronic equipments

- Electronic read - out equipments
- Multi media devices
- Data loggers
- Global Positioning System(GPS)

Methods of data collection using Complex electronic equipments

- Remote Sensing Satellite
- Aerial Photographs

Collection of Secondary Data for GIS

All secondary data represents primary data which has been converted into more accessible and processed form. The sources of secondary data at libraries, Government office, bookshops, private and public organizations etc. Secondary data which are more relevant to GIS are maps table and digital data. Mapped information is the major source of data for GIS.

Maps

Various types of maps are available. But following types of maps are important for fisheries.

- Hydrographic map
- Relief map
- Topographic map
- Political and administrative map
- District planning maps

Satellite Imageries and Aerial photographs

- Visible and reflective infrared Remote Sensing
- Thermal infrared Remote Sensing
- Microwave Remote Sensing

Non Spatial Data Source

It refers to numeric and textual data which has been presented in some logical and cohesive format. Important tabular data source includes census office, fisheries office, fisheries research institutes, academic research institutes, Fish marketing office, harbours and port authorities, meteorological office other than state and central government agencies and International organizations.

Database Preparation

In any information system database is the backbone. Without proper database it is not possible to make any information system. Geographic Information System has two types of data. These two data are entirely different in nature. They are :

1. *Spatial data* (latitude/longitude for Geo-referencing, the features on a map, e.g. water body spread area, administrative districts, road), and
2. *Attribute data* (descriptive data about features, e.g. productivity of pond, population of districts, length of road etc.)

Analogue maps (soil map, land use map administrative district map, agro-ecological zone map, wetland map etc.) or aerial photographs and satellite imageries are the sources of spatial data for creating GIS. Data input or data capture is the process of encoding analogue data in the form of maps, imageries or photographs into computer readable digitized form and writing data into the GIS database.

Data Storage and Retrieval

A GIS does not store maps. It stores data organized into database. The location data of different features (coordinates, topology) are generated during the digitization process. The attribute data of location are created separately. The GIS must provide the link between the location and attribute data. The relational database model is most suitable to ensure such linkage between the location and attribute data. The relational database model is most suitable to ensure such linkage and the database query language can be used to retrieve the data. Relational database concepts are therefore central to organizing and managing database in GIS. The specific format of data storage varies with the GIS software.

Advantages with GIS

- Consistency
- Accessibility
- Precision

- Accuracy
- Completeness
- Flexibility
- Refinability

Application of Remote Sensing and GIS in Inland Fisheries

Mapping of wetlands/Water body

Precise inland water resource information is vital for estimation of fish yields and production. Due to global change ad anthropogenic pressure either these resource shrinking or changing their shapes. Regular estimation of water area is very much essential to estimates the yield and production of the water resources. There are various steps involved in mapping of wetlands starting from geo referencing, base map development, image processing to tabulation. Following are the basic steps involved in water body mapping.

Acquisition of remote sensing Image

For any remote sensing related study it is very important to select the sensor, satellite and passing date of the satellite. Sensors should be selected on the basis of objective of the study. For the mapping of water body, it is required to select sensor which works in the visible range (0.4-0.7 μm). There are various sensors of Indian and other country's satellites, which work in visible range i.e. Pan, LISS-3, WiFS TM and OCM. For mapping bigger water body, LISS-3 sensor and for smaller, PAN is better. Water spread area of wetland varies temporally. For getting proper picture of water spread area, the mapping should be done more then one season in a year. For fisheries point of view, water spread area of two seasons should be recorded i.e. pre monsoon and post monsoon. For mapping or any remote sensing study the image should be cloud free. The remote sensing digital data of NRSA is pre-processed that includes radiometric and geometric correction.

Geo referencing of remote sensing Image

Geo referencing is the process which creates and store control information that relates raster cells or vector information to a coordinate system and map projection. The geo referencing of image can be done by various techniques. In the simple geo referencing, the coordinate system cell size and coordinate for corner or central cell location is known from reference materials supplied by the source of raster objects. The precise geo referencing can be possible with the help of topographic sheet (i.e. Survey of India topographic sheets).

Interpretation

Traditionally, visual interpretation methods have been followed for extracting information on various natural resources. Tone, Pattern, Texture Size shape shadow location and association are the key, which are very useful for interpretation of the image characteristic. Other image processing techniques such as supervised and unsupervised classification are also very much useful for rapid analysis of images.

National Remote Sensing Agency (NRSA) had carried out the mapping of all the wetland of India bigger then 56 ha with LISS II RS data. Central Inland Fisheries Research Institute (CIFRI) also working for delineation of water body for all the states of India using LISSIII and PAN data under Ministry of Agriculture project. The water bodies bigger then 0.5 ha had been mapped using LISSIII RS data.

After water spread area, water quality is most important factor for fisheries management. And for the fish production, planktons (i.e. chlorophyll) along with others are considered as an essential factor for the fish production.

Water Quality

Remote sensing applications are limited to measuring those substances or conditions with influence and change optical and / or thermal characteristics of surface water properties/ water quality. Suspended sediments, chlorophylls, dissolved organic matter and temperature are water quality indicator that can change the spectral and thermal properties of surface water are most readily measured by remote sensing techniques. (Jerry C. Ritchie. *et al.*).

Suspended solids in water produce visible change in the surface of water and in the reflected solar radiation, such changes in the spectral signal from surface water, captured by the satellite. (Richie.1990). There are significant relationship between total suspended sediment and the reflectance of Landsat TM digital data (Fay F. Macebo). Reservoir trophic state and Carlson trophic state index were measured in Te-Chi Reservoir in central Taiwan with the help of Landsat TM (TM1~TM4). (Cheng, 1988).There are significant correlation between LISS III RS digital data and total nitrogen, gross primery production and respiration.(Gupta, R.A. *et al.* 2003).

Plankton (Chlorophyll)

Central to the aquatic food chain is phytoplankton - microscopic plants that photosynthesize chemicals in water. This process depends on the chlorophyll content of plankton. Lakes and water bodies depend mainly on their watershed for nutrients and other substances to sustain biological activities. While these nutrients and substance are required for a healthy aquatic environment, an excess of these input leads to nutrient enrichment and eutrophication of the lake. Eutrophication of a water body is usually quantified in terms of concentration of the chlorophyll contained in the algal/ plankton. Plankton contains a pigment that strongly absorbs red and blue light. As plankton concentration increase, there is a corresponding rise in spectral radiances, peaking in the green. Upwelling masses of water (usually associated with thermal convection) containing phytoplankton take on green hues in contrast to the deep water with few nutrients. Remote sensing can be used to measure the chlorophyll concentration and patterns in water bodies. While measuring chlorophyll by remote sensing technique is possible, studies have also shown that the broad wavelength spectral data available on current satellites do not permit discrimination between chlorophyll and suspended sediments when suspended sediment concentrations are high due to the dominance of the spectral signal from suspended sediment (Ritchie et al. 1994). Recent research shows the relationship between chlorophyll-a and the narrow band spectral details at the red edge of the visible spectrum (Gitelson et al. 1994). Data have shown a linear relationship between chlorophyll-a and the difference between the emergent energy in the primarily algal scattering range (700-705 nm) and the primarily chlorophyll-a absorption range (675-680 nm). Laboratory and field studies using hyperspectral data have been used to develop algorithm to estimate green and blue green algae (Dekker et al. 1995). Hyperspectral data now available from several satellite platforms, should allow better discrimination between pigments thus allowing the identification of broad algal groups.

Land use / Land cover change

Associated land use/ land cover of water body affects the water quality of the water area. Susumu ogawa and others studied water quality in Okutama lake in Japan .The purpose of the study was to examine relationship between land cover change and water quality in river. They found water quality changed by land cover change and other factors (Susumu ogawa). Mangrove forest (dense, open), change in forest area and deforestation of mangrove can be assessed by remote sensing (Y. A. Hussain).

The long term management plan can be prepared based on the remote sensing data supported by ancillary data for habitat protection tourist attraction and eco development for the livelihood of the people. (Arun Kumar *et al.*). "Nal Sarovar" of Gujarat, India was monitored with the help of remote sensing for waterfowl habitat management (Tatu) .

Management of Resource Wise and Location Wise Data

Assessment of fisheries resources and estimation of production are the most required research area. Time series data of various resources (like estuaries, reservoirs floodplain wetland) at various level (national, state and district) are being considered for this purpose. Creating this data base in vector GIS has got many advantage in understanding not only temporal trends in fisheries growth but also gives spatial patterns. Further tool GIS gives will be useful in mapping these trends.

Now more than ever, decision makers at all levels need an increasing amount of information to help them understand the possible outcomes of their decision by integrating the data from different sources and types. Due to integration various statistical tools its give more liberty to decision maker and researchers to analyze the data.

Development of Inventory

In the fisheries sector of India, to develop a proper inventory is quite challengeable and Herculean task. The Indian estuaries have tremendous potential but production is quite below the predicted one. Due to lack of catch and inventory data the assessment of production became very difficult. By creating various thematic map based on biomass, salinity, mesh/Gear, craft and historical catch data its became easy to analyze and predict the production. Land use pattern, change in land use pattern, sediment deposition all can be monitored and quantified with the application RS and GIS. Even though this technology were used by Carrie Smith and others for linkage between change in estuarine vegetation and salinity. They found GIS and aerial photographs are very much useful for mapping of estuarine vegetation.

GIS is useful in assembling of wide ranges of data sources. Readily available sources of GIS data such as Coastline political boundaries are frequently used data. Many time Census data also used in GIS for analysis. Land use / land cover is either obtained from previous compilation or interpret from satellite data. Catchments boundaries based on DEM or delineated from relief maps. GIS also use available data to generate Information like rainfall and runoff.

Water body information system of West Bengal that was developed at CIFRI also used the RS data (LISS III) and GIS and found for preparing inventory of water bodies RS data and GIS are very much useful. This information system able to give the spatial information (latitude longitude nearest village block district water area and shape) of water bodies which have the water area more then 10 ha.

Monitoring of Mangrove Forest

Mangrove forests form one of the primary coastal ecosystems in the tropical and subtropical region of the world. They are biologically diverse and have therefore traditionally been utilized for food resources, firewood, charcoal, timber and other minor products. However, mangrove ecosystems are very sensitive and fragile. Deatruction of mangrove forest and change in species composition directly shows the impact on biodiversity. In India and abroad various scientific groups are using GIS remote sensing and aerial photographs for estimation and monitoring of mangrove forest. Yosif ali Hussain and others utilized the RS data (Landsat MSS, SPOT- XS and Landsat TM) and GIS for monitoring of mangrove forest of Indonesia (the delta of Mahakam River). Their study reveals that it is possible to monitor mangrove deforestation with reasonable accuracy using optical and radar satellite images and GIS. In inda Dr Sridhar Vadlapudi applied GIS and RS data for Identification and quantification of changes in mangrove forest near Kakinada bay and Atanu Kumar Raha applied same techniques for mentoring change in Sunderban Mangrove forest. Both of them found that GIS and RS data are very much useful monitoring and quantification of change in mangrove forest area. The same type studies were also conducted in Thailand and Viatenam and they found GIS is very much useful in Coastal zone environment management.

Brackish Water Aquaculture Site Selection

Brackish water aquaculture has tremendous potential due to ever increasing demand of prawns and export earnings. Aquaculture development and planning require comprehensive data on land use and water resources. The land use/wetland information has been used for evaluating quality of the surrounding coastal waters, saltpans, aquaculture ponds for feasibility of brackish water aquaculture. The areas under mangroves/marsh are of vital/ critical concern and were avoided for site selection.

The spatial distribution of mud/ tidal flat areas, which are most suitable from substrate condition point of view give an idea about potential available for brackish water aquaculture. IRS II data was used for preparing coastal land use maps by Space application center on 1: 50000 scale. This map shows various features like forest, lakes/ponds, mudflats, river, stream, High water line, Low water line, Mangrooves, etc. These maps along with other engineering, biological, meteorological and socio-economic related parameters were integrated using GIS for evaluating site suitability (Gupta *et al.* 1995, 2001). Site evaluation is carried out to determine whether a site is suitable or not and to appreciate about requirements to make the site suitable. This procedure had ensured development for brackish water aquaculture with minimum damage to the ecology of the area.

Assessment of Benthic Community Condition in Estuaries

A few papers showed the potential usefulness of GIS in assessing benthic Communities. Several application of GIS were showed that the capability to utilized diverse data, provision for use in statistical analysis and mapping visualization of the data.

GIS was also used to model data and to generate information for statistical analyses. In order to select coastal conservation areas in Tazmania, Australia, Edgar *et al.* (2000) utilized GIS coverage along with field-collected data to conduct a complex set of statistical analyses. These analyses divided the estuaries into similar classes, and then determined the least impacted estuary within each class. To determine how sediment chemistry and water chemistry affected benthic communities, Preston and Shackelford (2002) created collocated data for statistical analyses. ArcView Spatial Analyst was used to generate (interpolate) biodiversity values at a given spatial location where sediment chemistry was also available. In a related study Preston (2002) used the cumulative probability function of the interpolated biodiversity data to designate areas that were un impacted and others that were highly impacted. Comparison of different variables within these areas was used to gain some understanding of why benthic biodiversity was impacted. Paul *et al.* (2002) used variables generated in GIS to determine if there was a relationship between land use, point source pollution and sediment contamination (the primary determinant of benthic impact in this area). Sponseller *et al.* (2001) used GIS to subdivide their data and then examined how freshwater benthic invertebrates were affected by different variables at different spatial scales. This has relevance to estuaries since we don't know how important watershed effects are or at what spatial scale they may be important.

ERS SAR (Synthetic Aperture Radar) were used for monitoring and inventorying shrimp farms in Sri Lanka. SAR provides both timeliness and flexibility of its independence from weather condition on the ground. By the SAR data dykes of shrimp forms can also detect in one direction. Similarly if SAR data were used of descending direction the side dykes can also be trace out and it improve the mapping of shrimp forms.(C. Travaglia et. al.). Due to timeliness of SAR it is very useful to monitor illegal shrimp farms which are responsible for disturbing coastal ecology.

Aquaculture Site Selection and Development

Ross and others were used GIS for site selection for salmonid cage aquaculture in small bay. This study used a sequential screening process for different criteria to identified potential aquaculture site for cage culture. Depth, current velocity salinity dissolved oxygen, temperature and other factors (such as Infrastructure, topography and exposure) were used as simple overlay layer. Independent maps were prepared for above parameters using base topographic map for depicting their spatial characteristics. Different scores were assigned for parameters according to suitability for culture like depth less then 6m as 0 and 1 for more then 6 m similarly. They also shows that GIS can use to address biophysical, economic and social factors for decision support system.

Web GIS in Fisheries

Web-based GIS is becoming more and more common as time passes. GIS for anyone, anywhere, anytime - the Web enables geographic information to be shared and distributed quickly and cost effectively. Web GIS is a integration of Word Wide Web(WWW) and GIS technologies. In Web GIS, GIS server prepares maps and launches on web sites. In this technology, there are two things WebGIS server and map browser. Map browser sends the request to GIS server and GIS server creates the map and sends to the client. It is very useful tool for dissemination of information (Map based information) in large no of clients.

Other Application

Marcia S. Meixler *et. al* (1995) was developed a GIS model for the prediction of fish diversity at the stream reach level in the Allegheny River basin of western New York. They created habitat classification of each stream reach using the landscape attributes: stream size, habitat quality, and water quality. Fish species were associated with habitat and the species diversity of each habitat class was thus identified. Tests of landscape attributes revealed that stream size predictions were significantly correlated with observed data, while water quality and habitat quality predictions were less correlated.

Preservation of fresh water habitat is of utmost concern to salmon managers. Habitat degradation, due to anthropogenic impacts, has significantly reduced the number and size of streams considered suitable for spawning. In Scotland, GIS has been used to identify which streams are accessible to adult salmon returning to spawn. The Dee Catchment management planning Geographical information System project is a multi-agency approach to integrate catches management.

Suitable Habitat for near shore marine species is being studied with GIS. GIS can also be used for selection of site for development of aquaculture pond in coastal area. Sanjeev Srivastava, *et al.*, uses GIS for identification of optimum and degraded habitat for golden Mahaseer (*Tor putitora,* Hamilton).

Summery

The application of remote sensing is not a new tool for mapping of wetland and other natural resources. Previously the spatial resolution was very low, so the mapping of small water bodies was not possible. But after lunching of new satellite (OrbView and IKONOS, etc.) with high spatial resolution the mapping of water body less than 0.05 ha is possible. There are few parameters studied with the remote sensing but in future some more parameters can be correlated. Hyperspectral data now available from several satellites, allow better discrimination between pigments, thus allowing the identification of broad algal groups. Such information allows us to move away from empirical approaches now being used and develop algorithms that will allow us to use the remote sensing for assessment, monitoring and management of fisheries resources.

A GIS is computer-based tool for geographical analysis of information. It is not simply a digitized map, nor does it hold maps. It holds a database of spatial data and attribute of descriptive information about features on a map, which can be used to create desired maps. GIS has a significant role to play in the decision making process in fisheries. At various levels i.e. field, regional, national and global levels. This is the one of the important tool of information technology, which is very useful to fishery scientists / managers to arrive at better decisions by integrating research results with intensive location specific information on various biophysical and socioeconomic variables. This technology allows to use and handle a wider range of spatial database such as soil, hydrology, crop, fisheries, weather etc. and integrate with socioeconomic variable and with the new technologies that are being developed from time to time. Potential areas for growth in GIS application to fisheries management include real time analysis, stock assessment, spatial demand and supply analysis, inventory management, environmental impact assessment biodiversity degradation assessment and disaster management.

Current information technology trends and future GIS tools will provide a range of function embedded in various components that can be tailored for specific use. GIS enable application by incorporating software components supporting geographic manipulation and visualization that are needed for specific application for development of capture fisheries, aqua culture development, wetlands management, reservoirs management etc.

REFERENCES

Anji Reddy, M. (1990). Text book of remote sensing and geographical information systems (2nd editions), B S Publications Hyderabad, A.P

Arun Kumar and R. K. Chingkhei, Application of GIS & R. S. in formulating a management plan in Keibul Lamjao National Park Manipur" http://www.gisdevelopment/application/envoirnment/overview/env0006.htm

Atanu Kumar Raha, "Monitoring changes in Sunderban Mangrove Forest using RS/GIS" http://www.gisdevelopment.net/application/environment/wetland/envwm004.htm

Cheng, Ke Sheng and Lei, Tsu-Chiang (2000) : Reservoir trophic state evalution using Landsat TM data. Web address http://www.gisdevelopment/aars/acrs/2000/ts2/water0006pf.htm

Cheng, Ke Sheng Lei, Tsu-Chiang and Yeh, Hui-Chung (1998): Reservoir water quality Monitoring using Landsat TM Images and Indicator Kirging. Web address http://www.gisdevelopment/aars/acrs/1998/ts2/ts20004pf.htm

Dr Sridhar Vadlapudi, "Identification and quantification of changes in mangrove forest using Remote Sensing - A case study near Kakinada Bay, Andhra Pradesh, India" http:// www.gisdevelopment.net/ pplication/environment/wetland/mi03075pf.htm

Dekker, A.G., T.J. Malthus, and H.J. Hoogenboom. (1995) The remote sensing of inland water quality, pp. 123-142. In: Danson, F.M. and S.E. Plummer (Eds.), *Advances in Remote Sensing*. Chichester: John Wiley and Sons.

Edgar G. J, Barrett NS, Graddon DJ, Last PR. 2000. The conservation significance of estuaries: a classification of Tanjanian estuaries using ecological, physical and demographic attributes as a case study. *Biological Conservation*. 92:383-397.

Gitelson, A., M. Mayo, Y.Z. Yacobi, A. Paroarov, and T. Berman. (1994) The use of high spectral resolution radiometer data for detection of low chlorophyll concentrations in Lake Kinneret. *Journal of Plankton Research* 16, pp. 993-1002.

Gupta, M. C., Krishnarajan, V. P. and Nayak, S. 2001. Brackish water aquaculture site selection in coastal track of Cannanore (Kerala) using remote sensing and GIS techniques. *Jour. Ind. Soc. of Remote Sensing*, 29 (1&2):79-83.

Gupta M.C. et.al 1995. Brackish water aquaculture site selection using techniques of Geographical Information System(GIS) Scientific Note, Space Application Centre, Ahmedabad RSAM/SAC/CMASS/SN/ 08.95 56p

Gupta, R. A., Mondal, S. K., Nath D., Kumar D., Das D. and Sahu, S.K., (2002) "*Assessment of Water quality parameter using IRS 1C data*", pp: 191-195 in national symposium on "Fisheries enhancement in inland water challenges ahead", CIFRI Barrackpore, April 27-28, 2002.

Kazuyo Hirose, et.al "A case study of mangrove forest monitoring in Cangio district, southern Vietnam" http://www.gisdevelopment.net/aars/acrs/2002/pos2/078.pdf

Preston B. L. 2002. Spatial patterns in benthic biodiversity of Chesapeake Bay, USA (1984-1999): Association with water quality and sediment toxicity. *Environmental Toxicology and Chemistry*. 21:151-161.

Kasturirangan K., Aravamudan R., Deekshatulu, B.L., Joseph George, and Chandrasekhar, M. G. (1996). Indian remote sensing satellite (IRS)- The beginning of new era. *Current Science*, 70 (7): pp 495-500.

Narayanan, A. K. (1992). Overall strategy adopted at the government level to encourage conservation of important wetlands in India. In : Wetlands of India (Eds. K.J.S. Chatrath), Ashish Publishing House, New Delhi.

Marcia S. Meixler and Mark Bain. 1995. Use of GIS to predict stream habitats and fish diversity. FISHERIES GIS SYMPOSIUM 2000 American Fisheries Society August 20 - 24, 2000 St. Louis, MO. http://www.ecu.edu/org/afs/st_louis/ absgis/r950568419-95.htm

Pham Viet Cuong, et.al. Application of Remote Sensing imagery for investigation in the Haiphong estuarine and costal zone. http://www.gisdevelopment.net/aars/acrs/1989/F/ocean001pf.htm

Preston BJ, Shackelford J. 2002. Multiple stressor effects on benthic biodiversity of Chesapeake Bay: Implications for ecological risk assessment. *Ecotoxicology*, 11: 85-99.

Ritchie, J.C., F.R. Schiebe, C.M. Cooper, and J.A. Harrington, Jr. (1994). Chlorophyll measurements in the presence of suspended sediment using broad band spectral sensors aboard satellites. *Journal of Freshwater Ecology* 9, pp: 197-206.

Ritchie, J.C., and C.M. Cooper, Remote sensing techniques for determining water quality: application of TMDLs www.geo.tu-freiberg.de/studenten/Baikal_2004/remote_water_quality.pdf

Ross, L.G., Mendoza, E.A., Beveridge, M.C.M., 1993. The application of geographical information systems to site selection for coastal aquaculture: an example based on salmonid cage culture. *Aquaculture*, 112: 165–178.

Sponseller R. A., 2001. Relationships between land use, spatial scale, and stream macroinvertebrate communities. *Freshwater Biology*, 46: 1409-1424.

Sanjeev K. Srivastava, Uttam K.Sarkar, and Alphis G. Ponniah (1999) Arrangement of habitat information on a GIS platform to identify optimum and degraded areas of golden mahseer (*Tor putitora,* Hamilton) habitat First International symposium on geographic information system (GIS) in fishery sciences. Tom Nishida, Chunk E. Hollingworth and Patricia J. Kailola (Eds.) Seattle Washingaton USA; 2-4 March 1999

Tatu, Ketan, Parihar, J. S., and Kimothi, M. M. (1999). Remote sensing for Wetland Monitoring and Waterflow Habitat Management. A.P.H. Publishing Corporation, New Delhi.

Travaglia, C.,Kapetsky, J. M., and Profeti G.(1999). Inventory and monitoring of shrimp farms in Sri Lanka by ERS-SAR Data". http://www.fao.org/sd/eidirect/Elan0012.htm

Yousif Ali Hussin, Mahfud M. Zuhair and Michael Weir (1999) : Monitoring Mangrove Forests using Remote Sensing and GIS.

Web address http://www.gisdevelopment/aars/acrs/1999/ps5/ps5126pf.htm

Chapter **16**

Computer Based Advanced Tools for Soil-Water Management in Aquaculture

Subhendu Adhikari & Bibhudatta Mishra

Introduction

The major potential areas for inland aquaculture have already been explored in India. Additional production can only be achieved through successful manipulation of available resources that influence the productivity of various aquaculture systems. This can be achieved through maintenance of adequate levels of nutrients in the pond environment. Pond management in fish culture is mainly concerned with fertilization requirements and strategies with better management practices (BMP) of pond soil and water quality. BMP could be achieved through adoption of advanced technology, computer based diognostic/ decision tools, software packages for quick, accurate, authetic data management.

Advanced Technology for Integrated soil-water fertility Management

Routine chemical analysis of soil & water involves a number of operations such as pipetting, diluting, mixing and transfer solution to measuring instruments. The trends towards replacement of manual operation to handle large numbers of samples with automated system for routine analysis of water and soil parameters is observed. Automation is not restricted to chemical procedure, data capture by computer or a large system via LAN has reduced

Applied Bioinformatics, Statistics and Economics in Fisheries Research: pp. 217-231

New India Publishing Agency, New Delhi (India)

the need for manual computation and increased the reliability of analytical data. Listed below are some of the modern instrumentations for handling soil water parameters.

a. Colorimeter or spectrophotometer

The colorimeter or spectrophotometer measures the transmission of light of a particular wavelength. Beer-Lambert Law gives the relationship between transmission and concentration as

$$1=10\ e^{-kcd}$$

where 1 & 10 are the intensities of light transmitted and the incident light respectively, c is the concentration of the species (moles per litre) and k is the molar extraction coefficient for the particular wavelength, d is path length of the light

There are spectrophotometer used for nutrients *viz.* ammonium, nitrate, nitrite & phosphate analysis & results depicted in a computer or plotter and viewed through monitor with pre set standard curve for quicker analysis.

b. Atomic absorption spectrophotometer (AAS)

In AAS methods of chemical analysis, a portion of the sample is converted into atomic vapour by a flame and irradiated by the light from the source, whose emission lines are those of the metal being analysed. The absorption of the light by the atoms of the element in the vaporized sample is related to the concentration of the desired metal and is measured at a specific wavelength. The concentration of the metal in the solution is measured by comparison with absorbance measurements of standard of known composition. All AAS has a light source that emits resonance line radiation, a sampling flame (flame), monochromator that is used to isolate the absorbing resonance line, detector to measures the amount of absorption, electronic amplification of absorption signal and a read out system (strip chart recorder, digital display or a printer). Heavy metal guidelines for the protection of freshwater aquatic life has been given below:

Metal	Unit	Value
Arsenic (III)	Mg/L	<0.05
Cadmium		
Alkalinity> 100mg/L	Mg/L	0.006
Alkalinity < 100 mg/L	Mg/L	0.03
Chromium total	Mg/L	0.03

Contd...

Contd...

Cobalt	µg/L	0.9
Copper		
Alkalinity > 100 mg/L	Mg/L	0.006
Alkalinity < 100 mg/L	Mg/L	0.03
Iron (Fe) total	Mg/L	0.1
Iron (Fe) II	Mg/L	0.1
Iron (Fe) III	Mg/L	0.5
Lead	Mg/L	< 0.02
Manganese	Mg/L	< 0.01
Nickel	µg/L	< 50
Zinc	Mg/L	0.005

c. Inductively Coupled Argon Plasma atomic emission spectrophotometer (ICAP-AES)

Atomic emission spectroscopy, employing inductively coupled plasma (ICP) as the source of energy has most useful tool for simultaneous determination of major, minor & trace elements in all types of soil, water with higher sensitivity and emission stability. Two types of argon plasma are in use, one is energized by radio frequency a-c coupled to the gas through electromagnetic induction (ICP) where as the other is d-c excitation (DCP).

d. Gas Chromatography

Gas chromatography-mass spectrometry (GC-MS) used in determination of multiresidues analysis of pesticides in soil research. Field portable GC can be used for analysis of volatile organic compounds (VOCs) in water, soil, soil gas and ambient air. The C&N contents of soil could also be analyzed by GC.

e. Remote Sensing in water-soil Research

There are integrating remote sensing technology for the determination of water content of bare, cultivated mineral and organic soils. The development of models and technique for remote sensing technology in management and monitoring of aquaculture water and soil resources for better production is a newer approach.

Computer Based Diagnostic/ Decision Tools

There are diagnostic tools for predicting nutrient availability and water quality in aquaculture ponds. Implementation of precision farming as a strategy in aquaculture lead to a significant reduction in water requirement & environmental impact in fish culture practices. Following are some of the computer based diagnostic/decision tools usually used in aquaculture/ agriculture research.

1. QUEFTS: Quantitative Evaluation of the Fertility of Tropical Soil

   This model is a simple static model permits to calculate the yield of crop based on a number of soil parameters that are generally available. Although this model tested in agri-crop, the same may be applicable to aquaculture.

2. DESSAT- Decision support for agro-technology transfer
   http://www.icasa.net/dssat/index.html

3. APSIM-Agricultural Production system simulator
   http://www.apsru.gov.au/products/apsim.htm

4. ORD-Organic resource data base
   http://www.wye.ac.uk/Biosciences/soil

5. NUTMON-Monitoring nutrient flow and economic performance in tropic

   In this tool, nutrient flow could quantify, calculate and visualize and optimal fertilizer doses can be calculated.

   http://www.nutmon.org/index.htm

6. NuMass-Nutrient Management Support System
   http://ntdss.soil.ncsu.edu/NuMass

7. BASINS is a multi-purpose environmental analysis system that integrates a geographical information system (GIS), national watershed data, and state-of-the-art environmental assessment and modelling tools into one convenient package. http://www.epa.gov/waterscience/basins/

8. AQUATOX is a freshwater ecosystem simulation model. It predicts the fate of various pollutants, such as nutrients and organic toxicants, and their effects on the ecosystem, including fish, invertebrates, and aquatic plants. AQUATOX is a valuable tool for ecologists, water quality modelers, and anyone involved in performing ecological risk assessments for aquatic ecosystems. http://www.epa.gov/waterscience/models/aquatox/

9. PLOAD is a simple pollutant loading model. PLOAD estimates nonpoint sources of pollution on an annual average basis, for any user-specified pollutant, using either the export coefficient or simple method approach.

10. SWAT is a physically based watershed scale mode. It was developed to predict the impacts of land management practices on water, sediment and agricultural chemical yields in large complex watersheds with varying soil, land uses and management conditions over long periods of time.

11. WASP is a generalized framework for modeling water quality and contaminant fate and transport in surface waters. It is used primarily to investigate the water quality response to management actions, primarily point and nonpoint source load reduction. WASP can model biochemical oxygen demand and dissolved oxygen dynamics, nutrients and eutrophication, bacterial contamination, and organic chemical and heavy metal contamination. It is designed to permit easy substitution of user-written routines into the program structure. WASP can be applied in one, two, or three dimensions and it allows for graphical presentation of results. WASP is not part of the BASINS software package; however classes are occasionally offered at EPA Regional Offices and information will be provided on their availability. http://www.epa.gov/region4/water/tmdl/tools/wasp.htm

12. ARMS (Aquatic Realtime Management System) is a decision support system to aid managers and operators of surface water bodies (reservoirs, lakes, rivers, estuaries, coastal oceans). Its underlying philosophy is to provide an automated software system that requires minimal maintenance to monitor and forecast the conditions of surface water resources, and to notifyrelevant staff of current conditions on a regular basis.

    http://www.cwr.uwa.edu.au/services/models.php

13. Decision Support Systems in Hydrology
    http://in.groups.yahoo.com/group/dsshydrology/.

14. FISH FORUM: A web-based forum for questions and answers about fish
    http://www.fishbase.org/search.html.

15. SALINITY-L
    http://listserv.unl.edu/archives/salinity-l.html.

16. *WATERCOM:* provides an opportunity for exchanging commercial announcements associated with water resources engineering, hydrology, hydraulics and other water-related sciences.

    http://www.watercom.ca

Software Packages

1. Wateruse forcasting system-IWR-MAIN
2. Pond dynamic software- POND Version 3.0 focuses on providing a detailed view of pond aquaculture at both the individual pond as well as at the facility-level. Pond analysis is accomplished primarily by the use of simulation models combined with an economics package that can be used to generate enterprise budgets for a pond facility.
3. Land Data Management System (Land DMS)
4. Farm Field Information on-line Measurement Instrument

   This instrument, which mainly consists of a single-chip microcomputer (SCM), a keyboard, a LCD display, several sensors and a GPS receiver, was designed to synchronously integrate the signals from both sensors and GPS receiver
5. Integrated Crop and Soil Management Software
6. Aquaculture Decision Support Software
7. *DESTA*: a software tool for selecting sites for new aquaculture facilities
8. *SEDNET*-Sediment network modelling software

Bibliographic Research Database Related to Water Soil Chemistry

Agricola
Agris International
Aquatic Sciences & Fisheries Abstract (ASFA)
Cab Direct
Current Research Information System (CRIS)
Environment Abstract
Water Resources Abstract

Soil-water Management

Water Quality Management : Water is the most essential commodity in aquaculture production. For sustainable management of available surface and ground water there should be effective utilization of water, equity in water sharing and efficiency in water delivery.

pH

pH is a measure of the hydrogen ion concentration in water and indicates how much acidic or basic the water is. Water pH affects metabolism and physiological process of fish. pH also exerts considerable influence on toxicity of ammonia and hydrogen sulphide as well as solubility of nutrients and thereby water fertility. Guidelines for pH value in fish health management are given in Table 1 below:

Table 1. Effects of different pH levels on fish health.

pH level	Effect on fish health
4	Acid death point
4–6	Slow growth
6–9	Best for growth
9–11	Slow growth, lethal to fish over long period of time
11	Alkaline death point

Measures for Rectifying Alkaline and Acidic Water Bodies

Alkaline Waters

- Rapid fluctuations in pH caused by excessive phytoplankton populations may be rectified by ensuring good water management. Water body should have an alkalinity of more than 20 mg/l as $CaCO_3$.
- Application of acid forming fertilizers.

Acidic Waters

- Calcium carbonate ($CaCO_3$), calcium hydroxide ($Ca(OH)_2$), calcium oxide (CaO) or dolomite is used to rectify the acidic water bodies depending upon the pH.
- Salt water like sea water may be flushed through water bodies of coastal farms to neutralize acidity.

Dissolved Oxygen

The optimum dissolved oxygen (DO) content of pond waters should be in the range of 5 mg/l to saturation level for good growth of fish.

Table 2 : Guidelines for dissolved oxygen level in fish health management

Concentration	Effects
5.0 mg/l	optimum for normal growth and reproduction in tropical waters
1.0–5.0 mg/l	may have sub-lethal effects on growth, feed conversion and tolerance to disease
0.3–0.8 mg/l	lethal to many species if sustained for a long period

Oxygen depletion in water is rectified by the following aeration methods

- *Manual.* In this method, water surface is splashed with bamboo sticks. This helps in dissolving atmospheric oxygen in water.
- *Mechanical.* A diesel water pump is operated through method. Water is pumped out and simultaneously sprayed in again into the water body. This helps in dissolution of atmospheric oxygen.
- *Aerators.* Aerators are mechanical floating devices. Their rotating blades churn the water helping in dissolution of atmospheric oxygen in water. Depending upon the concentration of oxygen in waters, the number and placement of such aerators are determined.

Other steps taken to control the oxygen level are

- Care should be taken to feed fish in the afternoon or evening in heavily stocked pond systems as oxygen requirement in fish after feeding increases and dissolved oxygen is minimum in pond during early morning.
- Organic manure application in a water area should be done carefully as organic material consumes oxygen during decomposition. Therefore, the quality of manure to be applied without the risk of oxygen depletion can be calculated taking into consideration the availability of dissolved oxygen during the 24 hr period.
- During collapse of phytoplankton bloom, decomposition occurs and in the process oxygen requirements of micro-organisms increase. Thus, special care has to be taken during this time.
- Special care has to be taken as more oxygen are required with increasing temperature.

Ammonia

The total ammonia concentration in water comprises two forms, namely:

NH_3 = unionized ammonia (Free ammonia) and $NH4^+$ = Ionized ammonia

They maintain equilibrium as per the equation:

$$NH_3 + H_2O \ ? \ NH_4^+ + OH^-$$

The un-ionized ammonia fraction is more toxic to fish and the amount of the total ammonia in this form depends on the pH and temperature of the water. As a general rule, the higher the pH and temperature, the higher is the percentage of the total ammonia present in the toxic unionized form.

Table 3 : Guidelines for unionized ammonia level in fish health management

Concentration	Effects
0.02–0.05 mg/l	safe concentration for many tropical fish species
0.05–0.4 mg/l	sub–lethal effects depending on the species
0.4–2.5 mg/l	lethal to many fish species.

There are a number of measures to maintain safe ammonia concentration in pond water. Normally at high dissolved oxygen and high carbon dioxide concentration, the toxicity of ammonia to fish is reduced.

Some recommended measures to reduce the effects of ammonia are

- Aeration will increase the dissolved oxygen concentration and decrease the increasing pH thereby reducing toxicity.
- Healthy phytoplankton populations remove ammonia from water. Care should be taken while using fresh manure with high ammonia content. The manure should be dried to allow ammonia gas to escape before application to the pond.
- Biological filters may be used to treat water for converting ammonia to nitrite and then to harmless nitrate through nitrification process.
- A high quality feed that contains no more nitrogen (crude protein) and phosphorus than actually needed by fish should be used in ponds and also over–feeding should be avoided.
- Excessive liming should be avoided as it raises pH and high pH favors ammonia toxicity to aquatic animals.
- Water exchange can reduce ammonia concentrations in fish and prawn ponds. From both economic and environmental perspectives, water exchange should only be used when necessary.
- Formalin can be used to remove ammonia from fishponds.

Nitrite

Nitrite is an intermediate product in the biological oxidation of ammonia to nitrate, a process called nitrification. In most natural water bodies and in well maintained ponds, nitrite concentration is low. In water bodies with high organic pollution and low oxygen concentration, nitrite concentration may increase.

Table 4 : Guidelines for nitrite-N level in fish health management

Concentration	Effects
0.02–1.0 mg/l	sub-lethal effects depending on the species
1.0-10.0mg/l	lethal to many fish species

Measures to maintain safe nitrite level in water are

- Correct stocking, feeding and fertilization practices should be maintained. The ponds should be kept well oxygenated.
- Bio filtration is done through special filters by which biological conversion of nitrite to harmless nitrate occur.

Temperature

Temperature sets the pace of metabolism by controlling molecular dynamics (diffusibility, solubility, fluidity) and biochemical reaction rates. Under favorable conditions, the optimum temperature range for many coldwater and warm water fishes are 14–18°C and 24–30°C respectively. Water temperatures can be adjusted to optimum levels in controlled system such as hatcheries. It is difficult to adjust water temperature in large water bodies. Operation of aerator during calm and warm afternoon helps to break thermal stratification by mixing warm surface water with cool subsurface water. Planting of trees on pond banks to give shade will reduce stratification but at the same time, reduces the beneficial effects of wind mixing and restricts solar energy for photosynthesis.

Turbidity

Turbidity in cultural water is the resultant effect of several factors like suspended soil particles, planktonic organisms, humus substances produced through decomposition of organic matter, etc. Turbidity is measured by secchi disk visibility. Optimum secchi disk visibility in fish ponds is considered to be 40–60 cm. Turbidity resulting from plankton is generally desirable.

Guidelines for suspended soil particles value in fish health management are

Up to 10 000 mg/l Freshwater carps, *Tilapia* sp. and catfishes are tolerant to this level, however, the effect will depend upon the nature of the suspended particles.

Pond water turbid with suspended soil particles can be controlled by application of 500–1000kg/ha organic manure, 250–500 kg/ha gypsum or 25–50 kg/ha alum.

Hydrogen Sulphide

Freshwater fish ponds should be free from hydrogen sulphide (H_2S). Hydrogen sulphide is produced by chemical reduction of organic matter that accumulates and forms a thick layer of organic deposit at the bottom. Unionised hydrogen sulphide is toxic to fish, but the ions resulting from its dissociation are not very toxic. Guidelines for hydrogen sulphide value in fish health management are:

Table 5 : Guidelines for hydrogen sulphide level in fish health management

Concentration	Effects
0.01–0.5 mg/l	lethal to fish and any detectable concentration of hydrogen sulphide in water creates stress to fish
0.01–0.5 mg/l	prawn lose their equilibrium and create sub-lethal stress
0.1–0.2 mg/l	prawn die instantly.

Measures to rectify increase in hydrogen sulphide levels include

- frequent water exchange to prevent building up of hydrogen sulphide in the water body.
- when pH of water is increased by liming, the toxicity of hydrogen sulphide decreases

Alkalinity

Alkalinity refers to the concentration of bases in water and the capacity of water to accept acidity, i.e. the debuffering capacity. In most waters, bicarbonates and carbonates are the predominant bases.

Guidelines for alkalinity in fish health management are

- 300 mg/l : create stress to fish.
- 60–300 mg/l : ideal for fish.
- <60 mg/l : create stress to fish.

Low alkalinity can be rectified by treatment with lime.

Total Hardness

Contents of alkali earth metals, mainly calcium and magnesium constitute the total hardness of a water body. Guidelines for hardness value in fish health management are given below:

- 50 mg/l : satisfactory for pond productivity and helps protect fish against harmful effects of pH fluctuations and metal ions.
- <50 mg/l : creates stress to fish.

Measures for rectification of low hardness

- Ponds with low hardness can be treated with lime for rectification.

Carbon Dioxide

Carbon dioxide is present in the atmosphere in very small quantity. For this reason, in spite of its high solubility in water, its concentration in most water bodies is low. It occurs in water in three closely related forms, namely: (a) free carbon dioxide (b) bicarbonate ion (HCO_3^-) and (c) carbonate ion (CO_3^{2-}). The amount of each form presents, depends on the pH of water. For example, in neutral or acidic waters high concentrations of free carbon dioxide, i.e. the toxic form is frequently found.

Guidelines for carbon dioxide value in fish health management are

- 12–50 mg/l : sub–lethal effects include respiratory stress and development of kidney stones (nephrocalcinosis) in some species.
- 50–60 mg/l : lethal to many fish species with prolonged exposure.

Measures for controlling high carbon dioxide concentration include

- repeated aeration of water.
- increasing the pH of water by hydrated lime can control high carbon-dioxide concentration. Experiments have shown that 1.0 mg/l of hydrated lime can remove 1.68 mg/l of free CO_2.
- phytoplankton population and the organic loading in a water body should be regulated by correct stocking, feeding and fertilization.

Bottom Soil Management

Role of bottom soil in determining productivity of a pond is well documented. Production of various primary food organisms depends largely on the availability of different nutrients. Dynamics of availability of most of these nutrients, in turn, is determined by the condition prevailing in the bottom soil. Considering this significance, bottom soil is designated as the chemical laboratory of a pond. However, suitable soil quality problem is common in aquaculture ponds, and therefore, many methods are used for the purpose of improving pond soils.

Soil Texture

Many important physico-chemical properties affecting the fertility of fish ponds are influenced to a great extent by the relative proportion of the different size fraction of the soil. An ideal pond should not be too sandy to allow leaching of the nutrients or should not be too clayey to keep all the nutrients adsorbed in it. When the pond is constructed on sandy soils, then heavy doses of organic manure application is essential to control seepage loss of water. In general, the dose of raw or composted farmyard manure varies from 10 000–15 000 kg/ha/yr.

Soil Acidity

Soil may be acidic, alkaline or neutral. The ideal range of pH for fishpond soil is 6–8. Water passing over acid soil tends to be acidic with low alkalinity and hardness. High concentration of metal ions particularly aluminium and iron also may be present. Acidic ponds do not respond well to fertilization. Liming is the only way to improve water quality with acid soils. Recommended rate of application of lime at different soil pH has been given below:

Soil pH	**Dose of Lime ($CaCO_3$ kg/ha)**
4.0-4.5	1,000
4.5-5.5	700
5.5-6.5	500
6.5-7.6	200

Acid Sulphate Soils

Acid sulphate soils from mine spoils and coastal mangroves tide swamps contain high levels of pyrite (FeS_2, 1–6 percent). As long as sediments containing pyrite are submerged and anaerobic, they remain reduced and there is little change. However, as they are drained and exposed to the air, oxidation results and sulphuric acid is formed. Sulphuric acid reduces the pH of the water when pond is filled. In ponds, problems with acid sulphate soils usually originate in pond dikes. Pond bottoms are usually flooded and anaerobic, so sulphuric acid does not form. However, dikes dry and sulphuric acid formed during dry period enters pond with run-off water after rains. Acidity on dikes can be controlled by liming (0.5–1.0 kg/ m^2) and establishing good cover with an acid resistant grass species. A procedure for rapid reclamation of ponds with acid sulphate involves drying and filling of the soil to oxidize pyrite, filling the pond with water and holding till water pH drops to < 4 and then draining the pond, repeating the procedure until the pH stabilizes at a pH >5, followed by liming the pond with 500 kg of calcium carbonate/ha.

Bottom Soil Oxidation

Dissolved oxygen cannot move rapidly into water saturated soil, and pond soil becomes anaerobic below a depth of few mm. Aeration and water circulation are beneficial in improving bottom soil oxygenation, but the surface layer of soil may still become anaerobic in intensive fish culture ponds. When the redox potential is low at the soil surface (anaerobic conditions), hydrogen sulphide and other toxic microbial metabolites diffuse into the pond water. Sodium nitrate can serve as a source of oxygen for microbes in poorly oxygenated environment; the redox potential will not drop to a level enough for the formation of hydrogen sulphide and other toxic metabolites.

Some other Environmental Parameters

Nutrient Removal

It is possible to precipitate phosphorous from pond water by applying sources of iron, aluminium or calcium ions. These ions precipitate phosphate as insoluble iron, aluminium or calcium phosphates. Alum (aluminium sulphate) and ferric chloride are commercially available sources of aluminium and iron respectively. Alum is cheaper and more widely available than ferric chloride. Gypsum (calcium sulphate) is a good source of calcium, because it is more soluble than liming materials. Treatment rates of 20–30 mg/l alum and 100–200 mg/l gypsum have lower phosphorus concentration in pond waters. Alum is acidic and more suitable for use in waters with 500 mg/l of total alkalinity and above. Gypsum is better for use in low alkalinity waters.

Phytoplankton Removal

Algaecides are used to reduce the abundance of phytoplankton in intensive fish culture ponds. Copper sulphate pentahydrate ($CuSO_4.5H_2O$) is recommended for reducing phytoplankton abundance and the abundance of blue-green algae in particular. The usual recommendation is to apply a dose of copper sulphate equal to 1/100 of the total alkalinity. However, use of $CuSO_4$ in plankton removal is discouraged as excessive use of the chemical could cause human health and environmental problems. The best approach to phytoplankton control is to regulate nutrient inputs by moderate stocking and feeding rates, but it may be feasible to use alum or gypsum to precipitate excessive concentrations of phosphorus.

Sanitation of Ponds (Chlorination)

Hypochlorus acid and hypochlorite (free chlorine residuals) are the chlorine products which are used for disinfecting pond water. However, chlorination of water containing fish or prawn is both dangerous and unbeneficial. It is possible to sterilize bottoms of empty ponds and water in newly filled but unstocked ponds by applying chlorine products. When this is done, enough chlorine should be applied to overcome the chlorine demand and provide 1 mg/l or more of free chlorine residual. The residuals will detoxify naturally in a few days so that ponds can be stocked safety.

Water Exchange

There are reasons to exchange water in specific instances, such as to reduce salinity, to flush out excessive nutrients and plankton or to reduce ammonia concentrations. However, daily water exchange usually does not improve water quality in ponds, and pumping costs are a liability. Ponds are highly efficient in assimilating carbon, nitrogen and phosphorus inputs, which are not converted to fish or prawn flesh, but if water exchange is great, these substances are discharged from ponds before they can be assimilated. Thus, the pollution potential of aquaculture ponds increases as a function of increasing water exchange. From both economic and environmental perspectives, water exchange should only be used when necessary.

Conclusion

Now a days various softwares are available regarding pond soil and water quality management. Therefore, an understanding of chemical and biological conditions of pond soil and water through regular in-situ and modern techniques monitoring systems and adoption of efficient and careful management practices will lead to enhanced production of fish food organisms and thereby increase the growth and survival of fish.

Chapter **17**

Non-Invasive Methods of Fish Freshness Determination by Electronic Olfaction

Nabarun Bhattacharya, Bipin Tudu,

Rajib Bandopadhyaya & Devdulal Ghosh

Introduction

Of the five human sensory perceptions, the feelings of sight, hearing and touch have been simulated by artificial sensors quite successfully. Electronic eyes (digital cameras), electronic ears (microphones) and touch sensitive electronic devices are quite common nowadays. In comparison to these devices, the perception of smell and taste are relatively unexplored. Sensing of smell and taste is full of uncertainty and the prevalent technology is at its infancy. Electronic noses for smell and electronic tongues for taste or a combination of both have vast prospects in the field of standardization of quality of food and agro products. In India, proper use of these instruments can revolutionize the quality of the food and agro products, thereby increase the export worthiness of these products. In this article, we have tried to discuss applicability various electronic olfaction technologies like electronic nose and electronic tongue for fish freshness determination. A report on an experiment with electronic nose on fish samples has been presented.

About Fish Odour

Fish odour is due to a complex mixture of volatiles and it changes with storage duration. Human nose can detect such changes in volatile emission

Applied Bioinformatics, Statistics and Economics in Fisheries Research: pp. 233-245

New India Publishing Agency, New Delhi (India)

pattern under storage to some extent. But human evaluation suffers from difficulties like subjectivity, lack of reliability and repeatability, person to person differences etc. Electronic nose based measurements may be gainfully employed to objectively monitor smell of fish samples under storage. This paper gives a study on freshness determination of inland water fish using electronic nose. Also feasibility study of electronic tongue has been discussed in short.

As already stated, fish odour comprises of a very complex mixture of volatiles and their percentage presence varies with length of storage. Fresh fish odour is mainly contributed by:

- Long Chain of Alcohol and Carbonyls
- Bromophenols
- N-Cyclic Compounds

Spoilage of fish is caused by microbial formation during storage. Microbial growths nucleate generation of various chemicals like:

- Short Chain Alcohol and Carbonyls
- Amines
- Sulphur Compounds
- Aromatics
- N-Cyclic Compounds
- Few Acids

There are a number of instrumental methods by which fish quality may be analyzed. Some of such noteworthy methods are:

- Analytical Chemistry Methods (GC, HPLC)
- Firmness and Texture Inspection
- Impedance Measurement Method
- Platinum Electrodes to Detect Hydrogen Peroxide
- Enzyme Reactor Based Systems
- Amperometric Bio-Sensors for Determination of Amines

But such instrumental techniques suffer from a number of problems like:

- Time consuming analysis
- High cost
- Lack of portability
- Requirements of experts for handling
- Elaborate Sample Preparation

So there is a need for fast, non-invasive and real-time methodology for fish quality estimation and electronic nose based methodologies, as studied in this paper, may go a long way in this direction. Our limited survey on the applicability of electronic olfaction technology for fish quality determination reveals that this technology is although at its infancy, but intense research is going on all over the globe in this exciting area and considerable progress has already been made particularly on the study related to the freshness of sea water fish [1], [2].

Physiology of Human Olfaction

Odours are sensations that occur when compounds (called odorants) stimulate receptors located in the olfactory epithelium at the roof of nasal cavity. Odorants are hydrophobic, volatile compounds with a molecular weight of less than 300 daltons. Human can recognize and distinguish upto 10000 different substances [3] on the basis of their odour quality. Odorant receptors (OR) in the nasal cavity detect and discriminate among these thousands of diverse chemical traces. The olfactory region of each of the two nasal passages in humans is a small area of about 2.5 square centimeters containing in total approximately 50 million primary sensory receptor cells.

The olfactory region consists of cilia projecting down out of the olfactory epithelium into a layer of mucous which is about 60 microns thick. This mucous layer is a lipid-rich secretion that bathes the surface of the receptors at the epithelium surface. The mucous layer is produced by the Bowman's glands which reside in the olfactory epithelium. The mucous lipids assist in transporting the odorant molecules as only volatile materials that are soluble in the mucous can interact with the olfactory receptors and produce the signals that our brain interprets as odor. Each olfactory receptor neuron has 8-20 cilia that are whip-like extensions 30-200 microns in length. The olfactory cilia are the sites where molecular reception with the odorant occurs and sensory transduction (i.e., transmission) starts.

Researchers in olfactory study have shown that every neuron in the olfactory bulbs participates in the generation of olfactory perception. In other words, the salient information about the stimulus is carried in some distinctive pattern of bulbwide activity and not in a subset of specific neurons. In the absence of a stimulus, the pattern of activity across the olfactory bulb has "chaotic" characteristics. However, upon receiving a stimulus the chaotic behaviour rapidly assumes a cross-bulbar pattern. This pattern need not be the same each time for the same odour, but may change its characteristics depending upon the previous stimulus. This system allows for odorant conditioning, and also explains how we can be sensitive to odours we have never previously experienced [4].

Technology of Electronic Nose

Given below a block diagram of electronic nose (Fig.1). From the block diagram, it can be seen that an electronic nose system primarily consists of four functional blocks, viz., Odour Handling and Delivery System, Sensors and interface electronics, Signal processing and Intelligent pattern analysis and recognition. The array of sensors is exposed to volatile odour vapour through suitable odour handling and delivery system which ensures constant exposure rate to the sensors. The response signals of sensor array are conditioned and processed through suitable circuitry and fed to the intelligent pattern recognition engine for classification, analysis and declaration [5, 6].

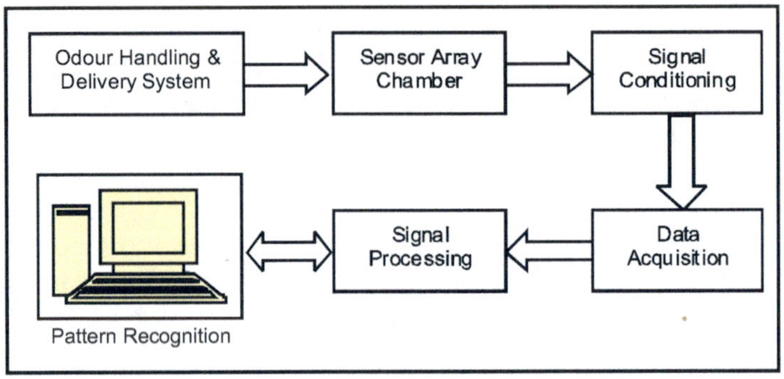

Fig. 1. Block Diagram of Electronic Nose

Odour Delivery using Headspace Sampling Technique: Fig. 2 below shows a schematic of conventional headspace sampling method. The headspace is the

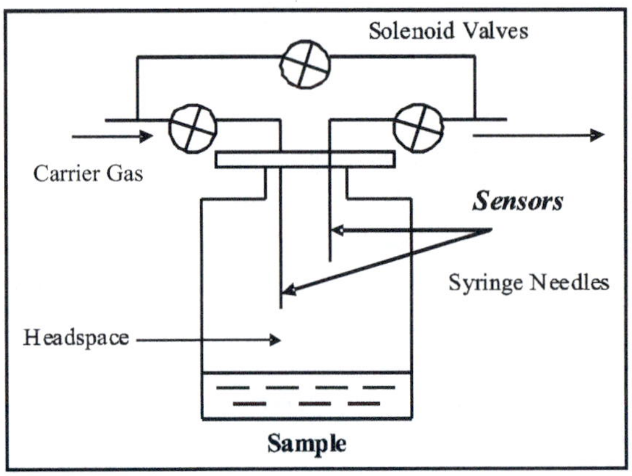

Fig. 2. Schematic of Conventional Headspace Sampling Method

space just above the liquid sample in a bottle. The carrier gas such as dry air is supplied at the inlet and the vapour evaporated at the liquid surface carried by the carrier gas is supplied to the sensors. Solenoid valves may be alternately switched so that sensors are exposed to pure carrier gas and liquid vapour alternately and difference in sensor output may be recorded.

Sensors: The most complicated parts of electronic olfaction process are odour capture and sensor technology to be deployed for such capturing. Any sensor that responds reversibly to a chemical in gas or vapour phase has the potential to be developed in an electronic nose format. Early electronic noses used either polished wires in contact with porous rods saturated with different electrolytes or thermistors coated with materials such as gelatin, fats or polymers. In the 1980's, advances were made with the appearance of chemically sensitive sensors and developments in electronics and computing [7, 8]. Table 1 shows the sensors and their measuring principles that can be used in electronic nose system instrument.

Table 1: Sensors used in Electronic Nose

Measuring Principle	Sensor Type	Common Materials Used
Chemo resistors	Metal oxides	Doped tin di-oxide, zinc oxide, tungsten trioxide
	Conductive polymers	Polypyrrole, polyaniline, polythiophene
	Conductive oligomers	Short chain length conductive polymers
	Conductive material loaded polymers	Carbon particles in non-conductive polymers.
	Self-organising films	Discotic liquid crystals
	Macrocyclic compounds	Phthalocyanines
Mass sensitive devices	Surface and bulk acoustic devices	Poly (siloxanes), charged polyelectrolyte, fluoropolymers, polyalkanes, hydrophilic polymers, calixarenes and other cavitands, chiral polymers, metal phosphonates, conducting polymers, lipids, self assembled monolayers.
Field effect devices	Metal insulator semiconductors or metal oxide semiconductors	Catalytic metals such as palladium, platinum and iridium, as well as other metals.

Contd...

Table 1. Contd...

Electrochemical devices	Oxidation or reduction catalysts	Precious metal and carbon with concentrated liquid electrolyte and counter electrode
Pellistors	Catalytic combustion devices	Platinum, palladium or alloys of the two.
Fibre optic systems	Polymer bids on fibre optic bundles	Fluorescent, solvochromic dyes to detect solvation of vapours in the polymer
Intrinsic molecular parameters	Mass spectrometer, ion mobility or pulsed spectroscopy instrumentation	

Out of all the above types of sensors, conducting polymer, metal oxide semiconductor and bulk acoustic devices are most commonly used in commercial electronic noses. Please see Fig. 3 for some commonly available electronic nose sensors [9].

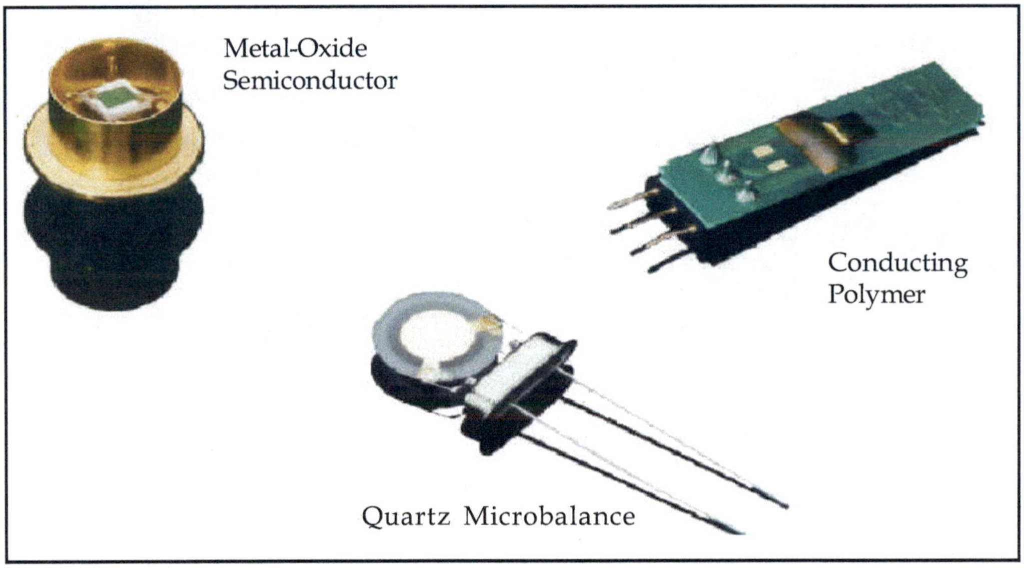

Fig. 3: Commonly Available Electronic Nose Sensors

Pattern Recognition Methods: A number of supervised and unsupervised pattern recognition techniques may be applied for electronic nose data classification [6, 9] and clustering. Few of such techniques are listed in the Table 2.

Table 2: Various Pattern Recognition Methods used in Electronic Nose

Statistical methods	Quantitative	Supervised	MLR, PLS
	Pattern Analysis	Unsupervised	PCA, CA
		Supervised	DFA, PCR
Biologically inspired methods	ANN	Unsupervised	SOM
		Supervised	MLP, PNN, RBF, LVQ
	Fuzzy Methods	Supervised	FIS, FNN, FCM
		Self-supervised	ART, Fuzzy ARTMAP
	Others	Self-supervised	GA
		Supervised	NFS, Wavelets

Electronic Nose Developed by C-DAC and JU

A customized electronic nose set-up (Fig. 4) has been developed by Centre for Development of Advanced Computing (C-DAC), Kolkata and Jadavpur

Fig. 4: E-Nose Set-up

University, which can be used in aroma characterization of fish. The electronic nose set-up developed is easy to install and can be operated and maintained by personnel with basic computer skills. The user-friendly software interface ensures easy and fast odour monitoring for fish during freshness measurement.

The electronic nose system consists of following accessories:

1. Gas sensor array comprises of eight gas sensors ftom Figaro Inc., Japan (TGS-832, TGS-823, TGS-831, TGS-816, TGS-2600, TGS-2610, TGS-2611 and TGS-2620).
2. Odour delivery system using headspace sampling technique.
3. PC.
4. NI data acquisition card.
5. Power cables, data cables.
6. Olfaction software.

Technical Details of the Electronic Nose Set-up

1. *Sample Vessel:* Specially designed sample holders made of glass has been used for sniffing of fish samples. The glass sample holders may be fixed to the instrument by simple threaded fitting.
2. *Cycles of Operation:* A sniffing cycle consists of automated sequence of internal operations, viz., Headspace Generation, Sampling, Purging and Dormancy before start of next sniffing cycle. The automated operation is controlled by the software supplied with electronic nose. Headspace generation ensures concentration of volatiles of fish samples before sampling and during sampling; the sensor array is exposed to a constant flow of volatiles through pipelines inside the electronic nose. During purging operation, sensor heads are cleared with blow of fresh air so that the sensors go back to their baseline values.The user may choose auto or manual mode of operation. In auto mode, following parameters are programmable on screen:

a. Headspace generation time.
b. Sampling time.
c. Purging time.
d. Intermediate interval between sample sniffing.

On completion of each sniffing cycle, sensor data will be automatically logged in the computer without operator intervention.

3. *Olfaction Software :* A user-friendly interactive software has been carefully designed which has got features like programmable sequence control, dynamic fermentation profile display, data logging, alarm annunciation, data archival, etc. Fig. 5 below shows typical sensor response curves while exposed to sample vapor.

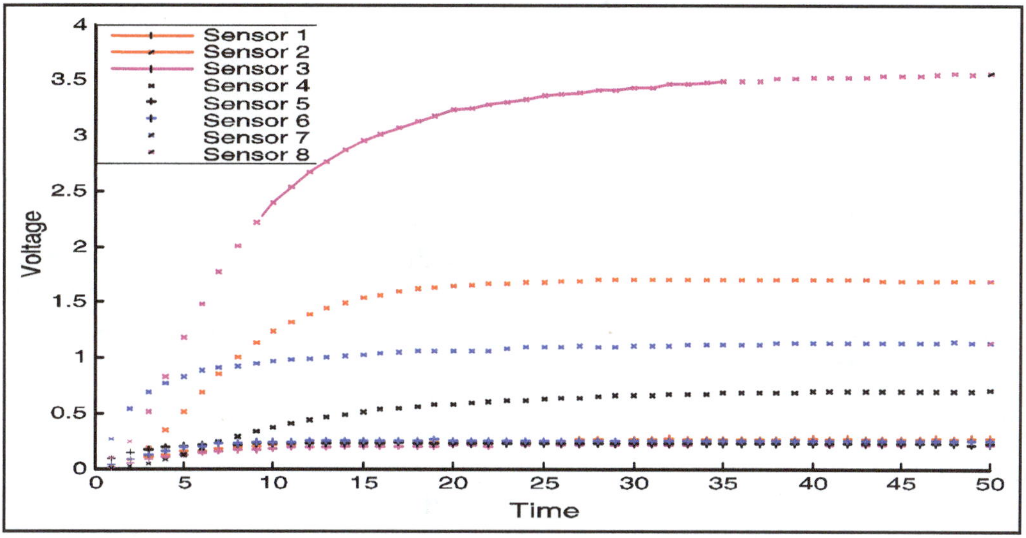

Fig. 5: Sensor Response Curves

Experimental Details

An experimentation for fresness study was carried out with *Labeo rohita* samples.

Sample Collection: CIFRI, Barrackpore provided following samples for electronic nose based experimentation as shown in Table 3.

Table 3: Sample Details

Type of Sample	Weight of Sample (in gm)	Remarks
Labeo rohita	50	Fresh fish (0 hours)
Labeo rohita	50	6 hours
Labeo rohita	50	12 hours
Labeo rohita	50	24 hours

Sample Preparation: The samples were taken from fish muscles and minced before allowing them to degrade as per times mentioned above. The samples were preserved at -20 ^{0}C before putting them for trials. Such preservation technique ensured no further microbial growth on the samples beyond the specified duration.

Experimental Settings: Following are the experimental settings for the study.

- Headspace generation: 100 Seconds
- Acquisition: 50 Seconds
- Purging: 100 Seconds
- Quantity used for each sample: 50 gms.

Data Analysis and Results

Principal Component Analysis: The multi-dimensional data matrix formed out of the signatures from the multi-sensor array have been subjected to Principal Component Analysis (PCA) [11]. The PCA plot obtained for 40 data vectors for each of the four samples is shown in Fig. 6 below. More than 99% significant information is available with the first two principal components (PCA1 and PCA2). It can be observed from the PCA plot that four distinct clusters have been formed for four types of samples tested with the electronic nose. The blue, red, green, and black indicates, 0 hrs, 6 hrs, 12 hrs, 24 hrs sample respectively. So, it is proved that the electronic nose has differentiated all the four samples very clearly.

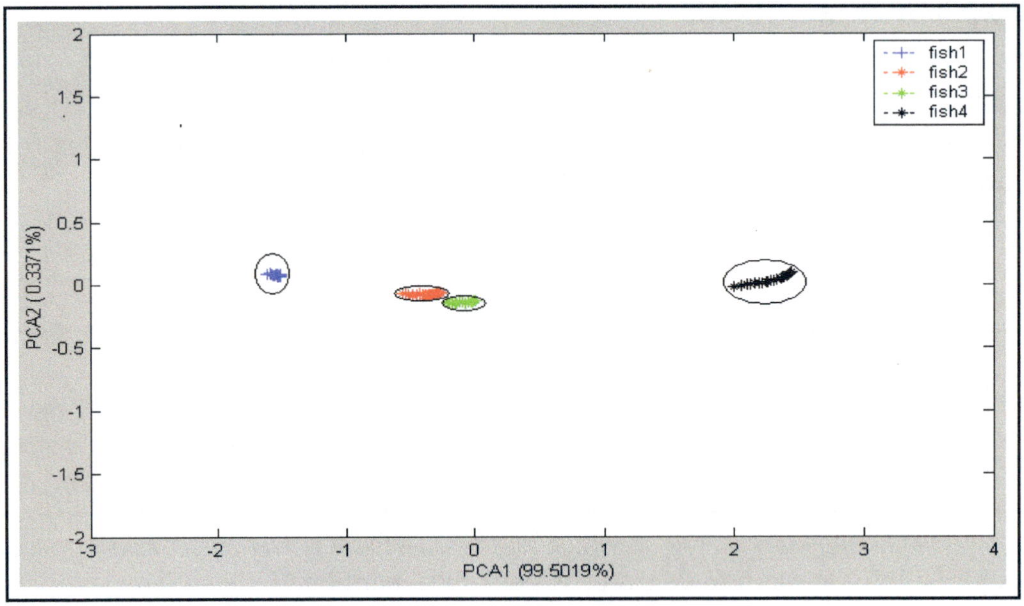

Fig. 6. PCA Clustering of Fish Samples

Back propagation Multilayer Perceptron (BP-MLP) Model: A three-layer Back propagation Multilayer Perceptron model with one input layer, one hidden layer and one output layer has been considered. The input layer has been fed with the output from the sensor array and the output layer has been configured to show the four classes of fish samples as shown in Fig. 7 [12]. For eight sensors in the multi-sensor array, the ANN has eight input nodes. The output layer of the network has been assigned with four (four classes of fish samples) nodes as shown. Convergence during learning process has been obtained with acceptable accuracy with only one hidden layer with 8 nodes. The BP-MLP model trained with the data obtained from the collected samples have been tested with unknown samples. The model has successfully predicted the grade of the fish and the accuracy of prediction is 100%.

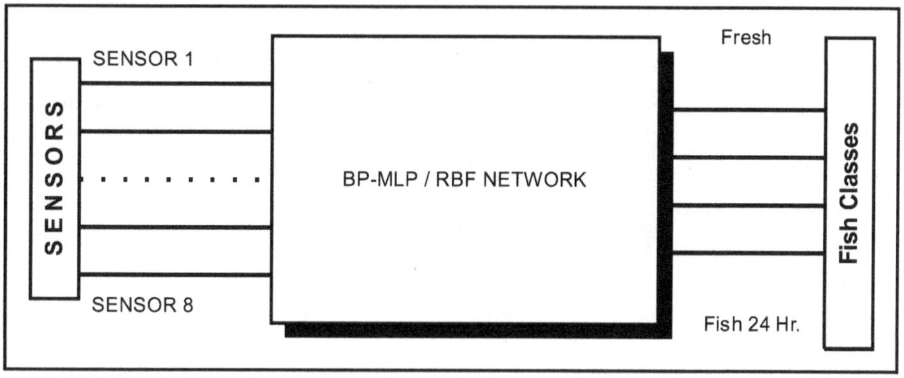

Fig. 7. BP-MLP Model

Electronic Tongue

Even though considerable progress has been made on the applicability of electronic nose for freshness evaluation of fish, taste sensors can also play a significant role. We could not perform the experiment on fish with electronic tongue or any taste sensor, but since this instrument shows a lot of promise and can be used with an electronic nose to have a broader perspective with higher sensitivity, it is noteworthy to describe briefly the basic principle. Block diagram of electronic tongue is shown in Fig.8. The electronic tongue system consists of a sensor and associated electronic equipment. Electrochemical method such as voltammetry or potentiometry is used in electronic tongues. The sensor response represents the total chemical and micro-organism content of the measured sample. The raw data is acquired in a computer through data acquisition device for further processing. The result is not necessarily the presence or absence of specific chemicals and/or bacteria with respective concentrations in the sample, but rather a measure of the total quality.

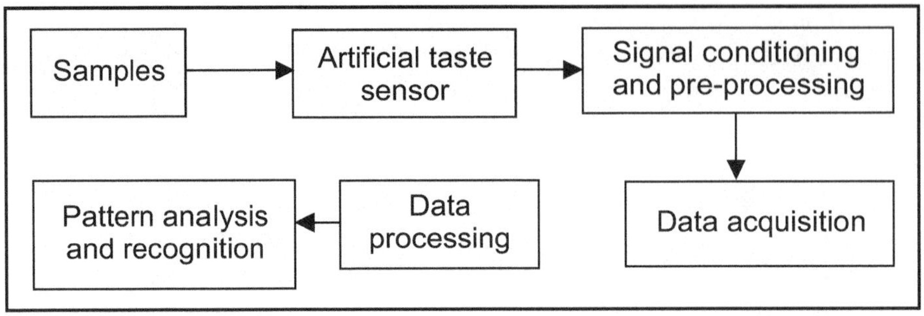

Fig. 8: Block Diagram of Electronic Tongue

The underlying technique in the electronic tongue is voltammetry. In this method a potential is applied at a working electrode while the resulting current between the working electrode and a reference electrode is measured. The input waveform is formed by pulses of varying amplitudes with a base potential in between. Different amplitudes are used to increase the sensitivity and improve the discrimination between samples. The reason for the improvement in the sensitivity is that all substances that are electrochemically active below the applied potential will add to the measured current. Applicability of such instrument in the field of flesh food, namely fish has been reported in [13].

Conclusions

The objective of the pilot level study was to explore the feasibility of application of electronic nose for prediction of degree of freshness of inland water fish. The results obtained from PCA as well as the neural network techniques demonstrate that e-nose may be successfully deployed as an effective tool for objective estimation of freshness of inland water fish.

The results definitely opens up numerous scopes of future research based on e-nose for quality assessment of fish. On the basis of the feasibility established as above, research may be taken up towards characterization of smell of different categories of fish. Also research efforts towards establishment of correlation of e-nose data with analytical instruments like High Profile Liquid Chromatograph (HPLC) as well as various socio-climatic variables related to fish production appear to be logical sequel to such pilot study.

REFERENCES

Kent, M., Oehlenschlager, J., Mierke-Klemeyer, S., Manthey-Karl, M., Kno¨chel, R., Daschner, F., Schimmer, O., 2004. A new multivariate approach to the problem of fish quality estimation. *Food Chemistry*, 87 (4), pp. 531-535.

Olafsdottir, G. Nesvadba, P., Di Natale, C., Careche, M., Oehlenschla¨ger, J., Tryggvado´ttir, S.V., Schubring, R., Kroeger, M., Heia, K., Esaiassen, M., Macagnano, A., Jorgensen, B.M., 2004. Multisensor for fish quality determination, *Trends in Food Science and Technology*, 15(2) : pp. 86-93.

Serby M.J. and K. L. Chobar, 1992. *Science of Olfaction*, Springer Verlag, New York.

Jinks, A. and D. G. Laming, 2001. The analysis of odor mixtures: evidence for a configurational process, *Physiol. Behav*, 72: 51-63.

Pearce T.C, S. S. Schiffman, H. T. Nagle and J. W. Gardner (Eds.), 2003. *Handbook of Machine Olfaction, Electronic Nose Technology*, Wiley - VCH, Wenham (FDR).

Gardner, J. W. and P. N. Bartlet, 1999. *Electronic Noses, Principles and Applications,* Oxford University Press, Oxford, U.K.

Gopel W, J. Hesse, J. N. Zemel (Eds.), 1997. *Sensors - A Comprehensive Survey,* VCH Verlagsgesellschaft GmbH, Weinheim, Germany.

Patranabis, D., 2003. *Sensors and Transducers*, Prentice Hall of India, New Delhi.

Ballantine, D. W., R. M. White, S. J. Martin, A. J. Ricco, G. C. Frye, E. T. Zellers, H. Wohltjen, 1997. *Acoustic Wave Sensors, Theory, Design and Physico-chemical Applications*, Academic Press, New York.

Bartlett, P. N, P. B. M. Archer and S. K. Ling Chung, 1989. Conducting polymer gas sensors Part I: fabrication and characterization, *Sensors and Actuators*, 19:125-140.

Bhattacharya N, and P. Kapur, 2004. Aroma Characterization of Black Tea Using Electronic Nose, Proceedings of 2004 International Conference on O-CHA (Tea), *Culture and Science*, Shizuoka, Japan.

Bhattacharya N, R. Bandyopadhyay, M. Bhuyan, A. Ghosh, and R. K. Mudi, 2005. Correlation of Multi-sensor Array Data with "Tasters" Panel evaluation for Objective Assessment of Black Tea Flavour, *Proc. of International Symposium* on Olfaction and Electronic Nose (ISOEN-2005), 13th to 15th April, 2005, Barcelona, Spain.

Andrey Legin Alisa Rudnitskaya, Boris Seleznev and Yuri Vlasov, 2002. Recognition of liquid and flesh food using an electronic tongue. *International Journal of Food Science & Technology*, 37(4): 375 - 385.

Chapter **18**

Nanotechnology and Its Impact on Aquaculture

Fahima Dilnawaz & Sanjeeb K. Sahoo

Nanotechnology: The Next Big Thing is Really Small

Nanotechnology can be considered as a brainchild of Richard Feynman, as the conceptual underpinings of nanotechnologies were first laid out in 1959 in his historical lecture, ' There is plenty of room at the bottom'. The term 'nanotechnology' was not used until 1974, when a researcher, Norio Taniguchi, at the University of Tokyo, Japan, used it to refer to the ability to engineer materials precisely at the nanometer level. The prefix 'nano' is derived from the Greek word 'drawf'. One nanometer (nm) is equal to one billionth of a meter or about the width of six carbon atoms or ten water molecules. A human hair is approximately 80,000 nm wide and a red blood cell is approximately 7000 nm wide[1]. Nanotechnology can be defined as the science and engineering involved in the design, synthesis, characterization and application of materials and devices whose smallest functional organization in at least one dimension is on the nanometer scale or one billionth of a meter[2-5]. It is a multidisciplinary field, which covers a vast, and diverse array of devices derived from Engineering, Physics, Chemistry and Biology and the cumulative efforts of all the disciplines is necessary for any significant outcome.

Applied Bioinformatics, Statistics and Economics in Fisheries Research: pp. 247-259

New India Publishing Agency, New Delhi (India)

Economic Impact of Nanotechnology

Nanotechnology has grown exponentially in recent years and it is anticipated that the economic impact of nanotechnology is growing by leaps and bounds. It has sparked a new industrial revolution and may emerge as one of the key technologies that shape the future economy. If Nanotechnology at maturity, achieves even a fraction of its promise, it will force the reassessment of global markets, Economies and industries on a scale never experienced before in human history. Nanotechnology, even if partially realized, over the next few decades has the potential to realign society, change business and affect economics at the structural level. Nanotechnology, in the years to come, will touch all aspects of economics: wages, employment, purchasing, pricing, capital, exchange rates, currencies, market demand and supply. In addition, nanotechnology may well drive economic prosperity or at the least will be an enabling factor in shaping productivity and global competitiveness [Table 1].

Table 1. Hot areas of nanodevelopment from a commercial viewpoint in the next 5 and next 10-15 years

Areas	5 years	10-15 years
Personal health diagnostics	6	17
Personalized medicines	3	9
Disposable displays	6	9
Artificial bone/tissue organs	3	8
Novel foods	0	6
Solar power	6	6
Retinal/ cochlear/ neural implants	0	6
Smart paints/pigments/coating	33	6
Security/crime detection	2	3
Smart Clothing	3	2
Prevention of congenital diseases	0	0
Robotic surgery	0	0
Microsatellites	2	0
Space medicines	0	0
Games/entertainment	5	0
Other/no response	31	28

Source: Meridian institute, Nano-Business Alliance

In recognition of the enormous scientific and commercial potential for nanotechnology, President Clinton established the National Nanotechnology

Initiative (NNI) in 2000, which is a multi-agency umbrella program to build, characterize and understand nanoscale devices. The program is slated to spend nearly $1 billion in fiscal year 2005 as compared to $464 million in 2001[6,7]. Moreover, the 'nanotechnology market' as a unified market was defined by the National Science Foundation in its 'Societal Implications of Nanoscience and Nanotechnology' report from March 2001, with the much publicized estimate of $ 1 trillion by 2015. The market growth for all nanotech products is estimated to be about 44% over the next 12-15 years. In anticipation of this economic impact, nanotechnology research programmes in several countries have increased substantially in recent years. In fiscal year 2005, federal agencies are slated to spend more than $1 billion on nanotechnology. They are attracting rapidly increasing investments from the government and from various business communities across many parts of the world. It has been estimated that the total global investment in nanotechnologies is currently around €5 billion, €2 billion of which comes from the private sector. In addition, the number of published patents in nanotechnology increased four fold from 531 patents 1995 is to 1976 patents in 2001. Thus, the advent of nanotechnology will have a profound impact on the US and the global economy [Table 2].

Table 2. Investment in Nanotechnology: A Global Technology

Name	Investment
United States	$3.7b until 2008, an annual investment twice that of the Human Genome project in its peak years
Europe	$1.7b in the Sixth Framework Programme FP6 (2002-2006)
Japan	$36b by 2010
China	$175m/year and a growth rate of 200%
Brazil	$25m (2004-2007)
India	$23m (2002-2007)
Taiwan	663m over six years
South Africa	$5m

Applications of Nanotechnology

Nanotechnology has been a boon to human society as its various applications offers solutions to the myriad of problems of the evolving society. A complete list of potential applications of nanotechnology is too vast and diverse to list but some of the major upcoming applications are briefly discussed.

(a) Nanotechnology in Health Care

(i) *Nanotechnology in Drug Delivery and Therapeutics:* Applications of nanotechnologies in medicine are especially promising and areas such as disease diagnosis, drug delivery targeted at specific sites in the body and molecular imaging are being intensively investigated and some products are undergoing clinical trials.[8-11] Formulating different drug delivery systems so that they maintain their stability and remain within their efficacious and safe target doses remains a great challenge. In addition, problems associated with higher doses of conventional medicines, immunogenicity, greater toxicity and non-specific delivery also remains an unsolved problem. Nanoparticles, an evolvement of nanotechnology, are increasingly considered as a potential candidate to carry therapeutic agents safely into a targeted compartment in an organ, particular cell or tissue. Nanoparticles are submicron-sized polymeric colloidal particles, with diameters ranging from 10-100nm, in which a therapeutic agent of interest can be encapsulated within their polymeric matrix or adsorbed or conjugated onto their surface[4,12,13]. Thus, nanotechnology based delivery systems could mitigate these problems by combining tissue or organ-specific targeting with therapeutic action. Actually, nanotechnology based localized drug therapy using sustained release drug delivery systems could be more effective because it can provide higher and prolonged drug levels in the target tissues without causing systemic toxicity.[12] It is also opening new therapeutic opportunities for a large number of agents that cannot be used effectively as conventional oral formulations, due to poor bioavailability.[14,15] Nanoparticles formulations provide protection for agents susceptible to degradation or denaturation in regions of harsh pH, and also prolong the duration of exposure of a drug by increasing retention of the formulation through bioadhesion.[16,17] In addition, recent advances in encapsulation and development of suitable animal models have demonstrated that micro and nanoparticles are capable of enhancing immunization.

(ii) *Nanotechnology in Gene Delivery:* Applications of nanotechnological tools in human gene therapy, which is a method for treatment or prevention of genetic disorders by correcting defective genes responsible for disease development based on delivery of repaired, or the replacement of incorrect genes could be introduced successfully for the delivery of the required gene.[18-21] Nanoparticles and microparticles formulated using poly-lactide-co-glycolide (PLGA) and polylactic acid (PLA) polymers are recently being investigated as a non-viral gene delivery system because of their sustained release characteristics, biocompatibility, biodegradability and their ability to protect DNA from degradation in endolysosomes.[12] The rapid escape of nanoparticles from the endolysosomal compartment could protect nanoparticles as well as the encapsulated DNA from the degradative environment of the endolysosomes.[22]

(iii) *Imaging and Molecular Diagnostics Revolutionized by Nanotechnology:* Advantages in nanotechnology also has a potential to transform the field of medicine as it offers novel opportunities for sensing clinically relevant markers, molecular disease imaging and tools for therapeutic intervention.[23] Quantum dots are semiconductor nanocrystals with unique optical and electrical properties and nanotechnologies already affords the possibility of intracellular imaging through attachment of quantum dots (QDs) or synthetic chromophores to selected molecules. Nanotechnology has also been identified as a field of great promise to detect and diagnose cancer. For example, semiconductor QDs can enable the detection of tens to hundreds of cancer biomarkers in blood assays, on cancer tissue biopsies or as contrast agents for medical imaging.[24,25] Nanotechnology is also being used for molecular diagnostics. Nanotechnology on a chip is one more dimension of microfluidic/lab-on-a-chip technology. For instance, when certain nanoscale particles are put to work as tags or labels, biological tests measuring the presence or activity of selected substances become quicker, more sensitive and more flexible.

(iv) *Cardiac, Orthopaedic and Dental Applications of Nanotechnology:* Nanotechnology also offers a broad platform in the field of cardiovascular sciences by offering tools to explore the frontiers of cardiac science at the cellular level. Nanotechnology based tools can be effectively used to treat the cardio-vascular diseases. These tools can be utilized in the areas of diagnosis, imaging and tissue engineering.[26] Moreover, it can also help in revealing the mechanisms involved in various cardiac diseases and facilitates the designing of atomic-scale machines by imitating or incorporating biological systems at the molecular level and the use of these newly designed nanomachines can have a paradigm shifting impact in the treatment of the dreaded cardiovascular diseases. It can also have an impact in the diagnosis and treatment of unstable plaques and in the management of other cardiovascular problems like calcification of valves. Thus, nanotechnology could be an effective treatment modality to achieve localized and sustained cardiac drug therapy for the prevention and cure of cardiovascular diseases. Nanotechnology is also fruitful for patients who suffer from debilitating bone fractures, as it provide an alternative platform with higher mechanical strength, enhanced bioactivity and resorbability.[27] Since bone is also made up of nanosized organic and mineral phases, nanostructure materials with size 1-100 nm can act as new and effective constituents of bone materials. Nanomaterials, nanopolymers, carbon nanofibers, nanotubes and nano-composites of ceramics will also lead to more efficient deposition of calcium-containing minerals on the implants. Furthermore, it is estimated that nanophase materials attract more proteins to their surface due to their altered surface energetics, brought about by higher exposed surface area and altered electron distribution as compared to conventional materials.[28,29] Thus, nanostructure materials represent an important growing area of research that may improve bonding between an implant and surrounding bone by increasing bone cell interactions and this will certainly aid in improving orthopaedic implant efficacy while drastically minimizing patient compliance and improving the quality of life of patients. The future of nanotechnology also looks promising with its roots gradually spreading into new, unexplored terrains. It will have future medical applications in the field of Nanodentistry. Nanodentistry will make it possible to maintain a near perfect oral health through the use of nanomaterials, biotechnology and nonorobotics and through this, it will be possible to provide high quality dental care to the millions of the world's population that currently receives no significant dental care.[30-33]

(b) Nanotechnology for Agriculture and Food Systems

Agriculture forms the backbone of third world countries. Research in agriculture has always dealt with improving the efficacy of crop production, food processing, food safety and environmental consequences of food production, storage and distribution. Nanotechnology, as a new enabling technology, has the potential to revolutionize agriculture and food systems, as it provides a tool to pursue these historically relevant goals.[34] Agriculture and food systems security, disease treatment delivery systems, new tools for molecular and cellular biology, new materials for pathogen detection and protection of the environment are examples of the important links of nanotechnology to the science and engineering of agriculture and food systems. In the present scenario, if a plant becomes infected with disease, it can be days, weeks or months before disease presence is detected by whole-organism symptoms, but as nanotechnology operates at the same scale as a virus or disease-infecting particle, it holds the potential for very early detection and eradication of the disease causing entity. It holds the possibility that 'smart' treatment delivery systems could be activated and that they would have the capacity to detect and treat an infection, nutrient deficiency, or other health problems long before macro symptoms appear. These 'smart' delivery systems can also have the capacity to monitor the effects of the delivery of pharmaceuticals, neutraceuticals, nutrients, food supplements, bioactive compounds, probiotics, chemicals, insecticides, fungicides, vaccinations etc. In addition, the exciting possibility of combining biology and nanoscale technology into sensors holds the potential of increased sensitivity and therefore the response time to sense potential problems is significantly reduced. Thus, nanotechnology could permit broad advances in agricultural research.

(c) Nanotechnology for Water Purification and Treatment

Nanotechnology has an enormous potential for creating cost-effective, simple and efficient tools for addressing water supply challenges, while preventing the creation of potentially toxic by-products. Nanotechnology could also play a major role in the purification process and current problems involving water quality could be greatly resolved or ameliorated using nanosorbants, nanocrystals, bioactive nanoparticles, nanostructured catalytic membranes and nanoparticles enhanced filtration.[35] Nano-engineered membranes could potentially lead to more energy-efficient water purification processes, notably in desalination by reverse osmosis and can also be used for desalination and detoxification of water. Photocatalysis degradation process by the use of nanoparticles has also gained popularity in waste water treatment process.[36] Photocatalysis is a reaction in which a chemical compound reacts in the presence of light and itself not being completely consumed in the reaction. Metal oxides like TiO_2, ZnO, SnO_2 etc. as well as sulphides like ZnS have were used for photocatalysis as these nanoparticles have efficient disinfectant rate.[37-39] In addition, nanosensors can be used for the detection of contaminants and pathogens. Other applications of nanotechnology include the use of nanospore zeolites, nanoporous polymers and attapulgite clays for water purification as well as the use of magnetic nanoparticles for waste water treatment and remediation. Additionally, nanotechnology-derived products that reduce the concentrations of toxic compounds can assist in the attainment of water quality standards and health advisories. Thus, nanotechnology plays a major role in waste water purification and treatment.

(d) Application of Nanotechnology in Aquaculture

Aquaculture is the cultivation of the natural produce of water (fish, shellfish, algae and other aquatic organisms). It is also known as aquafarming. Subsets of aquaculture include mariculture (aquaculture in the ocean), algaculture (the production of seaweed and other algae), fish farming (the raising of catfish, tilapia and milkfish in freshwater and brackish ponds or salmon in marine ponds), and the growing of cultured pearls. A considerable proportion of the harvest from capture fisheries is destined for the production of fish meal and fish oil, food rich in vitamins and fatty acids, which are primarily used by the feedstuff industry. Aquaculture thus remains the major means of maintaining current per capita "fish" availability as capture fisheries production available for human consumption has been increasing slowly for many years.[40]

This industry includes the rearing of aquatic animals and plants for and within public, private aquariums and research facilities, the production of bait fish, hatcheries and nursery rearing of stock intended for fisheries enhancement or restocking programs. Various species of carp top the list in terms of aqua cultural production. Most countries produce significant amounts of carp (China, India, and Europe). In North America, channel catfish farming is the largest aquaculture industry. Others of importance include trout, crawfish, and various species of shellfishes. In addition, many of the fish species taken are produced in hatcheries and reared to a size where they can be expected to have a good chance of survival before being released into the natural environment. The ornamental fish industry depends on animals caught in the wild and on those produced by aqua culturists and most of the bait minnows available in the marketplace come from fish farms. Seaweed culture as human food is a major industry, particularly in Japan and other Asian nations. Seaweeds are not only consumed by people as food, they are also a source of such chemicals as carrageenan and agar, which are utilized in everything from toothpaste and cosmetics to automobile tires. Other fish varieties such as, squid and cuttlefish are not being produced to any extent as human food, but they are reared as a source of giant axons for use in biomedical research. Moreover, an increasing number of potential pharmaceuticals are being identified from marine organisms. Culture of various species from a number of phyla, many of which have held little or no interest for aquaculture in the past, show promise as one means of meeting the demand for cancer-fighting and other types of drugs. Aquaculture also significantly affects the economy. It has been estimated that global aquaculture production will need to expand to 62 million tons by 2035 to maintain 1993 global average per capita consumption levels.[41] Therefore the modern aquaculture promises a 'blue revolution' akin to the agro industry's green revolution observed in the last century."

(i) *Cleaning Fish Ponds with Nanotechnology Devices:* Nevada-based Altair Nanotechnologies makes a water cleaning product for swimming pools and fishponds called 'NanoCheck'.[42] The compound is made such that it has a large surface area of 30 square meters per gram using a 40-nanometer spherical particle that acts like a sponge to absorb the phosphate. These 40 nm particles of the lanthanum-based compound which absorbs phosphates from water, prevents algae growth and does not cause sludge formation or filtration problems associated with lanthanum carbonate. Lanthanum is a transition metal that has been used to remove heavy metals and arsenic from drinking water. There is a potentially large demand for NanoCheck in thousands of commercial fish farms worldwide where algae removal and prevention is costly at present. This compound, instead of killing algae prevents their growth by removing the phosphate nutrient that the plant needs to survive. According to Altair, the company plans to expand its tests to confirm that its nanoparticles will not have any adverse effects on various fish species and the environment.

(ii) *Fishing with Nanotechnology: Fishing Lures* : Fishing lures typically use motion and color to attract the attention of fish, and they often feature decals and paint designed to reflect sunlight. These conventional lures are only capable of reflecting light in one direction. The ULVAC company (Japan) has formulated a new type of lure featuring a nanocoating applied with special vacuum technology to tackle the fishing industry.[43] The surface of the lure is colored and glossy-finished and then coated with a polyimide film with high light transmission formed by vapor deposition polymerization. The hue is determined by the effect of interference of light and adjusted based on the relation between the thickness of the polyimide film and the wavelength of light. The hue of the lure can be varied like the hologram depending on the angle from which it is seen. Further, it can be provided with a rainbow gradation by controlling the film forming conditions of the polyimide film. Better fishing results could be achieved with the use of a rainbow-colored lure. The surveys by ULVAC *et. al.* (2007) showed that when the lure is coated with a polyimide film so as to look pink and rainbow colored, the chance of catching fish is raised to about 2.6 and 3 times, respectively, compared to the case where a lure without a polyimide coating is used. This high-tech lure is called STROM. The lure appeals to freshwater fish such as trout, as well as to reef-dwelling saltwater fish such as rockfish.

(iii) *Nanotechnology in DNA Vaccines and the Increased Growth of Fish* : Modern aquaculture would have been impossible without use of injectable vaccines against certain diseases. However, the use of oil emulsions as adjuvant may cause major drawbacks in terms of unwanted effects. As stated in a recent review, it is evident that some fish possess unacceptable levels of side effects.[44] Adherations and skeletal deformations are the probable side effects associated with vaccines. Moreover, to replace the commonly used oil-based adjuvant, that cause severe side effects, it is obvious that a development of new adjuvant formulation based on new knowledge should start. By employing nanoparticle carriers (PLGA) of vaccine antigens together with mild inflammatory inducers (Toll-receptor ligands), one may achieve high levels of protection concomitant with a low level of vaccine-induced side effects. It is hypothesized that by using this novel concept, immune-prophylactic measures not only against bacterial but also certain viral diseases may be obtained.[45] There is an ongoing project at Norway (FISHVACCPLAT, 2006-2010) that has started production of PLGA particles containing tracer substances FITC analogues, LPS and recombinant IPNV VP2 peptides. The first results from *in vivo* injection experiments with FITC tracer, with the purpose to identify tissue localization of particles, will soon be available. The USDA (United States Department of Agriculture) is completing trials on a system for mass vaccination of fish using ultrasound. Nanocapsules containing short strands of DNA are added to a fishpond where they are absorbed into the cells of the fish. Ultrasound is then used to rupture the capsules, releasing the DNA and eliciting an immune response from the fish. This technology has so far been tested on rainbow trout by Clear Springs Foods (Idaho, US) - a major aquaculture company that produces about one third of all US farmed trout. The growth of fish has been boosted by the use of Iron Nanoparticles as is evident from the Russian Academy of Sciences have reported that young carp and sturgeon exhibited a faster rate of growth (30% and 24% respectively) when they were fed nanoparticles of iron.

Future Prospects and Conclusion

The multidisciplinary field of nanotechnology is making the science of the almost incomprehensively small devices closer to reality. The creation of new pharmacological compounds and novel treatment protocols are going to have a huge impact on the society and the genesis of nanotechnology can be traced to the promise of revolutionary advances across medicine, communications, genomics, robotics and human health care. Thus, in the twenty first century, nanotechnology may offer a vast number of breakthroughs and it holds the promise of delivering the greatest technological breakthroughs in history.

REFERENCES

1. Whitesides G.M., 2003. The 'right' size in nanobiotechnology. *Nat Biotechnol*, 21 (10): 1161-5.

2. Emerich DF and Thanos C.G., 2003. Nanotechnology and medicine. *Expert Opin Biol Ther.*, 3(4):655-63.

3. Parveen S and Sahoo, S.K., 2006. Nanomedicine: clinical applications of polyethtlene glycol conjugated proteins and drugs. *Clin Pharmacokinet*, 45(10):965-88.

4. Sahoo S.K. and Labhasetwar, V., 2003. Nanotech approaches to drug delivery and imaging. *Drug Discov Today*, 8(24):1112-20.

5. Sahoo, S.K., Parveen, S. and Panda, J.J., 2007. The present and future of nanotechnology in human health care. *Nanomedicine*, 3(1):20-31.

6. Roco, M.C., 2003. Nanotechnology: convergence with modern biology and medicine. *Curr Opin Biotechnol*, 14(3):337-46.

7. Roco M.C. and Bainbridge WS, 2002. Converging Technologioes for Improving Human Performance: National Science Foundation and Department of Commerce Report.

8. Li K.C., Pandit SD, Guccione S and Bednarski MD., 2004. Molecular imaging applications in nanomedicine. *Biomed Microdevices*, 6(2):113-6.

9. Wilkinson, J.M., 2003. Nanotechnology applications in medicine. *Med Device Technol*, 14(5):29-31.

10. Moghimi SM, Hunter AC and Murray JC., 2005. Nanomedicine: current status and future prospects. *Faseb J*, 19(3):311-30.

11. Shaffer C., 2005. Nanomedicine transforms drug delivery. *Drug Discov Today*, 10(23-24):1581-2.

12. Panyam J and Labhasetwar V., 2003. Biodegradable nanoparticles for drug and gene delivery to cells and tissue. *Adv Drug Deliv Rev*, 55(3):329-47.

13. Brannon-Peppas L and Blanchette JO., 2004. Nanoparticle and targeted systems for cancer therapy. *Adv Drug Deliv Rev*, 56(11):1649-59.

14. El-Shabouri MH., 2002. Positively charged nanoparticles for improving the oral bioavailability of cyclosporin-A. *Int J Pharm*, 249(1-2):101-8.

15. Hu L, Tang X and Cui F., 2004 Solid lipid nanoparticles (SLNs) to improve oral bioavailability of poorly soluble drugs. *J Pharm Pharmacol*, 56(12):1527-35.

16. Arbos P, Campanero MA, Arangoa MA, Irache JM., 2004. Nanoparticles with specific bioadhesive properties to circumvent the pre-systemic degradation of fluorinated pyrimidines. *J Control Release*, 96(1):55-65.

17. Arangoa MA, Campanero MA, Renedo MJ, Ponchel G, Irache JM., 2001. Gliadin nanoparticles as carriers for the oral administration of lipophilic drugs.

Relationships between bioadhesion and pharmacokinetics. *Pharm Res*, 18 (11) : 1521-7.

18. Ariga T., 2006. Gene therapy for primary immunodeficiency diseases: recent progress and misgivings. *Curr Pharm Des*, 12(5):549-56.
19. Rogers S., 1971. Gene therapy: a potentially invaluable aid to medicine and mankind. *Res Commun Chem Pathol Pharmacol*, 2(4):587-600.
20. Verma IM, Weitzman MD., 2005. Gene therapy: twenty-first century medicine. *Annu Rev Biochem*, 74:711-38.
21. Goverdhana S, Puntel M, Xiong W., 2005. Regulatable gene expression systems for gene therapy applications: progress and future challenges. *Mol Ther*, 12(2):189-211.
22. Panyam J, Zhou WZ, Prabha S, Sahoo SK, Labhasetwar V., 2002. Rapid endo-lysosomal escape of poly(DL-lactide-co-glycolide) nanoparticles: implications for drug and gene delivery. *Faseb J*, 16(10):1217-26.
23. Lin H, Datar RH., 2006. Medical applications of nanotechnology. *Natl Med J India*, 19(1): 27-32.
24. Wetts R, Kalaska JF, Smith AM., 1985. Cerebellar nuclear cell activity during antagonist cocontraction and reciprocal inhibition of forearm muscles. *J Neurophysiol*, 54(2): 231-44.
25. Gao X, Cui Y, Levenson RM, Chung LW, Nie S., 2004. In vivo cancer targeting and imaging with semiconductor quantum dots. *Nat Biotechnol*, 22(8):969-76.
26. Wickline SA, Neubauer AM, Winter P, Caruthers S, Lanza G., 2006. Applications of nanotechnology to atherosclerosis, thrombosis, and vascular biology. *Arterioscler Thromb Vasc Biol.*, 26(3):435-41.
27. Sato M, Webster TJ., 2004. Nanobiotechnology: implications for the future of nanotechnology in orthopedic applications. *Expert Rev Med Devices*, 1(1):105-14.
28. Webster TJ, Ergun C, Doremus RH, Siegel RW, Bizios R., 2000. Specific proteins mediate enhanced osteoblast adhesion on nanophase ceramics. *J Biomed Mater Res*, 51(3): 475-83.
29. Woo KM, Chen VJ, Ma PX., 2003. Nano-fibrous scaffolding architecture selectively enhances protein adsorption contributing to cell attachment. *J Biomed Mater Res* A, 67(2):531-7.
30. West J.L, Halas NJ., 2000. Applications of nanotechnology to biotechnology commentary. *Curr Opin Biotechnol*, 11(2):215-7.
31. Shi H, Tsai WB, Garrison MD, Ferrari S, Ratner BD., 1999. Template-imprinted nanostructured surfaces for protein recognition. *Nature*, 398(6728):593-7.
32. Sims MR., 1999. Brackets, epitopes and flash memory cards: a futuristic view of clinical orthodontics. *Aust Orthod J.*, 15(5):260-8.

33. Slavkin HC., 1999. Entering the era of molecular dentistry. *J Am Dent Assoc*, 130 (3) : 413-7.

34. Hsieh YH, Ofari JA., 2007. Innovations in food technology for health. *Asia Pac J Clin Nutr*, 16:65-73.

35. Savage N, Diallo MS.,2005. Nanomaterials and water purification: opportunities and challenges. *Journal of Nanoparticle Research*, 7:4-5.

36. Herrmann JM. 1999. Heterogeneous photocatalysis: fundamentals and applications to the removal of various types of aqueous pollutants. *Catalysis Today*, 53:115-29.

37. Bhatkhande DS, Pangarkar VG, Beenackers AACM., 2001. Photocatalytic degradation for environmental applications- a review. *J Chem Technol Biotechnol*, 77:102-16.

38. Li D, Haneda H., 2003. Morphologies of zinc oxide and their effects on photocatalysis. *Chemosphere*, 51:129-37.

39. Torres-Martinez LC, Nguyen L, Kho R, *et al.*, 1999. Biomolecularly capped uniformly sized nanocrystalline materials: glutathione-capped ZnS nanocrystals. *Nanotechnology*, 10:340-54.

40. Stickney RR., 1994. Principles of Aquaculture: John Wiley & Sons,Inc.

41. Stickney RR., 2000. Encylopedia of Aquaculture.

42. "Altair Nanotechnologies" Algae prevention Treatment confirmed Effective Testing. Market wire 2004.

43. Strom the original fishing lure. *ULVAC Technical Journal* 2007(66E).

44. Berg AB, O. Fjelldal,P.G, Hansen,T.Juell,J.E and Nerland,A., 2006. Animal welfare and fish vaccination-effects and side effects. *Fisken og Havet*, 9:1-43.

45. Reddy ST, Swatrz, M.A and Hubbel, J.A., 2006. Targeting dendritic cells with biomaterials: developing the next generation of vaccines. *Trends in Immunology*, 27:573-6.

Chapter **19**

Design and Development of Fisheries Data Mart in India

Anil Rai, K.K. Chaturvedi & P.K. Malhotra

Introduction

Technologies and tools that have developed in the recent information technology revolution can play a very vital role in overcoming most of the present problems related to agriculture and rural development. In order to achieve the goal of an economically sound and stable society through environmentally friendly systems of agricultural production, it is imperative to have a comprehensive data storage system. This will not only help in providing systematic and periodic information to the planners, decision-makers and developmental agencies, but also help to manage our limited natural resources efficiently. The development of a data warehouse is one of the recent tools in the field of information technology that can fulfill the demands of the information that is required in the field of agriculture research, management and education

In fact, the development of an integrated information systems and databases (non-spatial, spatial and bibliographic) in the field of agriculture research and education was recognized by the Indian Council of Agricultural Research (ICAR) review committee (1988). The development of a computerized satellite based information network (ICARNET) was recommended by the committee with on-line terminal connectivity to ICAR scientists and State Agriculture Universities (SAUs). In 1991, the development

Applied Bioinformatics, Statistics and Economics in Fisheries Research: pp. 261-292

New India Publishing Agency, New Delhi (India)

of a computer-based Agricultural Research Information System (ARIS) network was initiated by ICAR. The council also took up a NATP Mission Mode project entitled "Integrated National Agricultural Resources Information System (INARIS)" in 2001. In this a state of art a Central Data Warehouse has been developed at Indian Agricultural Statistics Research Institute (IASRI), New Delhi.

Agriculture in India

The Indian agriculture is highly diversified in terms of its climate, soil, field crops, horticultural crops, plantation crops, livestock resources, fisheries resources, water resources, etc. The diversity of its agricultural sector is both a bane and boon to the social, economic, and cultural bases of India's vast population. Moreover, the diversity among resources generates interactions among many different macro and micro factors, and is further complicated with the interdependencies that exist among these. These resources need to be evaluated, monitored, and allocated optimally for balanced and sustainable development of the country. To have a better understanding of the design of data warehouse for Indian agricultural resources (INARIS), it is necessary to have an overview of the administration, which supports the agricultural production and collection of information related to agricultural sector.

India is divided into states and union territories (UTs). Each state/UT is further divided into districts. Although, there are further divisions of these districts into tehshils or taluks, blocks, and finally to villages but in general a district is the basic unit of administrative set up in the country for all purposes. Information related to different sectors of agriculture is collected by various organizations throughout India. For example, the National Sample Survey Organization (NSSO) conducts national level agricultural surveys with the help of concern state/UT departments in coordination with Directorate of Economics and Statistics. Information related to horticultural crops is also collected by the National Horticulture Board and related state departments of horticulture. Similarly, there are many boards and organizations related to each sector of agriculture at national as well as state level. Each organization is entrusted with collecting information about different agricultural commodities, as per the mandate of their organization. Some of these organizations are specific to the region or state. Since, there is a large number of data collection agencies as well as equally diverse resources for which the information is to be collected, it is easy to visualize the heterogeneity of information from the agricultural sector. As stated earlier, the problem is compounded by the fact that there are no common standards that are applied in data collection process. Designing data warehouses to integrate the collected information from different sectors of agriculture poses a formidable challenge to any data warehouse architect. In order to use the information for planning and decision-making level, data have to be integrated and aggregated properly.

Data Warehousing

The data warehouse is an architectural construct that addresses the growing need of information for enterprise-wide data access. It functions as a core for decision support processing at the strategic/ managerial level, separate from day to day operational data. It is not software specific and can be used in any computing environment. The design and construction of data warehouse includes granularity analysis of data, partitioning of data, meta-data management etc. The data in a data warehouse can be seen as a set of materialized views derived from source data, where the source data can be relations in the operational database or other non-traditional data such as data files, legacy systems and document data. Since, the raw data normally changes over time, materialized views in a data warehouse have to be updated to ensure consistency with the source data. Ralph Kimball defines data warehouse as "a copy of information data specifically structured for query and analysis". Data warehousing is more than a database. Data warehousing involves the entire information delivery process i.e., from access and transformation of data from different operational sources through the process that makes it available for decision making, to the exploitation of the retrieved data via a range of decision support tools. The growing need for the data warehousing technology in recent years has stemmed because of the technology's importance in supporting decision support processing and analysis. A specific property of data warehouse which makes efficient application processing needs to summarize huge amounts of data. The growing trend in data warehouse architecture is to store the data both in the data warehouse and in several data marts, where each data mart contains the data pertaining to a particular domain of the organization's operations.

Data warehousing and data marts : A Data warehouse is a read only analytical database that is used as a foundation of a decision support system. In other words, a data warehouse is a repository of integrated information, available for queries and analysis. In this, data and information are extracted from heterogeneous sources as they are generated. This makes it much easier and more efficient to run queries over data that originally came from different sources. Data warehouse architecture comprises of

- Operational source systems
- A data staging area
- One or more conformed data marts
- A data warehouse database

An operational source system is developed to capture and process original business i.e. day-to-day transactions and is often referred as "mission critical" application processing. These systems are designed for data entry and scheduling the operations, not for reporting, but data can be populated in the warehouse from this system. Data staging area is where this raw (source) operational data is extracted, cleaned, transformed and combined so that it can be reported and queried by users. This area lies between the operational source systems and the user database and is typically not accessible to users. Data staging is a major process of data warehousing. The important concepts of this technology are given below:

Data mart: Data mart is a logical subset of an organizational data warehouse. The dimensional data marts are organized by a specific domain or subject area wise.

Data warehouse database: A data warehouse database contains the data that is organized and stored specifically for direct user queries and reports. It differs from an OLTP database, which is designed primarily for reads not writes. An OLAP application is a system designed for few but complex (read only) request. An OLTP application is a system designed for many but simple concurrent (and updating) requests. For creating a data warehouse environment, modeling techniques are required that can be used to help in understanding the business and identify the individual data marts. It is also useful for communicating the way data flows from one stage to the next on its way from source system to user.

Entity relationship modeling: This will be needed by the data staging team to understand how data is organized in each of the source systems. This is a standard system analysis technique used to build On-line Analytical Transaction Processing (OLTP) systems.

Multi-dimensional data modeling: It is primarily specialized technique used to design the central data warehouse and associated data marts.

Hierarchies: It describes the organizational structure and logical parent-child relationship existing within the data.

Granularity: It is the level of detail information within the data warehouse or data mart.

Fact tables: These are used to record actual facts and measures in specific sector. Facts are additive or non-additive numeric data items. Complex calculations are performed on these data items in OLAP process. Examples: - population, area, production etc.

Dimensional tables: Dimensional tables establish the context of the facts. Relatively small, denoralized lookup tables contain the description of the data generation process. Examples: - states-districts, species-breed, category-crop-product etc. Deciding how much historic data to keep, and how often the warehouse needs to be updated.

Schema for Dimensional Modeling: The schema for a dimensional model contains a central fact table and multiple dimension tables. A schema is a fact tables plus its related dimensional table. There are two types of schemas namely star schema and snowflakes that are usually followed in process of designing a data marts.

Star Schema: A schema is called a star schema if all dimension tables can be joined directly to the fact table. Here are the some basic steps for making the star schema:

- Denormalized the database design.
- Disable primary or foreign key constraints.
- Create bitmap indexes on every column of every table.
- Use bitmap indexes on columns with lengthy character data.
- Use bitmap indexes on columns with millions of different values.

Snow Flake Schema: A schema is called a snowflake schema if one or more dimension tables do not join directly to the fact table but must join through other dimension tables.

Integrated National Agricultural Resources Information System (INARIS): This Project was taken up as a sub-project under NATP. The mission set for this project was to design and develop a flexible Central Data Warehouse (CDW) of agricultural resources of the country at IASRI, New Delhi (lead center) and databases on different subjects at respective co-operating centers. The above project has been implemented with active collaboration and support from 13 other ICAR institutions, namely NBSSLUP Nagpur (for soil resources), CRIDA Hyderabad (for agro-meteorology), PDCSR Modipuram (for crops and cropping systems), NBAGR Karnal (for livestock resources), NBFGR Lucknow (for fish resources), NBPGR New Delhi (for plant genetic resources), NCAP New Delhi (for socio-economic resources), CIAE Bhopal (for agricultural implements and machinery), CPCRI Kasargod (for plantation crops), IISR Calicut (for spices crops), ICAR Research Complex for Eastern Region Patna (for water resources), NRC-AF Jhansi (for agro forestry) and IIHR Bangalore (for horticultural crops).

In all 59 databases on agricultural technologies generated by council, research projects in operation and related agricultural statistics from published official sources at least from the year 1990 onwards at the district level were integrated into this information system. Subject-wise data marts were created; multi-dimensional data cubes have been developed and published on Internet/ Intranet. The validation checks have been implemented wherever possible.

The information in this data warehouse available to the end-user in the form of decision support system in which all the flexibility of the presentation of the information, its on line analysis including graphic is inbuilt into the system. The system also provides facility of spatial analysis of the data through web using functionalities of Geographic Information System (GIS). Apart from this, subject wise information system has been developed for the general users. The user of this system has the access of subject wise dynamic reports through web. The facilities of data mining and generation of ad-hoc querying were also extended to limited users. Therefore, the dissemination of information from this data warehouse for different categories of users is through web browser with proper authentication of the users. The web site of the project is launched (www.inaris.gen.in) and the multidimensional cubes, dynamic reports, GIS maps and information systems are available to the users.

This project is viewed to strengthen the information system conceptualized by ICAR. Based on the interaction among the basic resources like soil, water, climate, animal and vegetation that form the prime components of the production system this data warehouse could help in determining the carrying capacity of the region. The project provided suitable opportunity on multi-disciplinary mode through enhanced linkages among research institutes and other development agencies by providing first hand information on problems and potential in production systems. This data warehouse is now being used with an ultimate aim of enhancing better quality of life of the farming community and society at large. The architecture of INARIS CDW is shown in Fig 1.

Fisheries Data Mart: The components of the fishery data mart which is a part of INARIS data mart have been decided giving more emphasis on user requirements and availability of information. Before finalizing on the components of the databases a sample of available data sets were grouped, arranged into different subheadings and prospects for further data availability were explored. The various components included are the collection, classification and thematic arrangement of information to the designed backend databases on fish genetic resources in the country. The information on freshwater fish technology and statistics, marine fisheries technology and statistics, brackish water fish technology and statistics, inland fisheries technology and statistics and fish museums were stored into various data bases. Further, information on fisheries research projects, fishery GIS resource map on geographic and water body scale habitats were also populated in appropriate formats. Therefore, the National fisheries database comprised of 16 databases on above subject areas. Together these databases provide information on the fisheries statistics, production and trade statistics, culture status and technologies used, thus establishing a fishery information system at the national level that could aid in rapid data access in minimal time.

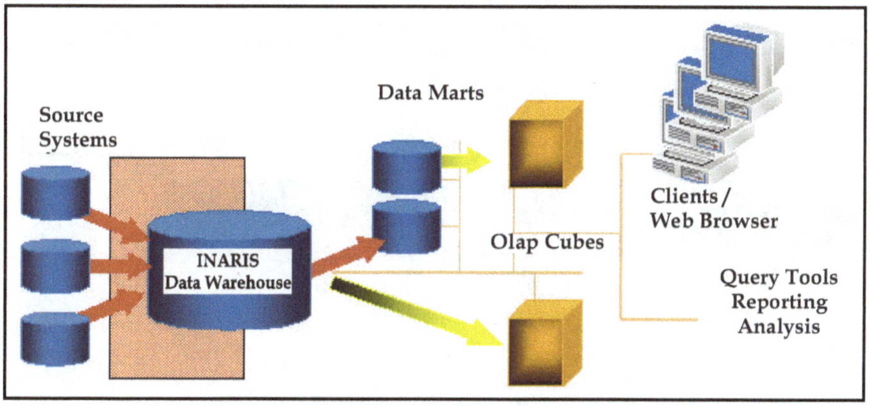

Fig 1. Architecture of CDW of INARIS

A prototype of the web enabled fishery database was prepared with the persistent efforts of all the fisheries institutions of Indian Council of Agricultural Research (ICAR), New Delhi. The features were again revised and the database was put on institutional intranet for feedback and further refinement. The databases were accessed under three major sub headings - Fish genetic resources, Fish technology and Fish statistics. Fish genetic database provides information on 2100 species of fishes reported from our country with their synonyms, fin formula, their latest taxonomic classification, habitats, morphology, fishery and distribution. Fish technology & Fish statistics database provides information on fishery production and export for both inland and marine sectors for the years after 1994, the areas under aquaculture / potential area in various states, the technologies used in aquaculture, the list and details of fish specimens kept at various museums of the country, the projects undertaken by the cooperating centers etc.

The freshwater fish distribution information collected from internet and published literature has been arranged on the digitized GIS maps on different levels like political boundary, region, basin level, watershed, river/drainage, which provides the distribution information in interactive mode can be extracted about a species on these scales. This fishery information system thus developed was then put on the Internet (Fig.2), institutional intranets. The project implementation resulted in the taxonomic validation of Indian fishes (in terms of literature) by comparing the synonyms and valid names. The availability of such a large volume of data in an easily retrievable form could facilitate rigorous work in the field of fish and fisheries. Overally the project was a new initiative to bring about socio-economic progress and environmental stability through integrated management of fish and fisheries with other relevant aspects.

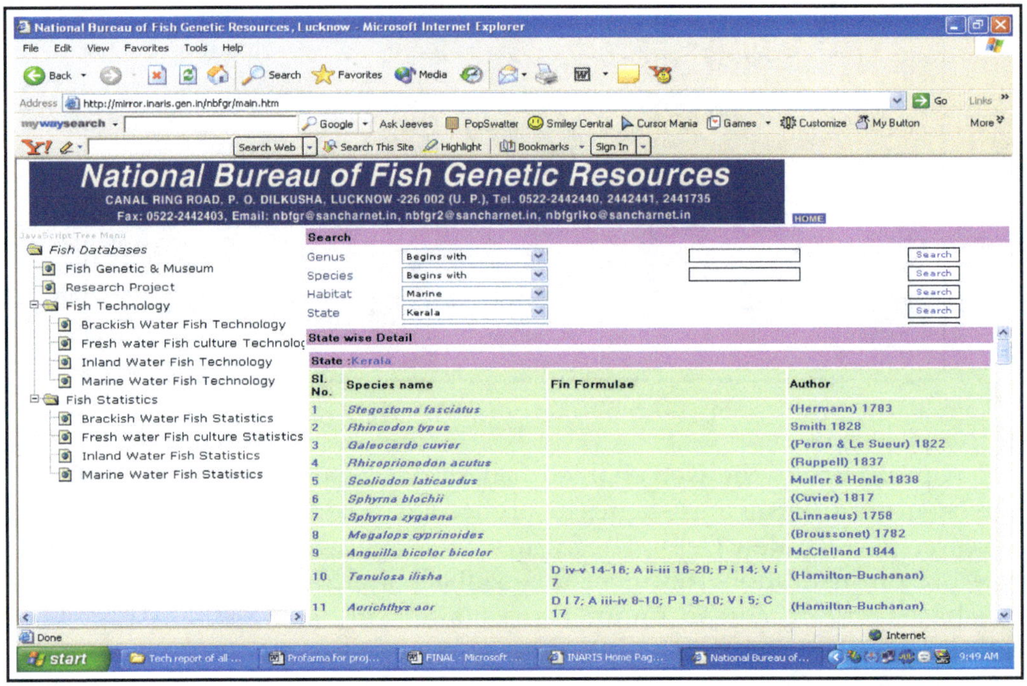

Fig. 2. Website of fisheries information system

Dimensional Modeling: A data warehouse is a "subject-oriented, integrated, time variant, non-volatile collection of data that is primarily used in organizational decision making" (Inmon, Bill 2005, Kimball, Ralph 2002). The data warehouse is maintained separately from operational databases. Since, data warehouse contain consolidated data, perhaps from several operational databases, over potentially long periods of time, they tend to be order of magnitude larger than operational databases. The organizational data warehouses are projected to hundreds of gigabytes or terabytes in size. The query mostly adhoc in nature can access millions of records and perform a lot of scans, joins and aggregates (Gupta, A. *et al.*, 1995, O'Niel, P. *et. al.* 1995, 1997). In this system query throughput and response time are more important than transaction throughput.

In case of complex analysis and visualization, the data in a data warehouse is typically modeled in multidimensionality. It requires some of the additional steps with OLTP systems. This type of storage is generally used for read only purpose. The updating will be done on periodic basis. This data storage is again converted into a form of multidimensional model, known as cube. If any user is retrieving only facts irrespective of any dimension, it is known as 0-D (Apex or Zero dimensional) cube. This type of cube will be meaningless at most of time. If it contains information specific to one field or dimension, then it is known as 1-D (One-Dimensional) cube. Similarly, if it contains two dimensions or fields, it is known as 2-D (Two-Dimensional) cube. For example, district wise area and production for different years. In the same way, if the table contains n-dimensions or fields for visualization of fact, it is known as multidimensional cube. These cubes can be designed by using the fact and dimension tables from the database. Since, these cubes are deployed on Internet for on-line analysis, these are also known as On-Line Analytical Processing (OLAP) cubes. In these cubes aggregations are pre-calculated and stored in multidimensional form. After deployment of these multidimensional cubes on the server, end user can perform his analysis and export the desired result to his desktop or computer in any form such as, MS-Word, MS-Excel, ASCII Text file or well-known Acrobat Reader (pdf) format etc.

Designing Multidimensional Cubes: Macro level planning and decision support processes in agriculture require access to data on different fisheries resources at varying levels of detail or aggregation. These and other requirements must translate to dimensions and fact tables of the OLAP cubes. Location, Time, and Commodity are few of the common dimensions that transcend all warehouse models. But Location and Time dimension poses biggest problem in integrating data from the varied sources within the agricultural sector.

The problem of integration basically boils down to the four most common dimensional issues as follows:

- Granularity of Location differs among different sources
- Granularity of Time varies among different sources
- Multiple overlapping time domains.
- Aggregation and disaggregation of information at different dimensional hierarchies.

These dimensional issues lead to the design issues of the fact table. Similar to business sectors such as retail, telecommunications, and others, the agricultural sector also uses the Location dimension extensively for its warehouse applications. In Indian agricultural sector, the Location dimension presents a number of interesting issues. Location, also known as the geography dimension, usually has a clearly defined hierarchical structure. In this hierarchy is determined by the administrative mechanisms implemented by Government of India. The highest grain level of hierarchy is the entire country (National level) followed by states. Further, each state is divided into districts, which are further divided into tehsils, blocks and villages respectively in the lower hierarchy. Generally, information related to agriculture is collected at district or above level. Information related to various census conducted in the country is available at village level which is the lowest grain level in this hierarchy. Another problem is encountered with respect to historical data. Some of the information such as production of some of the commodities is presently available at the district level but historical data is only available at the state level. Availability of resources, requisite need for information, and policies present at that time had an effect on the collection of information at any given level. There are following issues which are associated with creation of dimensions in the above situations in the development of OLAP cubes:

- How many levels are required for the location?
- How to integrate the information coming from different sources (organizations) at different grain level?
- How to define fact table for these dimensions?

Table: Location hierarchies at different levels according to collection of informartion

Hierarchical Level	Hierarchy Name	Example
Level one	National	Import and Export Statistics (monthly) International Prices (daily/weekly) Production of minor produce (annual) etc.
Level two	State	Production of major inland fish species (annual), National Account Statistics (annual), Information related to various agricultural development project at state level (annual) etc.
Level three	District	Production and area of principal crops (annual), land utilization statistics (annual) etc.
Level four	Village	The information related to land use (annual), information on different census such as human (decennial), livestock census (quin-qennial) etc.

From the above table it can be seen that there are three candidates for the lowest level, namely, the fourth level of hierarchy. Note that data from these dimensional attributes cannot be merged together. It is also not possible to represent them with a common name, as the other attribute information associated at this level is quite different.

Rule of aggregations for each of the fourth level hierarchy of the Location dimension to roll up to the next higher level of hierarchy is different. In case of villages it may be a simple aggregation whereas in case of agricultural market where the parameter is price, a simple aggregation will not work - a weighted average or any other suitable method might be applied, and in case of meteorological parameters one has to apply spatial models to interpolate or extrapolate the data at the District level. It may be further noted that agricultural market or weather stations are not available in every district of the country. Availability of agricultural markets in different states for a particular commodity depends on its area, production, and consumption. Also, not every state or every district in a state produces all commodities and accordingly markets are located. Hence, it may not be possible to aggregate the lower level data for the each district or in that sense for each state for the upper level of hierarchy. So in this case our hierarchical structure of the dimension will either collapse or provide misleading information to the user.

The integration of information coming from different sources, especially from various organizational sources, is also a big challenge in the design of the OLAP cubes. Data collection takes place at different levels (national, state, district, etc.) utilizing different methods (surveys, census, observations, etc.) and by different organizations each with its own formats, procedures and objectives. Further, definitions, concepts and purpose are likely to be different for different parameters. More importantly, each source and method contributes to different kinds of errors. In spite of these issues, if any information is available at the lower level it is relatively simple to aggregate (roll up) to the higher level by taking care of all above-mentioned factors. But when information is only available at the higher level and we have to disaggregate (drill down) to lower level, the task becomes difficult. Most of the information related to agriculture is collected through agricultural surveys or census, which are designed to elicit responses at the national or state level. The surveys collect information about agricultural products and resources from respondents at the state or national levels. The regional or lower level estimates cannot be obtained from these with reliable precision. Although, sampling strategies are employed in collecting data at these levels, the assumptions of sampling design do not hold well at lower levels. So, for handling this situation we need to have highly sophisticated statistical or mathematical modeling techniques. We must also provide estimates of errors /risk associated with the data so that users have acceptable levels of confidence with the information

Design of Fisheries Data Mart : In a data mart design, grain level of the fact tables is to be decided first (Kimball, 1996, 1998, Kimball & Ross 2002). In most data mart designs, decision depends on the level of detail at which fact should be addressed, namely, performance measure. Bonifati et al. (2001) discussed details regarding designing of data marts from a data warehouse. Data on fisheries resources do not have all available measures of fact at the lowest level of detail. Since, there are a large number of data collection agencies and equally diverse resources for which the information is being collected, it is easy to visualize the heterogeneity of information of this sector. The problem is further compounded by the fact that there are no common standards that are applied in data collection procedure. Designing a data warehouse to integrate the collected information poses a formidable challenge to any data mart architect in such situation. In order to use the information at micro/ macro planning and decision-making level (Chen R., et al., 2003) data were integrated and aggregated properly. Further, it is also important to capture information at its lowest level so that there is no loss of information due to its aggregation from lower level to higher level of the data flow hierarchy. Therefore, several fact tables have to be designed to capture information at its lowest level.

Source System: In case of fisheries resources, operational source system is a set of following relational databases in third normal form.

(a) Fisheries genetic resources
(b) Statistics of freshwater aquaculture
(c) Technologies of freshwater aquaculture
(d) Statistics of inland fisheries
(e) Technologies of inland fisheries
(f) Statistics of brackishwater fisheries
(g) Technologies of brackishwater fisheries
(h) Statistics of marine fisheries
(i) Technologies of marine fisheries
(j) Research project data base

The data is stored in these databases at various levels of locations. Fisheries genetic resources data base contains the information of various fish species and their characterization. Statistics database related to fresh water aquaculture contains information on production of various fresh water fish species. Technology database of fresh water aquaculture contains detail information of various technologies of production of this sector. Similarly, other data bases contains information on statistics and technologies related to marine fish species, inland fish species, and brackish water fish species etc.

In research project database, information related to research projects being taken up in Indian Council of Agricultural Research (ICAR), New Delhi was compiled. These databases were designed, developed and populated in Orcale 9i at National Bureau of Fish Genetic Resources (NBFGR), Lucknow. The information from these databases are extracted from the source system and after suitable transformation and aggregations stored in staging area of the central data warehouse. Various data marts of this central data warehouse are being populated in this staging area. Consistency checks were also applied while extracting information from the source system. The extraction process was applied using ODBC and providing user authentication information of databases.

Dimensions and hierarchies in the data mart: Five important major dimensions were identified for development of fish resource data mart i.e.(i) Species (ii) Location and (iii) Time (iv) Project and (v) Technology. Fish Species is a subject dimension. It consists of fish genetic species, fresh water species, inland fish species, marine fish species and its groups. Fish genetic species hierarchy provides information related to different kinds of various fish species and its groupings and its classification. Location dimension consists of District hierarchy and State Lookup. Fig. 3 shows different dimensions and hierarchies of fisheries data mart.

District hierarchy has two level state and district and provides logical structure for fact tables, which has grain data at district level. Time dimension has only year lookup as a hierarchy due to the fact that all information is available with respect to a year. The project dimension has hierarchy as the center which was responsible for conducting the research and the lowest level is name of the research project. Similarly, technology dimension has the lower level of the hierarchy as the name of the species and upper level is the description of the technology.

Data Staging Area: This staging area of the architecture lies between data warehouse and source systems. In this data mart, the staging has been extensively used to apply all consistency checks on each of the variables moving from source system to the data warehouse. All transformations for making data and code of the data mart consistence and uniform were performed in this region. Further, different processes of aggregation were also run here before moving the clean data in to the data mart of the warehouse.

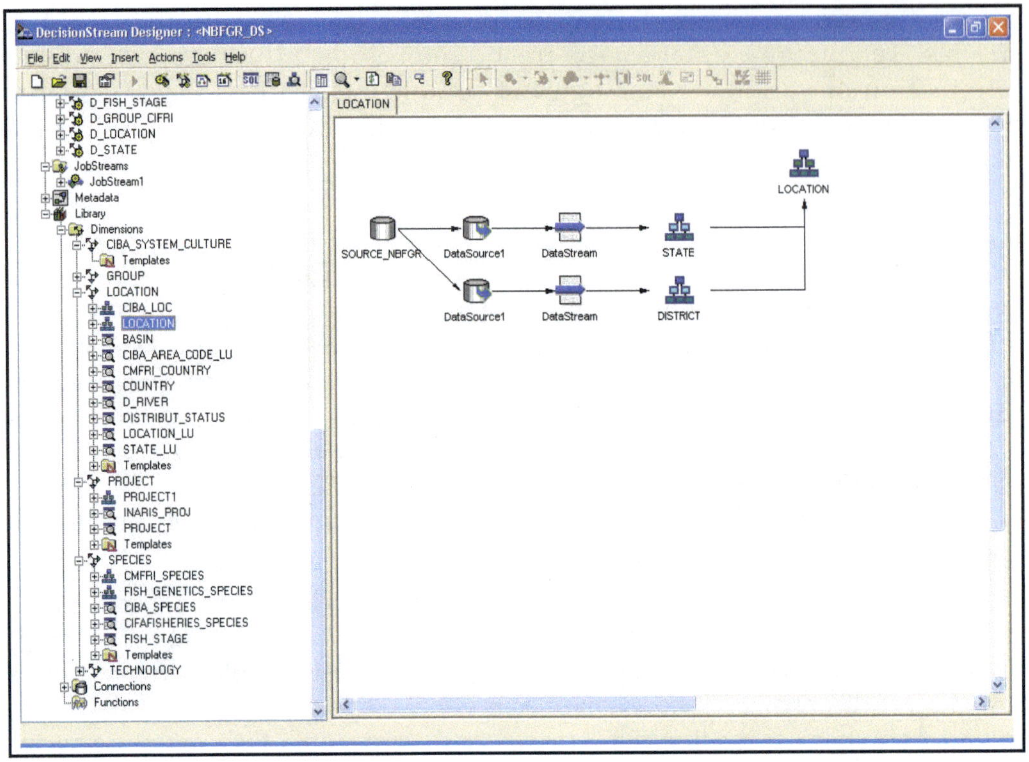

Fig. 3: Dimensions and hierarchies of Fisheries data mart

Facts Tables: Fact table are used to record actual facts and measures of the sector. Brief descriptions about these are provided below:

F_BASIN: This fact table consists of detail information related to water basin characterization of different animal species. Since, most the information is in textual form so there is no dimension associated with his fact. Fig 4 presents mapping process of this fact from source system to target data mart.

F_CIBA_COUNTR_STATISTICS: This provides year wise and country wise information about cultured area and production of brackish water fisheries at the international level.

F_CIBA_DISEASE_TREATMENT: This fact contains information related to various diseases and its management for brackish water aquaculture fish species.

F_CIBA_FARMER_AREA_ STATISTICS: This fact provides state wise information about the area of brackish water aquaculture and number of farmers for different holding size categories of the farmers for each year.

F_CIBA_FISH_TECH: This fact has information related to various technologies of brackish water aquaculture.

F_CIBA_FISHSTAT: This fact has district wise data about cultured area and production of brackish water fisheries for each year.

F_CIBA_SHRIMP_DIS_TECH: This fact has been created to provide specific information about the disease management in case of shrimp production.

F_CIBA_SPECIES_STATISTICS: This fact has national level data of production of various brackish water fish species.

F_CIBA_STATE_FISH STATISTICS: This fact provides state wise information about the production of various fish species brackish water aquaculture for each year.

F_CIBA_STATE_HATCHERY: This fact contains state wise information about the details of hatchery production for brackish water aquaculture.

F_CIFA_FISHERYSTATISTICS: This fact has detail state wise information about the production of fish species from fresh water aquaculture.

F_CIFA_TECHNOLOGY: This fact has information related to various technologies of fresh water aquaculture.

F_CIFRI_DISP_CATCH: This fact contains state wise information about total catch and its disposing for inland fish species of each year.

F_CIFRI_FISH_PROD_COUNTRY: This provides year wise and country wise information about production of inland water fisheries at the international level

F_CIFRI_PRODUCTIONSTATISTICS: This fact provides state wise information about the production of various inland fish species for each year.

F_CIFRI_SEED_PROD: This fact provides state wise information about the seed production of various inland fish species for each year.

F_CIFRI_TECHNOLOGY: This fact has information related to various technologies of inland fisheries.

F_CMFRI_EXPORT: This fact contains the information about exports of various marine fish species for each year.

F_CMFRI_FTECH: This fact has information related to various technologies of marine fish species

F_CMFRI_INARISPROJ_PROJ: This contains information about various research project taken up by the ICAR institute in the field of fisheries.

F_CMFRI_PRODNSTAT: It provides state wise and year wise catch statistics of marine fishes.

F_ECO_DISTRIBU_WATER_BODY: This fact contains information related to distribution of various water bodies in different ecological zones.

F_FISH_GENETICS: This fact contains detail information about various fish species along with its characterizations.

F_MUSEAM: This fact has the detail information about various international fish museums.

The process flow for designing and developing of fact table has been shown in the Fig. 4.

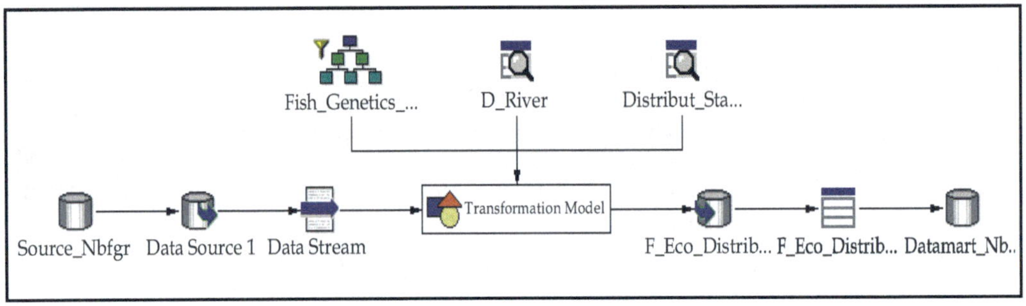

Fig. 4. Process flow for generation of fact table.

In order to generate fact table in a data mart the information from the source data is extracted and transformed in suitable form through transformation models. The various dimensions and look ups were logically linked with the information contents and then the fact table is created in the target data mart. After the establishment of complete logical design, the process is automated to populate the fact in the target data mart.

Designing data flow process of the data mart: The basic process of data warehouse requires the automation of data extraction, data transformation, data loading, exception/error handling, and logging/notification for the warehouse to function properly. Managing warehouse requires the coordination of various tasks involved in preparing the data marts. Once the data marts are created, performing these tasks is to be automated in a sequence or in parallel. Data warehouse administrator can be benefited from a well-constructed job control process that comprises the following:

- Build status notification (send email, if fails)
- Coordinates facts and shared dimension builds
- Data staging and cleansing prior to create a data mart
- Pre processing and post processing SQL
- Different arrival rates of source data etc.

The automation of this process has been obtained by using JobStream module which is a Cognos DecisionStream Designer component. It was used to coordinate groups of builds, processing instructions, conditions and SQL into an operational process. It was used to implement the job control processes needed to keep data warehouse or data mart functioning more smoothly. There is a node for each type of operation and operations are performed in a predefined sequence to maintain consistency of the system. The sequence or parallel nodes can be arranged and execute accordingly. The ultimate warehouse operation would run the regular load processes in the lights-out manner, that is, completely unattended. Ideally it is difficult to obtain but it is possible to get closer to this situation (Kimball et al, 1998). In the automation process first dimensional builds are run to create dimensions of the data mart. These dimensions are connected with a node at the center, which checks the consistency of the dimensional table. After execution of all dimensions facts are executed. Finally facts and dimension table are populated in the data base structure.

Development of On-Line Analytical Processing System (OLAP): In order to develop OLAP system for users data populated in fact and dimensional tables needs to be packed in a multidimensional cube structure. Multidimensional cube is a structure that organizes data on the basis of dimensions. Data access perspective from users point of view is to be modeled from dimensional tables. Hierarchies of dimensions represent logical flow and aggregations of the data from bottom to top and dis-aggregation from top to bottom. Building models for developing multidimensional cube mainly depends on users requirements. Generally development of data flow hierarchy on depends facts and its associated dimensions. These cubes are posted on webs in a ready to browse form for OLAP (Kambayashi *et. al.* 2004, Craig *et. al.* 1999, Scalzo *et. al.* 2003, Kimball R. 1996) analysis by the decision makers and can be accessed from anywhere by authenticated user with the help of web browser and internet connectivity. Fig.5 shows fisheries production at district level multidimensional cube in the form of *OLAP through Internet browser*. In this OLAP cube levels are All States, All groups, All Years and Measures. All States has state names as a top level and district as bottom level of data flow hierarchy. All groups has group names of various fish species. All Year has only one level i.e. years. Similarly, Measures has only production information.

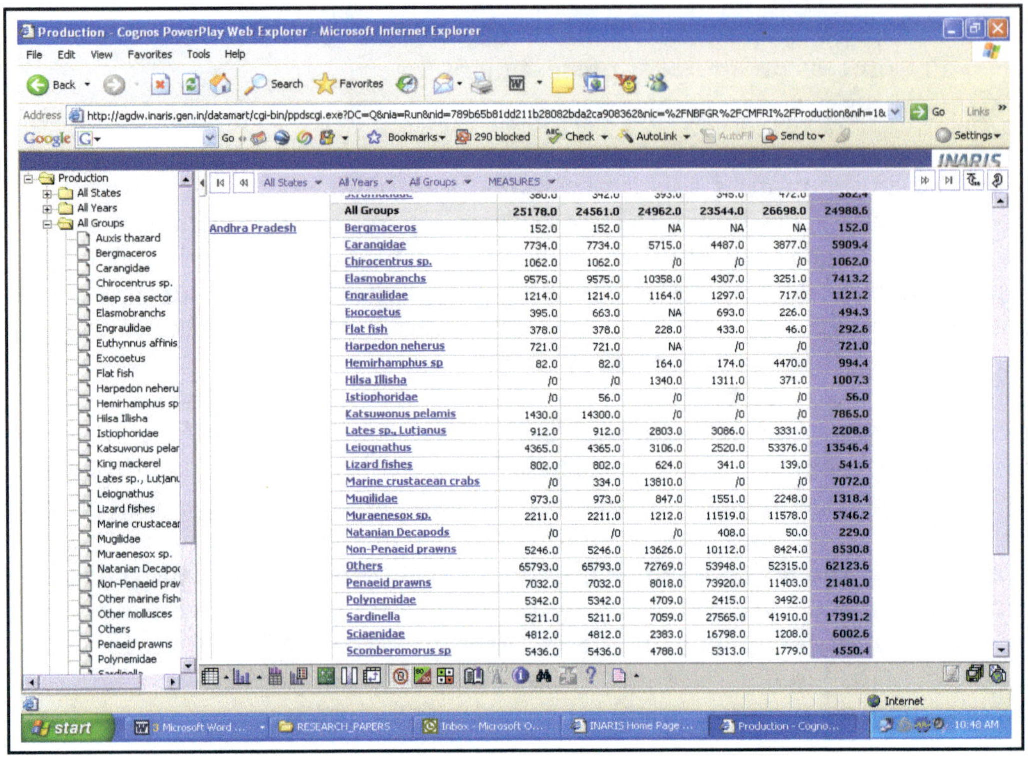

Fig. 5. Tabular presentation of information from OLAP

A multidimensional OLAP cube provides on-line decision support system. This online system has drag and drop option for creation of nested tables, drill up and drill down functionalities based on hierarchies of various dimensions. Number of simple calculation options on tabular data. Hide and show option to hide certain undesirable rows or columns to be displayed on the screen. Graphical representation options such as line, bar, charts three-dimensional graphs etc. (Fig.6) through single click of a button. Aggregations, segregations, slicing and dicing options are available with out any additional requirements.

The zero suppression option suppresses all rows/columns with no information. It also has 80/20 options, which display major rows/columns contributing to 80% of grand total of respective row/column. This may be used to identify major contributors in total. It has option of exceptional highlights in which well-defined exceptions in the tabular information can be highlighted with different text and cell colors. The history of a particular session of analysis is recorded and can be seen whenever is required by single click of a button. This will help user to look into list of operation performed by him in a particular session. The find and search option is also available for finding a particular information in tabular data of a cube. It also has help option to assist user in working with this decision support system. The most important option is the any table/graphs can be exported to number of format file through a single click of a button.

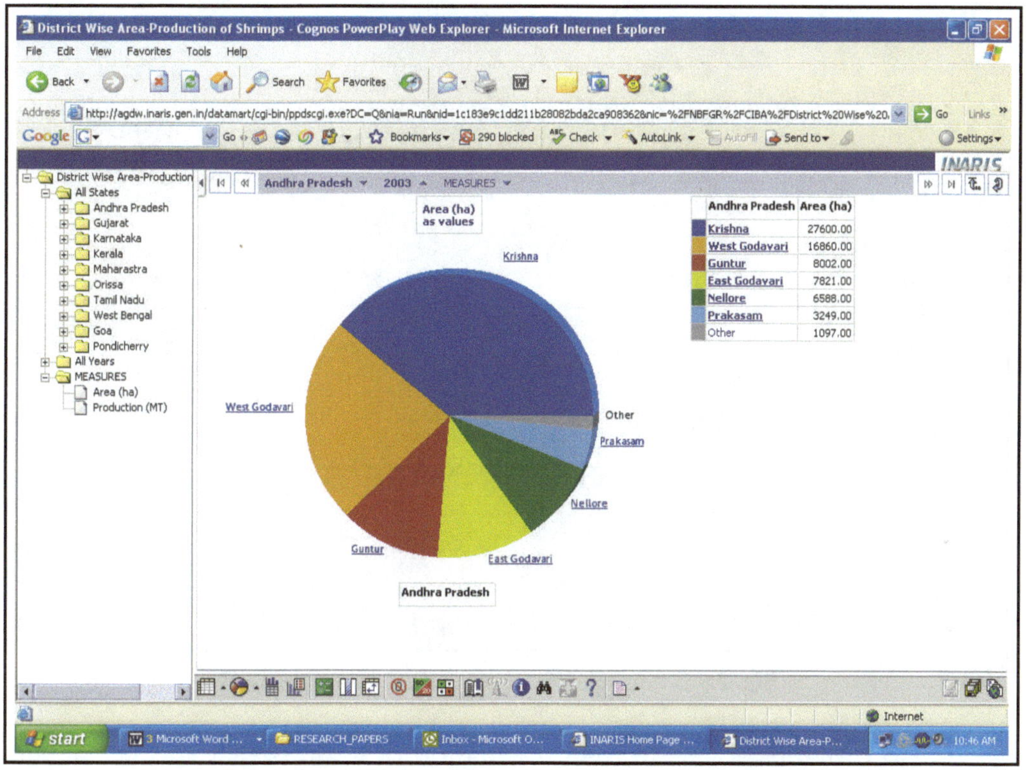

Fig. 6. Graphical represssentation of information from OLAP

Apart from these options it has functionalities of disseminating information through dynamic reports. In dynamic reports, a query is fired through web browser to the data warehouse database and preformatted reports are generated on-line in PDF/HTML format (Fig.7).

This decision support system also has functionality of on-line spatial analysis of information in the data mart. In this case user has access to web based Geographic Information System (GIS). In order to access this system user need not to install any GIS software. The authenticated user used to access this system using simple Internet browsers. This system helps in providing access to GIS technology to general user and decision makers with no extra cost and efforts.

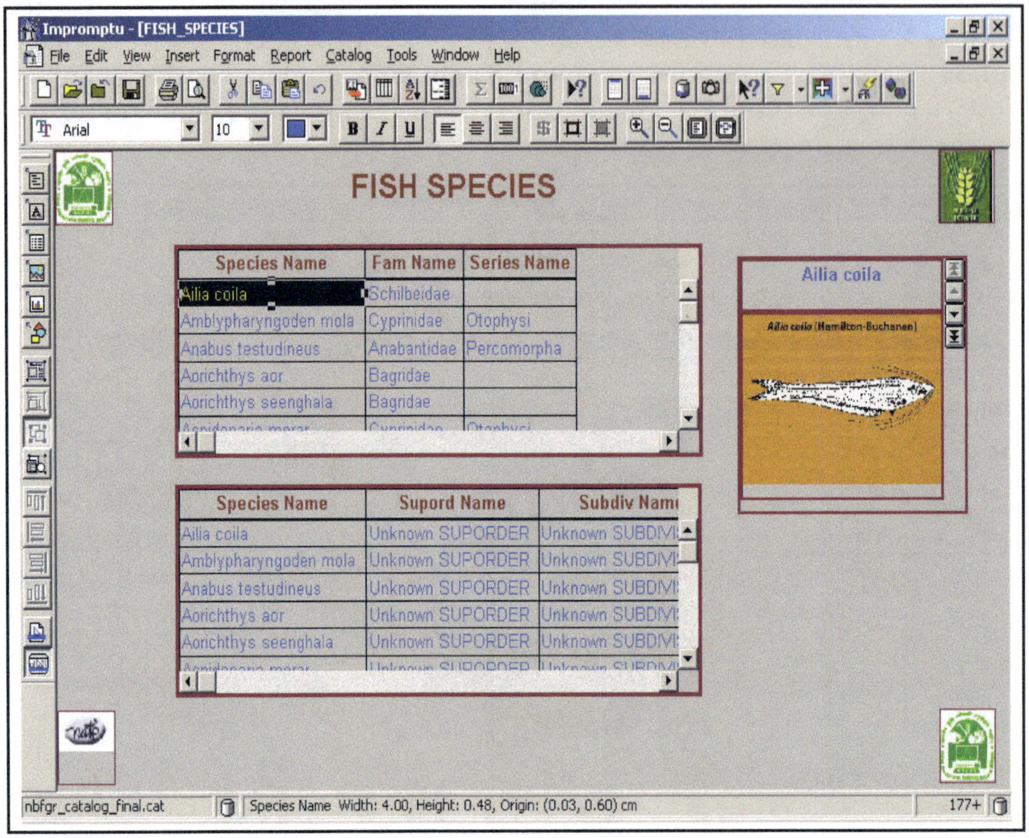

Fig. 7. Dynamic report from animal data mart

It also provides access to the information of various fish species and its characterizations through the web site developed under this project (Fig. 8). This web site has been launched at NBFGR, Lucknow.

Issues of selection of data mart/warehousing tools: The selection of data warehousing tools depends on number of factor such as type of data, amount of data, nature and type OLAP required by users, quality of data, heterogeneity of data sources etc.. The success of any data warehousing development process is highly related to selection of these tools. The requirement of tools varies from project to project. All developmental tools of data warehouse can be categorized in three categories i.e. (i) ETL tools (ii) OLAP modeling tools and (iii) reporting tools.

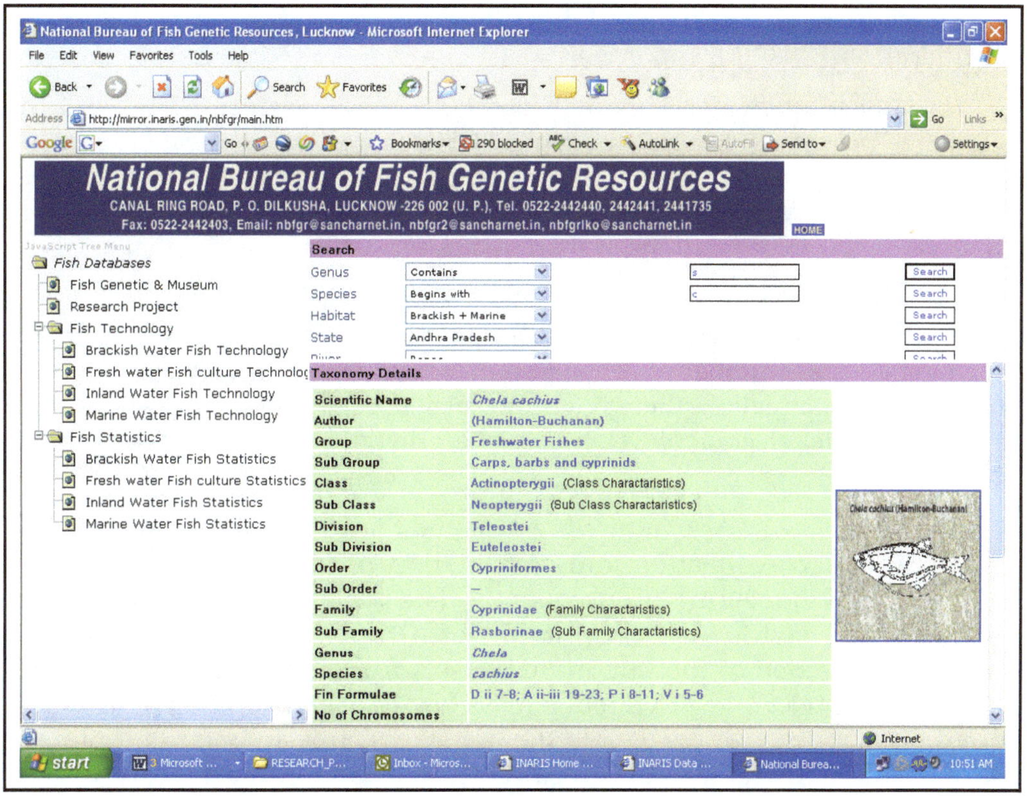

Fig. 8. User front-end of web GIS from fish data mart.

1. ETL tool Selection

Extraction, Transformation and Loading (ETL) is the most important component of data warehouse architecture. It is reported that 70% to 80% time of any project is consumed in implementing processes related to ETL. It is a backbone of any data warehouse project. Using these tools, data is extracted from the original source, transformed after cleaning in a suitable form and automatically loaded in a data warehouse at appropriate location. In case of ETL tool selection, the following points need to be considered:

- *Complexity of the data transformation:* The more complex the data transformation is, the more suitable it is to acquire an ETL tool.
- *Data cleansing needs:* It is best to procure a tool with strong data cleansing functionalities.
- *Data volume:* ETL tools must have the feature that can speed up data movement.

To evaluate ETL tools, the following characteristics must be taken into consideration

Functional capability: This includes both the 'transformation' and the 'cleansing' process. In general, the typical ETL tools are either meant for having strong transformation capabilities or having strong cleansing capabilities, but they are seldom very strong in both. If you know there are going to be a lot of different data transformations, it then makes sense to pick a tool that is strong in transformation.

Ability to read directly from your data source: In agriculture, data sets are heterogeneous as number of organizations are involved in collection of data. So, for each organization, there is a different set of data sources. It is advisable to make sure the selection of ETL tool connect directly to heterogeneous source data.

Metadata support: Agriculture is a vast and highly diversified subject. The responsibility for collection and compilation of data lies with different organization of the country. Consequently, data is available in different format by following various concepts and definitions. Therefore, metadata management is one of the important aspects for getting success in a data-warehousing project. The ETL tool plays a key role in generation of metadata because it maps the source data to the target, which is an important piece of the metadata. In fact, some organizations have come to rely on the documentation of their ETL tool as their metadata source. As a result, it is very important to select a metadata tool that works with overall metadata strategy.

2. OLAP Tool Selection

On-line Analytical processing tools enables developers to design and develop multidimensional model, which is finally published to web for query purpose. Selection of OLAP tool is quite critical from users perspective in development of decision support system. OLAP tools are oriented towards roll-up, drill-down, pivoting, slicing and dicing of the data. Generally OLAP tools in general can be classified as MOLAP (Multidimensional OLAP), ROLAP (Relational OLAP) and HOLAP (Hybrid OLAP).

MOLAP: In this type of OLAP, a cube is aggregated from data warehouse. When user generates a report request, the MOLAP tool can generate reports quickly because all data is already pre-aggregated within the cube.

ROLAP: In this type of OLAP, instead of pre-aggregating everything into a cube, the ROLAP engine essentially acts as a smart SQL generator.

HOLAP: It is Hybrid On-Line Analytical Process, which have combination of MOLAP and ROLAP functionalities.

Following criterion is to be followed for selection of OLAP tools

Ability to influence parallelism supplied by RDBMS and hardware: This would greatly increase the tool's performance and help loading the data into the cubes as quickly as possible.

Performance: In addition to force parallelism, the tool itself should be quick both in terms of loading the data into the cube and reading the data from the cube.

Customization efforts: More and more, OLAP tools are used as an advanced reporting tool. This is because in many cases, especially for ROLAP implementations, OLAP tools can be used as a reporting tool. In such cases, the ease of front-end customization becomes an important factor in the tool selection process.

Security Features: In general OLAP tools are geared towards a number of users, making sure people see only what they are supposed to see is important. OLAP tools should have a security layer that can interact with the common corporate login protocols.

Metadata support: OLAP tools aggregates the data into the cube and sometimes serves as the front-end tool, it is essential that it work with the metadata strategy which is selected.

3. Reporting Tool Selection

Data is useless if it is simply put idle in the data warehouse. Therefore, the presentation layer is very important for providing the information from data. Apart from OLAP tools, there should be a reporting tool, which can summarize the information in a desired format. Following criterion may be used for selection of reporting tools:

Data source connection capabilities: In general there are two types of data sources, one the relationship database and other is the OLAP multidimensional data source. Nowadays, chances are good that organization might want to have both. Many tool vendors offer both options, but upon closer inspection, it is possible that the tool vendor is especially good for one type, but to connect to the other type of data source, it becomes a difficult exercise in programming.

Scheduling and distribution capabilities: The reporting tool must have scheduling and distribution capabilities. Reports are scheduled to run, and the resulting reports are distributed to the users executives either by email or web publishing.

Security Features: The reporting tools, similar to OLAP tools, are geared towards a number of users, making sure people see only what they are supposed to see is important. Security can reside at the report level, folder level, column level, row level, or even individual cell level.

Customization: The reporting tool should provide an easy way to pre-set the reports to look exactly the way that adheres to the corporate standard, it makes the analysts jobs much easier, and the time savings are tremendous.

Export Capabilities: The most common export needs are to Excel, to a flat file, and to PDF, and a good report tool must be able to export to all three formats. A reporting tool must have the facility to export its information into common formats.

Issues of Selection of Schema for Designing a Data Mart: The snowflake schemas and star schema are dimensional models. The basic difference is in their physical implementations. Snowflake schemas support ease of dimension maintenance because they are more normalized. Star schemas are easier for direct user access and often support simpler and more efficient queries. The decision to model a dimension as a star or snowflake depends on the nature of the dimension itself, such as how frequently it changes and which of its elements changes. Very often it involves evaluating tradeoffs between ease of use and ease of maintenance. It is often easiest to maintain a complex dimension by snowflaking the dimension. By pulling hierarchical levels into separate tables, referential integrity between the levels of the hierarchy is guaranteed. However, it is important to present a simple and appealing user interface to target users who are developing ad hoc queries on the dimensional database. It may be better to create a star version of the snowflake dimension for presentation to the users. Often, this is best accomplished by creating an indexed view across the snowflake dimension, collapsing it to a virtual star.

Implementation Issues of the Data Mart/Warehouse: There are number of issues related to a data mart. Some implementation issues are described below:

1. Uniformity of information on a particular commodity collected by different organization

It has been observed that information about a particular commodity is being collected by different government/private organizations and there are wide differences. Even different government organization provides information which differs so widely that users loose their confidence. The basic reasons attributed to these differences are due to non-uniformity of the definitions and methodology. The methodology for data collection by these organizations depends on the mandate of the institution, purpose, technical capabilities, available resources etc. Sometimes, it has been observed that no objective scientific methodology has been adopted for data collection. The problem of differences in the statistics becomes more prominent due to lack of coverage of the information also. These problems exists almost every sector but it is more visible in the sector where national and international trading are extensive. Unfortunately, the data collection methodology is generally not accessible to the users so that one can try to bridge the gaps among these statistics. In case of development of the warehouse for this information selection of appropriate source is one of the major issue. If information of all the sources is populated in to the warehouse, it may create lot of confusion to the general warehouse users and decision makers.

2. Integration of the information with in a data mart using common dimensions

The integration of information related one data mart to another is a challenging task even sometimes not possible. The information of different parameters of a particular commodity is collected by different organizations following their own definitions and format. The grain level of information collected and published from these sources depends on their requirements and its demand on a particular period of time. For example INARIS data warehouse contains information related to agro-meteorology since 1950 on agriculture related climatic parameters associated with agro-meteorological stations but time grain level of recording previous years information on most of the parameters are weekly whereas presently it is recorded on daily grain level. In some of the cases the grain level of different fields with in the same data marts widely differs from each other for example in case of horticultural data mart of the INARIS data warehouse the imports and export statistics is available at location dimension at national level and time dimension at months level, the area and production of some of fruits and vegetables at location dimension of state and time dimension as agricultural year where as for other important fruits and vegetables for some of the states it is available at location dimension of district, the price statistics of different important fruits and vegetables are available at location dimension of agricultural markets with a district and time dimension of daily/weekly/monthly. There are only two possibility of integrating all the parameters in to one multidimensional cube. First by aggregating the parameters available at lower grain level to a minimum higher level of hierarchy and loosing the detail information, second is pooling all the parameters of the same level of hierarchical information in to different cube and forget about their integration. The statistical techniques of their interpolation or extrapolation may lead us to misleading conclusions.

3. Definition of aggregation rules in multidimensional cubes for drill-down and drill-up with in a hierarchy

The issues of defining rules for aggregations are very pertinent in the multidimensional cubes where drill-down and drill-up facility of decision support system is the most desirable feature. In ideal conditions this problem may exists when standard data quality checks and techniques of aggregations are applied to the data before releasing the same. This problem is more dominating in the historical data generated by different government organizations as in those days there were limited options for application of above techniques due to limited computing facilities. Probably, the best way to tackle this issue is to generate additional level in each of the hierarchical stages and accounts the differences of the data with the help of this psuedo generated level otherwise it may not be possible to have drill-up and drill-down in the cubes and the whole structure of the cubes will crash. The second very important issue in this area is that the rules of aggregation will be different for different measures over different dimensions and the selection of implementation of these rules should be done very carefully.

4. Data quality and data gaps in the historical information

The data quality is very important issue for design and development of successful data warehouse. If the data qualities of the databases are not good whatever efforts are done in design and development of a warehouse is a futile exercise. The developers will loose the confidence of their users in a very short span of time and support for these developments from the management of the organization will become weak. Again, it has been found that data gaps and data quality problems are more dominant in case historical data due to lack of awareness and importance of the information generated at that time. These problems in the past were more frequent due to lack of availability of proper technical and computing support in this area. It is understood that in this era of information revolution these problems are likely to reduce very fast. Unfortunately, the information related to agricultural sector in this area is more vulnerable as compare to other sector. The agricultural statistics are mainly collected by non-profit making government organizations and there are no direct visible losses due to data quality problems. Therefore, not much of attentions were paid in the past for data quality and data gaps. It has been seldom realized there are problems and in some cases huge losses to the country due to lack of proper planning by using these data.

Recently, the problem arises due to data quality and data gaps have gained importance due to computerization. Now it is easy to detect and correct handle these issue after computerization. The best solution of this may be to apply check programme while computerization and development of large databases where it is easy to detect and apply corrections.

5. Conformed dimensions for integration of information across the data marts.

Macro level planning and decision support processes require access to data on different resources (e.g., crops, livestock, fisheries, etc.) at varying levels of detail or aggregation. Information on the levels of production (demand and supply), price levels, and population and migration statistics is also expected from the systems. These and other requirements must translate to the dimensions and fact tables of the data warehouse. Location, Time, and Product are a few of the common dimensions that transcend all warehouse models. But Location and Time dimensions pose the biggest problems in integrating data from the varied sources within the agricultural sector. The problem of integration basically boils down to the four most common dimensional issues as follows:

- Granularity of Location differs among the different sources.
- Granularity of Time varies among the different sources.
- Multiple overlapping time domains such as Calendar Year, Agricultural Year and Financial Year.

Aggregation and de-aggregation of information at different dimensional hierarchies and the problem of integration can be visualized more clearly by looking in to the details of dimensions and cubes developed under different marts in this agricultural data warehouse. The information is usually collected as per the requirements by different organization mainly by surveys, census or administrative records. Also, the information may be available at different grain levels of time like daily, weekly, monthly or annual. Further, different definitions are followed for year depending on the commodity and the requirements such as calendar year, agricultural year and financial year. The information related to the indicators of performance is only available at one point on time. It can be seen by this example that information from these cannot be combined in cube, as integration of this information is not possible without disaggregating at the common minimum level. The combination of these information of various agricultural sector not only produces highly sparse cube but also leads to misleading decisions when aggregated to higher level with common aggregation rules. This problem of development, design, integration and implementation of the data warehouse can be solved by disaggregating the information coming from different sources at the lowest grain level wherever possible by using statistical or mathematical techniques before loading to the data warehouse. This can be done in the staging area of the warehouse. This will not only provide simplicity and flexibility to the data warehouse design but leads to better data mining and decision making. The decision support system built on this data warehouse will be definitely stronger as compare to the data warehouse, which is built without following the above procedure. Let us take a simple example of de-aggregation of census and survey information at the relevant information level. It may be possible due to the fact that survey information is at least available at the year level on time dimensional hierarchy whereas census information may be available after every 5-year or 10-year. Some of the information may not be available even after this time gap. So the census information can be disaggregated at least at the year level of time hierarchy. Similarly, survey information may not be available at district level of location hierarchy. It is generally available at state level or above whereas census information is always available at the lowest level of location dimensional hierarchy may be village. So the information of the census can be used for disaggregating the location hierarchy of the information coming from surveys. Both can be further used to lower the hierarchy of different relevant dimensions.

Issues Between Developer Organizations

1. Responsibility and control for development of data warehousing

The issue is whether the data warehousing development group should be a freestanding development organization or it will work under some apex body. These data warehousing systems should be protected from other external agencies as far as information storage and retrieval is concerned. Placing the data warehousing effort in a separate development group can lessen knowledge transfer and appreciation of how to make data warehouse safe.

2. Administration of data warehouse

The need to make data warehouse database structure changes can be relatively frequent, proliferating data marts, uncertainty about the usage pattern, and the nature of the data warehouse development can necessitate table and indexed changes. The data warehouse developers, concern about using the interest of data warehouse users, want changes made quickly and get quite frustrated being put on DBA backlog. On the other hand, DBAs often have knowledge about how to make database processing safe. Cutting the DBA organization of the data warehousing support loop can deprive the data warehousing effort of some valuable wisdom.

3. Cooperation from participating organization

Data warehousing efforts often bring problems if the feeder system organizations viz., ICAR Institutes/ SAUs/ other organizations feel that they are going to loose much as compared to the gain in the data warehouse development efforts. Sometimes, these organizations may be reluctant to help if they feel that the data warehousing effort is going to be an audit of their work.

4. Problem identification and rectification

Actually, the question often becomes whether (i) the feeder system should be fixed, or (ii) the feeder system should be left alone and the data in the warehouse should be fixed, or (iii) data should be fixed in data warehouse with the fixes fed back to the feeder system.

5. Data dissemination

A decision has to be taken about what kind of data should be made available to the different organizations, as some information is sensitive in nature and may create problems to the information generating organization as well as the data warehouse developers.

6. Data updation

Often there is a need for a time period to be decided so that processing systems are kept stable and changes made to the systems can be captured and fed into the warehouse. Sometimes, identification of the changes is not possible. A typical course of action is to compare a previous copy of the database with the current database. After changes are identified, a copy of the current database is made for comparison in the next processing cycle. Frequent changes of the information in the data warehouse may create confusion to the user organizations.

7. Data quality monitoring

Data quality is not a one-time concern while implementing data warehouse. In a complex feeder system it is not uncommon for previously undiscovered data quality problems that occur after the big push to clean data after the initial load of data warehouse is done. It is necessary to install procedures to regulate and audit the data quality. Many times it is not clear that who should have the responsibility for executing these procedures.

8. Rectification in data warehouse design

Small changes in the feeder system can have a major impact on the feed to the data warehouse. Many times conflicts arise among the system developers under the pressure from their users to make changes. A more complex situation is when a change is made in the system and it is not communicated to the DBA.

Issues Between Developers and Users

User issues can be especially thorny with data warehouses because unlike the developer organizations they have different kinds of queries to be made. Unless the data warehouses are tailored to their preferences, users may quickly decide not to use the data warehouse.

1. Control over databases

Many user departments have on their own developed databases that meet some of their key reporting needs. Often, these systems were built by user organizations on their own because the data warehouse developers were unwilling or unable to help the users. It is highly likely that when a data warehouse will subsume the function of these user-managed databases, users feel that these organizations cannot do a good job of fulfilling their needs.

2. Cooperation of users

Often part of the goal of a data warehouse is to automate the production of information that has been manually created by a user. Sometimes the user's identity is tied to the information and he/she feels threatened by the prospect of automation. This user's cooperation will be needed in the data warehouse development. Though dealing with this sensitive personal issue probably should be the responsibility of the user's management.

3. Needs of users

In the development of a data warehouse it should be clearly indicated that whether it is for highly technical and most demanding users or it will be used by a larger group of non-technical users as the needs of these two groups of users are quite different from each other.

Issues Between Developers and Other National/ International Organizations

There is a need to resolve the issues related to developers (NARS) and other National and International bodies at higher-level management. MoU has to be signed by the different organizations regarding the sharing of information and the usage of the data warehouse developed. Also, some of the officers are to be identified in these organizations for the proper coordination and monitoring of the information sharing process.

Need of Data Warehousing: The following are the important reasons for developing data warehousing in National Agricultural Research System (NARS):

- Improve the quality of research and planning
- Reduce duplication of research efforts
- Encourage dissemination of research findings
- Facilitate qualitative research supported by agricultural databases
- Help in the development of Decision Support Systems (DSSs)
- Used as effective tools for agricultural research and education planning
- Develop effective linkages with other national organizations in sustainable development
- Develop effective linkages with international organizations working in the same area
- Improve the efficiency of the National Agricultural Research System (NARS) by effectively managing its resources
- Useful for other national and international organizations in planning their related activities.
- Utilizing library resources of various organizations effectively

REFERENCES

Bonifati, A., F. Cattaneo, S. Ceri, A. Fuggetta and S. Paraposchi, 2001. Designing data marts for data warehouse. *ACM Transactions of Software Engineering & Methodology,* 10(4):452-483.

Chaudhuri S. and K. Shim, 1995. An Overview of Cost-based Optimization of Queries with Aggregates. *IEEE Data Engineering Bulletin.*

Chen, R., C. Chen & C. Cheng, 2003. A web-based ERP data mining system for decision-making. *Int. J. Comp. App. and Tech.,* 17(3):156-158.

Craig, R. S., J. A. Vivona & D. Bercovitch, 1999. Microsoft Technologies. John Wiley & Sons, New York.

Gupta, A. and I. S. Mumick, 1995. Maintenance of Materialized Views: Problems, Techniques, and Applications. *Data Eng. Bulletin,* Vol. 18 (2).

Inmon, Bill, 2005. Building the Data Warehouse. Fourth Edition, John Wiley, New York.

Kambayashi, Y., V. Kumar, M. Mohania & S. Samtania, 2004. Recent advances and research problems in data warehouse. *Lecture Notes in Computer Science*, 1552:81-92.

Kimball, R., 1996. The Data Warehouse Toolkit: Practical Technique for building data warehouses. John Wiley & Sons, New York.

Kimball, R., 1998. The Data warehouse lifecycle tool kit John Wiley & Sons, New York

Kimball, R. & M. Ross, 2002. The Data Warehousing Toolkit. John Wiley & Sons, New York

O'Neil P. and G. Graefe, 1995. Multi-Table Joins through Bitmapped Join Indices. Proceeding of SIGMOD Conference.

O'Neil P.and Quass D., 1997. Improved Query Performance with Variant Indices. Proceeding of SIGMOD Conference

Scalzo, B., 2003. Oracle DBA guide to data warehousing and star schema. Prentice Hall, New York

Zhuge Y., H. Garcia-Molina, J. Hammer and J. Widom, 1995. View Maintenance in a Warehousing Environment. *Proceeding of SIGMOD Conference.*

http://www.iasri.res.in

http://www.inaris.gen.in

Chapter **20**

Application of Softwares in Molecular Marker Data Analysis

P.K. Meher, A. Barat & P. Das

Introduction

The development of DNA technology over the last few decades has led to generation of huge amount of data from plant and animal genome. The main sources of markers are mitochondrial DNA (mt DNA), minisatellites, micro satellites and anonymous nuclear sequences assayed using highly specific PCR primers or by using the RAPD technique. A major contribution comes from molecular markers used for analyzing genetic variations. There has been the infusion of computers, softwares and online databases into the daily activities of many molecular geneticists. The identification of abundant, highly variable loci, detectable by a wide variety of techniques has led to a revolution in the way researcher view and analyses levels of variation. We will here discuss about current computer programs in this area to characterize the DNA profiles and data analysis for intra and inter population structure, phylogenetic inference and parentage determination.

In fish breeding, genetic markers can be used to investigate the genetic diversity present in the population to be used as brood stock. The breeding population can also be tested for Hardy-Weinberg equilibrium. The bands yielded from both dominant and codominant molecular markers such as

Applied Bioinformatics, Statistics and Economics in Fisheries Research: pp. 293-300

New India Publishing Agency, New Delhi (India)

Allozymes, RAPD, AFLP, RFLP, microsatellite etc. are converted into numerical data matrix to use in different genetic data analysis software. Plenty of software and computing materials are available in Internet and standalone Personal Computers. These are easily available and most of them are free to download from Internet. Here few selected programs are discussed, which are widely used globally.

Programmes

Various programmes are available now in the Internet. Some of these programmes are used for codominant and some are for dominant markers.

1. Genetic Data Analysis (GDA) version 1.1

The authors of this programme are Lewis and Zaykin (2000). GDA uses a data file format known as NEXUS. NEXUS files are plain text files, which can be viewed by any programme that is capable of displaying standard ASCII text files. These data files can be created by any word processor.

Analysis consists of a table of descriptive statistics, tests for Hardy-Weinberg and pairwise disequilibrium, linkage disequilibrium coefficients, estimation of F-statistics, bootstrapping over loci, genetic distances between populations and a dendrogram based on genetic distances.

Descriptive statistics include mean sample sizes over all loci, mean number of alleles per locus, the proportion of polymorphic loci, the mean number of alleles per polymorphic locus, the expected heterozygosity, the observed heterozygosity and an estimate of the fixation index (f_{is}). Tests can be performed for any departure from Hardy-Weinberg equilibrium at individual loci through an exact test and linkage disequilibrium will also be tested for all possible pairs of loci. Bootstrapping options include outputting a list of replicate estimate of f (=f_{is}), F (=F_{it}) and Theta (=F_{st}). Jackknifing across populations to obtain variances of F-statistics estimates for individual lociis also available. The genetic distance measures will be clustered using the UPGMA (Unweighted Pair Group Method using Arithmetic averaging) and a tree depicting these phenetic relationships among populations will be drawn.

2. GENEPOP version 3.4

GENEPOP is another population genetic software package authored by Raymond and Rousset (1995). This software runs on any Window processors including Pentiums II and III without any need for extended memory. It is not protected and available free of charge. Haploid or diploid codominant data of micro satellite markers can be analysed with this software.

This software estimates, i) exact tests for Hardy-Weinberg equilibrium, for population differentiation and for genotyping disequilibrium among pairs of loci, ii) F-statistics and other correlations, allele frequencies etc and iii) analysis of isolation by distance.

GENEPOP provides P-values for Hardy-Weinberg equilibrium, linkage disequilibrium and population differentiation by Exact tests, either by complete enumeration or by Markov chain methods. The U-test for heterozygosity excess or deficiency is also available. Global Hardy-Weinberg test across loci or across populations is another option for GENEPOP. *F*-statistics or *Rho*-statistics can be calculated for all populations or all pairs of populations. The matrix for F_{st} values can be used for isolation by distance by the ISOLADE programme using rank correlation coefficient.

GENEPOP allows the transformation (ecumenicism) of the data input file toward other formats required by other programmes, which is unique for this software.

3. POPGENE (version 1.30) (Yeh and Boyle,1996)

POPGENE is a user-friendly *Microsoft*® Window-based computer package for the analysis of genetic variation among and within natural populations. The current version is designed specifically for the analysis of co-dominant and dominant markers using haploid and diploid data. It performs most types of data analysis encountered in population genetics and related fields. It can be used to compute summary statistics (*e.g.*, allele frequency, gene diversity, genetic distance, *F*-statistics, multilocus structure, *etc.*) for (1) single-locus, single populations; (2) single-locus, multiple populations; (3) multilocus, single populations and (4) multilocus, multiple populations. This latest version also includes modules for quantitative traits. The software programme opens in two windows namely data display window and dialog boxes, it is menu driven and very easy to run by selecting the menus from the task bar. The popgene have provision for dominant and co-dominant marker levels, Each have Haploid or Diploid data analyses provision, each again have provision for 3 population levels like single population, groups or multiple population, at last estimation of single locus or multi locus genetic parameter can be assigned.

4. Arlequin

Laurent Excoffier of the Computational and Population Genetics Lab of the Institute of Zoology, University of Bern, Switzerland (laurent.excoffier@zoo.unibe.ch), Stephan Schneider, and David Roessli have released *Arlequin* version 2.001, a program for population genetics analysis. It does many kind of analyses for different kinds of molecular data of RFLP, AFLPs, minisatellites sequences, microsatellites, allozymes etc. It estimates gene frequencies, testing of linkage disequilibrium, and analysis of diversity between populations. It enables us to compute variety of genetic distance measures including Jukes and Cantor, the Kimura 2-parameter distance, and the Tamura-Nei distance, each of these with or without correction for gamma-distributed rates of evolution. It can also compute a Minimum Spanning Tree network. Arlequin has its interactive "front end" written in Java, and requires the Java Runtime Environment. The software and a PDF documentation file can be downloaded from http://lgb.unige.ch/arlequin/.

5. TFPGA (Tools for Population Genetic Analyses)

This is a Windows based programme for the analysis of allozyme and molecular population genetic data. Mark P. Miller developed this programme in 1997 from Northern Arizona University. This can analyse both haploid and diploid data and also dominant and co dominant. This software needs around 1 MB of hard disk space and any Pentium processor can do it.

The programme can estimate allele frequencies for diploid dominant markers by either as the square root of the frequency of the recessive genotype or by Taylor expansion. The analysis menus are descriptive statistics, F-statistics, Genetic Distance, Hardy-Weinberg, UPGMA, Exact Tests and Mantel Test.

Assignment Test

More information regarding population structure and gene flow can be available now with assignment test using highly variable microsatellite and RAPD marker. This test was developed originally by Paetkau et. al. (1995) for population differentiation. It was based mainly on frequency statistics using the likelihood of each genotype in each of the population. Individuals are correctly assigned to the population with the highest likelihood. The available softwares are as follows:

1. GENECLASS 1.0.02 (Cornuet et.al. 1999; http://www.ensam.infra.fr/URLB).

This programme involves both the frequency model and the Bayesian model with 'leave one out' option. The simulation - exclusion approach proposed by Cornuet *et al.* (1999) computes a probability that the individual belongs to a population by simulating 10000 genotypes and a p= 0.001. A threshold P-value is fixed according to the required certainty of exclusion. Each individual was assigned to the population in which it had the highest probability of occurring. When an allele in the individual to be assigned is absent from the population sample, the estimate of the corresponding allelic frequency is equal to zero, leading to a likelihood of zero and hence eliminating de facto this population. However, the allele in question may be rare in the population so that it was not represented in the sample (Cornuet *et al.* 1999). If the individual probability is lower than the decided threshold in a population, it considers that the individual does not originate from the population. Thus an individual was considered as correctly assigned to a population when it was excluded from all of the non-origin population (e.g. $P < 0.001$), but not from the true population of origin.

2. Structure (version 2) (Pritchard and Wen 2002; http://pritchard.bsd.uchicago.edu)

It is a model based clustering method using genotype data. The genotype data includes microsatellites, SNPs and RFLPs. The programme is divided into two steps. First step involves the estimation of number of population (K) and second step involves the assigning of individuals to each of the population with highest percentage of membership (q).

3. Assignment Calculator (Brzustowski 1998) http://www.biology.ualberta.ca/jbrzusto/Doh.php

This programme is an interactive one, therefore to run this software one has to log on to the website and use it. This calculates the assignment probabilities of the individuals to the most probable populations. That is, if individuals whose source population is unknown and we wish to find out the population in which their genotype are most likely, then this porgramme will help us. The input data can be directly imported from a Popgene file.

Genetic Distance- Similarity

Nei (1987) described the definition of genetic distance as " the extent of gene differences between populations or species that is measured by some numerical quantity". Various genetic distance measures have been used now for analysis of molecular marker data for the purpose of genetic diversity analysis. The commonly used measures of genetic distance using binary data are,(i) Nei and Li (1979), (ii) Jaccard's (1908) coefficient, and (iii) Social and Michener's (1958) simple matching coefficient.

Genetic similarity (GS) between individuals *i* and *j* is estimated according to the formula given by Nei and Li (1979):

$$GS_{ij} = 2N_{ij} / (N_i + N_j)$$

where N_{ij} is the number of bands common in individuals i and j and N_i and N_j are the total number of bands in individuals i and j, respectively, with regard to all assay units. Thus, GS reflects the proportion of bands shared between two individuals, and ranges from 0 (no common bands) to 1 (all bands identical).

Genetic dissimilarity (GD) is calculated as:
GD = 1 - GS.

Jaccard's coefficient (GD) takes into consideration only matches between bands-alleles that are present and ignores pairs in which a band or allele is absent in both individuals, where as Sokal and Michener's GD, a Euclidean measure of distance which takes into account of mismatches and matches, and gives equal weight to both in estimating genetic distance.

Phylogenetic Tree Construction

1) Genetic distance based
2) Genetic similarity based.

In case of co-dominant data like microsatellites, various allelic combination (genotypes) the genotypes can be determined by recording the presence (1) or absence (0) of these bands only. Two assumptions were made for the analysis of the RAPD data: I) markers from different loci did not migrate to the same position on the gel, II) each band was assumed to represent the dominant genotype at the locus, whereas lack of the same band in another individual was assumed to correspond to the alternative homozygous recessive genotype in the Hardy- Weinberg equilibrium (Lynch and Milligan, 1994).

This matrix data can be used to construct a UPGMA dendrogram using SAS (v. 8.0 or latest) or Quantity One (BIORAD) or Tree Programs from NTSYS-pc, v. 1.6 (Rohlf, 1992) or any one of the these software that has option for phylogenetic tree construction.

The major features of some of the programs

Feature	TFPGA	ARLEQUIN	GENEPOP	POPGENE	GDA
Diversity					
Observed heterozygosity	√	√		√	√
Expected heterozygosity	√	√		√	√
No. of alleles per locus		√		√	√
Proportion of polymorphic loci	√	√		√	√

Hardy-Weinberg equilibrium	√	√		√	√
Population structure					
F-statistics		√	√	√	√
AMOVA		√			√
Homogeneity	√		√		
Migration		√	√	√	
Linkage equilibrium					
Two locus	√	√	√		√
Multilocus					√
Genetic distance					
Nei's	√	√		√	√
Rogers'	√				
Pairwise FST	√	√			√
Clustering					
UPGMA	√			√	√
Neigbour-joining					√
Neutrality test		√		√	

REFERENCES

Brzustowski,J.1998. Assignment Calculator (http://www.biology.ualberta.ca/jbrzusto/Doh.php). Edmonton, Alberta: University of Alberta.

Cornuet, J, Piry, S. Luikart, G., Estoup, A., and Solignac, M., 1998. New methods employing multilocus genotype to select or exclude populations as origins of individuals. *Genetics*, 153, 1989-2000.

Excoffier, L. 2000. *Arlequin 2.000*, Windows 2000.

Felsenstein J. 1993. Phylip(Phylogeny Inference Package), version 3.5c. http://evolution.genetics.washington.edu/phylip. html

Jaccard, P.1908. Nouvelles researches sur la distribution florale. *Bull. Soc. Vaudoise Sci. Natl.* 44:223-270.

Lewis, P.O., Zaykin, D. 2000. Genetic Data Analysis: Computer program for the analysis of Allelic data: http://leis.eeb.uconn.edu/lewishome/software

Lynch, M.and Mlligan, B.G.1994. Analysis of population genetic structure with RAPD markers. *Mol.Ecol.* 3: 91-99.

Miller, M.P. 1997. Tools for population genetic analysis. TFPGA. Computer software distributed by the author.

Minch,E. 1996. Microsat. The Microsatellite distance program: http://human stanford. edu / microsat/microsat.html.

Nei, M. and Li. W.H. 1979. Mathematical modeling for studying genetic variation in terms of restriction endonucleases. *Proc. Natl. Acad. Sci.* USA. 76, 5269-5273.

Nei, M., 1987. *Molecular Evolutionary Genetics*, Columbia University Press, New York.

Paetkau, D. Calvert, W. Stirling, I. Strobeck, C. 1995, Microsastellite analysis of population structure in canadian poplar bears. *Molecular Ecology*, 4, 347- 354.

Pritchard, J. K. and Wen, W. 2002, Documentation for STRUCTURE, 2.0, available at http://pritchard.bsd.uchicago.edu

Raymond M and Rousset, F., 1995. GENEPOP (Version 1.2); Population genetics software for exact tests and ecumenism. *J. Heredity*, 86: 248-249.

Rohlf, F.J. 1992. NTSYS-PC: Numerical taxonomy and multivariate analysis system, v. 1.60. Exeter Software, New York.

Sokal, R.R. and Michener, C.D. 1958. A statistical method for evaluating systematic relationships. *Univ. Kansas Sci. Bull.* 38:1409-1438.

Yeh, F.C. and Boyle, T., 1996. POPGENE version 1.111, Microsoft Windows-base Software for Population Genetic Analysis.

Chapter **21**

Softwares Relevant in Fisheries and Aquaculture Research

A. Antony & A.K. Roy

Introduction

In this computer age software plays a vital role in speeding up the research activity. Software and databases help in storage, processing and accessing large volumes of information. Software helps to retrieve right information at the right time at the right place for the right purpose. The time of researchers is very precious. Instead of spending it in monotonous and tedious tasks, it could be used more effectively for new findings. With the advent of computer and its softwares, this problem could be overcome to a considerable extent.

Software is a set of instructions given to the computer to perform certain tasks. In this era of information technology, its essential to apply different softwares to speed up the research activity. It is therefore, utmost importance to have information about the different software already developed in the area of fisheries/aquaculture worldwide. Almost all the routine tasks which the scientists or the farmers are doing, could be automated in the areas like financial & economic analysis, development planning and project appraisal, facility design, management planning and production forecasting, facility construction, management decision support and record keeping, disease diagnosis and treatment, environmental regulation and aquaculture databases.

Applied Bioinformatics, Statistics and Economics in Fisheries Research: pp. 301-312

New India Publishing Agency, New Delhi (India)

FAO, World Fish Centre and other International Research Organisations have contributed software and database development in the area of aquaculture (Anon, 1999; ICLARM, 1998; FAO, 1999; Erist, 1998). In India also initiative to develop software and database has been taken up. Preliminary report on the availability of fishery related software and databases are available (Roy, 2000). In this communication, an attempt has been made to list softwares and databases that may be used in fisheries and aquaculture research and development/activities.

Softwares in Different Fields of Aquaculture

The following software's and databases are identified through contact with internatonal organizations and browsing different Internet sites in the field of fisheries and aquaculture which are divided into nine broad categories are presented below along with brief description of application of each.

(i) Disease Diagnosis and Treatment

Hawaii Aquaculture Module Expert System (HAMES): It is an expert system which aids in the diagnosis and treatment of tilapia disease and syndromes in both fresh and marine environment. Modules include macroscopic examination and water quality leading to disease diagnosis, followed by treatment procedures.

Fish Vet : Multi-media based diagnosis and treatment of fish diseases. It considers disease quality and nutrition parameters for diagnosis.

Diagnosis of shrimp diseases: It is an interactive CD ROM database on diagnosis of shrimp diseases. It covers the description and diagnosis of diseases of farmed shrimp. It guides the user on how to handle the shrimp from the moment it is collected through the different diagnostic techniques and the interpretations of the results. It covers all the major unknown diseases of penaeid shrimp.

(ii) Management Decision Making and Record Keeping

Shrimp farm management: Its for the management of a shrimp farm. ET&A specializes in software systems for the shrimp farm industry: Hatchery and grow-out SMT (Su Management Tool) system has been designed for a carefree transition from tedious "paper work management integrated computerized solution". The system integrates full multimedia features, complete internet based solution, interface with water proof hand held computers, touch screen, data acquisition and control.

AbaSim : A fisheries management simulation game of an abalone fishery. AbaSim simulates a reef in the Australian abalone fishery, allowing the user to vary effort, quota, and age at capture. It was designed for a wide audience to teach important fisheries management aspects relevant to all fisheries. It gives insights about how various management actions (age at first capture, fishing effort, and catch quotas) affect abalone landing, age structure and population density. Its numerous graphical screens allow the user to follow cohort breakdowns on an annual basis. This program has excellent documentation, which includes much additional material on fisheries management and the abalone fishery. It is extremely user friendly with very high education potential.

Microfish : A program for generating population estimates from removal data based on maximum-likelihood estimation theory.

Cagean: A program to analyze catch-at-age and effort data from multiple fisheries.

Genmod: A generalized age-structured population model. Its options include: creating, editing and printing data files that contain model parameters, determine equilibrium yield as a function of fishing effort, examine the short and long-term behavior of the model when recruitment is a random event, and to obtain optimum harvest policies.

Fishparm: Fishery model estimation by nonlinear least squares. FISHPARM performs nonlinear parameter estimation for 13 separate statistical models commonly used in fisheries. These models are: Von Bertalanffy growth function, Gompertz growth function, Beverton-Holt recruitment function, Gamma function, Shepherd recruitment function, Allometry equation, Power function, Exponential growth function, Logistic growth function, Quadratic function, Weibull cumulative distribution, Mesh selection or LD-50 by logistic model.

Fishcale 89-Disbcal89: Fishery Analysis Tools for length-frequencies, length-age frequencies, and back calculations. This is a commercial software. The Disbcal89 portion of this package is used to measure projections of bony parts with a digitizer, explore the relationship between body length and bony part size, and back-calculate body lengths. The Fishcale 89 portion performs cross tabulation and high-resolution graphics on fish length, weight, sex code and age. Other output includes length structure and population density tables and charts, CPUE and percent of sample by length interval, PSD, RSD and Gablehouse's length categories, age composition, mean lengths at age, year-class contribution to each length class, relative weights and percent deviation from relative weights for each length class.

Trout Dynamics: An age-structured population model, which simulates the effects of various angling regulations on fish populations. Trout Dynamics is a multiple cohort population simulator, designed to assess the complex rules and recreational goals encountered under "quality" fishing. It utilizes the kind of data typically collected from trout-stream fisheries (i.e., mark-recapture estimates using electrofishing gear), but can be used to assess fisheries for any species of fish if proper data is available. The model: 1) estimates the number of fish caught and released and the hooking mortality they sustain; 2) evaluates the effects of unconventional size limit regulations such as slotted or inverted limits; and 3) illustrates the effects of fishing by showing the expected population size of fish and the expected catch of fish by size (inch) group.

aquaCalc 1.0 : It is a calculator that can do many basic calculator designs and run an aquaculture facility for the monitoring and daily control of important parameters.

Promanager, Harvest Optimizer, Wizard, Water Wizard (for aquaculture farm management) : It is a multi-species, multi-stocking and multi-harvest program for each rearing unit on the production facility. It stores and analyses all of the critical production data on the farm, including stocking, water quality, feeding mortality and harvest information. It can deal with wide variety of species and site configuration. It is flexible in data handling, reporting and forecasting.

Aquadab: Aquaculture species database. Its intended for researchers and farmers who want to store and retrieve in a standard concise format basic information an research and trial on cultivable species.

Popdyn: Population dynamic database to store stock based information in specific format.

Fishy: It provides comprehensive, combined record keeping and decision support system for catfish production and freshwater fin fish.

Pond Class : It is a decision support system that provides guidelines for freshwater pond fertilization and liming. Nutrients requirement for pond are calculated using limnological principle.

Pond 4: It has the capability for pond level analysis, multiple fund definitions, multiple fish lot definitions, user definable model parameterization and selection, multi level simulation and viewing simulation results, parameter estimation for species, customisation, fertilizer, lime, feed and water requirement determination and enterprise budget analysis.

Ecopath I & II : Developed by ICLARM (presently World Fish Centre), Philippines help in simulation modeling of aquaculture in a system approach. Ecopath II is structured around a system of linear equations.

Carp Plus and Carp plus II : Developed by Aquas of Oregon, USA are ideal softwares for hatchery and growout management tool applicable to any fish species and capable of calculating fish feed, growth, number and harvest schedules in any IBM compatible PC with 64 KB RAM.

Samcalre 2.1 : The software facilitates calculation of fish haul, static treatment, flow through treatment, length weight relationship, dissolved oxygen solubility, 2 x 2 contingency tables, experimental designs, correlation, regression etc.

Fishy 3.2 : Related to management of fish production. It can simulate changes in fish population, drain ponds, add fingerlings and water quality.

Fisher's associate : It's a software that helps to make fishing maps that shows where you fished and what were caught and to understand their correlation.

Fastfish : Fish management software.

Fisher's Guide : The Fishing Software

FISHING EXPERT: Software for Fishing for Bass, Walleye, Trout, Musky and Panfish.It uses the expert system concept.

SmartFish Expert Systems (SES) : The SmartFish Expert System (SES) is an expert software system for aquaculture, that provides fish growers with powerful and accurate web-based farm management and analysis.

SmartControl : It controls and monitors projects for different aquaculture sites.

Generic EIAxpert system : Its an interactive editor tool for water resource development projects, that allow to maintain (and compare) multiple alternatives for each project;

(iii) Environmental Regulation

DSS for Mariculture Regulation: Uses a combination of models and quasi expert system methodology to assess environmental experts of fish in coastal environment.

Modeling the Trophodynamics of the Black Azor Sea Ecosystem: It is the study of ecological impact of introduced exotic species on native fish stock as illustrated by the mnemiopies - Anchovy interaction.

Fishing Net : FishNet Computer Software is a complete system for organizing fishing experiences, using detailed atmospheric and water conditions

Hydrotech : For water analysis in fishery. It is specially designed for river hydrologist, environmental scientist and fishery biologist. It provides a comprehensive set of tools for analyzing and managing daily or annual historical records of river discharge, temp, salinity, pH, etc

SURPH : Environmental Fishery Software

Neurocomputing Integration-Interaction Design Technology (NID-TECH) : It uses the neurocomputing and artificial intelligence techniques on a personal computer in areas such as aquaculture and agriculture – to save time and money, NID-TECH applications are Smart Conversion Tool, Smart Aquaculture and Fisheries Chemical Treatment Calculations, Smart Fish Die-Off Causes Identification, Smart Aquaculture Water Quality Management, Smart Aquatic Plant Identification, Intelligent Management of Aquafarms and Aerator Evaluation

(iii) Fishing Activity Simulation Tool(FAST)

It has been developed as an extension for the ESRI ArcView GIS Software (v.3.1) in combination with the Spatial Analyst v. 1.0 extension. It is based on two models focussing on the analysis of the spatial components of the fishing effort distribution.

MTBASE 1.1: The Model of Trophodynamics of Black - Azov Ecosystem - An ecology software package for the study of the impact of exotic species introduction on fish stocks

(iv) Nutrition

Nurtibase : Database on nutrition for fish

Maxims : A complete programme for estimating the food consumption of fishes from stomach content data and population parameters.

Prefer : Food and habitat preference evaluation program. Preference Assessment Program, compares and ranks food items among competing individuals. The software is adequately user-friendly, but the documentation is awkward. This software has good potential for the small audience of people interested in food preference comparisons and limited educational value.

Sam Calc : It aids fish culturists with calculation for fish transport, chemical treatments, weight-length relations and injection solubility.

Fish : It handles normal stock record keeping and feed.

Climprod : Experimental Interative software for choosing and fitting surplus production models including environmental variable - Various equations allow for the introduction of an environmental variable into surplus production models. CLIMPROD helps the user select the most appropriate model for a particular case, according to objective criteria.

ClimprodPlus : It is interactive software for choosing and filling surplus production models including environmental variables. Stock assessment by surplus production models incorporating an environmental variable. Its an expert system embedded within a comprehensive computational model and data.

SmartFeed : SmartFeed conveys food to a succession of fish growing units in the simplest way possible, providing an optimal feeding regime for each individual unit, with minimal maintenance and at an attractive price

(v) Facility Design and Management Planning

Aquafarm: It's a windows based decision support system for designing and managing a full range of finfish production facilities including extensive ponds, intensive water re-use farms and flow through system, it provides aquaculture expertise, simulation, decision support and data management function ability.

Pond 4.0 : It's a decision support system for warm water pond aquaculture. Simulation based, fish performance projections provide a comprehensive and integrated analytical frame work to compare alternative strategies for warm water fin fish production, including stocking, feeding, fertilization and water management practices. It is downloadable.

Samakia : Most powerful aquaculture software for shrimp farms Management

NID-Tech : Water quality management and environmental assessment, aquaculture and fisheries calculations. NID-TECH integrates neural networks, symbolic artificial intelligence, object and event-driven oriented languages, graphical user interface environments, hypertext, and multimedia.

(vi) Facility Construction

Aqua CAD: Software for designing ponds, provides capabilities like simple tagging, density from distances, community analysis package, species diversity, removal sampling, simply growth simply probity, dynamica the complete angler, etc.

Levee : Calculates earth moving requirements for pond construction, considering user supplied pond and levee direct familiarity with USA soil conservation service codes.

(vii) Development Planning and Project Appraisal

The Aquaculture Planning Simulator (APS): A database module based on two aquaculture information system of the FAO, SIPAM, and SIPAL. It is an analytical evaluation of project proposals by comparison of technical and economic parameters against similar project operation included in the database.

Aquaculture development Decision Support System (ADDSS): It is a DSS prototype for planning and developing aquaculture industry in a given region. Uses a combination of models and databases to provide guidance on the most appropriate species to culture, production technology and production intensity.

(viii) Financial Analysis and Economic Analysis

A guide to the financial analysis of shrimp farming: It demonstrates the economic feasibility to determine the financial success of the venture. To achieve this objective a financial analysis is conducted on hypothetical intensive shrimp farm on Texas coasts. It assist in the economic analysis of shrimp farm.

Aqua Sim PC : It is a financial management program incorporating uncertainty and adaptive practices.

Soft crab Education program: Educates potential investors in recirculating soft-shell crab shedding system. It generates income and cash flow statement for comprehensive financial analysis. It may be adapted to other type of shellfish operation.

Shellfish farm Spreadsheet and Economic Analysis Software Package: It provides estimates of profitability of user defined oyster farm

Beam 1 and 2 : Bioeconomic modeling of artisanal and industrial sequential shrimp fisheries. Beam 1 gives its simulated results by age groups. Beam 2 gives them by standard commercial categories as used in the shrimp industry.

Beam 3: It is bioeconomic simulation model of tropical shrimps fisheries using fixed or random recruitment.

Beam 4: It is an analytical bioeconomic simulation of spac structured multi-species and multi-fleets fisheries.

ASPIC: A Stock-Production model incorporating covariates for management of fish stock. This software estimates the parameter of a non equlibrium of logistic surplus-production model from one or more series of data on catch & effort

(ix) Statistical Software

Fishstat Plus: Software for fishery statistical time series. Data (downloadable) having time series structure potentially be stored and processed. This database contains the volumes of fish catches landed by country of capture, species. It contains the collection of statistics on aquaculture production of fish, statistics on the annual production of fishery commodities by country and commodity descriptions, statistics on the number of commercial and subsistence fisheries.

Conversion factor: Factors for the conversion of fish and fish product weight to live weight equivalent.

FAO/ICLARM stock assessment tool: Length frequency analysis and length based stock assessment. FiSAT was developed mainly for the detailed analysis of length frequency data, but also enables related analyses, of size at age, catch at age, selection and other data typically collected for tropical fish stock assessment.

Globe fish database: Fish marketing database

NASIS: Software for fishery survey data logging and analysis

Speciesdab: Global species database for fishery purpose

Spatial: Space time dynamics in marine fisheries

Thompson & Bell Yield Analysis: Thompson and bell yield analysis using excel sheet.

Fishers : This database contains the number of people engaged in fishing according to the working time devoted to the occupation, as national annual averages, from 1961 onwards. Starting with data for 1990, the database includes employment in aquaculture and separates inland and marine fisheries, on a gender disaggregated basis.

Fishery Fleet : This database contains statistics on fishery fleets (fishing and support vessels) for decked and undecked craft. Numbers of undecked vessels are provided by country and vessel type for powered and unpowered boats. For decked vessels, numbers and aggreagate GRT values are available by country and vessel type for classes of GRT, length and horsepower.

Importer: The register comprises name and address of producers, importers, exporters involved in trade of fishery products. The companies are searchable by country and by type of activity. The commodities covered are tuna, groundfish, salmon, shrimps, cephalopods and small pelagics.

Listal (List of Aquaculturists from Latin America and the Caribbean) : Listal Database is an up to date list of aquaculturists from Latin America and the Caribbean which allows the user to retrieve information on scientific and technical personnel as well as on the productive sector in the field of Latin American and Caribbean aquaculture. The purpose of this Programme is to produce a final publication, which consists of a list of aquaculturists by country and several indexes.

Popdyn (Population Dynamic Database): The POPDYN database allows the user to store stockbased information on the characteristics of fisheries systems.

(x) Bibliographical Databases

ASFA : Aquatic science and fishery abstracts (ASFA) is an abstracting and indexing service covering the world's aquatic science literature. Covers all aspects of marine, freshwater and brackish water organisms, environmental ocean technology, policy, non-living resources, aquatic pollution, environmental quality, aquaculture, marine biotechnology etc.

Cities under the sea : The CD ROM consisting of coral reefs in a spectacular and oration under sea cities of corel reefs

DIAS: It is a FAO database on introduction of aquatic species. It considered primarily only freshwater species of fish.

The FAO Fisheries Atlas: It contains the sample of the wealth of data, information and analysis available in FAO's fisheries department. It includes statistical data on fisheries production and trade, maps showing key elements of the fisheries environment and activities.

Fishbase: It is a research tool as well as reference and information sources on finfish. It contains more than 17,500 fish species. It introduces time series data on crititcal variable such as fish recruitment and adding some analytical tool for population dynamics.

Encyclopedia of Hydroponics gardening: It is a CD-ROM. It includes vibrant picture, informatics text and graphics and captivating video on hydroponics, growing techniques, plant culturing hydroponics systems, plants to grow, pest and diseases control and nutrient formula

NANSIS: Software for Fishery Survey Data Logging and Analysis - NANSIS is a survey Information System for logging, editing and analysis of scientific trawl survey data

ARTFISH: It has been developed as a standardized tool adaptable to most fisheries in the developing countries. Its design was driven by the need to provide users with robust, user-friendly and error-free approaches and computer software, and achieve the implementation of cost-effective fishery statistical systems with minimal external assistance.

Conclusion

With the hardware technology progressing in leaps and bounds and worldwide cyber networking infrastructure coming up softwares can play a major role by covering all the activities of aquaculture, which would be accessed freely as shareware from the web, from any corner of the world or could be procured or down loaded freely. This facility helps in exchanging the research activity among the researchers around the world. It helps to avoid redundancy of the same work.

The reported identified softwares and databases may be of immense help to the researchers and scholars.These softwares and databases help the scientists and scholars to bring out their additional potentials with quick access to the information.

REFERENCES

Anon (1996). AFS Computer User Section, Software Description AQUAFIC - Aquaculture calculations programme by, NID Technologies version 1. URL : http://www.fisheries.ogn:80/cus/library/cuslibm4.htm.

Christensen, V. and D. Pauly (1993). *A guide to the Ecopath II Software System* (V. 2. 1.), ICLARM, Manila, Philippines.

Erist (1998). Computer Software for Aquaculture: Oregon State University, USA. URL : http://www.scicentral.com

Erist (2007) FAO : www.fao.org

Frocse, R. and D. Pauly (1997). Fishbase 97. CD-ROM ICLARM, Manila, 256 p.

Jarre, A *et al.* (1990). A user manual for MAXIMS (SOFTWARE 4), ICLARM MANILA.

Roy, A. K. (2000). CD-ROM databses, softwares and listservers related to aquaculture research. *Proc. National Workshop-cum-Training on Bioinformatics and Statistics in*

Aquaculture Research held during Feb., 8-11 at CIFA, Bhubaneswar organised by *Bioinformatics Centre,* CIFA, pp. 20-27.

Roy, A. K. (2000). Electronic Publication in Fisheries Science (Abs.). The 5th Indian Fisheries Forum, held at CIFA organised by Asian Fisheries Forum and CIFA during January, 2000.

Chapter **22**

Networking Professionals through Discussion Forums on the Virtual Platform

Lakshmi Murthy

Introduction

Improvements in Information and Communication technology are revolutionizing the way people access information. These new technologies are helping to improve communication, disseminate information and facilitate sharing of knowledge and skills. Zijp (1994) notes, Information Technology is making both public and private sector agricultural information systems more accessible and more rapid in transmission. The increase in electronic information and communication technologies and in particular networked information is the most notable change to affect the working practices of information professionals in recent years. Digital technologies have led to the emergence of wide-area connectivity through the Internet, making information on agriculture available worldwide and on-demand. Information technology and electronic networking particularly can strengthen organizations by improving their knowledge base and ability to share information and experiences with partners in the field. Electronic networking not only helps in the exchange of the growing volume of information but also brings people together in partnerships and facilitates a joint programme of action on common themes.

Applied Bioinformatics, Statistics and Economics in Fisheries Research: pp. 313-321

New India Publishing Agency, New Delhi (India)

Internet: The Network of Networks

One of the major developments in the Information Age has been the advent of the Internet. The Internet - a network of networks as it is aptly called is millions of people connected together via millions of computers that are connected through a common protocol. Internet offers seamless access to a variety of information sources for education and research including electronic journals, reports, books, libraries and databases; facts and figures; software; entertainment; business and commerce related information, weather, news, electronic discussion forums, electronic publishing, online education etc. Internet is changing traditional ways of managing information by establishing new sources of information, new ways of accessing information and new methods of communication on a global basis. All the information is available to everyone on the network regardless of physical location of the resources or the user. National and International organizations, NGOs, Academic Institutions, etc. have hosted their homepages on the world wide web, and have started offering information on their organizations and operations enabling its sharing and wider dissemination. Databases on a wide variety of topics are accessible on the Internet. The net offers tremendous potential for sharing resources like documents, data and software and network services.

The Internet is impacting on various sectors like aquaculture, agriculture, health, industry, environment, telecommunication trade, etc. Wider access to the Internet, friendlier tools for using it, and a growing body of useful information are helping to make the Internet an essential tool for agricultural practitioners, advisors, consultants, and researchers. The Internet, apart from being a vast storehouse of information, is also a virtual meeting place. People can communicate through mailing lists, newsgroups, discussion forums, electronic bulletin boards etc. Listservs and Discussion Groups offer a platform for discussion at the global level, irrespective of location. It is possible to have instant access to many people in a particular field, and it is possible to send a question to a discussion group and get responses from different perspectives. The Internet provides an ideal platform for collaboration and partnerships between organizations and among professionals, with respect to information, programmes and projects, monitoring and evaluation, research, policy development etc. The Internet facilitates sharing of information among individuals irrespective of location, replacing the more traditional flow of information from experts to practitioners and thus facilitating an interactive information exchange. In order to keep pace with new developments, Internet connectivity alone is not enough. What is needed is active participation and contribution to the world's growing digital knowledge bank that has the potential to educate and reform in many sectors including agriculture. This paper presents one of the ways to facilitate this ie. through listservs or electronic discussion forums.

Electronic Discussion Forums

Listservs are email discussion groups that are categorized by the area of interest. Listservs allow members to be part of a group of practitioners through email messages. One can use the listserv to engage in collaborative work, discussion, or to make announcements. Discussion forums and listservs both offer opportunities to take part in online conversations with other professionals in the field.

A Discussion Forum is an online area where contributors can carry on debates and discussions by posting questions and observations and inviting responses from other participants. It is an open forum for exchange of information and ideas on specific topics, and to seek advice on various issues through electronic mail. To participate, one needs to subscribe to the discussion. There are a number of electronic discussion groups representing different subject domains. Most of these are based on academic topics, and offer researchers a place to discuss, share their concerns, experiences and lessons learnt and exchange views. It is a virtual meeting place for people interested in a certain issue, to establish communication among them. This meeting place is not a geographical location but the computer screen of each of the members of the forum. A subscriber to the discussion forum receives in his/her email account a copy of each message that has been sent to the forum and can send a message, which will reach every other participant.

Subject discussion groups in specific fields of research are a mechanism for discussing topics and exchanging experiences with others who are working in a similar subject area around the world and often result in valuable ideas. It is a good source for researchers to exchange experiences in their scientific research. There are a number of discussion groups in agriculture with members from different countries.

Discussion Forums bring together a larger and more diverse group of people than a physical conference or seminar would permit, and their duration, which is quite flexible unlike a conference or seminar at a physical location, permits sustained interaction and reflection on the chosen subject that would not be possible with a conference or seminar at a physical location. Contributions from knowledgeable experts can make the forum a resource base of expertise.

A forum may be moderated or unmoderated. In a moderated one, all messages go to a person acting as a moderator, who inspects each message and decides whether it should be posted to the group. The moderator filters inappropriate messages that are not relevant to the purpose of the forum. Moderation involves filtering inappropriate, irrelevant messages; editing messages and providing suggestions to participants about improving their messages; approving messages; stimulating the discussion if it tends to slow down, and providing direction.

An unmoderated list is one where a message sent gets redistributed to all other users automatically. Unmoderated discussions tend to work if the group is relatively small and members know each other. But when the number of members is larger, people are less familiar with each other, and the topic needs to be narrowed, moderated discussions are likely to produce better results. Moderation helps to organize and focus the discussion and thus ensure that members do not go in different directions.

Members of discussion forums have the option of subscribing or unsubscribing to a forum. Most forums offer the choice of receiving messages individually, or in a daily digest form, or only special notices from the moderator, or no mail (where they can check out the messages on the Forum website).

Setting up Discussion Forums

There are certain issues one needs to keep in mind before setting up a discussion forum:

- The desired goals and objectives, scope and intended participants for the discussion
- Time frame for the discussion.
- Who would serve as a moderator/facilitator? A successful discussion depends on a moderator who can organize and maintain the discussion, coordinate the discussion, keep the discussion focused, and capture lessons learned.
- Decide on the background materials for the discussion if the discussion is related directly to a study, report, conference, workshop, or project
- Identify the participants and resource persons who could be actively engaged in the discussion.
- Compile and share the results and lessons from the discussion.

The success of an online discussion depends on sustained commitment and involvement. Successful discussions are well thought out and designed from the beginning; receive a high level of attention from moderators/ facilitators; and involve knowledgeable participants.

Discussion Software

A range of discussion software is available for setting up Discussion forums, some of which are:

- *Lyris (www.lyris.com):* This is commercial discussion list software, used by a number of development organizations.
- *Web Crossing (www.webcrossing.com):* Web Crossing has many features including hosting discussions on their server and web applications like real-time chat. It is also available via World Crossing as a free, hosted service
- *Majordomo (www.greatcircle.com/majordomo/):* This is a free software, easy to install, simple to operate. Majordomo is a program, which automates the management of Internet mailing lists. Commands are sent to Majordomo via electronic mail to handle all aspects of list maintenance. Once a list is set up, virtually all operations can be performed remotely by email.
- *Collaborative Virtual Workspace (http://cvw.sourceforge.net/):* CVW is a collaboration software environment that provides a "virtual building" where teams can communicate, collaborate, and share information. CVW also enables virtual co-location through virtual rooms.

Setting up Discussion Areas Free

These are available free and can be downloaded from the web and are easy to set up.

- *Yahoo Groups (http://groups.yahoo.com):* Facility to create your own discussion list and archives.
- *Quick Topic (www.quicktopic.com)* provides a web forum that's also email-enabled. Participants can use just email to participate, or they can use the web board, or both. This includes a workspace area for joint editing/ commenting on word and HTML documents.
- *Community Zero (www.communityzero.com):* Community Zero is an interactive website that allows a group of people to communicate and exchange information over the Internet in their own private and secure

area. Within each area, called an online community, participants are provided access to a suite of tools that enable a group to effectively get organized, share knowledge and communicate.

- *World Crossing:* (www.worldcrossing.com) Forums, message boards, and chat rooms. It is customizable to match your website design, and offers an easy and free way to add forums, message boards or chat rooms to your Web site.

Discussion Forums in Aquaculture

Given here are some listservs and discussion forums in aquaculture, with details on the coverage and access.

AQUA-L (http://www.mi.mun.ca/mi/aac/aqual.html) is an aquaculture discussion group owned by the Aquaculture Association of Canada and maintained by the Fisheries and Marine Institute of Memorial University of Newfoundland. To Subscribe to AQUA-L: you may send a message to majordomo@killick.mi.mun.ca.

AquaBiz. The AquaBusiness Club (http://groups.yahoo.com/group/AquaBiz/). An online Aquarium/Aquaculture Business Club where, commercial aquarium fish breeders, home breeders, buyers, sellers, traders, importers, exporters, collectors, wholesalers, agents, Trans-shippers, transporters, equipment dealers, live and dry food breeders/manufacturers/dealers can buy and sell their fish, invertebrates, related goods and services. The group has over 300 members. Potential traders, sellers, buyers, breeders and others also can become members. Membership is open to the above-mentioned people/businesses in the USA, Europe, Asia and Worldwide.

Members can voice and discuss their business experience with regard to import/export and legal issues related to the aquarium/aquaculture trade. Members can also assist other members in providing useful information and assist them in market research etc.

Aqua-culture. (http://groups.yahoo.com/group/Aqua-culture/) Currently has over 330 members discussing and sharing information on all aspects of aquaculture.

Aquaculture (http://finance.groups.yahoo.com/group/aquaculture/). This e-group is formed for the benefit of aquaculturists, scientists seafood traders, equipment manufacturers, aquaculture product manufacturers and marketers and has 179 members. Those who are interested in Aquaculture can freely join this group and post messages, suggestions, questions, queries research papers.

AQUANIC : The Aquaculture Network Information Center (http:// aquanic.org) is a gateway to the world's electronic aquaculture resources. AquaNIC was conceived by the former USDA-Extension Service (currently Cooperative State Research, Education and Extension Service) as one of the nation's first network information centers to serve as a gateway to the world's electronic aquaculture resources. AquaNIC was created, through grants from USDA-Extension Service, the Illinois-Indiana Sea Grant College Program, and Purdue University Libraries. The goals of AquaNIC are to: provide access to all electronic aquaculture information at the national and international level; increase the quantity and quality of electronic information available to the aquaculture industry; provide self-paced aquaculture instruction to the aquaculture industry and obtain user input in directing AquaNIC services. The site hostsdiscussion groups around common species and systems.

Aquaponics (http://groups.yahoo.com/group/Aquaponics) Aquaponics offers innovative methods of growing aquatic and terrestrial plants and animals. Drawing on and adapting technology, knowledge and experience from many fields including, intensive recirculating aquaculture, controlled environment horticulture and hydroponics. The group has 330 members

AquaTT (http://www.aquatt.ie/) is an international foundation which provides project management and training services to support the sustainable development of Europe's aquatic resources. It is a network of over 650 co-operating university and industry members involved in the aquaculture sector in Europe. AquaTT gives access to PiscesTT site , a discussion forum on European Aquaculture.

FISHFOLK (Fisheries social science network). Fish folk at MIT is a list server devoted to fisheries issues of relevance to social scientists, fishermen, biologists, managers, government officials, conservation group members, attorneys, and other academics. Through discussions, members communicate lessons learned in one part of the globe that can help instruct others, and aid in sustaining the world's fisheries. One can mail to listserv@mitvma.mit.edu to subscribe to the list.

FISH-SCI : The Fish-Sci mailing list provides an international forum for the discussion of scientific ideas and events pertaining to fish and fisheries. Similar to FISHFOLK but focused on the natural, rather than social, science of fisheries. To join the Fish-Sci mailing list one may send an email message to listserv@segate.sunet.se. In the body of the message put subscribe fish-sci (your name).

OCEANF-F Ocean Farmers of America Forum OCEANF-F is a discussion list to support the open exchange of ideas and information on: 1) The engineering of ocean farming systems for survivability, economy, and improved productivity, 2) Mechanisms to encourage fishermen to expand into ocean farming, 3) Biological, husbandry, and environmental issues associated with ocean farming at exposed sites, 4) Regulation and ownership of ocean farms in the U.S. Exclusive Economic Zone, and 5) Livestock registration and ownership of biological inventory Mail to listserv@mitvma.mit.edu to subscribe to the forum.

WATER-FORUM A moderated discussion forum for water resources issues; including drinking water, water conservation, environmental chemistry, wetlands, wastewater, irrigation, recreational use, fisheries and wildlife use, aquaculture, coastal studies and oceanography, environmental and public health issues, contamination/remediation, computer modeling, climatologic, hydropower, and other relevant water resources topics. Members are from academia, industry, government, and the general public. To subscribe, mail to waterforum-subscribe@egroups.com.

Network of Aquaculture Centers in Asia-Pacific - Forums (http://www.enaca.org) NACA is dedicated to rural development and poverty alleviation through sustainable aquaculture and aquatic resource management. NACA seeks to improve rural income, increase food production and foreign exchange earnings and to diversify farm production. The ultimate beneficiaries of NACA activities are farmers and rural communities. NACA conducts development assistance projects throughout the region in partnership with governments, donor foundations, development agencies, universities and a range of non-government organizations and farmers. NACA supports institutional strengthening, technical exchange and the development of policies for sustainable aquaculture and aquatic resource management. Current member governments are Australia, Bangladesh, Cambodia, China, Hong Kong SAR, India, Indonesia, Iran, Korea (DPR), Malaysia, Myanmar, Nepal, Pakistan, Philippines, Sri Lanka, Thailand, Vietnam. The site hosts a number of discussion forums with farmers, scientists and government officials with experience in different fields as members.

Initiatives at India

The National Institute of Agricultural Extension Management (MANAGE) an organization under the Ministry of Agriculture, Govt. of India, based at Hyderabad has been involved in setting up and managing the following discussion forums:

1. Impact of Economic Reforms on Rural Livelihoods (as part of the Livelihood Options study of the Overseas Development Institute (ODI)

U.K.) (Now managed by ODI, UK). An informal email discussion and debate on the impact of ten years of economic reforms on rural economy, livelihoods and well being in India. The debate is open to development practitioners, policymakers and researchers with knowledge and opinion on this subject. It is an open space to pose questions and issues for discussion and provide opinions and commentary emerging from the length and breadth of India. (Software: Majordomo).

2. Indian Agripreneur. A platform on the web (http://groups.yahoo.com/group/indianagripreneur) to facilitate sharing of ideas, and exchange of messages among agripreneurs and with professionals in the field.

This is a virtual platform to enable agripreneurs, to communicate and share their experiences, innovations, interesting developments, problems, progress and any other issues with other agripreneurs, scientists and professionals. The idea is to create a network of agripreneurs, researchers involved in all aspects of the agricultural sector and eventually enable a sustainable delivery of products and services to the farming community. The forum set up on November 27, 2002, currently has over 500 members. Messages are exchanged among the members posting views, seeking solutions, offering advice and sharing experiences. (Yahoo Groups).

The response on these has been very encouraging. It would be a good idea to explore the idea of creating discussion areas to enable professionals in India to share their ideas and experiences on the virtual platform. Distance, which separates these institutions and the professionals, would no longer be a barrier to communication and knowledge sharing. Information provided in this chapter may be of great help for researchers in accessing and sharing knowledge in the field of aquaculture.

REFERENCES

Fisher, Catherine. Creating discussion areas, GDNet / Eldis at http://www.eldis.org.

Kim, Ron. Online Moderation. http://www.eldis.org

Zijp, William (1994). Improving the transfer and use of Agricultural information - A guide to information technology. World Bank Discussion Paper No.247. Washington: World Bank.

Chapter 23

Mapping of Fish Research in India

D.B. Ramesh & S. Mohanty

Introduction

Fish and the aquaculture products contribute a significant amount of animal protein to the diet of the people worldwide. Being highly nutritive they serve as valuable food supplement in diets lacking essential vitamins and minerals. Besides its role of food security, fish is a source of income for millions of people around the World. Though the trade is not prominent at Global level, in some countries its contribution to foreign currency earning is vital to the National Economy. India has a coastline measuring 7500 km and important aquaculture country in the World. Being the home for more than 10% of the global fish biodiversity, the country ranks third in the World in the total fish production with an annual fish production of about 6.4 million tones. Constituting over the 1% of the GDP, fisheries contributes to 5.3% of the agricultural GDP. The estimated foreign earning for the year 2006 was US$ 1365 million.

Objectives of the Study

Scientific performance is essentially a multidimensional concept, which cannot be measured by a single universal indicator. At best, there may be a number of imperfect or 'partial' indicators, each representing a different aspect of research performance, with varying degree of success. Nonetheless, publications in the refereed scientific journals constitute the most important

Applied Bioinformatics, Statistics and Economics in Fisheries Research: pp. 323-336

New India Publishing Agency, New Delhi (India)

indicator of research performance. Careful analysis of scientific output in the form of publications can provide deep insights for making inter-institution, inter-field and international comparison of research performance. During the past two decades, the assessment of scientific performance using scientometric techniques have gained wide interest for assessing the 'health of science', besides being used for strategic planning and decision-making in science and technology. The present study investigates India's research performance in the field of fish research for the period 1975-2006.

The study has been taken to quantify the research efforts of the Scientists and the Scholars of India in the field of Fisheries research in the World. For this purpose the Web of Science was taken for the study. The search word made using the word 'fish' from the topic field and India from the address field.

The following indicators have been used for the data analysis to find out the:

- Year-wise distribution of articles
- Document type
- Language wise distribution of publications
- Country wise distribution of publications
- Subject classification wise distribution of publications
- Institution-wise distribution of papers and patents published
- Source-wise (Journal) distribution of publications

Methodology

For this purpose online version of the Web of Science (WoS) database of Thomson Scientific is used. All the publications on the topic Fish research were downloaded for the purpose of data collection. Web of Science *offers access to journal articles in the sciences, social sciences and arts and* humanities. It contains over 100 years of valuable research, fully indexed and cross-searchable. It covers many journals back to 1900, including every issue of significant publications. Approximately 8,830 titles are indexed from 230 disciplines. Over 25,000 articles and 500,000 cited references are fully indexed and added to Web of Science every week. *WOS* delivers over 16 million full-text links to 10,000 journals from more than *Web of Science* 300 publishers. There were about 3250 pages published during the years 1975 to 2006 were downloaded and analysed.

Data Analysis

Details of the data analysis in the form of table are presented below.

(i) Year-wise distribution

Yearwise (1975-2006) distribution of the published papers are presented in the Table-1. The year 2006 has the highest of the publications (246nos) followed by 2005(245 nos.) showing an increasing trend in general.

Table 1: Year-wise distribution of the articles

Year	No. of Papers	Year	No. of Papers
2006	246	1990	78
2005	245	1989	55
2004	197	1988	59
2003	161	1987	58
2002	153	1986	65
2001	141	1985	54
2000	130	1984	70
1999	140	1983	69
1998	147	1982	60
1997	128	1981	79
1996	103	1980	79
1995	115	1979	69
1994	101	1978	58
1993	77	1977	43
1992	99	1976	25
1991	112	1975	34

(ii) Document Type

Classification according to document type of the publications is shown in Table 2. More than 88% of publications are articles followed by notes, reviews etc.

Table 2: No. of papers according to document type

Document Type	No. of Papers
Article	2895 (88.78%)
Note	193 (5.99%)
Review	71 (2.18%)
Meeting Abstract	60 (1.84%)
Letter	33 (1.01%)
Editorial Material	8 (0.25%)
News Item	1 (0.03%)

(iii) Language of Publication

Classification according to language of publication is presented in table 3. Most of the articles (99.78%) are published in English. Out of the 3250 publications only 6 articles are published in German and 1 in Chinese.

Table 3 : Language of publications

Language	No. of Papers
English	3243 (99.8)
German	6 (0.18%)
Chinese	1 (0.03%)

(iv) Productive Authors

Authorwise distribution of papers related to fisheries science is presend in table 4. Jana, B.B. of Aquaculture & Appl. Limnol. Res Unit, Department of Zoology, University of Kalyani, Kalyani, West Bengal, India has the highest no. of the papers (47 nos.) followed by the Bhattacharya, S. of Environm Toxicol Lab, Dept Zool ,Sch Life Sci, Visva Bharati University, Santiniketan, W Bengal (45 nos.); Pandian, T.J.,School of Biological Sciences, Department of Genetics, Madurai Kamaraj University, Madurai, Tamil Nadu India (43 nos.); Ray, A.K. of Visva Bharati Univ., santiniketan, India (43 nos.); followed by Sastri , K.V. of Department of Biosciences, Maharshi Dayananda University-124001, Rohtak, Haryana, India (35 nos.)

Table 4: Productive authors (> or =10 papers)

Author	No. of Papers	Author	No. of Papers
Jana, B B (Univ Kalyani, Kalyani 741235)	47	Gill, T S	23
Bhattacharya, S (Visva Bharati Univ, Dept Zool, Environm Toxicol Lab, Santini Ketan 731235)	45	Singh, B R	23
Pandian, TJ (Madurai Kamaraj Univ, Sch Biol Sci, Dept Genet, Madurai),	43	Raisuddin, S	23
Ray, AK (Visva Bharati Univ, Dept Zool, Fisheries Lab, Santini Ketan)	43	Mukherjee, SC	22
Sastry, KV (Maharshi Dayanand Univ, Dept Biosci, Rohtak)	35	Srivastava, AK	22

Contd...

Table 4. Contd...

Munshi, JSD	27	Venugopal, V	22
Karunasagar, I	26	Kumar, S	21
Dalela, RC	24	Mittal, AK	21
Joy, KP	24	Singh, A	21
Medda, Ak	24	Verma, SR	20
Jafri, AK	19	Dasmahapatra, AK	12
Sarkar, SK	19	Govindappa, S	12
Ayyappan, S	18	Johal, MS	12
Chakrabarti, R	18	Keshavanath, P	12
Kaviraj, A	17	Madhavi, R	12
Khan, MA	17	Pal, AK	12
Ojha, J	17	Parvez, S	12
Sahoo, PK	17	Philip, GH	12
Senthilkumaran, B	17	Sahu, NP	12
Das, BK	16	Sharma, S	12
Ghosh, S	16	Subhedar, N	12
Gopal, K	16	Das, M	11
Jeyasekaran, G	16	Ghosh, A	11
Kumar, R	16	Kaur, M	11
Singh, SK	16	Khangarot, BS	11
Tripathi, G	16	Kucheria, K	11
Kalavati, C	15	Murty, AS	11
Mahajan, CL	15	Pande, J	11
Oommen, OV	15	Pant, JC	11
Pandey, AK	15	Rao, JV	11
Garg, SK	14	Rao, KVR	11
Gopakumar, K	14	Roy, PK	11
Goswami, UC	14	Shankar, KM	11
Gupta, S	14	Singh, R	11
Kumar, A	14	Banerjee, TK	10
Mukhopadhyay, PK	14	Basu, S	10
Ram, RN	14	Begum, G	10
Sampath, K	14	De, NC	10
Srikar, LN	14	Haldar, DP	10
Tewari, H	14	Haniffa, MA	10

Contd...

Table 4. Contd...

Doke, SN	13	John, PA	10
Lakra, WS	13	Michael, RD	10
Mathavan, S	13	Mukherjee, D	10
Pati, AK	13	Ninjoor, V	10
Reddy, PM	13	Pandey, AC	10
Saha, N	13	Pandey, S	10
Sathyanesan, AG	13	Radhakrishnaiah, K	10
Shaffi, SA	13	Ramasamy, P	10
Shakila, RJ	13	Reddy, GVS	10
Srivastav, AK	13	Singh, D	10
Adhikari, S	12	Singh, RK	10
Banerjee, S	12	Srivastava, SK	10
Bose, AN	12	Vishwanath, W	10

(v) Subject Category

Categorisation of fisheries publications according to subject category reveals that Environmental Science (623) has the largest number of articles followed by the Fisheries (425 nos.), Marine and Freshwater biology (363 nos.), Toxicology (310 nos.) and Biochemistry and Molecular biology (268 nos.) etc. (table 5).

Table 5: Subject category (>20 papers)

Subject Category	**No. of Papers**	**Subject Category**	**No. of Papers**
Environmental Sciences	623	Genetics & Heredity	61
Fisheries	425	Nutrition & Dietetics	60
Marine & Freshwater Biology	363	Immunology	59
Toxicology	310	Agricultural Engineering	50
Biochemistry & Molecular Biology	268	Pharmacology & Pharmacy	47
Multidisciplinary Sciences	247	Energy & Fuels	43
Food Science & Technology	231	Parasitology	43
Zoology	226	Chemistry, Applied	42

Contd...

Table 5. Contd...

Physiology	151	Agronomy	38
Biology	142	Entomology	38
Biotechnology & Applied Microbiology	119	Agriculture, Multidisciplinary	34
Ecology	106	Anatomy & Morphology	34
Agriculture, Dairy & Animal Science	105	Public, Environmental & Occup. Health	30
Endocrinology & Metabolism	99	Chemistry, Multidisciplinary	28
Microbiology	98	Engineering, Environmental	28
Oceanography	95	Medicine, Research & Experimental	28
Veterinary Sciences	94	Medicine, General & Internal	25
Water Resources	72	Neurosciences	24
Cell Biology	65	Oncology	21
Biophysics	64		

(vi) Top Institutions

Identification of top institutions with more than ten publications shows that Banaras Hindu Univ., Varanasi is the top institution as per publication count by publishing 115 papers followed by CIFA (78) and Univ. of Kalyani, Kalyani (77 nos.); .Bhabha Atom Res. Centre, Mumbai (76 nos.) followed by others (Table-6).

Table 6: Top institutions (> or =10 papers)

Institution Name	**No. of Papers**	**Institution Name**	**No. of Papers**
Banaras Hindu University	115	Madurai Kamaraj University	65
Central Institute of Freshwater Aquaculture	78	University of Madras	58
Univ Kalyani	77	Bose Institute	56
Bhabha Atom Res Centre	76	Univ Kerala	54
Visva Bharati University	72	Aligarh Muslim University	53
		Cent Inst Fisheries Educ	53
		AIIMS	21

Contd...

Table 6. Contd...

Natl Inst Oceanog	53	Indian Inst Chem Technol	21
Univ Agr Sci Mangalore	50	Univ Hyderabad	21
Cent Inst Fisheries Technol	49	Vikram Univ	19
Ind Toxicol Res Ctr	48	Tamil Nadu Vet & Anim Sci Univ	18
Sri Venkateswara Univ	47	Cent Inland Fisheries Res Inst	17
Indian Inst Technol	46	Indian Inst Sci	17
Kumaun Univ	44	Berhampur Univ	16
Bhagalpur Univ	43	Govt Arts Coll	16
Univ Delhi	42	Bharathidasan Univ	15
Andhra Univ	41	Gauhati Univ	15
Annamalai Univ	39	Haryana Agr Univ	15
Osmania Univ	39	Manipur Univ	15
Cent Food Technol Res Inst	38	Coll Fisheries	14
Punjab Agr Univ	36	Jiwaji Univ	14
Indian Vet Res Inst	35	Natl Bur Fish Genet Resources	14
Jadavpur Univ	33	Panjab Univ	14
Kalyani Univ	33	Univ Bihar	14
Univ Agr Sci	33	Univ Lucknow	14
Univ Gorakhpur	33	Cent Inst Brackishwater Aquaculture	13
Gorakhpur Univ	31	Ctr Cellular & Mol Biol	13
Indian Council Agr Res	30	Dav Postgrad Coll	13
Univ Calcutta	30	HNB Garhwal Univ	11
Dav Coll	28	Jamia Hamdard Hamdard Univ	11
Ne Hill Univ	27	Jawaharlal Nehru Univ	11
Univ Rajasthan	26	Marathwada Univ	11
Nagarjuna Univ	25	Natl Inst Immunol	11
Sri Krishnadevaraya Univ	25	Tamil Nadu Agr Univ	11
Bharathiar Univ	23	Indian Council Med Res	10
Cent Leather Res Inst	23	Natl Inst Basic Biol	10
Cent Marine Fisheries Res Inst	23	Natl Inst Nutr	10
Cochin Univ Sci & Technol	22	Ravishankar Univ	10
Nagpur Univ	22		

(vii) Sourcewise (Journals) distribution of all fisheries Publications

Sourcewise list of distribution of the publications related to fisheries science is provided in the Table 7. Journal of Environmental Biology (207 nos.); Current Science (181 nos.); Aquaculture (113 nos.); Bulletin environmental Contamination and Toxicology (113 nos.) and Journal of Food Science and technology (87 nos.) are found to be the leading journals publishing fisheries papers.

Table 7: Source wise (Journals) distribution

Source Title	No. of Papers	Source Title	No. of Papers
Journal of Environmental Biology	207	Israeli Journal of Aquaculture-Bamidgeh	34
Current Science	181	National Academy Science Letters-India	33
Aquaculture	113	Journal of Biosciences	31
Bulletin of Environmental Contamination and Toxicology	113	Fish & Shellfish Immunology	29
Journal of Food Sci. and Technol. -Mysore	87	Chemosphere	28
Indian Journal of Animal Sciences	80	Fish Physiology and Biochemistry	25
Indian Journal of Marine Sciences	73	Proc. of the Ind. Acad. Of Sci.-Animal Sci.	25
Aquaculture Research	60	General and Comparative Endocrinology	36
Ecotoxicology and Environmental Safety	57	Journal of Fish Biology	36
Indian Journal of Experimental Biology	48	Acta Hydrochimica Et Hydrobiologica	35
Comparative Physiology and Ecology	46	Hydrobiologia	40
Indian Journal of Biochemistry & Biophysics	45	Journal of Advanced Zoology	40

Contd...

Table 7. Contd...

Bioresource Technology	40	Indian Veterinary Journal	22
Toxicology Letters	25	Acta Protozoologica	20
Aquaculture International	20	Comparative Biochemistry and Physiology B	20
Pesticide Biochemistry and Physiology	20		

(viii) Top Journals with Impact factor

Listed below in the table-8 top 13 journals with their Impact Factor (IF) values as per Science Citation Index (SCI). Aquaculture has the highest IF and published 113 papers in the areas of fish research.

Table 8: Top Journals with Impact factor

Title of the Journal	No. of Papers	IF value
Journal of Environmental Biology	207	0.197
Current Science	181	0.737
Aquaculture	113	2.081
Bull. of Environmental Contam. and Toxicology	113	0.505
Journal of Food Sci. and Technol.-Mysore	87	0.053
Indian Journal of Animal Sciences	80	0.064
Indian Journal of Marine Sciences	73	0.209
Aquaculture Research	60	1.051
Ecotoxicology and Environmental Safety	57	2.000
Indian Journal of Biochemistry & Biophysics	45	0.277
Bioresource Technology	40	2.180
Hydrobiologia	40	1.049
Journal of Advanced Zoology	40	0.054

The Impact Factor (IF) Value of a journal in a given year is defined as R / N

where, N= Total number of items (source items) published in the volumes of the journal belonging to a given block of the preceding years, R= Total number of citations received by these volumes in the given year. It is also referred as citation count.

Table 9: Collaboration with other countries

Country	No. of papers
India	3250
USA	120
Japan	50
Germany	42
UK	31
Scotland	26
Canada	17
Netherlands	16
China	12
France	11
South Korea	11
Norway	10
Australia	8
Philippines	7
Spain	6
Thailand	6
Bangladesh	5
Hungary	5

Table 10: City-wise distribution of papers

City	No. of papers
Kolkata	185
Mumbai	182
Chennai	164
Delhi	152
Hyderabad	136
Varanasi	114
Bhubaneswar	95
Lucknow	95
Cochin	84
Trivandrum	57
Bangalore	48

Table 11: States-wise distribution of papers

State	No. of papers
Uttar Pradesh	439
West Bengal	367
Tamil Nadu	323
Andhra Pradesh	275
Maharashtra	173
Delhi	152
Kerala	141
Orissa	124
Karnataka	112

Table 12: Country-wise distribution

Country	No. of papers
USA	42870
Japan	12251
UK	8676
France	7184
Germany	7331
Italy	4401
Sweden	3320
India	3250
China	2357
Russia	1936
Korea	1035

(ix) Highly cited Papers

The top ranked research papers in the area of fish research from India according to number times cited is as follows:

- Singh, RB (Med. Hosp. & Res. Ctr., Heart Res. Lab., Civil Lines, Moradabad), Niaz, MA, Sharma, JP, et al. Randomized, Double-Blind, Placebo-Controlled Trial Of Fish Oil And Mustard Oil In Patients With Suspected Acute Myocardial Infarction: The Indian Experiment Of Infarct Survival .4. *Cardiovascular Drugs and Therapy* 11 (3): 485-491 Jul 1997. *Times Cited: 201*

- Singh, RB (Med. Hosp. & Res. Ctr, Ct.r Nutr. & Heart, Moradabad), Dubnov, G, Niaz, MA, *et al.* Effect Of An Indo-Mediterranean Diet On Progression Of Coronary Artery Disease In High Risk Patients (Indo-Mediterranean Diet Heart Study): A Randomised Single-Blind Trial. Lancet 360 (9344): 1455-1461 Nov 9 2002. *Times Cited: 127*

- Gadgil, M (Indian Inst Sci, Ctr Ecol Sci, Bangalore), Berkes F, Folke C. Indigenous Knowledge For Biodiversity Conservation . *Ambio* 22 (2-3): 151-156 May1993.*Times Cited:115*

- Hader, DP, Kumar, HD (Banaras Hindu Univ, Ctr Adv Study Bot, Varanasi), Smith, RC, *et al.* Effects On Aquatic Ecosystems . Journal Of Photochemistry And Photobiology B-Biology 46 (1-3): 53-68 Oct 1998. Univ Erlangen Nurnberg, Inst Bot & Pharmazeut Biol, Staudtstr 5, D-91058 Erlangen, Germany. *Times Cited:110*

- Pandian, TJ (Madurai Kamaraj Univ, Sch Biol Sci, Dept Genet, Madurai), Sheela, SG. Hormonal Induction Of Sex Reversal In Fish . *Aquaculture* 138 (1-4): 1-22 Dec 15 1995. *Times Cited:106*

- The paper entitled "Randomized, Double-Blind, Placebo-Controlled Trial of Fish Oil and Mustard Oil in Patients with Suspected Acute Myocardial Infarction: The Indian Experiment of Infarct Survival .4". published by Singh, RB of Med. Hosp. & Res. Ctr, Heart Res. Lab., Moradabad along with Niaz, MA, Sharma, JP, *et al.* in the Journal "*Cardiovascular Drugs and Therapy* {11 (3): 485-491 Jul 1997 has received 201 citations".

Salient Findings

- The year 2006 has the highest of the publications (246nos) followed by 2005(245).

- More than 88% of papers are articles followed by notes, etc.

- Most of the articles (99.78%) are published in English.

- The most productive author as per the no. of the contributions is Jana, B.B. of Aquaculture & Appl. Limnol. Res Unit, Department of Zoology, University of Kalyani, Kalyani 741235, W Bengal, India with highest 47 papers.

- Journal of Environmental Biology (207 nos.); Current Science (181 nos.); Aquaculture (113 nos.); Bulletin environmental Contamination and Toxicology (113 nos.) and Journal of Food Science and technology (87 nos.) are the top ranked journals.

- Environmental Science has the large number of articles (623 nos.) followed by the Fisheries (425 nos.), Marine and Freshwater biology (363 nos.), Toxicology (310 nos.) and Biochemistry and Molecular biology (268 nos.)etc.
- Banaras Hindu Univ., Varanasi is the top institution as per publication count by publishing 115 papers followed by Central Inst. of Freshwater Aquaculture (78 nos.) ; Univ. of Kalyani, Kalyani (77 nos.); .Bhabha Atom Res. Centre, Mumbai (76 nos.); and Visva Bharati Univ. (72 nos.).
- Aquaculture has the highest Impact Factor (IF) and published 113 papers in the areas fish research
- The paper entitled "Randomized, Double-Blind, Placebo-Controlled Trial of Fish Oil and Mustard Oil in Patients with Suspected Acute Myocardial Infarction: The Indian Experiment of Infarct Survival .4". published by Singh, RB of Med Hosp & Res Ctr, Heart Res Lab, Moradabad along with Niaz, MA, Sharma, JP, et al. in the Journal "Cardiovascular Drugs And Therapy 11 (3): 485-491 Jul 1997 has received 201 citations.

Findings

- USA is the major collaborative country with India and published 120 papers followed by Japan with 50 papers.
- As per city-wise distribution of papers, Kolkata occupies top position by publishing 185 papers followed by Mumbai published 182 papers.
- Uttar Pradesh is the top ranked State and published 439 papers followed by West Bengal with 367 papers.
- USA has published 42870 papers followed by Japan with 12251 papers are the top ranked
- Countries as per as total numbers of papers published in the field of fish research in the world.

REFERENCES

Web of Science Online Database 1986-2006. http://www.isiknowledge.com/

Ramesh, D.B. Mapping of fish research in India – A study based on Scopus database. In: Bioinformatics and Statistics in Fisheries Research, Vol. III, A. K. Roy and N. Sarangi (Eds.), CIFA, 2007, pp. 510-526.

Ramesh, D. B. Research Trends in Aquaculture – a pilot study based on Agricola database. In: Application of Bioinformatics and Statistics in Aquaculture – National Workshop-cum-Training, February 4-7, 2003. Eds. K. Janaki Ram and A. Antony. Bhubaneswar, CIFA, 2003, pp. 25-29.

STATISTICS

Chapter 24

Statistical Tests Applied to Aquaculture Research

Rabi N. Subudhi

Introduction

We shall discuss in this section some of the fundamental Statistical Tests that may be applied in the field of research in *Aquaculture*. Introduction to tests of hypothesis, test procedures and the concept of decision making are also covered here. At the outset, stages of research process and different formulae are briefly enlisted, for a quick reference by the young researchers. Later, a detail discussion is made on various fundamental statistical tests, including one-way and two-way ANOVA.

Research Methodology

What is Research

A good researcher should first feel/ realize the importance/ relevance of the research, he is undertaking. It should be quite convincingly clear to him particularly on the issue of "why" such research.

Research is defined as a systematic, scientific method consisting of enunciating the problem, formulating a hypothesis, collecting data analyzing them and finally reaching at some conclusion/ solution. Every good research should necessarily follow/ pass through the definite stages/ steps of research, as listed in Box 1.

Applied Bioinformatics, Statistics and Economics in Fisheries Research: pp. 337-344

New India Publishing Agency, New Delhi (India)

Box 1 : Stages of "Research Process":

Define Research Problem
↓
Review Literature
(Last finding/theories/concepts)
↓
Formulate Hypothesis
↓
Design the Research
(Sampling designs)
↓
Collection of data
↓
Analysis of data
(Hypothesis testing etc.)
↓
Interpretation & Final report.

Statistical Hypothesis

Every research problem evolves around a few hypotheses. Researchers are concerned with two types of hypotheses: research hypotheses and statistical hypotheses. The research hypothesis is the conjecture or supposition that motivates the research. It may be the result of years of observation on the part of the researcher.

Research hypotheses lead directly to statistical hypotheses. Statistical hypotheses are stated in such a way that they may be evaluated by appropriate statistical techniques. The hypotheses that we shall focus on are statistical hypotheses.

Hypothesis is generally defined as a declaratory statement, that is testable. It is a statement about the population that we wish to verify on the basis of available sample information. It is also known as a provisional answer to the research problem under study that is tested empirically for its validity. One should be very careful while selecting/ constructing a hypothesis.

Criteria for a good Hypothesis is that

It should be i) conceptually in clear terms, ii) should be testable, iii) formulated in specific terms and iv) related to available techniques, facts and theories.

Many Hypotheses can exist for the same problem

We may find many possible provisional answers (each one to be called an alternative hypothesis) to the same problem. Only one of them is tested at a time for its validity against another. The hypothesis that is tested, is generally called *Null Hypothesis* (Ho) and the other, which usually gives a reverse statement, is called its *Alternative Hypothesis* (H_1).

We know that, neither hypothesis testing, nor statistical inference, in general, leads to the proof of a hypothesis. It merely indicates whether the hypothesis is supported or is not supported by the available data.

Box 2 : Classification of hypothesis testing

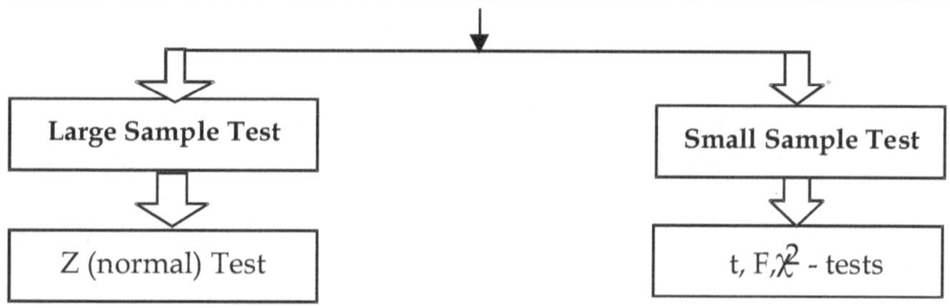

		Decision from Sample	
Actual (true) State		**Reject H_0**	**Accept H_0**
	H_0 True	Wrong: (Type-I Error) = α	**Correct Decision**
	H_0 False	Correct Decision	Wrong : (Type II Error) = β

As shown in the above table, at the time of sample testing (or, making decision based on sample information), we may come across two types of errors. These errors are other wise known as:

Type – I error: *Producer's Risk*, and

Type – II error: *Consumer's Risk.*

We may also note that a, the probability of type-1 error, is known as the level of significance of the test (= size of the Critical Region). And that,

β = Prob {Accepting H_o | H_1 is true}.

Here, (1-β) is called the *power of the test.*

It may be mentioned here that, we can't check/ reduce both α & β simultaneously.The usual practice is to control α at a predetermined low level and subject to this constraint on a (i.e. probability of type I error), we choose such a test that minimises β (or maximises the power function 1-β). Generally we choose $\alpha = 0.05$ or, 0.01, (i.e. 5% or 1% level of significance).

Test Criterion

In every test, we generally compute the value of (the ratio of the deviation of the statistic from its expected value, with its standard error, SE):

$$Z = \frac{|t - E(t)|}{S.E.(t)}$$

where t is the statistic, which is computed from sample values

The test statistic is some statistic that may be computed from the data of the sample. As a rule, there are many possible values that the test statistic may assume. The test statistic serves as a decision maker, since the decision to reject or not to reject the null hypothesis depends on the magnitude of the test statistic. If H_o is rejected, we conclude that H_1 is true. If H_o is not rejected, we conclude that H_o may be true.

3. Formulae for Tests

3.1 Large Sample Tests

By "Large Sample" we shall mean that the sample size is very large and that the population is NORMAL. Hence all the properties of Normal Distribution are to be applied in such a case. Let us enlist different possible cases and the corresponding test statistics as follows:

Cases

(i) Testing of Mean of a Normal Distribution with Known S. D. (s):

$$Z = \frac{|\overline{X} - \mu_0|}{\sigma / \sqrt{n}}$$

(ii) Testing of mean of a Normal Distribution with unknown S.D. (s):

$$t = \frac{|X - \mu_0|}{S / \sqrt{n}}$$

where

$$S = \sqrt{\frac{1}{n-1} \sum (X_1 - \overline{X})^2}$$

(S is called sample Standard Deviation)

(iii) (a) Test of equality of two means with known variances:

$$Z = \frac{\left|\overline{X}_1 - \overline{X}_2\right|}{\sqrt{\frac{\sigma_1^2}{n_1} + \frac{\sigma_2^2}{n_2}}}$$

$$(H_0 : \mu_1 = \mu_2)$$

(b) Test of equality of two means where variances are not known:

$$t = \frac{\left|\overline{X}_1 - \overline{X}_2\right|}{S\sqrt{\frac{1}{n_1} + \frac{1}{n_2}}},$$

where,

$$S = \frac{(n_1 - 1)S_1^2 + (n_2 - 1)S_2^2}{(n_1 + n_2 - 2)}$$

(iv) (a) Test of proportions (single sample case):

$$Z = \frac{|p_1 - p|}{\sqrt{pq/n}}$$

b) Test of equality of proportions of two samples:

$$Z = \frac{p_1 - p_2}{\sqrt{\frac{p_1 q_1}{n_1} + \frac{p_2 q_2}{n_2}}}$$

Small Sample Tests

t - Test

We use t-test, when:

(i) Sample size is 30 or less,
(ii) Population variance or standard deviation is unknown.

While testing hypothesis following assumptions are usually made:

(a) The population is normal (or approximately normal),
(b) Observations are independently drawn for the random sample,
(c) In case of 2 samples, population variances are assumed to be equal (for the test of equality of means).

Chi-Square (χ^2) Test

Areas of application:

(i) Test of dependence or association between two attributes

(ii) Test of goodness of fit

(iii) Test of homogeneity

Test statistic formula is given by:

$$\chi^2 = \sum_{i=1}^{n} \frac{(O - E)^2}{E}$$

This follows a χ^2 distribution, with (n-1) degrees of freedom.

F-Coefficient or Variance Ratio Test

This F-Coefficient is used to judge whether the difference between the two variances (i.e. between and within sample variances) is significant or just due to fluctuations of sampling. For this purpose we proceed on the Null-hypothesis presuming the difference between the two variances as insignificant. We then calculate F as stated above. We also look into the table value of F for the given degrees of freedom at certain level of significance. If the calculated value of F is either equal to or less than the table value of F, the difference is taken as insignificant i.e. due to chance and the null-hypothesis stands. But if the calculated value of F happens to be more than its table value, the difference is considered as significant (which means the samples could not have come from the same universe) and accordingly the conclusion may be drawn. The higher the calculated value of F is above the table value, the more definite and sure one can be about his conclusions.

Non-Parametric Tests

For testing of hypothesis, we mostly use some common tests (like large sample, t and F-tests) that estimates parameters of populations. Those are called *parametric tests*. But in some other cases we need not worry about the population parameters. Our test and result, both are about the sample observation/ function (which is called a 'statistic'). Such tests, as discussed below, are termed as *non-parametric tests*. Non-parametric tests are of course used for hypothesis testing. But it has also other extensive uses. Since it does not depend on the distribution of the parameter (of the population), there is no reference or comparison of tabulated value (like t-table or F-table). It checks the pattern of occurrence of items in the sample. It assumes randomness or uniformity of the items. It checks whether the distribution of items (or whether the fit of the distribution) is good or not, by using chi-square. There are several non-parametric tests. Most of those tests check whether the items (or data) of the same series are appearing randomly or not. That is, whether successive items are changing (higher or lower than previous item) randomly or not. Here we discuss only few such widely used tests as given under, viz. Sign Test and Rank Test.

Sign test

In this test, average of the items of the given series is found first (say, M). Then each item/value is deducted from this average or mean. Sign of such differences (X_i - M) is noted. Suppose we get 'r' plus signs and 's' minus signs. If there are 'n' items in the series, then $r + s \leq n$. In this case our null hypothesis is that the chance of any item or value exceeding M is 1/2. That is,

$$H_0 : P\ \{X>M\}=P=1/2.$$

Alternative hypothesis, $H_1 : P > 1/2$ (for one tailed test)

Or, $H_1 : P \neq ½$ (for two tailed tests).

In case there is zero difference, (when X = M) we have to ignore those cases. So, sample size will reduce to (r+s).

Wilcoxin signed Rank Test

Wilcoxin has suggested further improvement to this simple sign test. The suggested test is popularly known as *Wilcoxin signed Rank Test.*

Here, after finding the differences of individual items from mean, we have to find the ranks of combined series, for all the differences, (by taking absolute values of negative differences). Let T^- and T^+ be the sum of ranks of positive and negative differences respectively.

$$\text{So that, } T^+ + T^- = 1 + 2 \ldots.. + n = n\ (n+1)/2$$

For test we can check any of the values T^+ or T^- or (T^+ - T^-) or (T^+ + T^-)

While checking T^+, for large samples, normal distribution property, N~ (0,1), is assumed and the test statistic is given by the formula:

$$\frac{4T^+ - n(n+1)}{\sqrt{2n(n+1)(2n+1)/3}}$$

(Here too, zero difference cases are ignored.)

Run test (due to Wald - Wolfowitz)

A run is a sequence of identical letters (or sign) or by no letter at all Example: + -- ++ - -

In the above case, we find that there 4 runs in total; 2 of + and 2 of - signs.

If H_0 is true, then the number of runs (say r) will be large. If r is small, then H_0 is false.

We can convert any given series to a series of + - + etc by the following principle:

If the succeeding item is higher than the previous term, then a + sign is written, if it is less, then a - sign is written. Tie cases are to be ignored. Let us think of such a series of data (sequence of +, - signs), where there are 'm' + signs 'n'- signs and further that r is even (= 2d, say). We should expect 'd' runs of + and another 'd' runs of - signs in the series. For large sample cases,

$$\frac{r - E(r)}{SD(r)} = Z$$

where, $E(r) = \frac{2mn}{m+n} + 1$ and $Var(r) = \frac{2mn(2mn - m - n)}{(m+n)^2(m+n-1)}$

REFERENCES

Conover 1980, Practical Non-parametric Statistics, New York, John Wiley & Sons.

C. R. Kothari 2004, Quantitative Techniques (Vikas).

Daniel WW 2001, Bio-Statistics; John & Wiley.

ND Vohra 2007, QT in Management, 3rd Edn. Tata MGH

Panneerselvam 2000, Research Methodology, PHI Delhi

R Veerachamy 2005, Quantitative Methods, New Age Intl.

W.W.Daniel 2004, Biostatistics, John & Wiley, New York.

Chapter **25**

Some Statistical Procedures Useful for Analysis, Testing and Interpretation of Aquaculture Data

G.R. Maruthi Sankar

Introduction

A careful compilation and scrutiny of data of aquaculture variables is essential for appropriate statistical analysis and meaningful statistical inference and interpretation of results. In nature, many aquaculture variables are inter-related with each other and the changes in variables would often depend on different variables which are either correlated or independent. In some cases, the variables could also be intra-related over a period of time. For statistical analysis of data of aquaculture variables comprising of different physical growth variables like fish weight, body mass, length and bio-chemical growth variables like muscle protein, lipid, glycogen, protease etc. and variables like pond size, water and its quality, the environmental parameters, type of feed, chemicals used etc., appropriate statistical procedures of correlation and regression could be explored for statistical analysis, testing and interpretation of results. Different statistical procedures are used for analysis of aquaculture data with the objective of making an assessment of

- correlation between different inter-dependent aquaculture variables and their significance at a given point of time at one location or at different locations

Applied Bioinformatics, Statistics and Economics in Fisheries Research: pp. 345-363

New India Publishing Agency, New Delhi (India)

- relative changes in one or more variables at one or more locations over a period of time and their significance
- distribution parameters like mean, median, standard deviation, apart from coefficient of variation in the data of an aquaculture variable at a given location over a period of time
- testing of different treatment differences and their contribution to the the growth of an aquaculture variable based on analysis of variance (ANOVA) procedure, F-test and critical difference for testing differences between a pair of treatments at different levels of significance
- testing of differences of different variables, their relation, contribution to a dependent variable over a period of time
- efficient prediction of changes in a dependent variable through a set of either related or independent variables
- the relative contributions of a set of variables (rate of change) to a dependent variable through a statistical model of data over a period of time
- optimization analysis of different variables for maximizing a dependent aquaculture variable
- measurement of sustainability of growth of aquaculture variables over a period of time as a function of related variables

The details of different statistical procedures used for statistical analysis, statistical testing, statistical inference and interpretations of data of aquaculture variables are discussed in the paper.

The Fresh water Aquaculture experiments are conducted for testing differences between input treatments, environments, fish species etc., based on suitable statistical designs like completely randomized design, randomized block design, split-plot design, latin square design etc., using statistical principles of replication, randomization and local control. Based on a suitable sampling design, after experimentation, the observations on different parameters are carefully collected using samples taken from fish ponds. Suitable samples of observations (fish) are drawn from ponds based on simple random sampling, stratified random sampling, probability proportional to size sampling, multi-stage sampling and other types of sampling methods as discussed in literature (Cochran, 1972; Murthy, 1967). Since complete enumeration would be costly, time taking and has practical difficulties, one has to go for an efficient sampling procedure and collect observations on aquaculture parameters from time to time.

The principle of randomization is essential since it assigns an equal probability for all treatments while being allotted to experimental plots. Replication is essential for deriving an unbiased estimate of error; otherwise, no statistical test is possible for either comparing treatments or making an unbiased selection of superior treatments. Local control is the principle of dividing a field into homogeneous blocks, such that plots within a block are homogeneous and between blocks are heterogeneous. A suitable estimate of experimental error is derived for testing the treatment differences based on Snedecor's F-test which is computed as a ratio of treatment mean sum of squares and error mean sum of squares. The hypothesis of no difference between treatments (or null hypothesis) is rejected, if the calculated F-value is more than the critical F-value. The null hypothesis is not rejected, if otherwise. The F-test is carried out at a desired level of significance (usually at 5 or 1% level of significance or any level of experimenter's choice) by considering the degrees of freedom of treatments and error components (Gomez and Gomez, 1985; Henry and Scheffe, 1959). Based on a critical difference or least significant difference at a desired level of significance, the differences between treatments are statistically assessed and inferences are drawn about significance or otherwise.

The aquaculture data generated on different variables in more than one season could be pooled based on statistical criteria like homogeneity of variances, treatment differences in different years and their significance, and the contributions of different independent variables or explanatory variables to a dependent variable could be standardized. Efficient statistical models could be calibrated for the pooled data over years and predictions and optimization of variables could be made.

The statistical analysis of correlation of different aquaculture variables is carried out to assess the type of correlation (positive or negative) and depth of correlation which would range from –1 to +1. Positive correlation between a pair of variables would indicate that a change in variable would cause a positive change in another variable. Similarly, a negative correlation between two aquaculture variables would indicate that a change in one variable would cause a negative change in another variable. The changes could be significant or non-significant depending on the sample size, level of significance, magnitude or depth of correlation, practical importance of a variable and other things. This is essential to identify superior aquaculture variables based on a suitable test of correlations of variables. The analysis is a pre-requisite for screening variables, statistical modeling of fish growth, refining the predictability of a growth variable and assessment of different multivariable problems like inter-correlations of inputs of fish growth and environmental variables (problem of multicollinearity), errors-in-variables like measured and estimates of variables, serial dependence of residuals (difference between observed and expected values of variables) over a period of time like dependence of fish growth in 3rd week based on the actual growth occurred in 2nd week (problem of auto-correlation), extreme observations like outliers and influential observations (Hawkins, 1980) in both input variables and also growth of fish under different environments and other input and output aquaculture variables. The statistical models could be linear or non-linear depending on the type of relations of variables, univariate or multivariate, and could have fixed effects, random effects or mixed effects of variables which explain the changes in a dependent variable over a period of time. The details of different statistical procedures (Kapoor and Saxena, 1972) like measures of correlation, measures of regression, tests of significance, including problems of regression and statistical analysis of residuals are discussed in the following sections.

Measures of Correlation

The different measures of correlation computed for data of aquaculture variables are described in the following sections.

1. *Simple Correlation :* The simple correlation measures relationship between two aquaculture variables 'X' and 'Y'. It is a ratio of covariance of two variables to the product of standard deviations of the variables. It ranges from –1 to +1. The simple correlation 'r' could be computed as

$$r = \frac{\Sigma\, XY - (\Sigma\, X)(\Sigma\, Y)/n}{\sqrt{(\Sigma\, Y^2 - (\Sigma\, Y)^2/n)\, (\Sigma\, X^2 - (\Sigma\, X)^2/n)}}$$

   The simple correlation measures relation between any two-aquaculture variables like fish weight, fish length, age of fish etc. It indicates the association and depth of relationship between two variables. Higher the magnitude of correlation, better would be the relation between a pair of variables and vice versa. Negative correlation would indicate that a change in one variable would cause a negative effect on the other variable, while positive correlation would indicate that a change in one variable would have a positive effect on other aquaculture variable.

2. *Partial Correlation (first order) :* The partial correlation of first order measures the partial relationship between two aquaculture variables 'Y' and 'X1' keeping the effect of a third variable 'X2' as constant. When three variables are inter-related with each other, it would be difficult to assess the important variables for statistical modeling. In such situations,

an estimate of partial correlation between two variables keeping the effect of a third variable as a constant would be useful. The partial correlation is computated using simple correlations of pairs of variables. It ranges between -1 and +1. The partial correlation 'r' between two variables jointly with a 3rd variable could be computed as

$$r_{Y,X1.X2} = \frac{r_{YX1} - r_{YX2} + r_{X1X2}}{\sqrt{(1 - r^2_{YX2})(1 - r^2_{X1X2})}}$$

The partial correlation of first order measures the relation between fish weight and age of fish, keeping fish length as constant, etc. for all combinations of variables.

3. *Partial Correlation (second order) :* The partial correlation of second order measures a partial relationship between two variables 'Y' and 'X1' keeping the effects of two other variables 'X2' and X3' as constant. It ranges between -1 and +1. The partial correlation of second order 'r' could be computed as

$$r_{Y,X1.X2,X3} = \frac{(r_{YX1.X3}) - (r_{YX2.X3})(r_{X1X2.X3})}{\sqrt{(1 - r^2_{YX2.X3})(1 - r^2_{X1X2.X3})}}$$

The partial correlation of second order measures the relation between fish weight and length, keeping age of fish and number of fish in a pond as constant etc. for all combinations of variables. The partial correlation is useful for understanding the type of joint relationships among 4 different variables when two of them have a constant effect and are constantly related with the remaining 2 variables.

4. *Multiple Correlation :* The multiple correlation measures the correlation of a dependent variable 'Y' with a set of two or more independent variables 'X1 and X2' together. It is always positive and would range from 0 and 1. It is based on the simple correlations measured between all pairs of variables. In a 3-variable situation with one dependent and two independent variables, the multiple correlation 'R' could be computed as

$$R_{Y.X1,X2} = \frac{r^2_{YX1} + r^2_{YX2} - 2r_{YX1}r_{YX2}r_{X1X2})}{\sqrt{(1 - r^2_{X1X2}}}$$

The estimate of multiple correlation measures the correlation between a dependent variable like fish body weight with a set of two or more independent variables like age of the fish, pond size, quantity of water, quality of water etc. Higher the multiple correlation, better would be the relation of a set of independent variables with the dependent variable, and vice versa. The significance of multiple correlation would indicate that the dependent variable is significantly affected by the set of independent variables. The multiple correlation could be estimated for more than 3 variables also and interpreted accordingly.

5. *Rank Correlation* : For data of any pairs of variables, we can assign ranks for understanding the mutual growth and relation of variables over a period of time. The rank correlation measures correlation between ranks assigned to a pair of aquaculture variables instead of the actual observations of variables. Since the unit of measurement of variables would be different, it is always better to provide ranks to variables and determine correlation between variables. The estimate of rank correlation between a pair of variables is always positive and would range from 0 to 1. The differences in ranks of two variables are computed and assessed for closeness of ranks or otherwise between two variables. The rank correlation 'r' between the ranks of two variables could be computed as

$$r = 1 - \frac{6\Sigma d_i^2}{n\,(n^2 - 1)}$$

where 'd_i' is the difference in ranks and 'n' is the number of observations.

The rank correlation measures the correlation between ranks assigned to the variables like body weight, body mass index and length of fish etc., in aquaculture data. A higher rank correlation between a pair of variables would indicate either higher ranks or lower ranks attained by a pair of aquaculture variables over a period of time.

6. *Correlation Ratio* : Some times the relation between variables could be non-linear and we have to measure the curvi-linear relationship accordingly. The correlation ratio'h' is the appropriate measure of curvi-linear relationship when the relationship between two aquaculture variables is not linear. For example, if the body weight of a fish is increasing with its age and the growth rate is not consistent, and is non-linear over a period of time, then an estimate of correlation ratio could be computed. If the relationship is linear then h = r, otherwise h > r. The correlation ratio between two aquaculture variables 'Y' and 'X' ranges from –1 to +1 and could be given as

$$\eta^2_{YX} = [\Sigma (T_i^2 / n_i) - T^2 / N] / N \sigma_Y^2$$

where $T_i / n_i \quad = \quad \Sigma f_{ij} y_{ij} / n_i$

$T / N \quad = \quad \Sigma n_i y_i / \Sigma n_i$

The correlation ratio measures a curvi-linear relationship of variables like relationship between fish body weight and the quantity and ingredients of feed given, fish weight and fish length etc., over a period of time.

7. *Intra-class correlation* : The intra-class correlation means within class-correlation. Here both the aquaculture variables measure same characteristics. It is the correlation within a variable with respect to some common characteristic. For example, we may work out intra-class correlation between yields of plots in an agricultural trial. Suppose there are A_1, A_2, ...A_n families with k_1, k_2, ...k_n members, and let x_{ij} (I = 1,2,...n; j = 1,2, ...k_n) denote the measurement on the j th member in the i th family, then intra-class correlation can be given as

$$r = \frac{\Sigma k_i^2 (X_i - X)^2 - \Sigma\Sigma (X_{ij} - X)^2}{\Sigma\Sigma (k_i - 1)(X_{ij} - X)^2}$$

if $k_i = k$ (i.e., if all families have equal members), then

$$r = [1 /(k-1)][k \sigma_m^2 / \sigma^2) -1]$$

where s^2 denotes the variance of X and s_m^2 denotes the variance of means of families. The intra-class correlation ranges between - [1/(k-1)] and 1.

An estimate of intra-class correlation measures the correlation of a variable with in a group or a class with respect to the same characteristic. For example, intra-class correlation measures the relation of body weight of fish on a given day/week to its previous day/week for fish of a particular age group with in a given pond. We can compare the intra-class correlation of pairs of variables over a period of time and standardize for assessing the changes in variables from time to time.

Measures of Regression

1. *Simple Regression* : It measures the functional relationship between a dependent variable 'Y' and an independent variable 'X' with estimates of an intercept (α) and a slope (β). The estimates of α and β can be

negative, zero or positive. The linear regression will be useful to predict the changes in fish weight 'Y' as a function of either fish length or quantity of feed or time period or any variable 'X' that is related to the weight of a fish. The slope measures the rate of change in fish weight for a unit change in a variable 'X'. The model is based on the assumption that the relationship between the two variables in the model is linear. The linear regression is given as

$$Y = \alpha + \beta X$$

$$where\ \beta = \frac{\sum X_i Y_1}{\sum X^2_i} \quad \text{and} \quad \alpha = y - \beta x$$

In 'β', the numerator represents the covariance of 'X' and 'Y' and denominator represents the variance of 'X'. In 'α', y and x represent means of Y and X respectively. Based on above, we can compute simple regression equation of fish weight as a function of age of fish, quantity of feed and other inputs supplied, fish length etc. An estimate of correlation 'r' between dependent and independent variables will indicate the usefulness of the prediction model of 'Y' through 'X'. We can test the contribution of an independent variable to dependent variable based on statistical test of significance.

2. *Multiple Regression* : If a dependent variable 'Y' is a function of a set of independent variables 'X_i', then the estimates of regression coefficients of different variables (b) along with the intercept (a) are estimated using the matrix algebra. Here a variable 'Y' is influenced by a set of 'X' variables which have a different magnitude of correlation with Y and influence Y in a different manner. The multiple regression model of a dependent variable 'Y' through different independent aquaculture variables can be given as

$$Y = \pm \alpha \pm \beta_1 X_1 \pm \beta_2 X_2 \pm \beta_3 X_3 \pm \beta_4 X_4 \pm \text{---------} \pm \beta_n X_n$$

$$\hat{\beta} = (X' X)^{-1} X' Y$$

The regression coefficients 'b_i' can be negative, zero or positive and would measure the rates of change in 'Y' for an unit change in the independent variables. The regression coefficients could be assessed based on the magnitudes and their significance using a statistical test. The predictability of changes in fish weight could be determined by a measure of coefficient of prediction 'R^2' for a multiple regression model. The contributions of different aquaculture variables could be tested based on t-test at 5 or 1%

level of significance (Draper and Smith, 1998; Maruthi Sankar, 1986; Williams, 1987). The multiple regression models of fish weight can be computed through different input variables like quantity feed, number of fish at a given point of time, age of fish, quantity and quality of feed, pond size, water availability (depth), quality of water etc. The models can be iteratively calibrated and refined by identifying significant and important aquaculture variables that are useful for predicting the growth of fish over a period of time.

3. *Polynomial Regression* : If a dependent variable 'Y' is a function of linear, non-liner and other higher order effects of an independent variable 'X', then the polynomial regression is fitted to quantify the effects of a variable and its significance at different orders for prediction of 'Y'. The computation procedure of polynomial regression is similar to that of a multiple regression procedure. The nth order polynomial regression of 'Y' can be given as

$$Y = \alpha + \beta_1 X_1 + \beta_2 X_1^2 + \beta_3 X_1^3 + \beta_4 X_1^4 + ------+ \beta_n X_1^n$$

The polynomial regression coefficients can be negative, zero or positive and 'b_i'would measure the rates of change in 'Y' for a unit change in the ith independent variable. We can compute the polynomial regression models of fish weight through variables like quantity of feed, water in the pond, time period like weight on different days etc. Here dependent variable 'Y' is influenced by only independent variable 'X' but through different polynomial orders of 'X' like linear, quadratic, cubic, quartic, quintic, hexa etc. The regression coefficients could be tested based on t-test at 5 or 1% level of significance with n-k-1 degrees of freedom where k is the order of the polynomial. Based on the significance of the polynomial regression coefficients, we can decide about the order of a polynomial function for measuring the changes in a given aquaculture variable over a period of time.

4. *Principal Component Analysis* : If the aquaculture variables are positively or negatively and significantly correlated, a principal component model can be derived by computing different uncorrelated principal components as a function of different correlated variables. Each principal component could be expressed as a function of different aquaculture variables. The total variability in a set of data of different correlated variables could be assessed based on a principal component model. The dimensionality of data variables could be reduced to a few leading principal components for expressing the contributions of variables to

total variability in the data. Usually, the first two leading principal components would explain maximum variability of 70-80% in the data. The variables, which have a significantly higher loadings or weights on principal components, could be identified and inferred as dominant variables. A principal component model with correlated variables is postulated for calibrating a new set of uncorrelated principal components (Tabachnic and Fidell, 1983; Maruthi Sankar *et al.,* 2004; Johnson and Wicharn, 2001). The principal components are linear and are derived in the decreasing order of importance. Since the variables are related and measured on scales with a wide range, the correlations are used for determining principal components. The loadings of variables in a principal component model would range between –1 to +1. A variable is considered dominant if its loading is more than ± 0.3 on a principal component. However, we suggest more rigorous limits of loadings or weights exceeding ± 0.4 could be considered in aquaculture data for assessing the dominance of variables.

Tests of Significance

Different statistical tests of significance are used for testing the differences between (1) sample means and variances of aquaculture variables as observed on different days of sampling; (2) correlations of different pairs of variables; (3) contribution of different predictor variables (or independent variables) to a prediction variable (or a dependent variable) as measured by regression coefficients of different aquaculture variables. Based on a suitable statistical test procedure, the hypothesis of no differences (null hypothesis) would be rejected if the calculated value of the test statistic is more than the critical value at a desired level of significance (usually either 5% or 1 % level of significance). The null hypothesis is not rejected, if otherwise. Usually, based on a statistical test, the null hypothesis would be rejected at 5% or 1% level of significance. This would indicate that the treatments are significantly different from each other for influencing a dependent aquaculture variable. The different statistical tests of mean, variance, correlation, regression coefficient, coefficient of predictability and other statistics are described below.

1. *Testing significance in data of large samples :* If an aquaculture variable X is distributed as Normal with mean 'μ' and variance 'σ', then $Z = (X - \mu)/\sigma$ is distributed as Normal with mean 0 and variance 1. If $|Z| > 1.96$, then the sample mean is inferred to be significantly different from population mean at 5% level of significance. If $|Z| > 2.58$, then the sample mean is significantly different from population mean at 1% level of significance. If $|Z| < 1.96$, then there is no significant difference

between sample mean and population mean with the set of aquaculture data of variables.

2. *Testing single proportion :* Let P be proportion of success and Q = 1 - P be proportion of failure. If mean = n P and variance = n P Q, then Z = [X - n P] / (n P Q) is distributed as Normal with mean 0 and variance 1. The test and conclusions are same as given above. For example: proportion of sampled number of fish of a particular age group, or proportion of growth of fish with respect to time period etc.

3. *Testing difference of proportions :* This is a test for testing the difference between two sample proportions. Let p1 = X1 / n1 and p2 = X2 / n2. Let Mean (p1) = P1 and Mean (p2) = P2. Similarly, let Variance (p1) = P1 Q1 / n1 and Variance (p2) = P2 Q2 / n2

$$Z = \frac{(p1 - p2)}{PQ\left[(1/n1) + (1/n2)\right]}$$

Z is distributed as Normal with mean 0 and variance 1. The test and conclusions are same as given above.

4. *Testing a mean in small samples* **:** If X is distributed as Normal with mean m and variances, then Z = [(x - μ) / (σ/ sqrt (n))] is distributed as Normal with mean 0 and variance 1. The test and conclusions are same as above.

5. *Testing differences of two means :* If $\overline{X1}$ and $\overline{X1}$ are means and σ_1^2 and σ_2^2 are variances based on two samples with n1 and n2 observations of two aquaculture variables, then

$$Z = \frac{\overline{X1} - \overline{X2}}{\sqrt{(\sigma_1^2 / n1) + (\sigma_2^2 / n^2)}}$$

is distributed as Normal with mean 0 and variance 1. The test and conclusions are same as above.

6. *Testing difference of standard deviations :* If S_1 and S_2 are standard deviations of two samples with n_1 and n_2 observations from Normal distribution with variances σ_1^2 and σ_2^2 of two aquaculture variables, then

$$Z = \frac{S_1 - S_2}{\sqrt{(\sigma_1^2 / 2n_1) + (\sigma_2^2 / 2n_2)}}$$

is distributed as Normal with mean 0 and variance 1. The test and conclusions are same as above.

7. *Testing the difference between sample correlation (r) and population correlation (p)* by making Z - transformation can be given as

$$\frac{Z-Z^*}{\sqrt{(n-3)}}$$

$$where \quad Z=\log_e \sqrt{[(1+r)/(1-r)]}$$

$$Z^* = \log e \sqrt{[1+p)/(1-p)]}$$

Here the sample and population correlation coefficients are transformed to Z-transformation and the statistic is worked out. The test and conclusions are same as above.

8. *Testing the difference between two correlation coefficients 'r_1' and 'r_2'* by making Z - transformation can be given as

$$Z=\frac{Z_1-Z_2}{\sqrt{[1/(n_1-3))+(1/(n_2-3))]}}$$

$$Z_1=\log_e \sqrt{[(1+r_1)/(1-r_1)]}$$

$$Z_2=\log e \sqrt{[(1+r_2)/(1-r_2)]}$$

The differences between correlations would indicate whether the samples are homogenous or different at a given point of time, apart from indicating the type of relation. Here the two sample correlation coefficients are transformed to Z-transformation and the statistic is worked out. The test and conclusions are same as above.

9. *Testing the observed correlation 'r' between two variables against 'zero'* can be given as

$$t=\frac{r}{\sqrt{(1-r^2)/(n-2)}}$$

Here 'r' is the sample correlation between two aquaculture variables based on 'n' observations. A better correlation and its significance could be derived between aquaculture variables if the number of observations on fish growth, species and related variables are large compare to fewer observations. The calculated t-value is compared with the critical t-value at 5 or 1% level of significance with (n–2) degrees of freedom. The null hypothesis is rejected, if the t-calculated value is less than the t-critical value. The null hypothesis is not rejected, if otherwise. The test and conclusions are same as above.

10. *Testing the partial correlation 'r' between two variables keeping the effects of a third variable as constant* can be given as

$$Z = \frac{\log_e \sqrt{[(1+r)/(1-r)]}}{\left[1/\sqrt{(n-k-3)}\right]}$$

A significant partial correlation would indicate that the two variables have a significant relation when the effect of 3rd variable is constant. Higher the magnitude, better would be the partial relation of the 2 variables in the presence of a 3rd variable. The test and conclusions are same as above.

11. *Testing the regression coefficient (slope)* 'b' of an independent variable 'X' can be given as

$$t = \frac{\beta}{\sqrt{S^2{}_{Y,X} / X_i^2}}$$

where S^2 is the residual mean sum of squares. The significance of slope of X is dependent on the number of observations, magnitude of correlation with Y and magnitude of residual variance of aquaculture data. The test and conclusions are same as above.

12. *Testing the regression coefficient (intercept)* 'a' can be given as

$$t = \frac{\alpha - a^0}{\sqrt{S^2_{YX}\left[(1/n) + /(X^2 / X_i^2)\right]}}$$

where S^2 is the residual mean sum of squares. The intercept would indicate an adjustment coefficient in the model. Its significance would indicate the likely non-significance of the slope of an independent variables. However, this need not be the case always. The test and conclusions are same as above.

13. *Testing the homogeneity of regression coefficients (slopes) of 'k' sets of data* (or over different seasons) can be given as (Gomez and Gomez, 1985)

(null hypothesis) Ho: $\beta_1 = \beta_2 = \beta_3 = $ ---------- $= \beta_k$

$$F = \frac{(B-G)(\Sigma n_i - 2k)}{(B)(k-1)}$$

This is distributed as F with (k-1, Σn_i – 2k) degrees of freedom

G = D-E^2 / C

B = sum of 'residual sum of squares' of k sets

C = sum of 'corrected sum of squares' of X of k sets

D = sum of 'corrected sum of squares' of Y of k sets

E = sum of 'corrected sum of squares' of products of X and Y of k sets

The significance of F would indicate that either K sets or atleast a few of the K sets are significantly different from each other. Even based on Bartlett's test, the variances would be heterogenous. The number of observations would also be important for significance of F test over different sets of aquaculture data when pooled over a time period. The test and conclusions are same as above.

14. *Testing the predictability value (R^2) of a regression model with 'k' independent regressor variables* can be given as

$$R^2 = \frac{SSR}{\Sigma Y^2} = \frac{\beta_1 \Sigma X_1 Y + \beta_2 \Sigma X_2 Y + \text{———————} + \beta_k \Sigma X_k Y}{\Sigma Y^2}$$

where SSR is sum of squares due to regression

$$F = \frac{R^2}{(1-R^2)} X \frac{N-K-1}{K}$$

where N is number of observations and K is number of variables

The significance of F test would indicate that the changes in dependent variable could be significantly predictable through a set of independent aquaculture variables. For example, the fish growth and its weight could be significantly predictable through the quantitity of feed, nutrients in feed as well as water, temperature congenial to fish growth, and other related variables over a period of time. The test and conclusions are same as above.

15. *Testing the R^2 adequacy of a model with 'p' regressors compared to a model with 'q' regressors where p < q*

$$\frac{(R^2q - R^2p)}{(1-R^2q)/(n-q-1)} < qF$$

where F is with (q, n – q – 1) degrees of freedom

$$R^2p > R^2a$$

where $R^2a = 1 - (1 - R^2q)\ (1+d)$

$$d = \frac{qF}{(n-q-1)}$$

Here the number of regressor variables in the two models which are compared and the number of observations would decide about the significance of the test criteria. The level of significance could be either 5 or 1% level with degrees of freedom as stated above. The test and conclusions are same as above we can always calibrate models with different combinations of aquaculture variables and improve predictions iteratively by adopting Gauss forward, Gauss backward are step-wise regression procedures for identifying a suitable model for aquaculture data.

16. *Testing the sufficiency of a model with Residual Mean Square Ratio (RMSR) criterion:* This is used for testing the sufficiency of a regression model with 'p' regressors compared to a model with 'q' regressors (p < q) based on F-test and can be given as

$$F = \frac{RSS(p) - RSS(q)}{(q-p)\,RMSS)(q)}$$

where RSS is residual sum of squares; RMSS is residual mean sum of squares. This is distributed as F with (q – p, n – q –1) degrees of freedom. The sufficiency of a statistical model for predicting the changes in aquaculture data could always be tested using the standard Snedecor's F test based on analysis of variance. The test and conclusions are same as above.

17. *Percentage relative Efficiency (PRE) of a regression model 'A' compared to a regression model 'B'* can be given as

$$PRE(A) = \frac{\sigma^2{}_B(n+q+1)/n}{\sigma^2{}_A(n+p+1)/n} X100$$

where s^2_A and s^2_B are Residual mean sum of squares of models 'A' and 'B' models. The percentage relative efficiency of a model would indicate its superiority over another model, if PRE is more than 100%.

Problems of Regression

1. *Multicollinearity:* The problem of multicollinearity will occur when there are high and significant correlations between different aquaculture variables compared to the over all multiple correlation and predictability of a model. This will result in linear dependence of a variable on another variable and insensitive regression coefficients for prediction. The multicollinearity can be assessed by computing an estimate of c^2 and can be given as

$$\chi^2 = -n - 1 - (1/6)(2k + 5) \log_e (D)$$

where D = value of standardized determinant

n = number of observations

k = number of independent variables

Multicollinearity is the problem of high and significant intercorrelation between predictor variables compared to overall multiple correlation of dependent variable with a set of predictor variables. The linear dependence of regressor variables would lead to insensitive estimates of prediction, high standard errors and non-significant contributions of regressor variables to the dependent variable. In aquaculture data, the problem of multicollinearity would be severe in some situations where 2 or more variables are jointly influencing the dependent variable significantly over a period of time. We must identify the significantly correlated variables and drop the less important variable from a model for predicting the changes in a dependent aquaculture variable like fish weight and body mass index over a period of time.

2. *Examination of the residuals for identifying 'Outliers, high leverage and Influential'* observations, testing and deletion, and improving the predictability of models. The residuals can be examined in different forms as normalized, standardized, internally studentised and externally studentised residuals.

Normalised: $a_i = f(e_i, e'e) = e_i / \sqrt{e'e}$

Standardised: $b_i = f(e_i, \sigma) = e_i / \sigma$

where $\sigma = \sqrt{e'e/(n-k)}$

Internally studentised: $r_i = f(e_i, \sigma\sqrt{(1-p_{ii})} = e_i / \sigma\sqrt{(1-p_{ii})}$

Externally studentised: $r_i = f(e_i, \sigma_i\sqrt{(1-p_{ii})}) = e_i / \sigma_i\sqrt{(1-p_{ii})}$

where $\sigma_i^2 = \sigma^2_{(i)}\ (1 - p_{ii})$

$$\sigma_i^2 = \frac{Y_{(i)}(I - P_{(ii)})Y_{(i)}}{n-k-1}$$

In aquaculture data, some times a few observations may be unusual and would have a large residual when predicted based on a statistical model. We must identify the observations which are extreme and outlying compared to other observations and delete accordingly after testing for their influence on dependent variable. Similarly we must also identify variables which are extreme and would influence the dependent variable Y inconsistently over a period of time, test them for their influence and delete if they are either having inconsistent effect or no effect on Y.

3. *Analysis of residuals:* We must always carry out a detailed statistical analysis of residuals over a period of time. The residuals are difference between observed and expected values based on a model. Since the predictions are non consistent and uniform over a period of time, the changing influences of variables on a dependent variable, uncontrollable variables, there is a need for appropriate statistical test for residuals and improve the predictions based on a model iteratively. For example, if fish weight has a predictability of only 60%, this means 40% is unpredictable. The unexplained variation as measured by the residuals (difference between observed and expected values of a dependent variable) could be minimized by proper identification and deletion of extreme observations in the data of prediction (Y) and predictor (X) variables and recalibrating the model. The procedure could be iteratively carried out till there is no significant increase in the predictability estimates of a dependent variable (Cooper and Weeks, 1983; John Bibby and Helge Toutenberg, 1979; Draper and Smith, 1973). The residuals are model dependent and hence we must always improve the model iteratively with suitable number of observations. If sample size is too small, we get spurious correlation and regression coefficients which are inconsistent and carry a large residual and model would be highly insensitive. There is a need for suitable transformation of variables for reducing the variance in data of aquaculture variables so that the model will have a high predictability and is consistent over a period of time.

4. *Effects of variables on a model:* We must always examine the effects of independent variables on the dependent variable for their sensitivity

and significances, linear dependence and multicollinearity, lack of homogeneity of variances, randomness of independent variables, autocorrelation of variables, extreme nature of variables in its relationships with other variables, violation of the assumption of normality and other aspects. We must always adopt an appropriate statistical procedure for understanding the cause and effect relationships, contributions of variables, significance of variables and predictability of growth of aquaculture variables over a period of time efficiently.

5. *Normal Distribution:* The density function of normal distribution for a random variable 'X' can be given as

$$f(x) = \left[1/\sqrt{2\pi\sigma}\right] e^{-\left[(X-\mu)^2/2\sigma^2\right]}$$

(a) The curve is bell shaped and symmetrical about the line $x = \mu$

(b) Mean, Median and Mode would coincide

(c) As x increases numerically, f(x) decreases rapidly, the maximum probability occurring at the point $x = \mu$

(d) $\beta_1 = 0$ (Skewness) and $\beta_2 = 3$ (Kurtosis)

(e) Linear combination of independent normal varieties is also a normal variate

(f) Area property: $P(\mu - \sigma) < X < (\mu + \sigma) = 0.6826$

$P(\mu - 2\sigma) < X < (\mu + 2\sigma) = 0.9544$

$P(\mu - 3\sigma) < X < (\mu + 3\sigma) = 0.9973$

(g) x-axis is an asymptote to the normal curve

(h) The points of inflexion of the curve are given by $x = f(X) = [1/2]\ e^{-[1/2]}$

(i) Mean deviation about mean is $\sqrt{2} / \pi\ \sigma = (4 / 5)\ \sigma$

(j) Quartile deviation = $(Q3 - Q1) / 2 = (2 / 3)\ \sigma$

(k) $\mu_{2r+1} = 0\ (r = 0,1,2,....)$

$\mu_{2r} = 1.3.5.........(2r\text{-}1)\ \sigma^2\ r\ (r = 0,1,2,........)$

This implies $\mu 1 = 0$ $\mu 3 = 0$ $\mu 5 = 0$ ....(odd moments)

$\mu 2 = 1$ $\mu 4 = 3$ ....(even moments)

REFERENCES

Cochran, W. G., 1972. Sampling techniques. John Wiley & Sons, New York.

Cooper, R. A. and A. J. Weeks, 1983. Data models and statistical analysis. Heritage Publishers.

Draper, N.R. and Smith, H. 1998. Applied Regression Analysis. John Wiley and Sons, New York.

Gomez K.A. and A. A. Gomez, 1985. Statistical procedures for agricultural research. John Wiley and Sons, New York.

Hawkins, D. M., 1980. Identification of outliers. Chapman and Hill Publishers.

Henry Scheffe, 1959. The analysis of variance. John Wiley & Sons Inc., New York.

John Bibby and Helge Toutenberg, 1979. Prediction and improved estimation in linear models. John Wiley & Sons Inc., New York.

Johnson, R. A. and D. W. Wicharn, 2001. Applied multivariate statistical analysis. Prentice-Hall of India.

Kapoor, J. N. and H. C. Saxena, 1972. Mathematical statistics. S. Chand & Co Private Ltd., New Delhi.

Maruthi Sankar, G. R., 1986. On screening of regression models for selection of optimal variable subsets. *Journal of Indian Society of Agricultural Statistics*, 38 (2) : 161 - 168.

Maruthi Sankar, G. R., M. Vanaja, V. Maruthi, P. Raghuram Reddy and D. Narasimha Murthy, 2004. Selection of consistent plant traits for sunflower growth using principal component analysis. Helia, 27(41):113 - 122.

Murthy, M. N., 1967. Sampling theory and methods. Statistical publishing society.

Tabachnic, B. G. and L. S. Fidell, 1983. Using multivariate statistics. Harper and Row Publishers, New York.

Williams, E. J., 1987. Regression analysis. John Wiley & Sons Inc., New York.

Chapter **26**

Design and Analysis of Aquaculture Experiments Using Statistical Softwares

Ravi R. Saxena

Introduction

In almost all biological studies, it will be impossible to account for every individual in a population. Therefore, it is necessary to examine a sub-group of the total population and extrapolate from this to the whole population. The process by which the sub-group of the population is selected is sampling. Nearly, all statistical tests make a fundamental assumption that sampling of individual will be at random from all the individuals that could possibly be sampled. It is very important that timing is taken into account either by sampling strategy or by later analysis. Most population will be affected by season, time of day and local weather conditions.

What is considered here is the problem faced by an ecologist working in the field and trying to design a suitable sampling strategy. The use of the very powerful statistical technique; analysis of variance (ANOVA) is more common in the situation of a controlled experiment, where, we are analyzing the effects of different levels of a treatment (e.g. feed, fertilizer, temperature etc.) on some measured aspect of population, Then, to get a true estimate of the effect of the treatment, experimental design will be of paramount importance.

The appropriate design of an experiment is the key to successful analysis of a problem, for without the correct design, we will never have the right sort

Applied Bioinformatics, Statistics and Economics in Fisheries Research: pp. 365-384

New India Publishing Agency, New Delhi (India)

of data. The problem of sampling still exist in designed experiment, but the control of the system allows the experimenter to ensure that there are sufficient individuals to sample, that all factor combinations have the same number of observations.

Many of scientific investigation are based on the idea of hypothesis testing. The idea is to formulate a *hypothesis* (H_1) into a statement such as "male and female shrimps are of different sizes" followed by collection of appropriate data and then use the statistics to determine whether the hypothesis is true or not. However, it is not quite as simple as that. The statistical tests do not give a simple answer of true or not. First, we have to realize that every hypothesis will have an associated *null hypothesis* (H_0); most statistical tests use the null hypothesis.

Therefore, for these examples, the hypothesis (H_1) is: "male and female shrimps are of different sizes" and the associated null hypothesis (H_0) is "male and female shrimps are not of different sizes".

What a statistical test determines the probability that the null hypothesis is true (called the p-value). If the probability is low then the null hypothesis is rejected and the original or alternative hypothesis is accepted. The lower the probability (p-value) the more confident, we can be in the rejection of the null hypothesis but we can never be very sure, unless we have measured the whole population, that we are correct. It is a usual convention in biology to use a critical p-value of 0.05. This means that the probability of a null hypothesis being true is 0.05 (5%). Whenever a statistical test gives results with p-value less than 0.05, we reject the null hypothesis and accept the alternate hypothesis.

Unlike in agriculture application of designs of experiments in aquaculture experiments are very much limited mainly because of limitations of availability of experimental ponds. Survey of literature reveals the evidence of successful applications of experimental designs following the principles of randomization, replication and local control in brackishwater and sewage fed aquaculture (Roy *et al.*, 1985, 1986, 1995, 1997; Chakraborty *et al.*, 1981, 1985, 1987 and Karmakar *et al.*, 1986; Pillai et al., 1987, Rajyalakshmi *et al.*, 1981, 1982, 1986; and Mohapatra *et al.*, 2006). The findings of the study may be useful for planning, conducting, analysis and interpretation of freshwater aquaculture data particularly the extent of variation, replication and experimental error.

In the following section suggesting some commonly used experimental designs alongwith SPSS output. Step by step instructions are also given for analysis of data using the SPSS package.

One-way Analysis of Variance

Using analysis of variance (ANOVA) to determine whether just two groups have the same mean may seem like overkill. This may be the simplest use of ANOVA, but it still works and gives the same answer as the t-test. The t-test is restricted to two groups makes the use of ANOVA preferable in this situation because we do not have to learn a new test when we consider more than two groups.

ANOVA for three or more group makes the same assumptions as for two groups: that the data is continuous, at least approximately normally distributed and that the variances of each group/sample are homogeneous. These assumptions should be tested before the test is carried out. The null hypothesis being tested is that each set of data has the same mean. The way that ANOVA approaches this is to have a null hypothesis that the variation within groups is the same as variation between groups.

Example: A hypothetical research has data from an experiment conducted on growth performance in *Labeo rohita* fingerlings fed three experimental diets (Table 1) and wishes to determine weather there are any difference is mean growth between them. If there is a significant result (i.e. if $p<0.05$) this indicates that at least one pair of feed have different means, it does not tell which one.

Table 1. Data on growth parameters of *Labeo rohita* fed three types of feed.

Feed	Repetition	Final body weight (g)
Fishmeal	1	24.5
	1	23.4
	1	22.1
	1	25.3
	1	23.4
Mustard oil cake	2	26.4
	2	27.0
	2	25.2
	2	25.8
	2	27.1
Rice bran	3	25.5
	3	25.7
	3	26.8
	3	27.3
	3	26.0

Table 2. SPSS output of one way ANOVA

```
                    ------------- one way ----------
Variable        BODY WEIGHT
By Variable     FEED

                          Analysis of variance

Source          d.f.  Sum of square  Mean Square  F-ratio   F-Prob

Between group    2    21.5093          10.7547    11.8792   0.0014
Within group    12    10.8640           0.9053
Total           14    32.3733
```

If the data is analyzed using ANOVA on a statistical package the output will not be directly suitable for presentation in report. The ANOVA table that you would include in a write-up or publication should appear as something along the lines of the tables below:

Table 3. One Way Analysis of Variance (ANOVA)

Source of variation	d.f.	SS	MS	F	Prob.
Feed	2	21.5	10.8	11.9	0.0014
Residual (error)	12	10.9	0.91		
Total	14	32.4			

Steps Using SPSS: Under the 'Statistics' menu choose 'compare means' and then 'one way ANOVA….'. In the dialogue box that appears move 'body weight' into 'dependent list' box by first highlighting it then clicking on the appropriate move button. Next move 'feed' into the factor box. Now click the 'define range' button and then, assuming three groups have been labeled 1,2 & 3 , input 1 in 'minimum' and '3' in 'maximum' before clicking 'continue'. This will return to the first dialogue box. Click OK to run the test. The following will appear in the output windows:

This tells that the observed data for 'body weight' has been grouped using 'feed' and then goes on to a standard ANOVA table. As is always the case in one-way ANOVA, the top line of the table is the important one, this is the variation between groups. The second line is the variation within the groups that is being used as a comparison. In this case, the F-ratio is 11.8792. This value could be looked up on an F-table although SPSS gives the p value anyway, labeling it as F-prob. The value is 0.0014 indicating that at least two of the groups have means that are significantly different. There is three possible pair with three groups: 1&2; 1&3; 2&3. It is the post hoc test that tells which pairs are different from the other.

'Option' one of the assumptions of the ANOVA is that variances are equal. In the t-test in SPSS, this is tested automatically. In ANOVA, it is available under 'Option'. Just check the box labeled 'Homogeneity of Variance' and click Ok before running the test.

Two-way Analysis of variance (Without replication)

It may seem slightly odd that a statistical test that relies on the comparison of variation can be used when there is only one observation for each factor level combination but the test is perfectly valid. Experimenter (e.g. different concentrations) or naturally occurring (e.g. different sites) may set the two grouping variables. The assumptions of the tests are : that the data are continuous; are approximately normally distributed ; that the data would have about the same variance in each factor combination and that the grouping variables have at least two levels each that can be coded into integers.

Example

A hypothetical data of Hardness of soil on six different dates (coded with the letter A to F) at four different sites were collected. Factor 1 is date and Factor 2 is site. The hardness recorded at different sites and dates were (Table 4).

Table 4. Hardness of four different sites collected on six different dates.

Sites	Dates					
	A	B	C	D	E	F
S1	1130	1125	1350	1375	1225	1235
S2	1115	1120	1375	1200	1250	1200
S3	1145	1170	1235	1175	1225	1155
S4	1200	1230	1140	1325	1275	1215

Table 5. ANOVA - Two way form:

Factor	d.f.	SS	MS	F-ratio	P-value
Sites	3	11071	3690	0.797	0.515
Dates	5	59763	11593	2.58	
Error	15	69479	4631		

Note that there are 23 degrees of freedom in total in this example. There were 24 observations in the total data set, so that is correct. It is always sensible to check the number of degrees of freedom in the out to see if it matches to our expectations. There is no possibility of examining the interaction between the two factors, as there is no replication. The interaction is an important part of the power of analysis of variance where there is no replication.

In the example, the P value associated with the dates was close to critical 0.05 levels. In a report this could be highlighted in the text as follows: "There was no significant difference in the hardness between the sites ($F_{3,15}$ = 0.797, P>0.1) but there was an indication that hardness varied between dates ($F_{5,15}$ = 2.58, P = 0.071)". The subscripted numbers quoted with the F- ratios are the degrees of freedom.

Steps using SPSS: Input the observed data into a single column. Use two separate columns to code the observations with integers to represent the factors levels. Choose Statistics menu select 'ANOVA models' the 'simple factorial'. In the dialogue box move the observation into the 'dependent box and two factors into the 'factor(s)' box. For each factor, in turn, we must define the range of possible values of the factor levels. Select the first factor, click the 'define range' button then in the box, put the value of the lowest factor level in the 'minimum box and the highest in the 'maximum' box (in the example the factor 'site' would have a 1 in minimum an 4 in maximum. Click continues and returns to main dialogue box. Before carrying out the test, we must click on the 'option' button and bring up the option box. In the section marked 'maximum interaction' select 'none' then click 'continue'. If we do not do this step, the analysis will not provide any significance values. Once back to the original dialogue box and click OK to run test. The following output we will get (Table 6.).

Table 6. SPSS output of Two-way ANOVA.

HARDNESS
SITES
DATES

Unique sums of squares
All effects entered simultaneously

Source of variation	SS	DF	MS	F	Sig. of F
Main Effects	70833.333	8	8854.167	1.912	0.133
Site	11070.833	3	3690.278	0.797	0.515
Dates	59762.500	5	11952.500	2.580	0.071
Explained	70833.333	8	8854.167	1.912	0.133
Residual	69479.167	15	4631.944		
Total	140312.500	23	6100.543		

The output of first column give the name of the column containing the observation and then the factor names. The next two lines relates to the fact that the ANOVA is using standard assumptions. Then comes ANOVA table itself. The important lines are for the factor variables called Main Effects in SPSS (Site and Dates in the example). In the example, the p value for sites is well above the critical 0.05 level, so we accept the null hypothesis that all sites have the same hardness. However, the p-value for dates is 0.071, quite close to critical level. We still accept the null hypothesis, but should note the p-value is close to the critical level. In this output, it is best to totally ignore the lines labeled 'Main effect' and 'Explained' as they are produced by combining the values for the two factors.

Two-way Analysis of variance (with replication)

The situations where we have two factors and more than one observation for each combination of factor levels is a very common one in fisheries. If there are two factors then there are three null hypotheses associated with the test:

(i) that all levels of the first factor have the same mean

(ii) that all levels of the second factor have the same mean

(iii) that there is no interaction between the two factors.

If we have no interest in third null hypothesis then we might consider two separate one- way analyses.

Two-way ANOVA with replication is a very powerful statistical test with three null hypotheses. It is suitable when there are two independent ways of assigning the observations into groups and there is more than one observation per factor combination. Assumptions are the same as other ANOVA tests: the data are continuous (>30 possible values); at least approximately normally distributed and the variation is the same in each factor combination.

Example: A fishery scientist is investigating the effect of light on food intake on ornamental fishes. An experiment where fishes are placed in individual aquarium of identical size with controlled lighting and are given an excess of food. Fishes are sexed and four male and four female fishes are exposed to either 8 hours or 4 hours of light. Each fish is monitored for 7 days and its total food intake is recorded. Each fish is only used once (table 7).

Table 7. Raw data on male & female fishes exposed for different hours in light.

Length of light	Sex							
	Female				Male			
Long (8 hours)	78.1	75.5	76.3	81.2	69.5	72.1	73.2	71.1
Short (4 hours)	82.4	80.9	83.0	88.2	72.3	73.3	70.0	72.9

There are sixteen fishes used in total. Four were exposed to each factor combination. This is a balanced design. The output from the statistical packages varies somewhat. It is usually not sensible to just put the whole output into a report. For a two-way ANOVA there is a standard way of displaying the output from the test.

For the example here the output would be represented in the following way (Table 8):

Table 8: Two-way ANOVA with replication

Factor	d. f.	SS	MS	*F* ratio	*P* value
Length of light	1	42.3	42.3	8.00	0.015
Sex	1	316.8	316.8	60.00	< 0.001
Interaction	1	27.0	27.0	5.12	0.043
Error	12	63.4	5.28		

Steps using SPSS: From the 'Statistics' menu select 'ANOVA models' then 'Simple factorial...'. In the dialogue box move the observation column into the 'Dependent:' box and the two coded factor columns into the 'Factor(s):' box. We must then tell SPSS what the code numbers for the lowest and highest factor levels are. Select the factors in turn, click on the 'Define Range...' range and put the appropriate numbers into the '. Minimum' and 'Maximum' boxes. (If both factors have the same range of factor levels then we can select both before defining the range.) Click 'continue' to return to the main box and then 'OK' to run the test.

Using the example we would get the following output:

Table 9: SPSS output of Two-way ANOVA (with replication)

*** ANALYSIS OF VARIANCE * *

by LENG_LIGHT
SEX

UNIQUE sums of squares
All effects entered simultaneously

Source of Variation	Sum of Squares	DF	Mean Square	F	Sig of F
Main Effects LENG _LIGHT SEX 2-Way Interactions LENG_LIGHT * SEX	359.090 42.250 316.840 27.040 27.040	2 1 1 1 1	179.545 42.250 316.840 27.040 27.040	33.999 8.001 59.998 5.120 5.120	.000 .015 .000 .043 .043
Explained	386.130	3	128.710	24.373	.000
Residual	63.370	12	5.281		
Total	449.500	15	29.967		

16 cases were processed.
0 cases (.0 pct) were mi ssing.

In the example the *P* value for 'sex' is given as '0.000'; this is a highly significant result but should always be reported as $P< 0.001$. We can be sure that the two sexes have different food intake levels although the test does not tell us which sex eats more – that must be done by inspection of the data. The *P* value for the other factor 'leng_light' is also less than 0.05 but the value of 0.015 indicates that the effect is not as strong as that for sex. Finally the interaction of the two factors also has a *P* value less than 0.05 meaning, in this example, that the two sexes respond significantly different to length of light in the amount of food they eat. The direction of this effect can only be revealed by inspection of the mean values for each group.

Multi-factorial testing: If each observation can be assigned to groups using more than two different factors (ways of classifying the data), each of the factors can be divided into groups and assigned integers and, most importantly, the factors are all fully independent of each other then the data is suitable for multi-factorial testing.

If there are more than two factors and they are fully independent of each other then the design is said to be factorial. When there were just two factors there was scope for only one interaction term: factor A with factor B. As more factors are added to the analysis the number of interaction terms possible grows rapidly. For three factors there are four interaction terms (AxB, AxC, B x C and A x B x C), for four factors there are eleven.

Three-way Analysis of variance (without replication)

When time, space or resources are very limited then this design is quite common. If there are three independent ways of classifying the data into groups and there is only one observation for each factor combination then it is still possible to carry out an analysis of variance. The instructions for three-way ANOVA with replication can be followed but be aware that there is no way of calculating the interaction between the three factors (this is used as the 'error' by the calculation). This means that only two-way interactions should be allowed in the model.

Three-way Analysis of variance (with replication)

If there are three fully independent ways of dividing the data into groups and there is more than one observation for each factor combination then this design is appropriate. As with all analysis of variance the test assumes that the data is continuous, approximately normally distributed and that the variance is approximately equal for each factor level.

Example

The data collected from an experiment in which there are three independent factors: A (two levels coded 1 and 2); B (coded 1 and 2); C (three levels coded: 0, 1, and 2). There are two observations for each factor combination for a total of 24. Here is the data in table format (Table 10):

Table 10: Raw data of three factors

		A					
		1			2		
		B					
		0	1	2	0	1	2
C	1	112 116	115 102	187 175	141 101	121 157	189 186
	2	121 138	145 124	198 168	135 129	141 133	208 206

Output from statistical packages often has more information than is required. The ANOVA table included in a report or publication should be presented as follows (Table 11):

Table 11: Three way ANOVA

Source of Variation	DF	SS	MS	F	P
A	1	1218.4	1218.4	6.44	0.026*
B	2	6841.4	3420.7	18.07	<0.001***
C	1	590.0	590.0	3.12	0.103
A X B	2	327.0	163.5	0.864	0.446
A X C	1	1.0	1	0.006	0.942
B X C	2	58.3	29.2	0.154	0.859
A X B X C	2	282.3	141.2	0.746	0.495
Error	12	2271.5	189.3		
Total	**23**				

Steps using SPSS: From the 'Statistics' menu select 'ANOVA models' then 'Simple factorial...'. Put the observations into the 'Dependent:' box and the three factors into the 'Factor(s):' box. Define the range (highest and lowest coded integer) for each factor. Click 'OK' to run the test. Using the example, you get the following output: (Table 12).

Table 12. SPSS output of three way ANOVA

*** ANALYSIS OF VARIANCE * **

Recorded variable by A, B, C

UNIQUE sums squares
All effects entered simultaneously

Source of Variation	Sum of Squares	DF	Mean Square	F	Sig of F
Main Effects	8649.750	4	2162.438	11.424	.000
A	1218.375	1	1218.375	6.436	.026
B	6841.333	2	3420.667	18.071	.000
C	590.042	1	590.042	3.117	.103
2-Way Interaction	86.373	5	77.275	.408	.834
A X B	327.000	2	163.500	.864	.446
A X C	1.042	1	1.042	.006	.942
B X C	58.333	2	29.167	.154	.859
3-Way Interactions	282.333	2	141.167	.746	.495
A X B X C	282.333	2	141.167	.746	.495
Explained	9318.458	11	847.133	4.475	.008
Residual	2271.500	12	189.292		
Total	**11589.958**	**23**	**503.911**		

24 cases were processed. 0 cases (.0 pet) were missing.

The output confirms the test and then gives the name of the observation column and then the factor names. The next two lines show that the ANOVA is using standard assumptions. Then comes the ANOVA table proper. There is more information than you actually need here. The most important lines are for the factor variables, called 'Main Effects' "in SPSS (' A , B and C in the example) and their interactions (three possible two-way interactions and one three way). The 'DF' column gives the degrees of freedom which is one less than the number of factor levels for main effects and calculated by multiplication for interactions (e.g. for ' A and B' there are 2x1=2 d.f.). The sum of squares and mean square are given, mean square is sum of squares/ degrees of freedom. The last two columns give the important information. The *F* ratio is factor mean square/residual mean square. The *P* value, labeled 'Sig of F', is in the last column. In the example, the *P* value for 'C' is slightly above the critical 0.05 level so we accept the null hypothesis that there is no difference between C. However, the *P* values for 'A' and 'B' are both less than 0.05 so we reject the null hypotheses and accept the alternative hypotheses. None of the interaction terms has a *P* value even close to 0.05.

Nested or Hierarchical Designs

If the experiment or sampling design has only one 'random' factor at the top of the hierarchy and all other factors are 'nested' within it then the design is said to be a pure nested one. If the top level is 'fixed' and there are other factors nested within it the design is often called a 'mixed model'. There is no limit to the number of factors that can be nested in this way. The simplest case of a single fixed factor and one nested is considered first then a more complex one.

Two-level nested design ANOVA: Two-level nested design ANOVA is an analysis with two factors. The standard factor in this type of design may be 'fixed' or 'random'. There should then be a second, 'random' factor nested within the standard factor.

Method for Analysis of Data with Repeated Measurements

Many studies involve making repeated measurements over a period of time or space on each experimental unit, which may be a plot, fish or plant. For example, length measures at different growth stages or weekly gain in the weight.

When a character in an experiment is measured over time, the researcher is usually interested in examining the rate of change from one time period to another. For example, when a researcher measures weight of fish at different growth stages, interest is usually on the effects of treatment on the growth pattern (or the rate of change over time) based on weight rather than on the effects of treatment on weight at the individual stage. In other words, it is important to determine the interaction effect between treatment and stage of observation, but that cannot be done if the analysis of variance is obtained separately for each stage of observation. Hence, the common approach is to combine data from all stage of observation and obtain a single analysis of variance.

There are two types of repeated measurement experiments, first; each experimental unit receives the same treatment during the course of the experiment and second; each experimental unit receives a sequence of treatments over successive periods of time during the course of the experiments. Here, we will only consider the former type of repeated measurements experiment. The specific experimental design is not important since all methodologies covered are applicable to any design. To simplify the discussion, a simple experiment arranged in a Randomized Complete Block Design with t treatments occurring in b blocks is used. Data are taken at times t_1, $t_{2,\ldots,}$ t_r from each of the N = tb experimental units. An example of data from this type of experiments is shown in Table 13. It shows the weight of six species at 10, 20, 30, and 40 days.

Single Variable Procedures

Analysis could be carried out for each of the r times but with these analyses the change of treatment effect over time cannot be evaluated. Hence, what is needed in an analysis, which provides a test for the interaction between treatment and time? Two errors are frequently made in carrying out this analysis. The first arises from assuming an incorrect model and the second from assuming an incorrect covariance structure.

Incorrect model for repeated measures: The first common error in the analysis of repeated measures is to analyze the data as if time were a factor crossed with treatments in a factorial experiments. This incorrect analysis is shown in Table 14.

Such an analysis will lead to serious errors in the test of significance of the effects under study because comparisons between times are made on the same unit and therefore less variable than those between treatments. Further, they are sequential and hence may be auto- correlated.

The second error arises from assuming a split plot design which corrects the first problem with the previous design but not the second concerning the

Table 13: Weight (in gram) of six species of fish at 10, 20, 30, and 40 days

Species	Block	10 days	20 days	30 days	40 days
S1	1	26.2	37.0	62.5	62.0
	2	27.0	40.4	54.6	68.3
	3	26.4	38.2	60.8	64.0
	4	28.9	43.9	60.1	67.2
S2	1	27.9	39.5	58.9	69.5
	2	27.0	36.0	60.8	67.1
	3	26.9	41.0	61.6	70.6
	4	27.5	37.5	59.2	67.9
S3	1	29.4	38.7	60.7	70.1
	2	27.8	38.5	60.9	75.8
	3	28.1	38.7	59.2	70.8
	4	28.9	43.0	56.1	71.1
S4	1	28.9	39.5	61.9	64.9
	2	33.5	46.0	62.5	77.0
	3	24.1	41.0	53.5	67.8
	4	28.6	43.7	55.5	69.6
S5	1	25.8	44.1	61.6	66.8
	2	27.1	37.4	65.0	69.1
	3	28.8	39.5	63.4	70.8
	4	33.1	43.9	61.6	71.5
S6	1	28.8	45.0	70.0	76.4
	2	28.6	44.2	66.3	72.4
	3	32.9	52.9	62.9	83.2
	4	35.6	50.9	69.8	74.6

Table 14: Incorrect factorial analysis of repeated observations

Source of variation	Degree of freedom
Block	b – 1
Treatment	t – 1
Time	r – 1
Treatment x Time	(t – 1) (r – 1)
Error	(tr – 1) (b – 1)
Total	r N – 1

correlation between time periods. This error may be serious when the correlations between time periods are unequal. The split plot analysis for the data in table-13 is shown in table-3. Here, the treatments are considered as the main plots and the times as subplots.

Table 15: Split plot analysis of data in Table 13..

Source of variation	Degree of freedom	MS	F	P>F
Block	3	8.20	0.49	0.6967
Treatment	5	127.18	7.54	0.0010
Error (a)	15	16.87		
Time	3	8511.44	1217.79	0.0001
Treatment x Time	15	10.76	1.54	0.1237
Error (b)	54	6.98		
Total	**95**			

Under the assumption of equal correlation between time periods, the analysis in Table 15 may be performed.

Alternative Procedure of Analysis of Repeated Observations

An alternative analysis, which is valid no matter what auto-correlation pattern exists, is to partition the subplot section of the analysis using orthogonal contrasts. The choice of contrasts depends on the interest of the experimenter. If for example, the interest is to compare the growth responses over time, the use of orthogonal polynomials may be appropriate. The subplot part of the split-plot analysis may then be partitioned as in Table-16.

Table 16: Partitioning using orthogonal contrasts

Source of variation	Degree of freedom
Time linear (L)	1
L x Block	b – 1
L x Treatment	t – 1
Error (L)	(b – 1) (t – 1)
Time Quadratic (Q)	1
Q x block	(b – 1)
Q x Treatment	(t – 1)
Error (Q)	(b – 1) (t – 1)
Time Cubic (C)	1
C x Block	(b – 1)
C x Treatment	(p – 1)
Error (C)	(b – 1) (t – 1)

The advantages of this analysis are that it examines directly interactions of interest between the treatments and the repeated measure and the different contrasts are independent (because of orthogonality) so that the different tests are also independent and have the correct level.

However, there is a question as to whether to use a pooled error term [Error (b) of the original split plot analysis] or a partitioned (individual) error terms in table-16. Many Scientists have recommended that the individual errors be routinely used because they yield better control of type I error and of independent statistical tests. The pooled error may be used under the unlikely conditions of homogeneity of partitioned errors, and of poor power of partitioned errors.

Example of Analysis by Orthogonal Contrasts

Step 1. Determine the orthogonal contrasts to be used from tables. For example the orthogonal polynomials to be used for four equally spaced time periods are the following:

Total	1	1	1	1
Linear	-3	-1	1	3
Quadratic	1	-1	-1	1
Cubic	-1	3	-3	1

The last row 'Total' is included to provide the main plot part of the analysis.

Step 2. Evaluate each linear function for each of the experimental units. For the data in Table-13, the results are in Table-17.

Table 17: Evaluated linear functions for data in Table-13.

Species	Block	Total	Linear	Quadratic	Cubic
S1	1	187.7	132.9	-11.3	-40.7
	2	190.3	138.1	0.3	-1.3
	3	189.4	135.4	-8.6	-30.2
	4	200.1	131.1	-7.9	-10.3
S2	1	195.8	144.2	-1.0	-16.6
	2	190.9	145.1	-2.7	-34.3
	3	200.1	151.7	-5.1	-18.1
	4	192.1	142.9	-1.3	-24.7
S3	1	198.9	144.1	0.1	-25.3
	2	203.0	166.4	4.2	-19.2
	3	196.8	148.6	1.0	-18.8
	4	199.1	139.7	0.9	2.9

Contd...

Table 17. Contd...

S4	1	195.2	130.4	-7.6	-31.2
	2	219.0	147.0	2.0	-6.0
	3	186.0	143.6	-2.6	6.2
	4	197.4	134.8	-1.0	5.6
S5	1	198.3	140.5	-13.1	-11.5
	2	198.6	153.6	-6.2	-40.8
	3	202.5	149.9	-3.3	-29.7
	4	210.1	132.9	-0.9	-14.7
S6	1	220.2	167.8	-9.8	-27.4
	2	211.5	153.5	-9.5	-22.5
	3	231.9	160.9	0.3	20.3
	4	230.9	135.9	-10.5	-17.7
Total			3471.0	-93.6	-406.0

Step 3. Run analysis of variance for each function (column of table-17) using the basic design of the experiment. For our example, the basic design is RCBD and the resulting ANOVAs are the following:

Table 18 : ANOVA for RCBD

Source of variation	Degree of freedom	Sum of squares			
		Total	Linear	Quadratic	Cubic
Block	3	98.49	733.12	89.67	986.73
Treatment	5	2543.76	1052.06	235.92	998.69
Error (a)	15	1012.26	756.09	216.86	3539.74

Step 4. Divide each of the sums of squares with the sum of the squares of the contrast coefficients. For our example, the divisors are 4, 20, 4 and 20 for total, Linear, Quadratic and Cubic, respectively. The resulting ANOVA table is as follows:

Table 19 : ANOVA showing linear, quadatic and cubic some of squares

Source of variation	Degree of freedom	Sum of squares			
		Total	Linear	Quadratic	Cubic
Block	3	24.62	36.65	22.41	49.33
Treatment	5	635.94	52.60	58.98	49.93
Error (a)	15	253.06	37.80	54.21	176.98

The sums of squares for the main plot part of the analysis are given under the total column. The mean squares and F-values are computed in the usual manner.

Step 5. Compute the Linear, Quadratic and Cubic sums of squares from the totals in Step 2 by squaring these values and dividing them with the product of the total number of observations and the sum of the squares of the contrast coefficients. For our example:

Linear SS = $(3471)^2$ / (24) (20) = 25099.66
Quadratic SS = $(-93.6)^2$/ (24) (4) = 91.26
Cubic SS = $(-406)^2$/ (24) (20) = 343.40

Step 6. Enter the sums of squares from steps 4 and 5 to table-16. The resulting ANOVA is as follows:

Table 20 : Final ANOVA showing linear, quadatic and cubic effect.

Source of variation	Degree of freedom	Sums of squares	Mean squares	F	Prob.
Linear (L)	1	25099.66	25099.66	9958.93	0.000
L x Block	3	36.65	12.21	4.84	0.015
L x Treatment	5	52.60	10.52	4.17	0.014
Error (L)	15	37.80	2.52		
Quadratic (Q)	1	91.26	91.26	25.24	0.000
Q x block	3	22.41	7.47	2.069	0.148
Q x Treatment	5	58.98	11.79	3.264	0.034
Error (Q)	15	54.21	3.61		
Cubic (C)	1	343.40	343.40	29.104	0.000
C x Block	3	49.33	16.44	1.394	0.283
C x Treatment	5	49.93	9.98	0.846	0.538
Error (C)	15	176.98	11.79		

Step 7. If a time x Block effect is not significantly different from its corresponding error and the error degrees of freedom is small, say 6 or less, then the sums of squares of the two effects can be pooled and used as error term. In our example, the error degree of freedom is sufficiently large; hence, no pooling is necessary.

Step 8. The result shows that the linear, quadratic and cubic effects of time are significant and the linear and quadratic effects vary significantly from one species to the other.

REFERENCES

Bryman, A. and D. Cramer, 1997. Quantitative Data Analysis with SPSS for Windows: a guide for Social Scientists, Routledge, London.

Campbell, R. C., 1989. Statistics for Biologist, 3rd Edition, Cambridge University Press, Cambridge.

Chakrabarty, N.M., A. K. Roy, D.D. Halder, R.K. Chakrabarti, 1987. Effect of Rearing Density on the Growth and Survival of the Fish *Liza parsia. Environ. Ecol.* 5 (1): 60-64.

Chakrabarty, N.M., H.C. Karmakar, A. K. Roy. 1985. Observations on the Effect of Supplementary Feed on Growth and Survival of Grey Mullet *Liza parsia* (Hamilton) fry in brackish water nursery ponds at Kakdwip. *Proceedings of the Symposium on Coastal Aquaculture* held at Cochin from January 12 to 18, 1980. Part 3: Finfish Culture. *Marine Biological Assoc. of India*, Cochin, India no. 6: 812-817

Charkrabarty, N.M., A. K. Roy, D.D. Halder. 1981. Studies on the Effect of Supplementary Feed, Fertilizer, and Fertilizer alone on the Growth and Survival of *Liza parsia. Ecol. Environ.*, 13(2): 92-99

Clarke, G. M., 1994. Statistics and Experimental Design. An Introduction for Biologists and Biochemist, 3rd Edition, Edward Arnold, London.

Cole, J.W.L. and Grizzle, J.E. (1966). Application of multivariate analysis of variance to repeated measurement experiments. *Biometrics*. 22, 810-28

Crowder, M.J. and Hand, D.J. (1990) . Analysis of repeated measures. Chapman & Hall. London.

Dytham, C., 1999. Choosing and Using Statistics - a biologist guide. Blackwell Science Ltd. USA.

Karmakar, H. C., A. K. Roy and D. D. Halder. 1986. On the Efficacy of Some Natural Feed for Successful Nursery Rearing of *L. calcarifer* by Using a 3 x 3 Factorial Experiment. *J. Indian Soc. Coastal Agric. Res.*, 4(2): 133-134.

Mohapatra, P. K. J., S. K. Prusty and C. K. Mukherjee, 2006. Modeling of Multiobjective Freshwater Aquaculture. *In: Proc. National workshop on Bioinformatics and Statistics in Aquaculture Research* held at CIFA, Kausalyaganga during 7-10 February 2006, pp.264-275.

Parker, R. E., 1979. Introductory Statistics for Biologist, 2nd edition. Studies in Biology no. 43. Edward Arnold, London.

Pillai, S.M., P.U. Verghese, P. Ravichandran, A. K. Roy. 1987. Effect of Thyroxine on Growth and Moulting in *Penaeus monodon* Fabricius. *Indian J. Anim. Sci.*, 57 (3): 241-245.

Rajayalakshmi, T., S.M. Pillai, A. K. Roy.1981. Observations on the Interaction of Aeration and Stocking Density on Growth and Survival of the Post-larvae of *Penaeus monodon* Fabricius. *J. Inland Fish Soc., India,* 13(2): 87-91.

Rajyalakshmi, T., S. M. Pillai, A. K. Roy and P. U. Verghese.1986. Feed and Nutrition of Prawn *Penaeus monodon*. *Environment & Ecology*, 4(3): 415-422.

Rajyalakshmi, T., S.M. Pillai, A. K. Roy, P.U. Verghese. 1982. Studies on Rearing of *Penaeus monodon* Fabricius in Brackish Water Ponds Using Pelleted Feeds. *J. Inland Fish. Soc. India* 14(1): 28-35.

Roy, A. K., A. K. Datta, Apurba Ghosh and S. K. Mukhopadhaya.1995. Experimental study on the extent of variation in aquaculture experiments. *In:* P. C. Thomas (eds.), *Current and Emerging Trends in Aquaculture.* Daya Publishing House, New Delhi, pp. 45-51.

Roy, A. K., A. K. Datta, S. K. Mukhopadhyaya and Apurba Ghosh. 1997. Contribution of Exotic Carps in net Production under varied level of Composite Carp Culture. *In:* K. K. Vass and M. Sinha (Eds), *Proceedings of the National Seminar on Changing Perspectives of Inland Fisheries,* March 16-17, 1997, Inland Fisheries Society of India, Barrack pore. pp. 129-133.

Roy, A. K.1985. Evaluation of Supplementary Feeding on the Growth and Survival of Grey Mullet *Liza tade* (Forsskal). *J. Indian Soc. Coastal Agric. Res.*, no. 6: 807-811.

Roy, A.K.; H.C. Karmakar, N.M. Chakraborti, D.D. Halder. 1986. Studies on the Effect of Rearing Density on the Growth, Survival and Net Yield of *Liza tade* (Forskal). *Indian Soc. Coast Agric. Res.* 4 (1): 43-55.

Scheiner, S. M & J. Gurevitch, (Eds.), 1993. Design and Analysis of ecological experiments. Chapman & Hall, New York.

Watt, T. A., 1993. Introductory Statistics for Biology Students. Chapman & Hall, London

Chapter 27

Some Standard Sampling Techniques and Use of Auxiliary Information in Fishery Surveys

Ajit Kumar Das & Rajendra Nath Das

Introduction

The method of obtaining the data for each and every unit belonging to the population or universe, which is the aggregate of all units of a given type under consideration, is known as complete enumeration survey or complete census. Examples of units are family, a period of time such as an hour, day, fishing village, fishing household, fishing establishment, fishing boat, fish landing centres etc. The type of unit to be considered would depend on the purpose of the view. The population may be finite or infinite. It is obvious that for any statistical investigation complete enumeration of the population is rather impracticable. For example, if we want to have an idea of the total catch of fish (monthly) in India, we will have to enumerate all the fishers who catch fish in the country, which is rather a very difficult task. If the population is infinite, complete enumeration is not possible. In addition, if the units are destroyed in the course of inspection, 100% inspection though possible, is not at all desirable. However, even if the population is finite or the inspection is not destructive, complete enumeration is not taken because it is time-consuming, expensive, and as well as it needs a large amount of resources in terms of trained personnel. Hence, we take the help of sampling. A sample is a finite subset of units in a population. If the population is homogeneous

Applied Bioinformatics, Statistics and Economics in Fisheries Research: pp. 385-420

New India Publishing Agency, New Delhi (India)

with respect to the characteristic under study, it hardly matters how we draw a sample. If, however, the units constituting the population vary considerably in respect of the characteristic under study, the method by which we draw a sample plays a critical role and it is necessary to pay due attention to the selection of the sample and drawing inference there from . Since data are collected from only a small part of the population, expenditures are smaller than if complete census is attempted. For the same reason the data can be collected and summarized more quickly with a sample than with a complete census. This is a vital consideration when the information is urgently needed. Since personnel of higher quality can be employed and given intensive training and since more careful supervision of the field work and processing of results become feasible when the volume of work is reduced, a sample may produce more accurate results than the kind of complete enumeration. Sample Survey has generally greater scope as compared with complete census. Since sample survey saves in time and money, it is possible to have a thorough and intensive enquiry because information that is more detailed can be obtained from a small group of respondents. We should keep in mind that proper care should be taken in the planning and execution of the sample survey, otherwise the results obtained might be inaccurate and misleading.

We now discuss the basic concepts and definitions of the theory of sampling.

1. Population and Parameter

A Collection $U = \{U_1, U_2, U_3, \ldots\ldots\ldots\ldots, U_N\}$ of N (given) well-defined, identifiable and observable objects under which certain valid conclusions are to be drawn is called a finite population. N is called the population size. The objects U_i's are called the sampling units. The list of all sampling units with identification numbers 1,2,3,.........,N is called the sampling frame.

Let y be a character defined on U and be the value of the character y for Y_i th unit of the population U (i=1, 2, 3 ...., N). Thus

$$U = \{U_1, U_2, U_3, \ldots\ldots\ldots\ldots, U_N\}$$

$$\underset{\sim}{y} = \{Y_1, Y_2, \ldots\ldots\ldots\ldots, Y_N\}$$

For example, the unit may be a day and the characteristic may be the catch of fish at a fishing center or the unit be a open water/ fresh water culture center and the characteristic may be the number of fishing / culture households engaged in a particular district or the unit be a fish pond and the characteristic may be the area of fish pond etc.

Any function of the values of all the population units is known as a parameter. The parameters of particular interest are population mean, population total, population variance, population co-efficient of variation, population ratio and population proportion defined by

$$\overline{Y} = \frac{1}{N}\sum_{i=1}^{N} Y_i, \quad Y = N\overline{Y}, \quad \sigma_y^2 = \frac{1}{N}\sum_{i=1}^{N}(Y_i - \overline{Y})^2, \quad C_y = \frac{\sigma_y}{\overline{Y}}, (\overline{Y} \neq 0)$$

$R = \frac{\overline{Y}}{\overline{X}} = \frac{Y}{X}, \; (X \neq 0)$ *and* $P = \frac{N_A}{N}$ respectively where N_A is the number of the units in the population possessing some attribute A.

Random sample and Statistic: A sample is a small or a representative part of the population. Now the question is: How is this small part chosen?

There are different procedures of selecting a sample such as

(i) Purposive or Judgment sampling
(ii) Probability or random sampling
(iii) Mixed sampling.

In purposive sampling, the sample is selected with definite purpose in view and the choice of the sampling units depends entirely on the discretion and judgment of the investigator.

In probability or random sampling, the sample is selected in such a manner that each unit in the population has a predetermined probability of selection.

In mixed sampling, the samples are selected partly according to some laws of probability and partly according to a fixed sampling rule. (No assignment of probabilities).

Purposive sampling and mixed sampling are seldom used in large scale surveys because it is not possible to get valid estimates of the parameters under consideration and of their sampling errors due to the risk of bias in subjective selection and the lack of information on the probabilities of selection of the units.

So we only confine to random or probability sampling because it is must for valid inferences and for valid estimates of sampling errors.

Let a random sample of size n be drawn from the population according to a well defined probability law and the sample value are $(Y_1, Y_2, \ldots\ldots\ldots\ldots, Y_N)$.

Any function of the sample values is known as a Statistic.

For example, the sample mean = $\bar{y}=\frac{1}{n}\sum_{i=1}^{n}y_i$ sample mean square = $s_y^2=\frac{1}{n-1}\sum(y_i-\bar{y})^2$ sample coefficient of variation = $\frac{s_y}{\bar{y}}$ and sample proportion $=p=\frac{n_A}{n}$ etc. where n_A is the number of units in the sample possessing some attribute A.

Unbiased estimator and mean square error

Parameters are unknown. We have to determine or estimate a parameter based on a suitable statistic. A statistic is called an estimator when it is used to estimate a parameter. The estimator is a random variable as its value differs from sample to sample due to sampling fluctuations. There are multiple choices of the samples that can be drawn from the population. Let the number of possible samples of size n that can be drawn from a finite population of size N is M_0. The particular value, which the estimator takes for a given sample, is known as an estimate. For each of these samples we can compute t_i, the value of an estimator t of the population parameter θ and P_i is the probability of getting i_{th} sample (i = 1, 2, ... M_0). Thus, we can have the following sampling distribution of the estimator t:

$$\frac{t\,|\,t_1\ \ t_2\ \ldots\ldots\ldots\ldots\ t_{M_0}}{P(s)\,|\,P1\ P2\ \ldots\ldots\ldots\ P_{M_0}};\sum_{i=1}^{M_0}P_i=1,\ P_i\geq 0 \text{ for all i}.$$

The expected value or the average value of the estimator t is given by

$$E(t)=\sum_{i=1}^{M_0}t_iP_i$$

The estimator t is called an unbiased estimator for θ if E (t) = θ. If E (t) ≠ θ then t is called a biased estimator and the bias of t is B (t) = E (t) - θ. In estimating a parameter θ by an estimator t, t_s - θ is called the estimation – error resulting, for the sample s, in the estimation of θ. Any convex function G of (t_s - θ) is called a loss function and E (G) = $\sum_s G_s P_s$ is called the expected loss or risk of estimation by t_s. A commonly used loss function is the squared error $(t_s - \theta)^2$ and the expected loss is known as mean square error (mse). The mse of an estimator t of θ is

$$M(t) = E(t-\theta)^2 = \sum_s (t_s - \theta)^2 P_s$$

The variance of t is defined by

$$V(t) = E\left[t - e(t)\right]^2 = E(t^2) - \left[E(t)\right]^2$$

$$= \sum_s t_s^2 P_s - \left(\sum_s t_s P_s\right)^2$$

In general, one may observe that M (t) = V (t) + B^2 (t). If t is unbiased for θ then M (t) = V (t). If t_1 and t_2 are two estimators for the same parameter θ then t_1 is said to be uniformly better or simply better than t_2 if $M(t_1) \le M(t_2)$ for all $\underset{\sim}{y}$ with strict inequality holding true for at least one value of $\underset{\sim}{y}$.

2. Some Standard Sampling Techniques and Estimators

We mention the following some standard sampling techniques:

1. Simple Random Sampling - with replacement (SRSWR)
   and without replacement (SRSWOR)
2. Probability proportional to size sampling with replacement (PPSWR)
3. Stratified sampling
4. Systematic Sampling
5. Cluster sampling
6. Two-stage sampling
7. Stratified two-stage sampling.

Simple Random Sampling

Simple random sampling is a method of drawing a sample from a population such that each and every unit in the population has an equal probability of being included in the sample.

From a population of N units we select one unit by giving equal probability 1/N to all units. This is best done with the help of random numbers. We make a note of the unit selected and return it to the population. If this operation is performed n times, we get a simple random sample of n units, selected with replacement (SRSWR). If, however, this procedure is continued till n distinct units are selected and all repetitions are ignored, a simple random sample of n units, selected without replacement (SRSWOR), is obtained. The later procedure is exactly same as retaining the unit selected and selecting a further unit with equal probability from the units that remain in the population.

The use of random numbers for drawing a random sample from a finite population may be illustrated by means of an example.

Example: A list of 42 landing centres during a month with the number of boats landing (y) is given below:

S. No.:	1	2	3	4	5	6	7	8	9	10	11	12
y:	2	12	10	8	9	11	3	8	10	12	7	9
S. No.:	13	14	15	16	17	18	19	20	21	22	23	24
y:	4	5	7	12	14	8	10	0	5	6	10	12
S. No.:	25	26	27	28	29	30	31	32	33	34	35	36
y:	14	15	2	5	15	17	8	9	12	13	14	5
S. No.:	37	38	39	40	41	42						
y:	6	7	8	12	11	9						

We shall select two random samples of 10 landing places (i) with replacement and (ii) without replacement from the above population.

The landing places are already serially numbered from 1 to 42. Now N=42 is a two-digit number.

Method-1 (Direct Approach)

We shall use a two-digit random number table since N=42 is a two-digit number. We shall consider two-digit random numbers from 01 to 42, rejecting the numbers greater than 42 and 00. We shall start from any number of the random number table. We can move in any random way we like, it can be vertically downward or upward, to the right or to the left. We start at random from a number in the 12th column and 1st row and move vertically downwards. The numbers selected from the random number table (given in Appendix) are given in the following table:

Random No. taken from the table	Serial No. as selected	
	SRSWR	SRSWOR
13	13	13
51	-	-
57	-	-

Contd...

31	31	31
38	38	38
17	17	17
84	–	–
41	41	41
68	–	–
84	–	–
58	–	–
56	–	–
80	–	–
95	–	–
20	20	20
26	26	26
51	–	–
40	40	40
33	33	33
31	31	–
78		–
71		–
72		–
21		21

The random samples of size 10 with replacement and without replacement consist of the observations

4,8,7,14,11,0,15,12,12,8 and, 4,8,7,14,11,0,15,12,12 ,5 respectively.

Method-2 (Remainder Approach)

In the above procedure, a large number of random numbers is rejected. Hence, a commonly used device i.e., remainder approach, is employed to avoid the rejection of such large numbers. The greatest two-digit number, which is multiple of 42, is 84. We consider two-digit numbers from 01 to 84, rejecting the numbers greater than 84 and 00. By using two-digit random numbers as above, we prepare the following table:

Random No. taken from the table	**Remainder when divided by 42**	**Serial No. as selected**	
		SRSWR	**SRSWOR**
13	13	13	13
51	9	9	9
57	15	15	15
31	31	31	31

Contd...

38	38	38	38
17	17	17	17
84	42	42	42
41	41	41	41
68	26	26	26
84	42	42	
58	16		16

The random sample of size 10 with replacement and without replacement consist of the observations

4,10,7,8,7,14,9,11,15,9 and, 4,10,7,8,7,14,9,11,15,12 respectively.

When n units are selected from a population of N units with SRSWOR, the number of possible samples is $\binom{N}{n}$. Here order of drawing is ignored. The probability of selecting any one of these possible samples is $1 / \binom{N}{N}$

When n units are selected from a population of N units with SRSWR, the number of possible samples (order of drawing to be taken account) is N^n and the probability of selecting any one of these possible samples is $1/N^n$.

It is noted that the probability of drawing any unit at any draw is same for SRSWR and SRSWOR and equal to 1/N.

That is $P_{i_r} = \frac{1}{N}$ for all i=1,2,...,N; r =1,2,...,n.

Suppose a simple random sample of n units is drawn from a population of N units and the sample values are (y_1, y_2, ... y_n).

Estimation of population mean (or total): We want to estimate the population mean $\overline{Y}$. In this case $\overline{y}$ (sample mean) is an unbiased estimate of $\overline{Y}$ i.e. E ($\overline{y}$) = $\overline{Y}$, and the estimate of variance of sample mean is given by

$$v(\overline{y}) = \frac{N-n}{N} . \frac{s_y^2}{n} = (1-f)\frac{s_y^2}{n}$$

where $s_y^2 = \frac{1}{n}\sum_{i=1}^{n}(y_1 - \overline{y})^2$ is the sample mean square and $f = \frac{n}{N}$ is called the sampling fraction.

In case of SRSWR,

$$\hat{\bar{Y}} = \bar{y} \text{ and } v(\bar{y}) = \frac{s_y^2}{n}$$

It is noted that the population variance (σ_y^2) is estimated by s_y^2 and $\left(\frac{N-1}{N}\right)s_y^2$ for SRSWR and SRSWOR respectively.

If population size is large i.e. N $\rightarrow \infty$ then f = 0 and estimate of variance of sample mean is equal to $\frac{s_y^2}{n}$ for both SRSWR and SRSWOR.

Similarly unbiased estimator of population total (Y) is $(\hat{Y}) = N\,\bar{y}$ with its estimated variance is $v(\hat{Y}) = N^2\ v(\bar{y})$.

Estimation of population proportion: Sometimes we may be interested to estimate the population proportion P. Out of a sample of n units, let a_1 be the number of units possessing some attribute a, then the sample proportion $p = \frac{a_1}{n}$ is an unbiased estimator for P and estimated variance of the sample proportion is given by $v(p) = \frac{p(1-p)}{n-1}$ in case of SRSWR , In addition, $v(p) = (1-f)\frac{p(1-p)}{n-1}$ in case of SRSWOR.

Estimation of Population ratio: We have a population of N units. With the ith sampling unit, we associate numerical values Y_1 and S_1 respectively of two variables.

Thus the population is

$$Y_1, Y_2, \ldots\ldots\ldots, Y_i, \ldots\ldots\ldots, Y_N$$
$$X_1, X_2, \ldots\ldots\ldots, X_i, \ldots\ldots\ldots, X_N$$

X and Y denote population totals. For X $\neq$ 0, the ratio $R = \frac{\bar{Y}}{\bar{X}} = \frac{Y}{X}$ is defined. Now we want to estimate R based on simple random sampling of size n. An estimator of R is $\hat{R} = \frac{\bar{y}}{\bar{x}}$

Examples of this kind occur frequently when the sampling unit comprises a group or cluster of elements and our interest is in the population mean per element. Suppose in a population there are N families and for each family X_i and Y_i are the values of the character x and y respectively where x and y are the size and the weekly expenditure on food respectively. Now we want to estimate the mean weekly expenditure on food per person. Ratios also appear in many other applications, for example, the ratio of loans for buying fish boat to total loans in a bank or the ratio of fish cash to total area (hectares) on a water unit.

The estimator $\hat{R}$ is usually a slightly biased estimate of R. The bias $E(\hat{R})$ will, however, for most variables met in survey practice be negligible in relation to σ ($\hat{R}$), if n is "sufficiently large" where $\sigma^2(\hat{R})=E\left[\hat{R}-E(\hat{R})\right]^2$. The estimated standard error of $\hat{R}$, assuming sample size n large, is

$$est.S.E.(\hat{R})=\frac{\sqrt{1-f}}{\sqrt{n}\overline{X}}\sqrt{\frac{\sum_{i=1}^{n}y_i^2-2.\hat{R}\sum_{i=1}^{n}y_ix_i+\hat{R}^2\sum_{i=1}^{n}x_i^2}{n-1}}$$

$$=\frac{\sqrt{1-f}}{\sqrt{n}\overline{X}}\left[s_y^2+\hat{R}^2s_x^2-2.\hat{R}.s_{xy}\right]$$

If $\overline{X}$ is not known, the sample estimate $\overline{x}$ is substituted in the denominator. Clearly, we have the following cases:

for X_i = 1 for all i, R = $\overline{Y}$, the average of y per sampling unit;

for $X_i = \frac{1}{N}$ for all i, R = Y, the population total of y.

For the special cases R = $\overline{Y}$ and R = Y respectively, E ($\hat{R}$) = R.

PPS Sampling with replacement

Whenever the units vary in size, simple random sampling is not an appropriate procedure as no importance is given to the size of the unit. Such ancillary information about the size of the units can be utilized in selecting the sample so as to get more efficient estimators of the population parameters.

Now, we shall discuss probability proportional to size (PPS) sampling procedure. There are two methods of selection

(i) Cumulative total method

(ii) Lahiri's method

Cumulative Total Method

Let the size of the ith unit be X_i (i=1,2,...,N), the total being $X = \Sigma X_i$. We associate the numbers 1 to X_1 with the first unit, the numbers (X_1+1) to (X_1+X_2) with the second unit and so on. A number R is chosen at random from 1 to X and the unit with which this number is associated is considered as selected. Clearly, the ith unit in the population is being selected with a probability proportional to X_i. If a sample of size n is required, the procedure is repeated n times with replacement of the units selected. This procedure of selection is termed cumulative total method, since this method needs cumulative of the sizes.

Lahiri's Method

Lahiri (1951) suggested a method of PPS selection, which does not require a cumulation of the sizes at all. It consists in selecting a number at random from 1 to N and noting down the unit with the corresponding serial number, provisionally. Another random number is then chosen from 1 to M, where M is the maximum of the sizes of the N units or some convenient number greater than the maximum size. If the second random number is smaller or equal to the size of the unit provisionally selected, the unit is selected into the sample. If not, the entire procedure is repeated until a unit is finally selected. For selecting a sample of n units with PPSWR, the above procedure is repeated till n units are selected.

An example: A district has 10 different Rural Development Programmes consisting of 76,50,30,26,40,50,34,27,60,90 water units, respectively.

We shall select a sample of four Rural Development Programmes with the replacement method and with probability proportional to the number of water units in the programmes.

In cumulative total method, the first step in the selection of programmes is to form cumulative totals as shown below:

S. No. of R.D.P.	Size (X_i)	Cumulative size	Numbers associated
1	76	76	1–76
2	50	126	77–126
3	30	156	127–156

Contd...

4	26	182	157 – 182
5	40	222	183 – 222
6	50	272	223 – 272
7	34	306	273 – 306
8	27	333	307 – 333
9	60	393	334 – 393
10	90	483	394 – 483

To select a programme, a random number not exceeding 483 is drawn with the help of a random number table. Suppose the random number thus chosen is 343.

It can be seen from the cumulative totals that 9th programme is selected corresponding to the random number 343. Similarly, we select three more random numbers. Suppose these numbers are 472,212 and 059. Then the programmes selected corresponding to these random numbers are 10th, 5th and 1st respectively. Hence, a sample of 4 programmes selected with probability proportional to size contains the 9th, 10th, 5th and 1st programmes.

Now we select a sample of 4 programmes using Lahiri's method. In this example, N=10, M=90. First we have to select a random number from 1 to 10 and a second random number from 1 to 90. Referring to the random number table, the pair is (8, 23). Here $23 < X_8 = 27$. Hence, 8th unit is selected in the sample. We choose another pair (2, 48). Here $48 < X_2 = 50$. Thus 2nd unit is selected in the sample. We choose another pair (7, 55). Here $55 > X_7 = 34$. So the pair (7, 55) is rejected. Choosing two pairs, we can have (10, 70), (6, 38). Here 10th and 6th units are selected in the sample. Hence, the sample consists of the programmes with serial numbers 8,2,10 and 6.

We consider a population of N units and Y_i be the value of the characteristic under study for i^{th} unit of the population (i=1,2,…,N). Suppose further that $P_i = X_i / X$ be the probability that the ith unit is selected in a sample of one such that $SP_i=1$. Suppose a sample of n units is drawn with PPSWR. Further, let (y_i, p_i) be the sample observation and the initial probability of the unit selected at the ith draw, i=1,2,…,n. An unbiased estimator for the population total Y is given by

$$\hat{Y}_{PPS} = \frac{1}{n}\sum_{i=1}^{n}\frac{y_i}{p_i}$$

with it's estimated sampling variance

$$v\left(\hat{Y}_{PPS}\right)=\frac{1}{n(n-1)}\sum_{i=1}^{n}\left(\frac{yi}{pi}-\hat{Y}_{PPS}\right)^2$$

Stratified Sampling

In stratified sampling, the population of N units is first divided into sub-populations of N_1, N_2, ... N_L units respectively. These subpopulations are non-overlapping and together they comprise the whole of the population, so that $N_1 + N_2 +$ .... $+ N_L = N$ the subpopulations are called strata. An appropriate estimator for the population as a whole is obtained by suitably combining the stratum-wise estimators of the characteristic under study. The division of the population into strata is usually done in such a way that the strata are homogeneous within themselves with respect to a suitably chosen auxiliary variable termed stratification variable. Examples of stratification criterion used in biological and agricultural experiments are soil fertility pattern, stress level, tiller size, sex of fish, and different species of fish and so on. One of the advantages of stratification is that it is possible to use different sampling techniques in the different strata.

Let N_h be the number of individuals in the hth stratum with mean $\overline{Y}_h$ and variance σ_h^2 and $\sum_{h=1}^{L} N_h = N$.

One may be interested to estimate the population mean $\overline{Y}=\sum_{h=1}^{L} w_h \overline{Y}_h$ and the population proportion $P=\sum_{h=1}^{L} W_h P_h$ where $w_h=\frac{N_h}{N}$.

Let us take a simple random sample of size n by selecting n_h individuals from h^{th} stratum such that $\sum_{h=1}^{L} n_h = n$ and the samples are drawn independently from stratum to stratum. Unbiased estimators for the population mean $\overline{Y}$, the population total Y and the population proportion P are given by

$\hat{\overline{Y}}=\overline{y}_{st}=\sum_{h=1}^{L} w_h \overline{y}_h$, $\hat{Y}=N\,\overline{y}_{st}$ *and* $\hat{P}_{st}=\sum_{h=1}^{L} w_h p_h$ respectively and their estimated variances are given by

$$v(\overline{y}_{st}) = \sum_{h=1}^{L} w_h^2 (1-f_h)\frac{s_h^2}{n_h},\ v(N\,\overline{y}_{st}) = N^2 v(\overline{y}_{st})$$

$$\text{and } v(\hat{p}_{st}) = \sum_{h=1}^{L} w_h^2 (1-f_h)\frac{p_h(1-p_h)}{n_h - 1}$$

where $f_h = \frac{n_h}{N_h}, \overline{y}_h = \frac{1}{n_h}\sum_{j=1}^{n_h} y_h,\ s_h^2 = \frac{1}{n_h - 1}\sum_{j=1}^{n_h}(y_{hj} - \overline{y})^2$ and p_h is the sample proportion in the hth stratum.

An estimator for the population ratio $R = \frac{\overline{y}}{\overline{x}}$ is given by

$$\hat{R} = \frac{\overline{y}_{st}}{\overline{x}_{st}}$$

$$\text{and } v\left(\hat{R}\right) = \frac{1}{\overline{x}_{st}^2}\sum_{h=1}^{L} w_h^2 \frac{1-fh}{n_h}\left[s_{y_h}^2 + \hat{R}^2 s_{x_h}^2 - 2.\hat{R}s_{xyh}\right]$$

$$\text{where } Sxy_h = \frac{1}{n_h - 1}\sum_{j=1}^{n_h}(x_{h_j} - \overline{x}_h)(y_{h_j} - \overline{y}_h).$$

Allocation of Sample to Strata: The sample sizes n_h can be determined in the following ways:

1. Equal allocation

In this allocation, $n_h = \frac{n}{L}$ for all h i.e. the sample sizes for all the strata are equal.

2. Proportional allocation

One may use the allocation $n_h = n\frac{N_h}{N} = nw_h$ that will be called N-proportional allocation.

3. X-Proportional allocation

In case measures of size X_h are available for all units in the population, the sample size n_h may be found as a proportion of $X_h = \sum_{j=1}^{N_h} X_{h_j}$ rather than of N_h. This allocation will be called X-proportional. In this case $n_h = \frac{nX_h}{X}$

4. Optimum allocation

The sample sizes may be selected to minimize V ($\overline{y}_{st}$) for a specified cost of taking the sample. The simplest cost function is of the form

$$\text{cost} = C = c_0 + \sum_{h=1}^{L} c_h n_h$$

where c_0 is an overhead cost and the cost per unit c_h may vary from stratum to stratum. In this case

$$n_h = (C - c_0)\frac{w_h S_h / \sqrt{c_h}}{\sum w_h S_h \sqrt{c_h}}$$

This is known as optimum allocation. If $c_h = c'$ i.e. if the cost per unit is the same in all strata then

$$n_h = n\frac{w_h S_h}{\sum w_h S_h}$$

This allocation is known as Neyman allocation.

It is noted that if $S_h = S$ for all h then this allocation reduces to proportional allocation.

Systematic Random Sampling

A sampling technique in which only the first unit is selected with the help of random numbers and the rest get selected automatically according to some pre-designed pattern is known as systematic random sampling. It is also referred to briefly as systematic sampling. Suppose N units of the population are numbered from 1 to N in some order. Let N = nk, where n is the sample size and k is an integer, and a random number less than or equal to k be selected and every k^{th} unit there after. The final sample is called every k^{th} systematic sample and such a procedure termed linear systematic sampling.

If N≠nk and every k^{th} unit be included in a circular manner till the whole list is exhausted, it will be called circular systematic sampling.

Systematic sampling is simple and it is widely used in various types of surveys i.e., in census work, forest surveys, milk yield surveys and in fisheries etc. because in many situations it provides estimates more efficient than simple random sampling.

Linear Systematic Sampling (LSS)

Case-1 [N=nk]

Let a random number r from 1 to k be selected. The selected random number r is known as the random start and k is called the sampling interval. In this procedure, the sample comprises the units r, r+k, r+2k,....,r+(n-1) k. The technique will generate k systematic samples with equal probability.

Case-2 [N≠nk]

In this case, the present sampling scheme will give rise to samples of unequal size. k is taken as an integer nearest to N/n. Then a random number is chosen from 1 to k and every k^{th} unit is drawn in the sample. Under this condition, the sample size is not necessarily n and in some cases it may be (n-1).

Circular Systematic Sampling (CSS)

The disadvantages of LSS regarding the actual sample size being different from that required and the sample mean being a biased estimator of the population mean when N is not a multiple of n can be overcome by adopting a device, known as circular systematic sampling. This device consists in choosing a random start from 1 to N and selecting the unit corresponding to this random start and thereafter every k^{th} unit in a cyclical manner till a sample of n units is obtained, k being the integer nearest to (N/n). That is, if r is a number selected at random from 1 to N, the sample consists of the units corresponding to the numbers

$$\left.\begin{array}{l}\{r+jk\}, \text{ if } r+jk \le N \\ \text{and } \{r+jk-N\}, \text{ if } r+jk > N\end{array}\right\} \quad j = 0,1,2,\ldots, (n-1).$$

In selecting n units from a population of N (=nk) units with k as the interval, there are k possible samples and let $\overline{y}_r$ be the sample mean of the rth possible sample (r = 1,2,..., k). Then the sample mean $\overline{y}_{sy}$ is an unbiased estimator of the population mean $\overline{Y}$ It can also be shown that in case of circular systematic sampling, the sample mean is an unbiased estimator of the population mean.

An unbiased estimate of the variance is not available for a systematic sample with one random start because a systematic sample is regarded as a random sample of one unit. Some biased estimators are possible on the basis of a systematic sample. If two or more systematic samples are available, an unbiased estimate of the variance of the estimated mean can be made.

Instead of selecting a single sample of size n, if it is selected in the form of m independent systematic samples each of size t (integer) such that n = mt then unbiased estimate of $\overline{Y}$ is given by

$\hat{\overline{Y}}_{sys} = \frac{1}{m}\sum_{i=1}^{m}\overline{y}_i \text{ where } \overline{y}_i = \frac{1}{t}\left(\sum_{\alpha=1}^{L} y_{i\alpha}\right)_r$ with the estimated variance as

$$v\left(\hat{\overline{Y}}_{sys}\right) = \frac{1}{m(m-1)}\sum_{t=1}^{m}(\overline{y}_i - \hat{\overline{Y}}_{sys})^2$$

The above sub-samples are called interpenetrating sub-samples defined by Mahalanobis (1946).

Cluster Sampling

In cluster sampling the whole population is divided into a number of clusters each consisting of several sampling units. Cluster size may definitely vary from cluster to cluster. Then some clusters are selected at random out of all the clusters. The advantages of cluster sampling from the point of view of cost arise mainly due to the fact that collection of data for nearly units is easier, faster, cheaper and more convenient than observing units scattered over a region.

The cluster sampling is useful in many practical surveys where listing of population units is not available but the same may be available for small segments of the population. For example, in a crop survey in a district, list of plots in the district cannot be available but list of villages may be easily available. Here, villages will be treated as clusters. In a rice field experiment, list of panicles in the field is not available but list of hills may be available. In such case hills will be considered as clusters.

Suppose a finite population is divided into N mutually exclusive clusters having M_i units in the ith cluster (i=1,2,..., N). Let Y_{ij} be the value for j^{th} unit in the ith cluster (j=1,2,…., Mi; i=1,2,…., N). Let $\overline{Y}$ be the mean for i[th] cluster and $\overline{Y}$ be the overall population mean where

$$\overline{Y} = \sum_{i=1}^{N} \frac{M_i \overline{Y}_i}{N\overline{M}}, N\overline{M} = \sum_{i=1}^{N} M_i$$

Suppose n clusters are selected with SRSWOR and all the units in the selected clusters are surveyed. An unbiased estimator of $\overline{Y}$ is given by

$$\overline{y}_c = \frac{1}{n} \sum_{i=1}^{n} \left(\frac{M_i}{\overline{M}} \right) \overline{y}_i$$

where $\overline{Y}_i$ is the mean of the ith sample cluster and the estimated variance is

$$v(\overline{y}_c) = \frac{(1-f)s_b^2}{n}$$

$$\text{where } s_b^2 = \frac{1}{n-1} \sum_{i=1}^{n} \left(\frac{M_i}{\overline{M}} \overline{y}_i - \overline{y}_c \right)^2 \text{ and } f = n/N.$$

Unbiased estimators for the population total Y and the overall population proportion P are given by

$\hat{Y}_c = \frac{N}{n} \sum_{i=1}^{n} M_i \overline{y}_i \ \ and\ \ \hat{P}_c = \frac{1}{n} \sum_{i=1}^{n} \left(\frac{Mi}{\overline{M}} \right)_{pi}$ respectively where p_i is the proportion of units belonging to the specified category in the ith sample cluster and their estimated variances are

$$v\left(\hat{Y}_c\right) = \left(N\overline{M}\right)^2 v(\overline{y}_c)$$

$$\text{and } v\left(\hat{P}_c\right) = \frac{(1-f)}{n(n-1)} \sum_{i-1}^{n} \left(\frac{M_i}{\overline{M}} p_i - \hat{p}_c \right)^2$$

An estimator for $R = \frac{\overline{Y}}{\overline{X}}$ is given by $\hat{R}_c = \frac{\overline{y}_c}{\overline{x}_c}$ and

$$v\left(\hat{R}_c\right) = \frac{1-f}{n} \cdot \frac{1}{\overline{x}_c^2} \left[S_{yb}{}^2 + \hat{R}_c^2 s_{xb}{}^2 - 2\hat{R}_c S_{xy_b} \right]$$

$$\text{where } \overline{x}_c = \frac{1}{n}\sum_{i=1}^{n}\left(\frac{M_i}{\overline{M}}\right)\overline{xi}$$

$$\text{and } s_{xyb} = \frac{1}{n-1}\sum_{i=1}^{n}\left(\frac{M_i}{\overline{M}}\overline{x}_i - \overline{x}_c\right)\left(\frac{M_i}{\overline{M}}\overline{y}_i - \overline{y}_c\right)$$

Two-stage Sampling

Two-stage sampling has been found to be very useful in practice and this procedure is being commonly used in large-scale surveys. Two-stage sampling consists in first selecting the cluster and then selecting a specified number of elements from each selected cluster. In such sampling designs, clusters which form the units of sampling at first stage are called the first stage units (fsu) or primary sampling units (psu) and the elements within clusters are called second-stage units (ssu). The main advantage of this sampling procedure is that, at the first-stage, the frame of fsu's is required, which can be prepared easily. At the second-stage, the frame of ssu's is required only for the selected fsu's. This design is more flexible as it permits the use of different selection procedures in different stages.

For example, if we are interested in obtaining a sample of households with pisciculture as principal occupation from a particular block, the first stage units may be villages and the second stage units will be households in the villages. Since the samples are selected in stages, it is possible to draw samples of units after properly utilizing the information collected in the preparation of sampling frame at that stage. In selecting villages as fsu's, one may stratify villages according to their sizes and may use the stratified PPSWR-sampling procedure. Again, while preparing lists of households belonging to each selected village, one may collect some easily available information such as size of households, water area of the households, type of dwelling etc. Such information may be suitably used for the drawing of samples at the second stage. The efficiency of two-stage sampling can be greatly enhanced by proper use of such information.

Suppose the population consists of N fsu's, the ith fsu comprising of M_i ssu's. The n fsu's are selected at the first stage and from the ith selected fsu, m_i ssu's are selected. Let Y_{ij} be the value of the character y on j^{th} ssu in the ith fsu in the population, $j=1,2,\ldots,M_i$; $i=1,2,\ldots,N$.

Let

$$M_0 = \sum_{i=1}^{N} M_1 \text{ = total number of ssu's in the population;}$$

$\overline{M} = \frac{M_0}{N}$ = average number of ssu's per fsu in the population

$Y_i = \sum_{j=1}^{M_i} Y_{ij}$ = population total for the ith fsu;

$\overline{Y_i} = \frac{Y_i}{M_i}$ = population mean for the ith fsu;

$Y = \sum_{i=1}^{N} Y_i$ = population total;

$\overline{Y} = \frac{Y}{N}$ = population mean per fsu;

$\overline{\overline{Y}} = \frac{Y}{M_0} = \frac{\sum_{i=1}^{N} M_i \overline{Y}_i}{\sum_{i=1}^{N} M_i}$ = population mean per ssu

$S^2{}_b = \frac{1}{N-1}\sum_{i=1}^{N}\left(\overline{Y}_i - \overline{\overline{Y}}\right)^2$ = population variance between fsu-means;

$$S'^2_b = \frac{1}{N-1}\sum_{i=1}^{N}\left(\frac{M_i \overline{Y}_i}{\overline{M}} - \overline{\overline{Y}}\right)^2$$

$S^2{}_{w_1} = \frac{1}{M_i - 1}\sum_{J=1}^{M}\left(Y_{ij} - \overline{Y}_i\right)$ = population variance for the i_{th} fsu;

$S^2{}_w = \frac{1}{N}\sum_{i=1}^{N} S^2{}_{w_i}$ = population variance within fsu's.

Let y_{ij} be the value of the character y on the jth ssu in the ith selected fsu in the sample, j=1,2,..., m_i; i=1,2,..., n. Let

$$y_i = \sum_{j=1}^{m_i} y_{ij}$$

$$\overline{y}_i = \frac{y_i}{m_i}$$

$$\bar{\bar{y}} = \frac{\sum_{J=1}^{n}\sum_{J=1}^{m_i} y_{ij}}{\sum_{i=1}^{n} m_i} = \frac{\sum_{i=1}^{n} m_i \bar{y}_i}{\sum_{i=1}^{n} m_i} = \text{sample mean per ssu.}$$

$$S_b^2 = \frac{1}{n-1}\sum_{i=1}^{n}(\bar{y}_i - \bar{\bar{y}})^2$$

$$S_b^{/2} = \frac{1}{n-1}\sum_{i-1}^{n}\left(M_i\bar{y}_i - \frac{\sum_{i=1}^{n} M_i\bar{y}_i}{n} \right)^2$$

$$S^2{}_{w_i} = \frac{1}{m_i - 1}\sum_{j=1}^{m_i}(y_{ij} - \bar{y}_i)^2$$

$$S^2{}_w = \frac{1}{n}\sum_{i=1}^{n} s^2 w_i$$

$f_1 = \frac{n}{N}$ = sampling fraction at the first stage of sampling;

$f_{2_i} = \frac{m_i}{M_i}$ = sampling fraction for the ith fsu at the second stage of sampling

Simple Random Sampling without Replacement at both Stages

Suppose a sample of n fsu's is selected with SRSWOR and from the ith selected fsu a sample of m_i ssu's is selected for each i = 1,2,..., n again with SRSWOR. Unbiased estimators for Y and $\bar{\bar{Y}}$ are given by

$$\hat{y}_i = \frac{N}{n}\sum_{i=1}^{n} M_i\bar{y}_i = \frac{1}{n}\sum_{i=1}^{n} NM_i\bar{y}_i \qquad .....(1)$$

and $$\hat{\bar{\bar{Y}}}i = \frac{1}{n}\sum_{i=1}^{n}\left(\frac{M_i}{\bar{M}}\right)\bar{y}_i \qquad .....(2)$$

respectively and unbiased estimators of $V\left(\hat{Y}_t\right)$ and $V\left(\hat{\bar{\bar{Y}}}_t\right)$ are

$$v\left(\hat{Y}_t\right)=\frac{N^2(1-f_1)}{n}s_b'^2+\frac{N}{n}\sum_{i=1}^{n}\frac{M_i^2(1-f_{2i})}{m}s_{w_i}^2 \qquad ......(3)$$

$$\text{and } V\left(\hat{\bar{\bar{Y}}}_t\right)=\frac{v(\hat{Y}_t)}{(N\bar{M})^2} \qquad .....(4)$$

respectively.

It may be noted that for calculating $\hat{Y}_t$ and v $(\hat{Y}_t)$ the value of $\bar{M}$ is not required. If the value of $\bar{M}$ is known in advance, $\hat{\bar{\bar{Y}}}_t$ and v $(\hat{\bar{\bar{Y}}}_t)$ can be obtained from (6.2) and (6.4). In practice, M_0 is not known and is estimated by

$\hat{M}_0 = N\sum_{i=1}^{n}\frac{M_i}{n}$ and an estimator of $\bar{\bar{Y}}$ is given by

$$\hat{\bar{\bar{Y}}}_{tR}=\frac{\hat{Y}_t}{\hat{M}_0}=\frac{\sum_{i=1}^{n}M_i\bar{y}_i}{\sum_{i=1}^{n}M_i} \qquad ......(5)$$

The estimated variance of $\hat{\bar{\bar{Y}}}_{tR}$ is given by

$$v\left(\hat{\bar{\bar{Y}}}_{tR}\right)=\frac{1-f1}{n\bar{M}^2_s(n-1)}\sum_{i=1}^{n}M^2{}_i\left(\bar{y}_i-\hat{\bar{\bar{Y}}}_{tR}\right)^2+\frac{1}{nN\bar{M}_s^2}\sum_{i=1}^{n}\frac{M_i^2(1-f_{2i})s_{w_i}^2}{m_i} \qquad ....(6)$$

$$\text{where } \bar{M}_s=\sum_{i=1}^{n}\frac{M_i}{n}$$

If in particular, M_i = M, m_i = m $\forall$ i,

$$\hat{Y}_t=\frac{NM}{n}\sum_{i=1}^{n}\bar{y}_i \quad and \quad \hat{\bar{\bar{Y}}}_t=\frac{1}{n}\sum_{i=1}^{n}\bar{y}_i$$

and (6.3) and (6.4) reduce to

$$v\left(\hat{Y}_t\right) = N^2M^2\left[\frac{(1-f_1)}{n}s_b^2 + \frac{f_1(1-f_2)}{nm}s_w^2\right]$$

$$v\left(\hat{\overline{\overline{Y}}}_t\right) = \frac{(1-f_1)}{n}s_b^2 + \frac{f_1(1-f_2)}{nm}s_w^2$$

respectively.

An estimator of population ratio $R = \frac{Y}{Z}$ is given by

$$\hat{R} = \frac{\sum_{i=1}^{n} M_i\overline{y}_i}{\sum_{i=1}^{n} M_i\overline{z}_i} \qquad .....(7)$$

and estimated variance is

$$v\left(\hat{R}\right) = (1-f_1)N^2\sum_{i=1}^{n}\frac{(M_i\overline{y}_i - \hat{R}M_iz_i)^2}{n(n-1)\hat{Z}^2} + \frac{N}{n\hat{Z}^2}\sum_{i=1}^{n}\frac{M_i^2(1-f_{2i})}{m_i(m_i-1}\sum_{j=1}^{m_i}\{y_{ij} - \overline{y}_i - \hat{R}(z_{ij} - \overline{z}_i)\}^2$$

$$\text{where } \hat{Z} = \frac{N}{n}\sum_{i=1}^{n} M_i\overline{z}_i \qquad .....(8)$$

In large scale surveys, it is desirable to select the fsu's with replacement, since it enables us to get an estimate of the sampling variance of the estimator without having to calculate separately the estimates of the within and between components. Suppose the sampling is done with SRSWR at the first stage and with SRSWOR at the second stage, the estimates of Y and $\overline{\overline{Y}}$ given in (1) and (2) remain unbiased and unbiased variance estimators of $\hat{Y}_t$ and $\hat{\overline{\overline{Y}}}_t$ are given by

$$v\left(\overline{Y}_t\right) = N^2\frac{s'^2_b}{n} = \frac{N^2}{n}\frac{1}{n-1}\sum_{i=1}^{n}\left(M_i\overline{y}_i - \frac{\sum_{i=1}^{n} M_i\overline{y}_i}{n}\right)^2$$

$$= \frac{1}{n(n-1)} \sum_{i=1}^{n} (NM_i \bar{y}_i - \hat{Y}_t)^2$$

$$\text{and } v\left(\hat{\bar{\bar{Y}}}_t\right) = \frac{v(\hat{Y}_t)}{(N\bar{\bar{M}})^2}$$

Self-Weighting Design

A sample design which provides a single common weight to all sampled observations in estimating the population mean, total etc. is called a self-weighting design and the corresponding estimator is called a self-weighted estimator. The estimator (1) is not generally self-weighted. If $\frac{m_i}{M_i} = \lambda$ (a constant), i.e. a constant proportion of ssu's, is sampled from each selected fsu, then

$\hat{Y}_t = \frac{N}{n\lambda} \sum_{i=1}^{n} \sum_{j=1}^{m} y_{ij}$ becomes self-weight. Again $\lambda = \frac{\sum_{i=1}^{N} m_i}{\sum_{i=1}^{N} M_i} = \frac{N\bar{m}}{M_0}$

$$\text{where } \bar{m} = \frac{1}{N} \sum_{i=1}^{N} m_i$$

$$\text{so that } \hat{Y}_t = \frac{M_0}{nm} \sum_{i=1}^{n} \sum_{j=1}^{m_i} y_{ij}$$

In large-scale surveys, it is always desirable that the weight is the same for every unit so that (i) tabulation becomes simpler, (ii) analysis becomes easier, and (iii) cost is minimized. Another advantage of the self-weighting design is that it gives a constant sample size from each selected fsu.

Example

In a survey for estimating the impact of a comprehensive development project for the fish-farmers (FFD Project) in West Bengal, 8 villages were selected out of 47 in a block in the district of Burdwan by SRSWOR and a sample of households was selected by SRSWOR from each selected village. The following table gives the total number of households with pisciculture as principal occupation (M_i) and water-area (in Satak) (household size) of the sampled households in each selected village.

Sampled Village No.	M_i	Water-area in Satak (household size)
3	8	50(3), 33(7), 57(12)
7	6	66(4), 92(6)
13	9	66(18), 50(8), 33(5)
17	11	33(7), 45(12), 60(7), 55(10)
21	13	99(6), 66(18), 50(10), 67(7)
31	13	60(5), 155(7), 77(8), 80(9)
37	10	78(10), 33(8), 230(11)
41	12	78(12), 66(7), 60(10), 77(8)

Source: FFD-Survey in West Bengal, 1991, I.S.I.

We want to estimate (a) the total water-area of the block, (b) the mean water-area per fish-farming household when it is given that $\bar{M}=12$ and $\bar{M}$ is not known and (c) the per-capita water-area in a fish-farming household.

Let Y and Z be the total water-area and the total number of persons in the fish-farming households in the block. In this example, N=47, n=8.

i	M_i	m_i	$\bar{y}_i$	$M_i\bar{y}_i$	$\bar{z}_i$	$M\bar{z}_{ii}$
1	8	3	46.67	373.36	7.33	58.67
2	6	2	79	474	5	30
3	9	3	49.67	447.03	10.33	93
4	11	4	48.25	530.75	9	99
5	13	4	70.5	516.50	10.25	133.25
6	13	4	93	1209	7.25	94.25
7	10	3	113.67	1136.70	9.67	96.67
8	12	4	70.25	843	9.25	111
	82			5530.34		715.84

(a) $\hat{Y}=\frac{N}{n}\sum_{i=1}^{n} M_i\bar{y}_i=32490.75$

(b) $\hat{\bar{\bar{Y}}}_t = \frac{\hat{Y}}{N\bar{M}} = 57.61$

When $\bar{M}$ is not known,

$$\hat{\bar{\bar{Y}}}_{tR} = \frac{\sum_{i=1}^{n} M_i \bar{y}_i}{\sum_{i=1}^{n} M_i} = 67.44$$

(c) $\hat{Z}_t = \frac{N}{n}\sum_{i=1}^{n} M_i \bar{z}_i = 4205.56$

$$\hat{R} = \frac{\hat{Y}_t}{\hat{Z}_t} = \frac{32490.75}{4205.56} = 7.73$$

We can also find the estimated variances of the above estimators using the expressions (3), (4), (6) and (8) respectively.

Stratified Multi-Stage sampling

The most common design in surveys is stratified multi-stage sampling. The population of first stage units is divided into strata; within each stratum a SRSWOR of first-stage units is selected and each of the selected fsu is further sub-sampled.

We shall consider an unbiased estimator of the population mean in stratified two-stage sampling.

Let the population be divided into L strata.

Let N_h = no. of fsu's in the hth stratum of the population

$$N = \sum_{h=1}^{L} N_h$$

Mhi = no. of ssu in the ith fsu of the hth stratum, i=1,2,......... $\frac{NW}{1}$

$M_{ho} = N_h \bar{M}_h = \sum_{i=1}^{NH} M_{hi}$ = total number of ssu in the hth stratum

Y_{hij} = Value of the characteristic on the jth ssu in the ith fsu of the hth stratum

$$\overline{Y}_{hi} = \frac{1}{M_{hi}} \sum_{j=1}^{M_{hi}} Y_{hij}$$

$$\overline{Y}_{h..} = \frac{1}{M_{ho}} \sum_{i=1}^{N_h} \sum_{i=1}^{M_{hi}} Y_{hij} = \frac{1}{N_h \overline{\overline{M}}_h} \sum_{i=1}^{N_h} \sum_{j=1}^{M_{hi}} Y_{hij}$$

$$\overline{Y} = \overline{y}.... = \frac{\sum_{h=1}^{L} M_{ho} \overline{Y}_{h..}}{\sum_{h=1}^{L} M_{ho}} = \sum_{h=1}^{L} \lambda_h \overline{Y}_{h...}, \quad \lambda_h = \frac{M_{ho}}{\sum_{h=1}^{L} M_{ho}} = \frac{M_{ho}}{M_o}$$

n_h = no. of fsu in the sample from h^{th} stratum

$$n = \sum_{h=1}^{L} n_h$$

m_{hi} = no. of ssu in the sample from the ith selected fsu of the hth stratum

$$\overline{y}_{hi} = \frac{1}{m_{hi}} \sum_{j=1}^{m_{hi}} y_{hij}$$

$$\overline{y}_{hs} = \frac{1}{\overline{\overline{M}}_h n_h} \sum_{i=1}^{n_h} \frac{M_{hi}}{m_{hi}} \sum_{j=1}^{m_{hi}} y_{hij} = \frac{1}{n_h} \sum_{i=1}^{n_h} \frac{M_{hi}}{\overline{\overline{M}}_h} \overline{y}_{hi}$$

An unbiased estimator of $\overline{Y}$ is

$\overline{\overline{y}}_{w2} = \sum_{h=1}^{L} \lambda_h \overline{y}_{hs}$ with its estimated variance

$$v(\overline{\overline{y}}_{w2}) = \sum_{h=1}^{L} \lambda_h^2 v(\overline{y}_{hs})$$

where $v(\overline{y}_{hs}) = \left(\frac{1}{n_h} - \frac{1}{N_h}\right) s_{hb}^2 + \frac{1}{n_h N_h} \sum_{i=1}^{n_h} \left(\frac{1}{m_{hi}} - \frac{1}{M_{hi}}\right) \left(\frac{M_{hi}}{\overline{\overline{M}}_h}\right)^2 s_{hiy}^2$

$$s_{hb}^{2} = \frac{1}{n_h - 1}\sum_{i=1}^{n_h}\left(\frac{M_{hi}}{\overline{M}_h}\overline{y}_{hi} - \overline{y}_{hs}\right)^2$$

$$s_{hiy}^{2} = \frac{1}{m_{hi} - 1}\sum_{j=1}^{n_h}\left(y_{hij} - \overline{y}_{hi}\right)^2$$

Some commonly used Estimation Procedures

The important methods of estimation are

i) Ratio method of estimation
ii) Product method of estimation
iii) Regression method of estimation
iv) Difference method of estimation

The customary estimators discussed in previous section can be improved through these estimation procedures provided the required information is available. Let y be the variable under study and the estimates for mean $\overline{Y}$ or total Y of y are wanted. Let the population mean $\overline{X}$ or the total X = N$\overline{X}$ of another variable x be known. The variable x is called as auxiliary variable, ancillary variable or supplementary variable. The information on x may be used to obtain better estimators for Y, $\overline{Y}$, σ_y^2, C_y etc. compared to their customary estimators.

Some examples of y and x variables may be given as follows:

a) y: fish catch on a day ; x: number of boats bringing fish
b) y: fish catch on a day ; x: area of water unit
c) y: number of trees ; x: area under orchards

The discussion will be related to simple random sampling of n sampling units drawn from a population of N units.

Let Y_i and X_i be the value of the characteristic under study and ancillary characteristic respectively for the ith unit of the population. Suppose the population mean $\overline{X}$ is known.

Ratio and Product method of estimation: The univariate ratio and product estimators for the population mean $\overline{Y}$ are given by

$$\overline{y}_R = \frac{\overline{y}}{\overline{x}}.\overline{X} = \hat{R}\overline{X} \; and \; \overline{y}_p = \overline{y}\left(\frac{\overline{x}}{\overline{X}}\right) \text{ respectively}$$

$$\text{where } \hat{R} = \left(\frac{\overline{y}}{\overline{x}}\right)$$

These estimators are biased and the estimates of their MSE's are given by

$$\hat{M}(\overline{y}_R) = \frac{1-f}{n.}\left[s_y^2 + \hat{R}^2.s_x^2 - 2.\hat{R}s_{xy}\right]$$

$$\text{and } \hat{M}(\overline{y}_p) = \frac{1-f}{n.}\left[s_y^2 + \hat{R}^2.s_x^2 - 2.\hat{R}s_{xy}\right]$$

It is known that given any auxiliary variable x, we may decide whether to use this variable for obtaining a ratio estimator or a product estimator as an improvement over $\overline{y}$ according as $\rho > \frac{1}{2}\left(\frac{C_x}{C_Y}\right) or \; \rho < -\frac{1}{2}\left(\frac{C_x}{C_Y}\right)$ when R is positive and vice-versa when R is negative where ρ is the population correlation coefficient between y and x. If ρ is between these two values, then it is better to use the conventional estimator $\overline{y}$ since in this case it is more efficient than the ratio and product estimators.

Regression and difference method of estimation: If b is the least-square estimator of the population regression coefficient β of y on x then $\overline{y}_{lr} = \overline{y} - b(\overline{x} - \overline{X})$ is a regression estimator for $\overline{Y}$ with its estimated mse (for large samples) $v(\overline{y}_{lr}) = \frac{1-f}{n} s_y^2(1-r^2)$ where r is the sample correlation coefficient between x and y.

If a constant λ is used in $\overline{y}_{lr}$ instead of b , we get the difference estimator

$$\overline{y}_D = \overline{y} - \lambda(\overline{x} - \overline{X})$$

with its estimated variance

$$v(\overline{y}_D) = \frac{1-f}{n} \left[s_y^2 + \lambda^2 s_x^2 - 2\lambda s_{xy}\right]$$

In practice λ is taken to be the good guessed value of $R = \frac{Y}{X}$ or $\beta = \frac{\sigma_{yx}}{\sigma_x^2}$.

It is noted that the regression estimator $\bar{y}_{lr}$ is always more efficient than the usual mean estimator $\bar{y}$ and the ratio estimator $\bar{y}_R$ unless ρ = 0 and R = β respectively.

Use of auxiliary information in sample surveys

In most of the survey situations the auxiliary information is always available. It may either be readily available or may be made available without much difficulty by diverting a part of the survey resources. The customary sources of obtaining relevant auxiliary information on one or more variables are various census data, previous surveys, pilot survey etc. The auxiliary information may infact be available in various forms.

The use of auxiliary information in estimating the population total or mean was recognized during the decade 1940 – 1950 themselves. It is now well recognized that if used intelligibly, the auxiliary information increases the efficiency of the estimators. Broadly speaking , the auxiliary information may be used in three basic ways.

i) At pre-selection or designing stage: for example in stratifying the population or constructing the clusters on the basis of a related character.

ii) At Selection stage: for example in selecting the units with varying probabilities, based on an auxiliary character, with or without replacement.

iii) At estimation stage: for example in constructing ratio, regression, difference or product type estimators.

The information may also be used in mixed ways. The use of auxiliary information on a single character to improve upon the customary estimator of the mean or total of a single principal character has been widely discussed in the sample survey literature and a host of estimators have been put forward (See, e.g. Cochran (1977), Raj (1968)). Some of the estimators are discussed in Section 3. All the estimators given in Section 3 are defined for the estimation of $\bar{Y}$ when the population mean of an auxiliary character is known.

In case the required auxiliary information is not readily available and collection of information on x is not quite expensive, the usual technique is to adopt the double sampling (two-phase sampling) procedure . The technique consists in taking a larger sample of size n′ to estimate the population mean $\overline{X}$ while a sub-sample of size n is drawn from n′ to observe the characteristic under study. Several estimates of the population mean $\overline{Y}$ can be made.

A preliminary random sample, WOR, of size n′ is taken and the information on x is collected. A sub-sample of size n is drawn, WOR, from the preliminary sample and information on y is measured.

The ratio estimator for $\overline{Y}$ based on double sampling is given by

$$\overline{y}_{Rd} = \frac{\overline{y}}{\overline{x}}\overline{x}^{1}$$

where $\overline{y}$, $\overline{x}$, are sub-sample means for y and x respectively and $\overline{x}'$ is the sample mean of x in the first sample . It is noted that $\overline{y}_{Rd}$ is biased and the estimator of $V(\overline{y}_{Rd})$, to the first order of approximation, is given by

$$v(\overline{y}_{Rd}) = \left(\frac{1}{n'} - \frac{1}{N}\right)s_y^2 + \left(\frac{1}{n} - \frac{1}{n'}\right)\left[s_y^2 + \hat{R}^2.s_x^2 - 2.\hat{R}.s_{xy}\right]$$

$$\text{where } \hat{R} = \frac{\overline{y}}{\overline{x}},\ s_{yx} = \frac{1}{n-1}\sum_{i=1}^{n}(y_i - \overline{y})(x_i - \overline{x})$$

A regression estimator for $\overline{Y}$ based on double sampling may be defined by

$$\overline{y}_{ld} = \overline{y} - b(\overline{x} - \overline{x}')$$

where $b = \frac{s_{yx}}{s_x^2}$ is the least square estimate of the regression co-efficient β in the population. The estimator is a biased and the estimate of V ($\overline{y}_{ld}$) to the first order of approximation, is given by

$$v(\overline{y}_{ld}) = \left(\frac{1}{n'} - \frac{1}{N}\right)s_y^2 + \left(\frac{1}{n} - \frac{1}{n'}\right)s_y^2(1 - r^2)$$

For the comparision of double sampling strategy with single phase sampling strategy based on cost consideration we refer to Cochran (1997).

In many situations, several auxiliary variables may be available. For example, in a village land survey in the costal area, the information on the characters such as area of the village, population of the village, number of reservoirs, number of fishing household, number of fishing boat etc. may be easily obtained through the past census data and may be used to improve the customary estimator for the mean or total of the related principal character (study variable).

The use of information on several auxiliary variable $\underset{\sim}{x} = (x_1, x_2, \ldots\ldots\ldots\ldots x_q)'$ for estimating the population mean of a principal variable y has also been considered by several authors, e.g. Olkin (1958), Raj (1965) etc. If the study variable y is positively correlated to q auxiliary variables $\{x_i\}$, then the multivariate ratio estimator of $\overline{Y}$ is given by

$$\hat{\overline{Y}}_{MR} = \sum_{i=1}^{q} w_i \left(\frac{\hat{\overline{Y}}}{\hat{\overline{X}}_i} \right) . \overline{X}_i$$

where $\hat{\overline{Y}}$ and $\{\hat{\overline{X}}_i\}$ are the estimators of $\overline{Y}$ and the means $\{\overline{X}_i\}$ of the q auxiliary variables and $\sum_{i=1}^{q} w_i = 1$. If y is negatively correlated to q auxiliary variables $\{x_i\}$, then the multivariate product estimator of $\overline{Y}$ is given by

$$\overline{Y}_{MP} = \sum_{i=1}^{q} w_i \left(\hat{\overline{Y}} . \hat{\overline{X}}_i \right) / \overline{X}_i$$

If some of the auxiliary variables are highly positively correlated and the others highly negatively correlated with the study variable y, then one can take a linear combination of ratio and product estimators.

For simple random sampling $\hat{\overline{Y}} = \overline{y}$ and $\hat{\overline{X}}_i = \overline{x}_i$, i = 1, 2, ..., q.

Our problem is to estimate the population mean $\overline{Y}$ of the study variable y when supplementary information is available on two auxiliary variables x_1 and x_2. The multi-variate ratio estimator is

$$\hat{\overline{Y}}_{MR} = w.\frac{\overline{y}}{\overline{x}_1}.\overline{X}_1 + (1-w)\frac{\overline{y}}{\overline{x}_2}\overline{X}_2 = w.\overline{y}_{R1} + (1-w).\overline{y}_{R2}$$

where w is a constant and $\overline{y}_{Ri} = \frac{\overline{y}}{\overline{x}_1}.\overline{X}_i = \hat{R}_i \overline{X}_i$

Estimates of mses of $\overline{y}_{R_1}$ and $\overline{y}_{R_2}$ are given in Section 3. The estimated value of

$$Cov(\overline{y}_{R_1}, \overline{y}_{R_2}) = \frac{1-f}{n}\left[s_y^2 + \hat{R}_1.\hat{R}_2.s_{x_1x_2} - \hat{R}_1.s_{yx_1} - \hat{R}_2.s_{yx_2} \right]$$

The estimated optimum value of w is

$$\hat{w}_{opt} = \frac{\hat{M}(\overline{y}_{R_2}) - C\hat{o}v(\overline{y}_{R_1}, \overline{y}_{R_2})}{\hat{M}(\overline{y}_{R_1}) + \hat{M}(\overline{y}_{R_2}) - 2.C\hat{o}v(\overline{y}_{R_1}, \overline{y}_{R_2})}$$

and the estimated minimum value of $M\left(\hat{\overline{y}}_{MR}\right)$ is then

$$\frac{\hat{M}(\overline{y}_{R_1})\hat{M}(\overline{y}_{R_2}) - \left[C\hat{o}v(\overline{y}_{R_1}, \overline{y}_{R_2})\right]^2}{\hat{M}(\overline{y}_{R_1}) + \hat{M}(\overline{y}_{R_2}) - 2.C\hat{o}v(\overline{y}_{R_1}, \overline{y}_{R_2})}$$

Similarly one can define Multivariate product estimator for simple random sampling.

In the situations where the population means $\overline{X}_1, \overline{X}_2, ..., \overline{X}_q$ are unknown the usual double sampling procedure is used to collect the required information and to define the estimators for $\overline{Y}$.

Generally fisheries surveys are multivariate, i.e. the variables investigated are more than one (revenues, costs, employment etc.). In the situations where several principal variables are to be observed the usual problem in such multipurpose surveys is to estimate population means or totals of several study variables simultaneously. In this situation, also the information on the auxiliary characters may be used to improve the customary estimators of the means or totals of the related principal characters. For example, in a freshwater culture in fish pond survey conducted in a district, the multiple characteristics such as area of pond, fish seed stocked, fertilizers and pesticides used, number of culture household, total fish catch etc. are observed for each unit. Such surveys, called as multivariate surveys or multi-purpose surveys have vast area of application in various statistical enquiries and investigation. Undoubtedly, there is a good scope of improvement in the practice related to planning of sample survey and choice of sampling units, sampling procedure, estimation procedures and methods of data collection, statistical analysis and interpretation of data. There is also a good scope of improving quality of data and increasing reliability of sample surveys in the field of Inland water fisheries.

Some Application of Different Sampling Schemes in Aquaculture

A simple problem arises to take samples from large catches of fish caught by commercial. It is often impossible to take data on all the specimens captured by experimental means or commercial catches, whether on shipboard or at landings, or elsewhere. Since the total mass of specimens at hand on a given day (say) is regarded as a sample, selection of smaller groups for examination is usually called sub- sampling. In sub -sampling, the objective is to obtain data free from any systematic bias due to the sampling procedure. Commonly sub-sampling has been resorted to for data on length, weight, sex, scale or bone specimens for age determination, stage of sexual maturity, stomach contents and other similar characteristics.

Stratified random sampling is particularly useful when the large sample is already classified into groups i.e. catches in successive net hauls, sex of fish, different species of fish, different size groups etc. that are known or suspected to differ in respect to the character being examined. Each of these groups is then a stratum to be sampled independently by the system of random numbers. The mean values of length, for example in each stratum can be estimated. Then these estimates can be combined by giving suitable weights to yield the estimate for the whole population. The weights are taken as the size of the stratum. When variances within strata are equal, proportional stratified sampling typically minimizes variability.

Now the problem is to estimate the total catch of fish on a particular day of 12 hrs. (6 am to 6 pm) at a particular fishing center on the coastal area. Here the unit is an hour and the characteristic (y) is the catch of fish. Also the number of boats (x) landing every hour was counted. Here x may be taken as an ancillary variable. If all x_is are known then the sample may be drawn with probability proportional to x. Then the unbiased estimator of Y is given by $\sum_{i=1}^{n}(y_i / p_i)/n$ where $p_i=x_i/X$. Here auxiliary character (x) is used in selection stage as well as estimation stage. But if a random sample is drawn then the usual unbiased estimator of Y is $N\overline{y}$ and the ratio estimator of Y is $Y_R = \overline{y}/\overline{x}$ X. But in this case information on x is used only in estimation stage but not in selection stage.

Another problem is to estimate N the number of fish in a population i.e. in a pond. At first a sample of size m fish is taken from the population and they are marked or tagged and returned to the population. A second sample of size c fish is secured from the population and the number of marked fish counted are designated r. Then the estimated value of N is (mc/r).

This estimator is based on the assumption that the proportion of marked fish in a random sample is the same as the proportion of a known number of marked fish in the population. This method is referred to as the Petersen method. It is noticed that the recapture sample *i.e.* the second sample of size c is drawn without replacement. The variance of the estimated population number is estimated by $\hat{N}^2(\hat{N}-m)(\hat{N}-c)/mc\ (\hat{N}-1)$.

The chapter with illustrated examples will help researchers to adopt right type of sampling methodology in fisheries survey.

REFERENCES

Cochran, W.G. (1977). Sampling Techniques. John Wiley and Sons, New York.

Mahalanobis, P.C. (1946). Recent experiments in statistical sampling in the Indian statistical Institute. *Jour. Roy. Stat. Soc.*, A, 109, 325-378.

Mukhopadhyay, Parimal (2000). Theory and Methods of Survey Sampling. Prentice- Hall of India Pvt. Ltd., New Delhi.

Murthy, M. N. (1967). Sampling Theory and Methods. Statistical Publishing Society, Calcutta.

Olkin, I., 1958. Multivariate ratio estimation for finite populations. *Biometrika*, 45:154-165.

Raj, Des (1965): On a method of using multi-auxiliary information in sample surveys. *JASA*, 60:270-277

Raj, Des, 1968. Sampling Theory. McGraw Hill Book Co., New York

Sukhatme, P. V., Sukhatme, B. V., Sukhatme, S. and Ashok, C. (1984). Sampling theory of Survey with applications. IOWA State University Press, U.S.A

Appendix Contd...

APPENDIX

Random Numbers Table

39 65 36 63 70	77 45 85 50 51	74 13 39 35 22	30 53 36 02 95	49 34 88 73 61
73 71 98 16 04	29 18 94 51 23	76 51 94 84 86	79 93 96 38 63	08 58 25 58 94
72 22 56 28 11	72 65 71 08 86	79 57 95 13 91	97 48 72 66 48	09 71 17 24 89
70 17 25 99 76	89 37 20 70 01	77 31 61 95 46	26 97 05 73 51	53 33 18 72 87
37 48 60 82 29	81 30 15 39 14	48 38 75 93 20	06 87 37 78 48	45 56 00 84 47
68 08 02 80 72	83 71 46 30 49	89 17 95 88 29	02 39 56 03 46	97 74 06 56 17
14 23 98 71 47	70 52 85 01 50	01 84 02 78 43	10 62 98 19 41	18 83 99 47 99
40 08 96 21 44	25 27 99 41 28	07 41 08 34 66	19 42 74 39 91	41 96 53 78 72
78 37 06 08 54	63 61 62 42 29	39 68 95 10 96	02 24 23 00 62	56 12 80 73 16
37 21 34 17 68	68 96 83 23 56	32 84 60 15 31	44 73 67 34 67	91 15 79 74 58
14 29 09 34 04	87 83 07 55 07	76 58 30 83 64	87 29 25 58 84	86 50 60 00 25
58 43 28 06 36	49 52 83 51 14	47 56 91 29 34	05 87 31 06 95	12 45 57 09 09
10 43 67 29 70	80 62 80 03 42	10 80 21 38 84	90 66 35 03 09	43 12 74 49 14
44 38 88 39 54	86 97 37 44 22	00 95 01 31 76	17 16 29 56 63	38 78 94 49 84
90 69 59 19 51	85 39 52 85 13	07 20 37 07 61	11 16 36 27 03	78 86 12 04 95
41 47 10 25 62	97 05 31 03 61	20 26 36 31 62	68 69 86 96 44	84 05 48 46 45
91 94 14 63 19	75 89 11 47 11	31 51 34 19 09	79 57 92 37 59	14 93 87 81 40
80 06 54 13 66	09 19 94 06 19	98 40 07 17 71	22 45 44 84 11	24 62 20 42 31
67 72 77 63 48	84 08 31 55 38	24 33 45 77 08	80 45 67 93 82	75 70 16 08 24
59 40 24 13 27	79 26 88 86 30	01 31 60 10 39	53 58 47 70 93	86 81 56 59 38
05 90 35 89 95	01 61 16 96 44	50 78 13 68 33	37 68 53 37 31	71 26 35 03 71
44 43 80 69 89	46 62 05 14 82	90 71 50 05 62	77 79 13 57 44	59 60 10 39 66
61 81 31 96 82	00 57 25 60 59	40 72 60 18 77	55 66 62 65 11	08 99 56 64 57
42 88 07 10 05	24 98 65 63 21	47 21 61 88 32	27 80 30 21 60	10 92 35 36 12
37 94 30 05 39	28 10 99 00 27	12 73 73 99 12	49 99 57 96 82	96 88 57 17 91
68 83 19 76 16	94 11 68 84 26	23 54 20 86 85	23 86 66 99 07	36 37 34 92 09
87 16 59 61 81	43 63 60 61 61	65 76 36 95 90	18 48 27 45 68	27 33 65 30 72
91 43 05 96 47	55 78 99 95 24	37 54 85 78 78	01 48 41 19 10	35 19 54 07 73
84 97 77 72 73	09 62 06 65 72	87 12 49 03 60	41 15 20 76 27	50 47 02 29 16
87 41 60 76 83	44 88 96 07 80	83 03 88 38 96	73 70 66 81 90	30 56 15 48 59

Chapter **28**

Biometrical Techniques in Aquaculture Experiments

Ritu R. Saxena & Ravi R. Saxena

The need for better analytical tools to analyze the data in aquaculture leads to major developments in statistics. These tools play a significant role in the analysis and interpretation of data. Improvement in carp culture through genetic techniques is the only alternative for increasing the carp production and its quality when area of pond is fixed. For better hybrids or efficient strains for higher production it is essential to collect the data on the genetic diversity among the different carp species on the basis of traits. Therefore, sound statistical procedures are needed in the design of experiments as well as in the analysis of the data. This involves, the use of variability studies and the use of multivariate statistical techniques like correlations, path coefficients, D^2 analysis and other classificatory analysis.

Methods for Assessment of Variability

Range

It is the difference between maximum and minimum value. The main feature is to assess the crude measure of variability in breeding populations.

Standard deviation

It is the square root of the arithmetic mean of the squares of the deviation measured from the mean. In other words, it is the square root of the variance.

Applied Bioinformatics, Statistics and Economics in Fisheries Research: pp. 421-429

New India Publishing Agency, New Delhi (India)

It is less affected by fluctuation of sampling and its value is always definite. It is based on all the observations of a sample.

$$\text{S.D.} = \text{Sqrt} \left(\Sigma X^2 - \{(\Sigma X)^2 / N\} / N - 1\right)$$

$$\text{or S.D.} = (\text{variance})^{1/2}$$

Variance

It is defined as the average of the squared deviation from the mean or it is the square of the standard deviation. It is an effective measure of variability, which permits partition of variation into various components.

$$\text{Variance} = \{\Sigma X^2 - ((\Sigma X)^2 / N) / N - 1\}$$

Coefficient of variation (CV)

The ratio of standard deviation of a sample to its mean expressed in percentage is called as coefficient of variation. It is used for the study of relative variation. Simple measures of variability can be worked out from both un-replicated and replicated data. In breeding, PCV, GCV, ECV are estimated from the corresponding variances. The outcome can be interpreted in such a way that if a sample in which CV is higher would have greater variation than the one, which has lower. In other words, when CV is high the sample is less consistent or more variable and when the sample is low, it is more consistent or less variable.

$$\text{C.V. (\%)} = (\text{S.D.} / \bar{x}) * 100$$

where, C.V. = Coefficient of variation

S.D. = Standard deviation

$\bar{x}$ = Mean

Standard error

It is the measure of the mean difference between sample estimate of mean (X bar) and the population parameter (μ). It is estimated by dividing the estimates of standard deviation by the square root of number of observations in the sample, and is denoted by SE.

$$\text{S.E.} = \text{S.D.} / (N)^{1/2}$$

where, S.D. = Standard deviation

N = Number of observations

Test of Significance

The statistical procedures, which are used to decide whether the differences under study are significant or non-significant, are known as tests of significance. For practical purposes, sample may be regarded as large if it is more than 30 and small if it is equal or less than 30. In breeding examples, significance of Z test or t test between the selected sample and base population indicates effectiveness of selection.

For comparison of two means following tests are suitable:

1. *Z-test:* sample size is large
2. *t-test:* sample size is small

The calculated value of Z or t is compared with their table value at appropriate degrees of freedom. If the calculated value is grater than table value, it is considered as significant and vice versa. In breeding, significance Z or t test between the selected sample and base population indicates effectiveness of selection.

Comparison of several means

The significance of difference among several means is tested by F test. It requires F value, which is based on analysis of variance.

The calculated value of F is compared with table value of F at desired level of significance and appropriate degrees of freedom. If calculated value is greater than the table value its is considered as significant and vice versa.

Analysis of Variance

The statistical procedure, which separates or splits the total variation into different components, is known as analysis of variance. It is useful in estimating the components of variance. It provide base for test of significance and helps in partitioning of total variation into two main parts viz., variation between species and variation within species. It permits estimation of three types of variances

i. Genotypic variance: $(\sigma^2 g) = (MSt - MSe) / r$
ii. Phenotypic variance: $(\sigma^2 p) = \sigma^2 g + \sigma^2 e$
iii. Environmental variance: $(\sigma^2 e) = MSe$

These different types of variances estimate three different coefficients of variability:

1. PCV (%) = $(\sigma^2 p / \bar{x}) * 100$
2. GCV (%) = $(\sigma^2 g / \bar{x}) * 100$
3. ECV (%) = $(\sigma^2 e / \bar{x}) * 100$

where, $\sigma^2 p$ = Phenotypic variance

$\sigma^2 g$ = Genotypic variance

$\sigma^2 e$ = Environmental variance

$\bar{x}$ = mean

MSt = Mean sum square due to treatment

MSe = EMS = Error mean sum of square

r = Number of replication

If the value of genotypic coefficient of variation (GCV) is higher than PCV, it indicates that there is little influence of environment on the expression of character. Selection for improvement of such character will be rewarding. If the value of PCV is higher than the GCV, it means that the apparent variation is not only due to genotypes but also due to the influence of environment. Selection of such traits sometimes may be misleading. If the value of ECV is higher than the PCV and GCV, it indicates that environment is playing a significant role in the expression if such character. Selection for the improvement of such character will be ineffective.

Heritability and Genetic Advance

Heritability and genetic advance are important selection parameters. Heritability estimates along with genetic advance are normally more helpful in predicting the gain under selection than heritability estimates alone. The ratio of genotypic variance to the phenotypic variance or total variance is known as heritability. It is expressed in percent. Thus, heritability is the heritable portion of phenotypic variance. It is a good index of the transmission of characters from parents to the offspring. Heritability is of two types:

Broad sense heritability (h^2_{bs})

It is the ratio of genotypic variance to the phenotypic variance or total variance. It is calculated from total genetic variance, which consists of additive, dominance, and epistatic variance.

$$h^2_{bs} = (Vg / Vp) * 100$$

where, Vp and Vg = Phenotypic and genotypic variance respectively.

Narrow sense heritability (h^2_{ns})

It is the ratio of additive or fixable genetic variance to the total or phenotypic variance. It is calculated from the following formula:

$$\text{Heritability} = \tfrac{1}{2}\, D / Vp$$

where, D = additive genetic variance, Vp = phenotypic variance

Genetic advance as percent of mean

Improvement in the mean genotypic value of selected species over the parental population is known as genetic advance. It is the measure of genetic gain under selection. The success of genetic advance depends upon three main factors:

- Genetic variability
- Heritability
- Selection intensity

$$GS = (k)\ (Vp)\ \tfrac{1}{2}\ (Vg/Vp)$$

or

$$GS = (k)\ (h^2) * SD\ (P)$$

where, GS = Genetic advance under selection

k = selection differential (2.06)

S.D. (P) = Phenotypic standard deviation

h^2 = heritability

*Genetic advance as percent of mean = {GS/ mean} * 100*

High heritability accompanied with high genetic advance indicates that most likely the heritability is due to additive gene effects and selection may be effective. High heritability accompanied with low genetic advance is indicative of non- additive gene action. The high heritability is being exhibited due to favorable influence of environment rather than genotype and selection for such traits may not be rewarding. Low heritability accompanied with high genetic advance it reveals that the character is governed by additive gene effects. The low heritability is being exhibited due to environmental effects. Selection may be effective in such cases. Low heritability with low genetic advance it indicates that the character is highly influenced by environmental effects and selection would-be ineffective.

Genetic Divergence

Genetic divergence plays a significant role because hybrids between lines of diverse origin generally display a greater heterosis than those between closely related parents. Genetic diversity arises due to geographical separation or due to genetic barriers to cross ability.

The number of clusters represents the number of groups in which a population can be classified on the basis of D^2 analysis. The distance between two clusters is the measure of the degree diversification. The greater the distance between two clusters the greater the divergence and vice versa. The genotype falling in the same clusters are more closely related than those belonging to another cluster. In another words, the genotypes grouped together in one cluster are less divergent than those, which are placed in a different cluster. It provides information about relationship between two clusters.

Methods for Indirect Selection

Correlation coefficient

It refers to the degree and direction of association between two or more than two variables. It is represented by r. A positive value of r shows that the changes of two variables are in the same direction i.e. high value of one variable are associated with high values of other and vice versa. When r is negative the movements are in opposite direction i.e. high value of one variable is associated with low values of other.

$$r_{xy} = \text{Cov}(X, Y) / \{V(X) * V(Y)\}^{1/2}$$

The main features of correlation coefficient are:

- The significance of correlation is tested with the help of table value at n-2 degree of freedom.
- It's independent of the unit of measurement.
- It's value lies between-1 and +1
- They are of three types, viz., simple or total, partial and multiple correlations.

The association between any two variables is termed as simple correlation or total correlation or zero order correlation coefficient. It is of three types:

(i) *Phenotypic correlation:* The association between two variables, which can be directly observed, is termed as phenotypic correlation. It includes both genotypic and environmental effects, and therefore, differs under different environmental conditions.

(ii) *Genotypic correlation:* The inherent or heritable association between two variables is known as genotypic or genetic correlation. This type of correlation may be either due to pleiotropic action of genes or due to linkage or more likely both. If the association between two traits (whether positive or negative) remains the same in the parental populations as well as segregating populations, it means the association is due to pleiotropy. This type of correlation is more stable and is of paramount importance for a breeder to bring about genetic improvement in one character by selecting the other character of a pair that is genetically correlated.

(iii) *Environmental correlation:* This type of correlation is entirely due to environmental effects. In other words, this is due to error variance. This is of less importance, as it is not heritable and stable.

All the three types of correlations viz., simple, partial and multiple correlation coefficients can be estimated from both unreplicated and replicated data. But phenotypic, genotypic and environmental correlations can be estimated from replicated data only.

From un-replicated data:

$$r = [\Sigma XY - (\Sigma X * \Sigma Y) / N] / (\Sigma X^2 - (\Sigma X)^2 / N * \Sigma Y^2 - (\Sigma Y)^2 / N)$$

where, N = Number of observations

X&Y = Two variables

From replicated data:

- ANOVA for all the characters included in the study.
- ANCOVA for all the character combinations
- Using respective variances and co variances three types of correlations are worked out as-

$$r_p = P\,Cov\,(X,Y) / \{PV\,(X) * P\,V\,(Y)\}^{1/2}$$
$$r_g = G\,Cov\,(X,Y) / \{GV\,(X) * G\,V(Y)\}^{1/2}$$
$$r_e = E\,Cov\,(X,Y) / \{EV\,(X) * E\,V(Y)\}^{1/2}$$

where, rp = Phenotypic correlation coefficient

rg = Genotypic correlation coefficient

re = Environmental correlation coefficient

Correlation may interpreted as if the value of r is significant, the association between two characters is high. If the value of r bears negative (-) sign it means that increase in one character will lead to decrease in second and vice versa. If the value of genotypic correlation coefficient (rg) is higher than the phenotypic correlation coefficient (rp), it means that there is strong association between these two characters genetically, but the phenotypic value is lessened by the significant interaction of environment.

Path Analysis

Path analysis is simply a standardized partial regression coefficient, which splits the correlation coefficient into the measures of direct and indirect effects. It measures the direct and indirect contribution of independent variables on dependent variable. Wright first developed the concept in 1921. Path analysis is of three types:

- Phenotypic path
- Genotypic path
- Environmental path

The phenotypic path splits the phenotypic correlation coefficient into the measures of direct and indirect effects, while genotypic path and environmental path is estimated from the genotypic and environmental correlation coefficients, respectively. Simple path can be worked out both from replicated and unreplicated data, but phenotypic, genotypic and environmental path can be estimated from replicated data only. The estimation of path analysis is as follows :

- Estimation of variances and covariances for all the characters and their combinations, respectively.
- Calculation of all possible simple correlation coefficients among independent and dependent variables, which is equal to n (n-1)/ 2, where n is the number of variables.
- Then path analysis is carried out. It consists of three steps, viz., calculation of (a) direct effects, (b) indirect effects and (c) residual effects.

Before estimation of direct effects a line diagram known as path diagram is constructed. It is constructed with the help of simple correlation coefficients among various characters under study. The dependent variable is kept at one side and all independent variables on the other side.

Results can be interpreted as follows if the correlation between yield and character is due to direct effect of a character, it reveals true relationship between them and directs selection for this trait will be rewarding for yield improvement. If the correlation is mainly due to indirect effects of character through another component trait, indirect selection through such trait will be effective in yield improvement. If the direct effect is positive and high, but the correlation is negative, in such situation direct selection for such trait should be practiced to reduce the undesirable indirect effect. If the value of residual effect is moderate or high, it indicates that besides the character there are some other attributes, which contributes for yield.

Methods for the Choice of Parents and Breeding Procedures

Hybridization is the most important technique for breaking barriers and evolving varieties having built-in high potential. The selection of suitable parents for hybridization is one of the most important steps in a breeding programme. Selection of parents on the basis of phenotypic performance alone is not a sound procedure since phenotypically superior lines may yield poor recombinants in the segregating generations. It is therefore, essential that parents should be chosen on the basis of their genetic value. There are several biometrical techniques for the evaluation of varieties or strains in terms of their genetic make up, viz., diallel, partial diallel, line x tester etc.

Methods of Assessment for Varietal Adaptability

Varietal adaptability to environmental fluctuations is important for the stabilization of production over years and regions. Stability analysis is useful in the identification of adaptable genotypes over changing environments. There are different types of stability analysis models for this purpose, such as Finley and Wilkinson model, Eberhart and Russell model, Perkins and Jinks model and Freeman and Perkins model.

REFERENCES

Dudley, J. W. and Moll, R. H. (1969). Interpretation and use of estimates of heritability and genetic variances in plant breeding. *Crop Sci.* 9 :257-262

Falconer, D.S. (1981). Introduction to quantitative genetics, Longman Group Ltd. London.

Kempthorne, O. (1957). An introduction to genetic statistics. John Wiley ans Sons, New York.

Singh R.K. and Choudhary B.D. (1977). Biometrical Methods in Quantitative Genetic Analysis. Kalyani Publishers, New Delhi.

Chapter **29**

Statistical Programs in Quantitative Genetics and Selective Breeding Studies of Fish

Kanta Das Mahapatra

Introduction

India possesses rich fish germplasm resources, accounting for about one-tenth of the 20,000 and odd species of fish known in the world. The Ganga network of rivers in the North, the Brahmaputra and the Mahanadi in the East, the Sutlej, the Narmada & the Tapti in the West and, the Godavari, the Krishna and the Cauvery in the South are very rich sources, harbouring bulk of the important fish fauna. (Jhingran, 1982).

The major carps of India which are also referred to as Gangetic carps are the most important among the commercial fish species and form the major component of the aquaculture system in the country (Jhingran, 1982). Since the last two to two and a half decades, there was a steady progress in carp farming in India due to the availability of seed of the major carps and the development of different packages of culture technologies (Chaudhuri *et al.* 1975).

Apart from the already existing different aquaculture practices in the country such as extensive, intensive and semi-intensive farming systems, the availability of huge quantities of organic wastes from plants and animals gives scope for integrated fish farming.

Applied Bioinformatics, Statistics and Economics in Fisheries Research: pp. 431-446

New India Publishing Agency, New Delhi (India)

Exploitation of genetic potentials of these carps has been initiated seriously from early 1980s particularly through genome manipulations and later through selective breeding from 1990s, to add to the already developed culture technologies, for further enhancement of fish production in the country.

Selective breeding in increasing production level is well established in agriculture and animal husbandry. Today the high yielding crops and land animals are totally depending on genetically improved domesticated breeds. This has not been true for aquaculture. Proper exploitation and utilization of genetic potential is lacking in aquaculture. Less than 5% of the total output of the aquaculture production is coming from improved breeding progamme. Aquaculture species are thus genetically much closer to their wild counter part than the land animals and plant species (Gjedrem, 2005). During the last few years it has been well documented that high selection response can be obtained in fish as well as in shellfish for economic important traits like growth, disease resistance, flesh quality etc.

Carp occupies very important position in freshwater aquaculture. 95% of world production of carps is coming from Asia only. Annual growth rate of carps is 11% and most of the low-income groups consume the carps. Very little genetic improvement work has been carried out in carps. Targets for genetic improvement of carps are complex because of the diversity of species, farming system, socioeconomic scenario and longer generation period. However, various commercially important traits can be improved through selective breeding in carps.

Taking all these into considerations, a project on genetic improvement of Rohu, particularly for better growth performance through selective breeding has been initiated at the Central Institute of Freshwater Aquaculture (CIFA), Bhubaneswar, during the year, 1992 in collaboration with the Institute of Aquaculture Research (AKVAFORSK), Norway. The main aim of the project was to develop a national fish-breeding plan and to provide the right seed producing procedures to the hatchery managers in the country, leading to quality seed production. Different statistical program can be used to analyze data in an effective manner. Some of the programs are discussed here which can be applied in quantitative genetics study to determine different parameters used in the selective breeding programme.

Selective Breeding

The potentials of genetics have already exhibited promising trends in aquaculture too. The correct breeding procedures followed through selective breeding in the case of Atlantic salmon and rainbow trout in Norway, Channel catfish in USA and Nile tilapia in Philippines are the standing examples in this regard (Gjedrem, 2005). The recent selective breeding programme of rohu (*Labeo rohita*) in India too is another example of the kind (Reddy *et al.*, 2002).

Before success of induced breeding through hypophysation in late fifties rivers were the main source for seed collection of Indian major Carps. With the introduction of induced breeding technique, hatcheries have been able to produce enough quantity of carp seed. The hatcheries in India hardly follow any genetic norms to produce carp seed. A limited number of brood fishes are used repeatedly for successive generations. As a result, the quality of carp seed is showing negative effect of inbreeding i.e. slow growth rate, disease proneness etc.

Inbreeding

Inbreeing in an infinitely large population is defined as the mating of individuals that are more closely related to each other than individuals mating at random within a population (Gjedrem, 2005). Inbreeding occurs due to mating of closely related individuals. Genetically inbreeding leads to homozygosity. Almost all individuals carry deleterious recessive genes, which are hidden in heterozygous state. Related individuals are likely to share common genes and probability of pairing of deleterious recessive genes gets enhanced with the increase of closeness between the parents. Due to which, inbreeding depression occurs to the population with decreasing growth efficiency, disease resistance, survival etc. Inbreeding depression is measured as the average performance differences between an inbred population and the base population.

Effective Population Size

Effective population size is one of the most important concepts in the management of a population in that it gives an indication about the genetic stability of the population. It depends upon several factors such as total number of breeding individuals, sex ratio, mating system and variance of family size. The effective population size can be calculated by following formula,

$$N_e = 4N_f X N_m / N_f + N_m$$

Where N_e = Effective population size

N_f = Number of female brood fishes used for seed production

N_m = Number of male brood fishes used for seed production

Effective population size is inversely related to inbreeding

$$\Delta F = 1/2N_e$$

$$\text{So} \quad \Delta F = 1/8N_f + 1/8N_m$$

Where ΔF= Rate of inbreeding per generation

N_e= Effective population size

N_f= Number of female brood fishes used for seed production

N_m= Number of male brood fishes used for seed production

Therefore, to improve the genetic status of any population, hatchery managers should decrease inbreeding rate and increase effective population size.

Crossbreeding

Cross breeding is a well-known method of genetic improvement, which also has application in aquaculture. Crossbreeding is mating between species, breeds, populations, strains or inbred lines. The main objective in crossbreeding is to exploit non-additive genetic variance (heterosis or hybrid vigor).

Selective breeding plays vital role to improve genetic status of fish in a positive direction, which has already been demonstrated in case of Salmon at Norway, Tilapia at Philippines and Rohu at CIFA, India. The aim of selection is to identify and select parents for the next generation the individuals whose progeny, as a group, have the highest possible additive genetic merit for the trait or traits in question.

Natural Selection

Fitness of an individual can be determined by its contribution to the next generation. This is also know as adaptive value or selective value. Natural selection select the fitness as the charcter. This fitness also changes with change of environment. The animal with highest fitness in the changed environment, will produce at a higher level and have a higher survival rate than less fit animal. The effect of natural selection over generation is to establish a population adopted to the new environmental condition. Thus adoptability is a response of a population rather than of individual. If the change in the environment condition represent large changes and take place rapidly then a population may be lost or destroyed due to non adaptiveness of the population to the changed environment and new condition.

Artificial Selection

Artificial selection was created by man to improve genetic status of a population.

Objectives of a selective breeding programme is to change the average performance of the targeted trait i.e. growth rate, disease resistance, better flesh quality, feed conversion efficiency etc. of the population in a favorable direction. Artificial selectiob can be performed in several ways .

1. Stabilizing selection, where selected phenotype are around the mean and both the extremes are discarded. Example: fat content of a fish where high and low fat content are less desirable.
2. In the diversifying selection both the extremes are selected through which subpopulation could emerge.
3. Directional selection is the most common type of selection form used in aquaculture sector. Selected individuals are taken from positive direction with high ranking animals to produce better offspring every generation (Gjedrem, 2005)

Estimation of Phenotypic and Genetic Parameters

The properties of a population that can be observed for a metric trait are based and estimated from the phenoptypic measurement on individuals of a population measured in metric units i.e. g, kg, cm, day etc. The measurement in metric unit on an individualis quantitative value is called as phenotypic value. Estimation of phenotypic and genetic parameters are essential in the design of genetic improvement programs. They include phenotypic and genetic variances and covariances and their linear functions such as heritability and phenotypic and genetic correlations. They enable the prediction of genetic gain and of correlated responses in genetic improvement programs. Selective breeding procedure is based on principle of quantitative genetics. Which indicated that phenotype of an individual, which can be measured or scored, could be partitioned into two components. One attributable to the influence of genotype i.e. the particular assemblage of genes possessed by the individual and other one attributes to the influence of environment i.e. all non-genetic components.

$$\text{So,}\quad P = G + E$$

Where P = Phenotype of an individual

G = Genotype

E = Environmental (Non-genetic) component

Quantitative phenotype exhibit continuous variation, the only way to study them is to analyze the variance that exists in a population. The phenotypic variance (V_P) that is observed for a quantitative trait is the sum of the genetic variance (V_G) and (V_E) and the interaction that exists between the genetic and environmental variance (V_{G+E}).

$$\text{So, } V_P = V_G + V_E + V_{G+E}$$

Genetic variance is the component of interest in selective breeding program. V_G is further subdivided in to three component i.e. additive genetic variance (V_A), dominance genetic variance (V_D) and the epistatic genetic variance (V_I).

$$V_G = V_A + V_D + V_I$$

V_A, V_D and V_I differ from each other according to their mode of inheritance. Dominance genetic variance is the variance that is due to the interaction of the alleles at each locus. Because of this, V_D cannot be inherited, it is created a new in each generation. Since it is the interaction between alleles at each locus so V_D is a function of the diploid state as alleles occur in pairs. During meiosis, homologous chromosomes and allelic pairs are separated during reduction division and the chromosome complement reduced to half. Therefore, V_D is not transmitted to next generation.

Epistatic genetic variance (V_I) is due to interaction of alleles between two or more loci. Epistatic interaction occurs across loci so it is also not transmitted from parents to offspring. This is also created a new in each generation.

V_A or additive genetic variance is the additive effect of genes. It is the sum of the effects of each allele that is responsible for phenotype. It does not depend on specific interaction or combination of alleles so it is not disrupted due to meiosis so additive genetic variance transmitted from parents to offspring. It is transmitted in a reliable and predictable manner. V_A is also called the variance of breeding value.

Heritability: Heritability describes genetic component that is not disrupted by meiosis. The proportion amount of phenotypic variance (V_P) that is controlled by V_A is called heritability (h^2). It is important to know the size of the heritability when planning a breeding program as well as when predicting the response to selection or individual breeding values. The heritabilty can vary between 0 and 1. Heritability in a broader sense can be defined as the ratio of the total genetic variance σ^2 G to the total phenotypic variance σ^2 P. However since σ^2G also contains dominance and epistatic effects which do not respond to selection, heritability in the broader sense is not often used. Heritability in narrow sense can be defined as the ratio of the additive genetic variance σ^2 A and the phenotypic variance σ^2 P.

$$h^2 = \sigma^2 A / \sigma^2 P$$

Once you know the heritability, response of selection can be predicted as

$$R = S * h^2$$

Where R = response to selection,

S = selection differential (difference between offspring and parent generation)

h^2 = heritability of the trait

Heritability is not a general and static characteristic of a breeding population. It is only relevant to the data used for estimation or the population from which the data was sampled. When environmental conditions are changed and or if a population responds to selection, the heritability may change.

Selection Methods

Several selection methods are available for obtaining additive genetic variance. The methods differ with respect to which type of relatives that provide information used for the selection decisions. The objective of all the methods is to maximize the probability of correct ranking of animals with respect to their breeding value, an estimate of each individual ability for producing high / low performing offspring. The breeding value of an individual cannot be estimated on basis of the phenotypic value of the trait (s). In fish such records are usually obtained from the individual itself (Individual selection), full and half sibs (family selection) or from all three sources of information (combined selection).

Breeding Objectives and Selection Indices

Selection indices have two main uses such as it combines information from different relatives for population trait and it combines the genetic merit for different traits in an aggregate breeding value. The information regarding different traits may vary widely. Some coming from an animal relative and some from the animals own performance for traits which, are expressed once or repeatedly during its life. These factors make selection a complicated and uncertain procedure. In addition, fluctuating, vague and sometimes erroneous ideals often cause the improvement resulting from selection to be much less than could achieve if these obstacles were not overcome.

Steps in Designing Breeding Programme

Following steps may be considered in designing breeding program

- Production system
- Breeding system and breeds
- Formulation of objectives of the system and calculation of economic values (weight)
- Development of selection criteria and estimation of selection parameters
- Design of animal evaluation system
- Design mating for selected animals
- Design system for expansion
- Monitoring and comparing alternative programs.

The most important decision in any animal improvement program is the choice of breeding objective. A badly chosen objective will lead the genetic progress in the wrong direction. There may be conflicts to the fragmented nature of production chain. Interest of the breeders, multiplier and producers may differ. The breeding objectives or the aggregate genotype is about where to go. But the selection index is about how to get there. The selection index is a linear function of recorded characters used as a basis for selection. It is derived so that the correlation between the index and the aggregate genotype are maximized. Such a linear function includes characters of the animals and its relatives to predict and improve the aggregate genotype. The overall economic merit can be thought of as a breeding value of a new trait.

While developing breeding objectives, the starting point should be the determination of an equation for economic benefit derived from a population. Genetic improvement typically takes place in a very small fraction of the population, which is multiplied and disseminated, to the production systems. Breeding objectives should be defined for the production system. Instead of using specific relatives, parameters may be estimated from the complete pedigree.

Trait of selection can be of four types i.e. Production traits, Morphometric traits, Reproductive traits and Quality traits.

Among all traits, growth rate occupies the top in the list of preference.

Role of population size in selection programs: Some of the considerations we often make about genetic improvement assume we are working with populations of infinite size but in reality, we are always working with population of finite size. If the population size is of finite then rate selection response declines, the maximum response is reduced and the selection limit is reached faster. The inbreeding coefficient increases and that may have undesirable effects. Rate of inbreeding can be calculated from effective population size.

$$F=1/2\ Ne \text{ after 't' generations } F_t = 1-(1-(1/2Ne)\ t$$

Response per generation $R= I\ h\ 6_A$

Under inbreeding σ_A is reduced by

$$\sigma_{A-F} = \sigma_A\ (1\text{-}F)^{\ 0.5}$$

Aim of any selection is to increase frequency of favorable alleles. Chance of missing out in favorable alleles increases with reduced Ne. When all are fixed, we are at limit. It is accepted that rapid inbreeding is more harmful than gradual mild inbreeding. Incidence of genetic defect increases. For a rate of inbreeding of ~ 1% per generation 50 brood fish pairs or more are required.

For analyzing all the data for the selective breeding program different statistical programs are used. The accuracy of any breeding program depends on the accuracy of analysis of genetic merit of the individuals. Some of the statistical programs are described in the present paper.

Selective Breeding of Rohu Program

For the first time in India, selective breeding was carried out to genetically improve one of the most preferred carp species (Rohu). Under the program a range of selective breeding techniques (production of fullsib groups, individual identification by PIT tags, communal pond rearing, estimation of breeding value and ranking of individuals in different year classes) were used. In five generations of selection for greater harvest weight an average 17 % genetic gain per generation was obtained. On farm trials have proven the strain's superiority and it is being disseminated to farmers.

Base population and mating procedure. Rohu fry or fingerlings were obtained from five rivers of India, the Ganga, the Yamuna, the Brahmaputra, the Gomati and the Sutlej and raised at CIFA for 2 to 3 years until they were sexually mature (Reddy *et al*, 2002). CIFA farm (local) rohu was also added to the base population. The fry or fingerlings collected were under quarantine for about two weeks in cement cisterns. After quarantine, fingerlings were marked by fin clipping, M-procaine blue dye marking or by combinations of both. They were stocked in communal ponds for rearing until they attain sexual maturity.

The technique of breeding followed for ovulation of females and sperm collection (milting) from males. A synthetic hormone (Ovaprim) was used to obtain the gametes. To date 13 year classes of fullsib and half sib families have been produced with six complete generations A nested mating design (two males nested within a female or vice versa) was used. Approximately 60 full sib families were produced during each year class.

In order to estimate the magnitude of heterosis for harvest body weight two 3 by 3 diallel crosses were made as described in Table 1. The local stock was common to both diallel crosses (Gjerde *et al,* 2002).

Table 1. Rohu stock included for two diallele crosses

Male parent	Female parent				
	Ganga	**Yamuna**	**Local**	**Brahmaputra**	**Sutlej**
Ganga	X	X	X		
Yamuna	X	X	X		
Local	X	X	X	X	X
Brahmaputra			X	X	X
Sutlej			X	X	X

Tagging: Passive Integrated Transponder (PIT) tags were used to individually identify the fingerlings from different families. 10 to 15g fingerlings were found quite suitable for tagging (Das Mahapatra *et al.,* 2001).

Test environment for rearing: Tagged fish were stocked in three communal ponds of 0.1ha each under mono-culture and two 0.4 ha ponds under poly-culture practice. In poly-culture, rohu was stocked along with Catla and Mrigal in the ratio of 1.2 : 1 : 1, respectively. This practice was continued until 1997. After 1997, only mono-culture was practiced, as a high correlation between the performance in mono and poly-culture was observed.

Statistical analysis: Editing of the data and basic stastistical analysis were performed using SAS (SAS Institute Inc., 1990). For each year class full sib, half sib and individual data were considered for breeding value estimation. Individuals with average breeding value were chosen as controls and in every generation they were compared with the selected (greatest breeding value) individuals.

Field testing: Field trials were arranged in several parts of India, namely 1.Rahara, West Bengal/ 2.Ludhiana, Punjab/ 3. Vijayawada, Andhra Pradesh and 4. State Fisheries Department, Orissa. In 2005 and 2006 demonstrations were also arranged at farmers' ponds in different states of India. A comparison of improved rohu was made with local rohu and with the control group from the research center.

Dissemination of improved rohu: Initially dissemination of improved rohu was carried out in three states (Andhra Pradesh, Orissa and West Bengal). CIFA, the research center, is where the nucleus resides, and provides improved rohu seed to the multiplier units. Multiplier units are raising the brood fishes, producing seed and distributing the seed to the fish farmers. (Gjerde *et. al,* 2005)

Characterization: Genetic characterization of the six base population stocks indicated a wide variation within each stock. The variation within stocks was greater than that between stocks (Reddy *et al.*, 2002).

Estimation of genetic parameters: Total heterosis for each of the six crosses was low or negative, whereas average heterosis was low and mostly not significantly different from zero. It was concluded that genetic improvement of rohu by crossbreeding would be futile (Gjerde *et al.*, 2002).

At tagging, the heritability (h^2) for body weight was very low whereas the environmental effect common to full sib families was high. Final sampling heritability was moderate.

The phenotypic and genetic correlations between sampling and final sampling weights were 0.92 (± 0.01) and 0.98 (±0.01), respectively, indicating that they were virtually the same trait.

Realized selection response: Each generation a control was created by mating average individuals. The progeny of these was compared with that of selected parents in the selection line. Over five gene rations the average response to selection per generation bose final body weight was 17 per cent.

Field trials: Field trials at different parts of India the improved rohu showed significantly higher growth rate in comparison to the control and local rohu. A growth rate superiority up to 75 per cent was observed by improved rohu over local rohu. In farmers' fields, the evaluation of the fifth generation of improved rohu showed 98 to 128 per cent higher growth rate than the local rohu.

Statistical Analytical System (SAS)

SAS software is a combination of a statistical package, a data base management system and a high level programming language. It can be used to produce a variety of statistical analyses.

Use of SAS in Rohu selective breeding program-case study: In the rohu selective breeding program, individual tagging was done to identify fishes of different fullsib families in the communal ponds. After fishes attained taggable size i.e. 10-15 g they were tagged with Passive Integrated Transponder (PIT) tags and stocked in communal rearing ponds for further grow out experiment. Before tagging, initial body weight of fishes was noted and during final sampling final individual body weights were noted.

Adjustment of Fixed Effects

All recorded data need to be carefully edited to eliminate errors during recording. Special attention was paid to the outliers (too small or too large value for trait or an ID) and individual having same ID.

As selection index procedure is used to calculate breeding value, the data need to be preadjusted for fixed effects i.e. pond. Multiplicative adjustment is applied in the rohu-breeding program.

Estimation of Breeding Value

After adjustment of pond effect, corrected body weights are being considered for the estimation of breeding values. It is estimated using following formula (Reddy *et al*, 1999).

$$\hat{I} = b_1 (P_i - P_{fs}) + b_2 (P_{fs} - P_{hs}) + b_3 (P_{hs} - P_{pop})$$

Where

$\hat{I}$ = Predicted breeding value

P_i = Adjusted phenotypic record on the individual

P_{fs} = Adjusted phenotypic mean of n-full sibs, P_i included in P_{fs}

P_{hs} = Adjusted phenotypic mean of mn halfsibs, P_i included in P_{hs} and m is the number of dams nested to each sire or vice versa

P_{pop} = Adjusted population mean

b_1 b_2 b_3 = the weights given to each source of information

Other Stastical Packages

Apart from SAS other packages can also be utilized for analysis of selective breedy data.

The Mixed Procedure

The mixed procedure fits mixed linear models (models with both fixed and random effects). A mixed model is a generalization of the standard linear model used in the GLM procedure. One can analyze the data with several sources of variation instead of just one.

Some features of MIXED model are as follows,

Covariance structures, including simple random random effects, compound symmetry, unstructured data

GLM type grammar using MODEL, RANDOM and REPEATED statements for model specification and CONTRAST, ESTIMATE AND LSMEANS statement for inferences

Appropriate standard errors for all specified estimable linear combinations of fixed and random effects and corresponding t- and *F* test.

Capacity to manage unbalanced data

Ability to create a SAS data set corresponding to any printed table.

The Inbreeding Procedure

The inbreed procedure calculates the covariance or inbreeding coefficients for the pedigree. PROC INBREED is unique in that it handles vary large populations. The inbreed procedures has two modes of operation. One mode carries out analysis on the assumption that all the individuals belong to the same generation. The other mode divides the population into non-overlapping generations and analyzes each generation separately assuming that the parents of individuals in the current generation are defined in the previous generation. Proc inbreed also compute average of the covariance or inbreeding coefficient within sex categories if the gender of individuals is known.

Strain Comparison and Heterosis Estimation

The diallel cross is an efficient design for evaluating the performance of purebred and crossbred stocks or lines. In complete diallel cross, each stock is crossed with itself and all other stocks. For each species that may be stripped artificially, the following mating procedure should be followed to obtain an equal contribution of genes from each cross and from each individual within a cross. For each stock, milt from a given number of sires is used to fertilize with an equal number of eggs from a given number of dams of the same and of each of the other stocks. Thus both crosses between the pure stocks and all possible cross combinations between stocks including reciprocal are produced.

A complete design may provide estimate of additional genetic effect of each stock, overall heterosis, general heterosis for each stock and specific heterosis for each cross. Following formula can also be used to determine above-mentioned aspects.

P_n = mean phenotype of all purebreds

$X_{n\,(n-1)}$ = mean phenotype of all crossbreds

Pj/i = mean phenotype of one purebred

= $P_n + gj/I + g^m\, j/i$ (here g^m is the maternal effect)

$h^1..$ = average heterosis for all crosses

= $\Sigma\, h_{ij} = X_{n\,(n-1)} - P_n$

h^1_{ij} = average heterosis for reciprocal cross

= $X_{ij} + X_{ji} - P_i - P_j)/2$

h^1_j = average strain heterosis

= $\Sigma h^1_{ij}/(n-1)$

g^m_j = $\Sigma\, (X_{ij} - X_{ji})\, /\, n$

g_{ij} = average individual effect of each strain

= $P_j + P_n - g^m_j$

Selection Index Program (SIP)

Selection index theory plays an important role in teaching animal breeding. It can be used to determine the expected response to selection. The department of Animal breeding and the department of informatics have worked on the development of user-friendly program to calculate selection indexes. Johan Van Arendonk developed the program. The program SIP can calculate many selection indexes. These selection indexes may be based on economic value as desired gain or on a combination of both. The program is written in Turbo vision. The program can create a number of temporary files, which can be deleted after leaving the program. The program will start in DOS promt. The important steps in operating SIP is as follows, first we have to set trait dialog box.

Then we have to set r_p and r_g boxes.

Next, we have to set group dialog box which includes additive genetic relationship and number of observation.

Then we have to set genetic relation dialog box

Common environment box is the next to be filled in.

Sources of index is marked next

Breeding goal is set next

Finally, through calculation box index can be calculated.

Genetic Parameter Estimation in Excel (GPEX): GPEX runs in Excel 97 or higher versions. Size is about 60 MB for analysis of about 5000 observations. User is not required to have advanced programming skills. To some extent, it could be a good alternative for using top of the line statistical packages in estimation of genetic parameters. It can handle unbalanced data and handle 48 full sib groups with maximum 150 progeny per full sib group. It can calculate genetic parameter like intra-class correlation, additive genetic, environmental and phenotypic variance, covariance, correlation, and heritability. It can also predict breeding values based on calculated genetic parameters from data and accuracy can be calculated.

PEST: Pest is a software package for multivariate prediction and estimation. It covers fixed, random and mixed model. Pest read raw data and translates class codes like month name integer or character identification of animal into internal representation. Identification can be up to 16 characters long. A parameter file contains commands to pest. It consists of input data description, the statistical model, output, data transformations etc. Pest is written in FORTRAN 77. It consists of around 15000 lines of source code in 250 routines. PEST provides very general and flexible strategies to solve mixed model equations. Its objective has not been to maximize memory requirements. Together with the large amount of program code PEST does require a certain minimum amount of processing capability.

ASREML: ASREML estimates variance component under a general mixed model by restricted maximum likelihood (REML). Its scope covers genetic, multivariate, repeated measures, spatial and multi-environment analyses. It uses the average information algorithm and sparse matrix technique to efficiently solve large mixed models. The user interface is basic and assumes a good understanding of the models that can be fitted; the results may need to be imported into another statistical / reporting program for further processing. ASREML enables limited testing of some fixed effect in the model. ASREML is available in complied form of MSDOS, Windows 95/ NT. We need an ASCII editor to prepare the data and parameter file before running ASREML. Base name is in the name of as command file. Output file names are generated from the input name by changing file extension from as to ars which is primary output file summaries the data, iteration sequence, the final variance parameters and solutions for fixed effects pin is an input file required for predicting means and functions of the variance components when the P option is specific pvs is the report produced with P option. In ASREML blank space in the data set can be taken care by "*" mark.

Conclusion

Determination of breeding value i.e. genetic merit of the individual is very important in selective breeding studies. Success in selective breeding program depends on the correct ranking of individuals according to its genetic merits. Different statistical packages can be utilized for this. However, SAS proved to be very effective programs for that. Other programs like ASREML is also equally effective and can be utilized for determining different parameters for selective breeding studies.

REFERENCES

Chaudhuri, H. P.R.Sen, N.G.S.Rao and S.Jena, 1975. A new high in fish production in India with record yield by composite fish culture in freshwater ponds. *Aquaculture* 6:343-355

Das Mahapatra,Kanta, Bjarne Gjerde, P.V.G.K.Reddy, M.Sahoo, R.K.Jana, J.N.Saha and M.Rye. 2001. Tagging: on the use of Passive Integrated Transponder (PIT) tag for identification of fishes. *Aquaculture Research* 32:47-50.

Das Mahapatra, K., B.Gjerde, J.N. Saha, P.G.G.K.Reddy, R.K. Jana and M. Sahoo, 2003. Realised genetic gain for growth in rohu (*Labeo rohita*). ISGA VIII, 9-15 November, Puerto Varas, Chile:116

Gjedrem, T, 2005.*In*: Selection and breeding programs in Aquaculture. Springer, Netherland publication.:89-110

Gjerde, B., P.V.G.K.Reddy, K. Das Mahapatra, R.K.Jana, J.N.Saha, P.K.Meher, M.Sahoo, S.Lenka, P.Govindswammy and M.Rye 2002.Growth and survival in two complete diallele crosses with five stocks of Rohu carp (*Labeo rohita*). *Aquaculture*,209 (1-4): 103-115.

Gjerde, B., K.Das Mahapatra, R.K.Jana, J.N.Saha and M.Rye 2005. Selective breeding improves Indian carp culture. *Global Aquaculture Advocate*, 8 (1):55-56

Jhingran, V.G, 1982. Fish and Fisheries of India Hindustan Publishing Corporation, India.

Reddy, P.V.G.K., B.Gjerde, K.D.Mahapatra, R.K.Jana ,J.N.Saha, M.Rye and P.K.Meher, 1999. Manual on Selective breeding procedures for Asian carps, CIFA and AKVAFORSK Publication.

Reddy, P.V.G.K., B.Gjerde, S.D.Tripathy, R.K.Jana, K.D.Mahapatra, S.D.Gupta, J.N.Saha, M.Sahoo, S.Lenka and P.Govindswamy, M.Rye and T. Gjerdem. 2002. Growth and survival of six stocks of rohu (*Labeo rohita*) in mono and polyculture system. *Aquaculture*, 203 (3-4) : 239-250.

Chapter **30**

Quantitative Methods in Limnology Research

Snehasish Mishra

Introduction

Information Technology (IT) is a relatively-recent gift of Science and Technology for the mankind, revolutionising even the basic perception of people. It is a boon to the ever-growing complexities of life in each sphere. Be it simple household chores to the complex and humanly impossible task of advanced space research, *Bioinformatics* is the application of Information Technology in the field of bio- or life-sciences. The advantages from this fusion of two sciences are many. Bioinformatics in particular, and information technology revolution in general have been a grace to today's biological researchers, particularly the groups involved in experiments at micro level, be it microorganisms (such as bacteria) of microbiologists or macro-molecules (of micro/nano size, such as DNA/RNA) of geneticists/molecular biologists or the nanotechnologists. For example, computer software package *IMPAC* (Interactive Modeling Package for Carbohydrates) has its applications in the modeling of 3-D structure of carbohydrates. *MYCIN* is another package popular amongst medical doctors for diagnosis of bacteriological diseases and to prescribe treatments.

Information Technology as a Boon in Limnological Research

Carrying out any sort of research work without the help of Information Technology in recent times is not only cumbersome and time-taking, but also

Applied Bioinformatics, Statistics and Economics in Fisheries Research: pp. 447-454

New India Publishing Agency, New Delhi (India)

in certain cases impossible. To cite an instance, during data recording, analyses and certain other lengthy processes, Information Technology makes the work lighter.

Limnology of current times has diversified into many fields of specialised research. For example, aquatic and environmental microbiology studies are a field of specialisation in limnology which have further diversified to aspects of biological nitrogen fixation; nitrification and biofiltration; biofertilisation; microbial processing and recycling of organic matters; biological treatment of organic wastes for use as aquaculture inputs; optimisation of aquatic bioenergetics; selective enrichment of microbial biota, pathways of bioaccumulation and biomagnification of pesticides, weedicides etc.; isolation and use of natural bioactive compounds such as probiotics; characterisation of microbial fish diseases and control, to cite a few.

Aquaculture research involves monitoring intricately intertwined complex processes of pond dynamics in managed ponds and is sensitive to tens of environmental and operational parameters governing pond productivity. Management of large amount of data for statistical analyses and also to obtain information for planning and operation of culture system could be eased through use of various tailor-made computer programmes. In developed countries (*e.g.*, Israel and Norway) where commercial aquaculture and research is cent percent technically driven, feeding, movement, location of fish shoal etc. are monitored round-the-clock by computerised sensors.

Collection of primary data, conversion of the information into convenient formats, transmission of the information and its presentation, storage and retrieval etc. make the cornerstone of Information Technology. During all these, a micro-/molecular-biologist is strongly benefited by utilising information or computer technology, apart from researchers from other branches of fisheries. This is because they deal with the living organisms or molecules (such as DNA/RNA) of very small size at organism levels. In each of their experimental samples these researchers are likely to obtain voluminous data of their interest which only a suitable computer system can handle and analyse properly to produce foolproof results.

1) *Data Recording*

Accurate knowledge of fish stock performance requires detailed record-keeping on attributes *viz.*, water temperature, water pH, transparency, dissolved oxygen, sediment pH, chemical quality of water, feeding schedule, fish biomass and its increase and number of fish in the pond etc (depending on the purpose and priority with which data is constructed). The major components which affect pond productivity, and hence whose record-keeping is of primary importance are,

Pond site	:	area, soil status, agro-climatic conditions etc.
Pond water	:	depth, physico-chemical conditions etc.
Fish	:	stocking size, number, composition etc.
Nutrition	:	feed composition, feed intake etc.
Pond management	:	feeding schedule, soil/water sampling, sample-netting etc.

2) *Data analyses*

A computer is an useful gadget for any sort of data analysis, from simple to complex. There are standard packages available with software vendors (and some tailor-mades with concerned research institutes of repute), such as SAS, LOTUS 1-2-3, SPSS etc. But, the advent of Microsoft's Windows (as Microsoft Office) during the early-to-mid nineties, which contains a spread-sheet package christened as *MS-EXCEL* revolutionised data analysis job. It provides the advantages of tabulation, statistical analysis, graphical representation and final documentation of a given set of data on single run (on one spread-sheet).

Limnology is the study of ecology and environment of aquatic system. As is evident, ecology and environment are the two essential components of limnological study. Discussed in detail below are examples of some quantitative methods followed in these two separate components of limnological research, where information technology contributes greatly:

i) *Ecology*: In ecological studies, the various interactions of biotic and abiotic components are taken into consideration. In a pond ecosystem, microorganisms and other small bottom dwellers (benthos) play a major role in such interactions. And many of these studies pertain to cycling of various natural elements, such as, Carbon, Nitrogen, Phosphorus and so on. While studying Carbon cycling, as an instance, the fate of carbonic (organic) inputs into the pond system is tracked. One of the parameters studied under this is the nutrient leaching rates from the detritus, litter and/or lignocellulosic organic-inputs. The nutrient leaching rates are determined through a popular procedure known as 'litter bag method'. Through this method, initially the dry-matter loss with reference to time is estimated by taking the initial and final dry-matter weight into consideration. The difference between the initial and final dry weights gives the measure of decomposition in terms of total dry matter loss. Further, some quantitative procedures are followed involving certain mathematical calculations, such as the loss rate coefficient or decay rate constant:

$$W_t = W_0 \text{ X } \exp(-kt)$$

or
$$k = \frac{\log W_0 - \log W_t}{t}$$

where, W_t = the final residual mass after a time interval t (in days)

and W_0 = the initial mass of the organic input

As for example, the loss rates of three lignocellulosic waste materials, water hyacinth (*Eichhornia crassipes*), floating water fern (*Salvinia cucullata*) and rice straw were observed as a case study by the author and coworkers. The three substrates, due to their recalcitrance, were subjected to initial solid substrate fermentation (SSF) and then, both the raw and processed forms were subjected to natural recycling in the pond system. From the initial and final weights of the substrates after each fortnight for a period of three-and-a-half months, the loss rates were calculated as:

	RAW SUBSTRATE			PROCESSED SUBSTRATE		
Days	**WH**	**RS**	**FWF**	**WH**	**RS**	**FWF**
15	2.96	1.89	0.37	3.40	1.69	0.72
30	2.00	1.15	0.20	2.14	1.57	0.61
45	1.45	0.88	0.27	1.47	1.25	0.46
60	1.17	0.81	0.28	1.17	0.96	0.44
75	1.09	0.81	0.27	1.10	1.04	0.37
90	0.93	0.76	0.29	0.93	0.91	0.54
105	0.84	0.67	0.29	0.82	0.79	0.53

WH: water hyacinth; RS: rice straw; FWF: floating water fern

From this, the mean loss rates were calculated to be 1.49, 0.99, 0.28, 1.57, 1.17 and 0.52, respectively. Taking the initial and final mass of the six (three raw and three processed) substrates into consideration, the loss rate coefficient was estimated based on the above-mentioned formulae as:

	RAW SUBSTRATE			PROCESSED SUBSTRATE		
Days	WH	RS	FWF	WH	RS	FWF
15	0.0206	0.0096	0.0033	0.0171	0.0127	0.0010
30	0.0116	0.0061	0.0029	0.0132	0.0120	0.0009
45	0.0105	0.0065	0.0032	0.0118	0.0083	0.0022
60	0.0074	0.0036	0.0001	0.0076	0.0057	0.0014
75	0.0105	0.0054	0.0013	0.0097	0.0087	0.0007
90	0.0087	0.0046	0.0032	0.0089	0.0083	0.0009
105	0.0082	0.0054	0.0034	0.0078	0.0101	0.0015

WH: water hyacinth; RS: rice straw; FWF: floating water fern

ii) *Environment*: With the ever-increasing living matter on planet Earth and further urbanisation, there is an increased pressure to maintain a delicate environmental balance. Hence, eco-sustainability of an enterprise is at stake and under scrutiny time-and-again. The primary casualties have been the land, air or water-polluting industries. The industries which pollute the water bodies range from intensive aquaculture industries to various food and commodity-processing industries which utilise the water resource to a greater extent, needless to say about the ever-problematic urban sewage wastes. With increased environment consciousness and concern, following the Environmental Impact Assessment (EIA) regulation has become mandatory for these polluting or potentially polluting industries. One easy and obligatory method adopted to study the level of pollution in such industrial effluents is the bioassay. In simple words, *Bioassay* means assessing the potential toxicity of a medium (preferably aquatic) with reference to a test organism. Bioassay is also a potential tool employed to ascertain the permissible levels of various farm-application chemicals such as, pesticides, weedicides and insecticides.

A range-finding bioassay is conducted at the beginning before the test bioassay. In such studies in aquatic systems, a test organism (of interest, preferably fish) is exposed to the collected effluent. In various bioassays concerning aquatic ecosystem study, many microorganisms such as zooplankton are also used as test organisms. When the study is conducted in a standing collected effluent, it is referred to as *static bioassay*. Bioassay could be short- or long-term. Further, cumulative percentages of death after 06, 12, 24, 48, 72 and 96 hours after stocking the test organism upon which the bioassay is being conducted are noted. The data are then processed using Probit Analyses for determination of LC_{50} and 95% upper and lower (fiducial) limits of the LC_{50}, for which following formulae are used:

$$Slope(S) = \frac{\frac{LC_{84}}{LC_{50}} + \frac{LC_{50}}{LC_{16}}}{2}$$

$$\text{Confidence limit } (f_{LC_{50}}) = S^{2.77/\sqrt{N}}$$

where, N = total number of test organisms in each concentration whose expected results are between 16 and 84%, and 2.77 is a constant

95% fiducial limits: Upper limit = LC_{50} X $f_{LC_{50}}$

$$\text{Lower limit} = \frac{LC_{50}}{f_{LC_{50}}}$$

In one of such studies, dairy effluent was obtained from a nearby dairy processing plant and was subjected to *Wolffia* bioremediation for 15 days and a bioassay was conducted for four days with rohu (*Labeo rohita*) as the test organism in raw as well as the treated effluents. The result in terms of LC_{50}, fiducial limits and confidence limit in both raw and bioremediated effluent were estimated to be :

Exposure period (in hours)	LC_{50} (%)	95% fiducial limits		Confidence limit (%)
		upper limit	lower limit (%)	
RAW				
6	27.8	30.0	25.8	1.08
12	26.5	28.9	24.3	1.09
24	26.2	28.3	24.3	1.08
48	26.2	28.3	24.3	1.08
72	25.8	27.6	24.1	1.07
96	25.5	27.3	23.8	1.07
TREATED				
6	84.0	87.0	81.1	1.04
12	82.0	86.0	78.2	1.05
24	81.5	88.2	75.3	1.08
48	83.5	97.2	71.8	1.16
72	76.0	87.2	66.2	1.15
96	73.5	81.7	66.1	1.11

From these probit results and the logit graph drawn with the help of these data, the 96-hour LC_{50} of raw and treated dairy effluent on *L. rohita* were found to be 25.5% and 73.5%, respectively. The other conclusions those could be drawn are, the LC_{50} (median lethal concentration) decreased gradually with the increase in exposure time from 06 to 72 hours; and the treated effluent was less toxic than the raw one, the toxicity reducing three-folds.

The 50% lethal concentration values are plotted against time for obtaining the toxicity curve. Such studies not only provide information on the extent of pollution in the industrial effluent but also further give clues for its judicious utilisation as a resource upon treatment/dilution. It could also be inferred from the above study that, the effluent could be utilised as potential pond manure with least toxicity effect with a dilution factor of 10 (*i.e.*, one-tenth of the 96-hour LC_{50} approximate threshold concentration).

Other Facets of Bioinformatics in Limnology

Apart from these, aquatic and environmental microbiology (limnological) research could be greatly benefited in many ways through IT tools and technique. Some far-reaching visionary concepts in this regard are, development of suitable software packages for easy and step-wise identification of microorganisms based on their biochemical characterisation, fatty acid contents etc.; data-base preparation of the various aquatic microorganisms according to their taxonomic position (phylogeny); tailor-made package development for identification of various microbial DNA fragments (or bands) according to their molecular weights; developing models to study the *in situ* activities of the specific microbial groups (*viz.*, nitrogen-fixers, nitrifiers, cellulolytes, amylolytes, phosphatase-producers, phosphate-solubilisers, pectinolytes, ureolytes etc.); pond-site (*in situ*) round-the-clock (perpetual) monitoring of water quality and sediment nutrient status with respect to microbial activity; designing packages to ascertain/forecast the effect of application of specific probiotic microbial groups such as, biofertilisers, biofilters, bioremediators, bioaugmenters, biosensors etc. on pond productivity.

Besides, limnological studies involve a lot of other procedures and quantitative methods which involve arduous mathematical calculations. To cite a few, total plankton counts in a pond, carbon conversion efficiency in the pond food chain, estimation of the actual levels of various physico-chemical parameters (in ppm, mg/l etc.).

Conclusion

Lastly, the economy of commercial aquaculture, the ultimate aim of aquaculture research, is a direct measure of its productivity, quality of products and its ability to cope with environmental sustainability. Information Technology or the silicon-chip revolution shall play a major role when aquaculture grows out of its infancy and becomes a large-scale, highly productive industry with globally-competitive quality and pricing. Needless to say, aquaculture quantitative genetics (precisely, selective breeding projects), molecular biology and aquaculture productivity (environmental and aqua-health management) researches shall benefit of it immensely.

REFERENCES

Ayyappan, S., Olah, J., Raghavan, S.L., Sinha, V.R.P. and Purushothaman, C.S., 1986. Macrophyte decomposition in two tropical lakes. *Arch. Hydrobiol.*, 106: 219-231.

Kowal, N.E., 1969. Effect of leaching on pine litter decomposition rate. *Ecol.*, 50: 739-740.

Mishra, S., 2000. Microbial ecology and productivity of fish ponds applied with processed lignocellulosic wastes. Ph. D. Thesis, Utkal University, Bhubaneswar: 226pp.

Mishra, Snehasish, S.K. Barik, S. Ayyappan and B.C. Mohapatra, 2000. Fish bioassays for evaluation of raw and bioremediated dairy effluent. *Bioresource Technology*, 72(3): 213-8

Mohapatra, B.C. and Rangarajan, K., 1995. A Manual On Bioassays in Laboratory and their Techniques. CMFRI Spl. Publ., 64: 75pp.

Olsen, J.S., 1963. Energy storage and the balance of producers and decomposers in ecological systems. *Ecol.*, 44: 322-331.

Reish, D.L. and Oshida, P.S., 1987. FAO Fisheries Technical Paper, No. 247: 62pp.

Chapter **31**

Analysis of Aquaculture Data Using Computers

P. K. Satapathy & A. Antony

Introduction

Data collection, compilation and analysis of Aquaculture data has a major role to play in Aquaculture research. The time spent in collecting, compiling, analysing statistical data and taking appropriate timely action creates a gap in aquaculture. Therefore timely action is very much essential. This gap can be reduced by taking help of computers in collection, compilation and analysis of statistical data which will save time as well as manpower and will help in taking appropriate action in appropriate time. Electronic spreadsheet has a major role to play in this context even if the user is having little or no programming knowledge. The most widely used office automation package is one of the components which consists of spreadsheet application, in which accounting, balance sheet calculations, statistical data analysis, graphical presentations, etc. are possible with few key strokes and mouse operation only. EXCEL has several capabilities which include opening of a workbook; entering and editing data; building formulas to calculate values; managing list of data; formatting data; creating a chart; saving a workbook; opening and saving files from other spreadsheets; linking documents from other spreadsheets, data analysis, etc.

Collection of Aquaculture Data

Data collection manually takes a considerable amount of time for statistical investigators. Preparation of schedules/questionnaires can be

Applied Bioinformatics, Statistics and Economics in Fisheries Research: pp. 455-471

New India Publishing Agency, New Delhi (India)

possible using MS-EXCEL spreadsheets which can be used for data collection by preparing spreadsheets. Data validation is possible with programmable automated sheets developed in MS-EXCEL helping to reduce errors and data redundancy. These sheets can be sent by e-mail to investigators spread over different parts of the country. These filled in automated sheets can be transmitted to the desired location for further compilation, analysis, interpretation and generation of information reducing the time of postal delay. Online data collection is also possible using server/web based application programmes by which one can enter the desired website with user based authentication system. This way instant updation of information is possible with little or no time spent. With wide spread of lap top/palm top computers the questionnaires/schedules can be filled at the data collection site itself. By connecting it to internet uploading of information to central investigator is also possible instantly. This way there is reduction of considerable amount time spent in transmission of information and reduces typographical error mostly committed by data entry operators.

Starting Microsoft Excel: Double-click the Microsoft Excel icon in the program manager.

The Microsoft Excel Window appears. If necessary, enlarge the window by clicking maximize button in the upper-right corner of the window. A new file can be created by File - Save/Save As or by clicking save button and giving appropriate file name. Clicking the Open button can open an existing file. Clicking the Open button is equivalent to choose the Open command from the File menu. Many of the commands on the menu bar have equivalent toolbar buttons. Either the command or the button can be used depending on preference. In the Directories box, double-clicking of the directory should be done to go to other directories.

Most of the work done in Microsoft Excel is on a worksheet. A worksheet is a grid of rows and columns. Each cell is the intersection of a row and a column and has a unique address or reference. For example, the cell where column B and row 5 intersect is cell B5 Cell references are used while writing formulas or referring to cells. At first select the cell or cells you want to work with, and then enter data or choose a command. Selected cells appear highlighted on your screen. The active cell is the cell in which data is entered when you start typing. Only one cell is active at a time. A heavy border shows the active cell.

To change the active cell: Move the mouse pointer into the respective cell and click.

To scroll through the worksheet: Move the pointer to the down arrow in the vertical scroll bar on the right edge of the window and then click. The worksheet scrolls down one row. Click the scroll bar area below the scroll box. The worksheet scrolls down one screen. Move the pointer into the scroll box. Hold the mouse button down, move the pointer back to the top of the vertical scroll bar and then release the mouse button. This is called dragging.

Entering Formulae: Place the cursor in the cell where you want to write the formula. Write the formula starting with equal (=) sign and then press enter key.

Formulas with Data

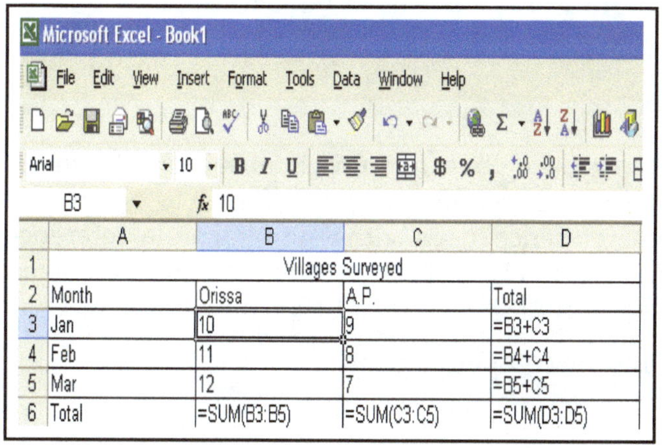

	A	B	C	D
1	Villages Surveyed			
2	Month	Orissa	A.P.	Total
3	Jan	10	9	=B3+C3
4	Feb	11	8	=B4+C4
5	Mar	12	7	=B5+C5
6	Total	=SUM(B3:B5)	=SUM(C3:C5)	=SUM(D3:D5)

Result

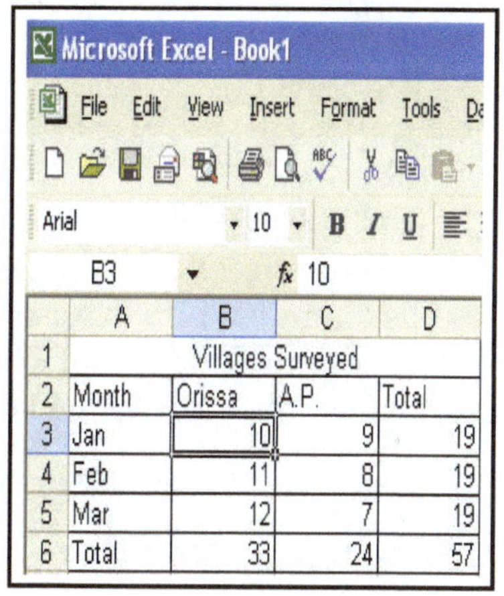

	A	B	C	D
1	Villages Surveyed			
2	Month	Orissa	A.P.	Total
3	Jan	10	9	19
4	Feb	11	8	19
5	Mar	12	7	19
6	Total	33	24	57

To copy the formula to the other columns drag the fill handle (by pointing the crosser at rightmost corner below the cell where + sign appears) from B9 through C9 and D9 and formula changes automatically in respective cells.

	A	B	C	D	E	F
1	CENTRAL INSTITUTE OF FRESHWATER AQUACULTURE					
2						
3	Fish production from various ponds (in kg)					
4						
5		Rohu	Catla	Mrigal		
6	Pond1	200	500	200		
7	Pond2	100	300	300		
8	Pond3	300	100	800		
9	Total	600	900	1300		

=sum(b6:b8) =sum(c6:c8) =sum(d6:d8)

If you combine several operators in a single formula, Microsoft Excel performs the operations in the order shown below:

-	Negation
%	Percent
^	Exponentiation
* and /	Multiplication and divison
+ and -	Addition and subtraction
&	Text joining
= < > <= >= <>	Comparison (<> stands for not equal to)

If a formula contains operators with the same priority, Microsoft Excel evaluates the operators from left to right. If you want to alter the order of evaluation, use parentheses to group expressions in your formula.

Ex :- 2+4*5 = 22

(2+4)*5 = 30

There are three types of references relative references, absolute references, and mixed references. In relative references while copying formulas, references to cells change while in absolute reference (used with $ sign) formula, reference is absolute and does not change. Mixing of both references is called mixed references. ex: relative (A1), absolute ($A$1), mixed (A$1), mixed ($A1).

There are three types of reference operators:

- Range (colon): B2:D2
- Union (comma): B4, D4
- Intersection (space): B7:D7 C6:C8

Sorting Data in a list: Select a single cell in the list. Choose Data - Sort

Sheetwise Compilation of Data

Collected data can be put sheetwise for further compilation and summarization. Suppose, there was a collection of state-wise/district-wise data. If state-wise and district-wise summarization of information is required, it can be possible by creating as many sheets as required and putting information on these sheets and thereafter a summary sheet can be programmed to have state-wise/district-wise summary of collected data.

Ex : Data collected from 3 districts of Orissa and 2 districts of Andhra Pradesh for trainings collected. A total of 5 sheets can be created by inserting sheets in workbook of Excel. These sheets can be renamed appropriately like OR01, OR02, OR03, AP01 and AP02. Then data entry in respective sheets can be performed. Finally one more sheet can be created naming TOTAL. The formulas in sheet TOTAL can be as follows :

	A	B
1	=OR01!A1+OR02!A1+OR03!A1+'AP01'!A1+'AP02'!A1	=OR01!B1+OR02!B1+OR03!B1+'AP01'!B1+'AP02'!B1
2	=OR01!A2+OR02!A2+OR03!A2+'AP01'!A2+'AP02'!A2	=OR01!B2+OR02!B2+OR03!B2+'AP01'!B2+'AP02'!B2
3	=OR01!A3+OR02!A3+OR03!A3+'AP01'!A3+'AP02'!A3	=OR01!B3+OR02!B3+OR03!B3+'AP01'!B3+'AP02'!B3
4	=OR01!A4+OR02!A4+OR03!A4+'AP01'!A4+'AP02'!A4	=OR01!B4+OR02!B4+OR03!B4+'AP01'!B4+'AP02'!B4
5	=OR01!A5+OR02!A5+OR03!A5+'AP01'!A5+'AP02'!A5	=OR01!B5+OR02!B5+OR03!B5+'AP01'!B5+'AP02'!B5

Analysis of Data

Step-1 Select Tool from Menu bar.

Step-2 Select Data Analysis from the Pull Down Menu. (If it is not available, it can be made available by Tools - Add-Ins - Analysis TookPak)

Step-3 Select required Analysis from the Analysis Tool box

Step-4 Select OK.

Step-5 Type the data range in Data box and other information needed for analysis.

Step-6 Select finish.

The output obtained from step-6 can be saved by opting File - Save. Selecting the range of worksheet to be printed and opting File - Print, can do the printing.

The Analysis ToolPak includes ANOVA: Single-Factor ANOVA: Two-Factor with Replication; ANOVA: Two-Factor without Replication; Covariance; Correlation; Descriptive Statistics; Exponential Smoothing; F-Test; Two-Sample for Variances; Histogram; Moving Average; Random Number Generation; Rank and Percentile; Regression; t-Test: Paired Two-Sample for Means; t-Test: Two-Sample Assuming Equal and anequal Variances; t-Test: z-Test: etc.

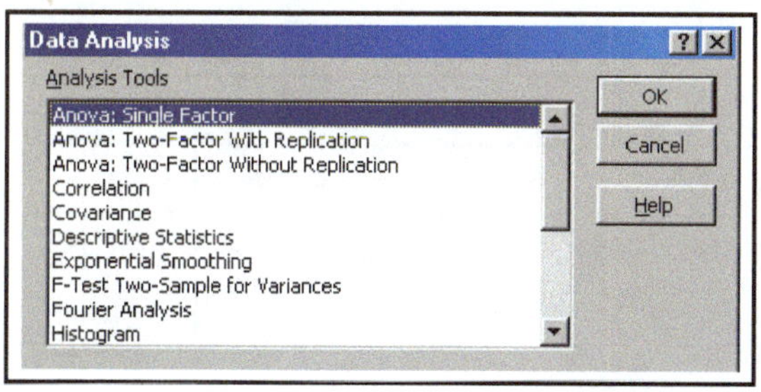

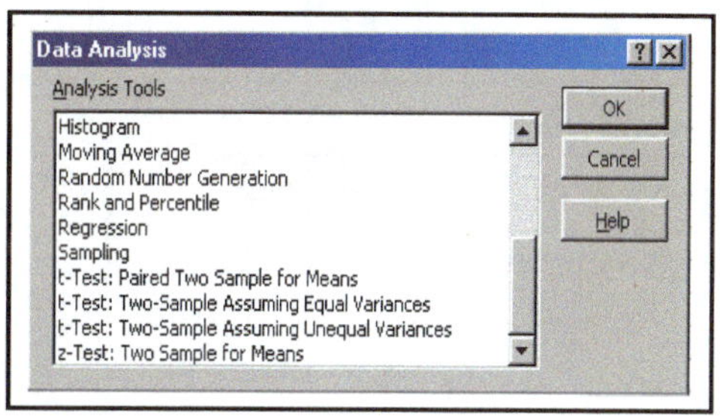

ANOVA: Single-Factor: The single-factor ANOVA tool performs a simple analysis of variance, which tests the hypothesis that means from several samples are equal. The data arrangement and analysis options should be done in the following manner:

Group1	Group2	Group3
60	55	61
68	56	63
71	36	65
75	32	64
66	36	61
70	60	68
65	78	63
66	66	61
55	55	64

ANOVA: Single Factor

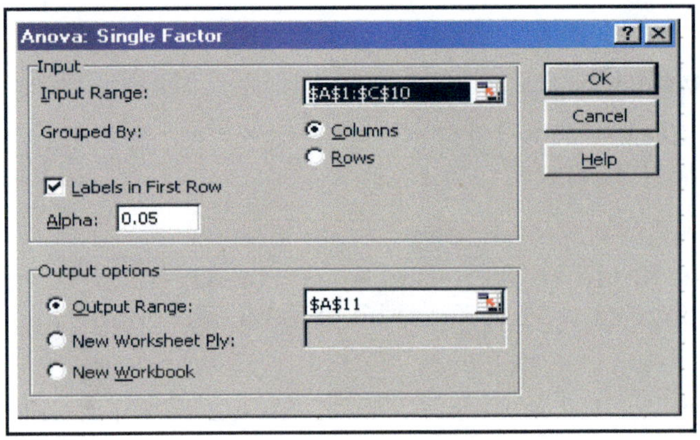

SUMMARY

Groups	Count	Sum	Average	Variance
Group1	9	596	66.22222	35.44444
Group2	9	474	52.66667	234.75
Group3	9	570	63.33333	5.25

ANOVA

Source of Variation	SS	df	MS	F	P-value	F crit
Between Groups	917.6296	2	458.8148	4.997176	0.015334	3.402832
Within Groups	2203.556	24	91.81481			
Total	3121.185	26				

ANOVA: Two-Factor with Replication: The two-factor with replication ANOVA tool performs an extension of the single-factor ANOVA that includes more than one sample for each group of data. The arrangement of data should be as follows:

	Group1	Group2	Group3
Expt-1	60	55	61
	68	56	63
	71	36	65
	75	32	64
Expt-2	66	36	61
	70	60	68
	65	78	63
	66	66	61

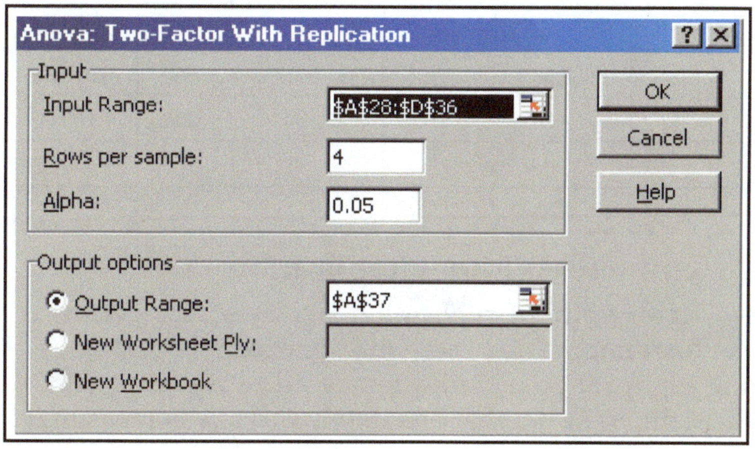

ANOVA: Two-Factor with Replication						
SUMMARY	Group1	Group2	Group3	Total		
Expt-1						
Count	4	4	4	12		
Sum	274	179	253	706		
Average	68.5	44.75	63.25	58.83333		
Variance	40.33333	156.9167	2.916667	167.7879		

Contd...

Expt-2						
Count	4	4	4	12		
Sum	267	240	253	760		
Average	66.75	60	63.25	63.33333		
Variance	4.916667	312	10.91667	97.69697		
Total						
Count	8	8	8			
Sum	541	419	506			
Average	67.625	52.375	63.25			
Variance	20.26786	267.4107	5.928571			
ANOVA						
Source of Variation	*SS*	*df*	*MS*	*F*	*P-value*	*F crit*
Sample	121.5	1	121.5	1.380682	0.255295	4.413863
Columns	986.5833	2	493.2917	5.605587	0.012809	3.554561
Interaction	349.75	2	174.875	1.987216	0.166033	3.554561
Within	1584	18	88			
Total	3041.833	23				

ANOVA: Two-Factor Without Replication

The two-factor without replication ANOVA tool performs two-factor ANOVA that does not include more than one sample per group. The data in the following input range includes temperature readings, in Celsius, taken from a pond at different depths. The depth variable is measured in feet. The readings were taken on four different dates.

Depth	Date 1	Date 2	Date 3	Date 4
1	23	24.2	24.8	25.2
2	22.2	23	24	24.5
3	22	22.5	22.8	23
4	21.2	22	22.3	22.5
5	18.4	20	21	22
6	13.5	15	17	18
7	9.7	10	13	15
8	5.9	6.5	8	9
9	5.7	5.9	6.5	7
12	5.5	5.7	6.1	6.5
15.5	5.3	5.5	6.9	6.2

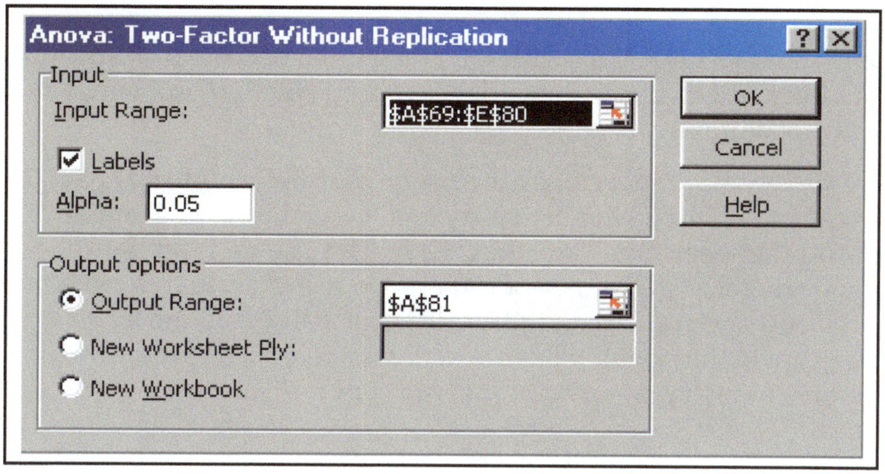

ANOVA: Two-Factor Without Replication

SUMMARY	Count	Sum	Average	Variance
1	4	97.2	24.3	0.92
2	4	93.7	23.425	1.055833
3	4	90.3	22.575	0.189167
4	4	88	22	0.326667
5	4	81.4	20.35	2.356667
6	4	63.5	15.875	4.0625
7	4	47.7	11.925	6.4225
8	4	29.4	7.35	1.99
9	4	25.1	6.275	0.349167
12	4	23.8	5.95	0.196667
15.5	4	23.9	5.975	0.529167
Date 1	11	152.4	13.85455	58.43873
Date 2	11	160.3	14.57273	63.36818
Date 3	11	172.4	15.67273	59.62618
Date 4	11	178.9	16.26364	60.46655

ANOVA

Source of Variation	SS	df	MS	F	P-value	F crit
Rows	2402.421	10	240.2421	434.8274	1.3E-29	2.16458
Columns	38.62	3	12.87333	23.30015	5.53E-08	2.922278
Error	16.575	30	0.5525			
Total	2457.616	43				

Covariance and Correlation: The following illustration shows an input range that could be used in either a covariance or a correlation problem. The data is arranged in columns.

The Covariance tool returns the average of the product of deviations of data points from their respective means. Covariance is a measure of the relationship between two ranges of data. The use of Covariance tool is to determine whether two ranges of data move together; that is, whether large values of one set are associated with large values of the other (positive covariance), whether small values of one set are associated with large values of the other (negative covariance), or whether the values in the two sets are unrelated.

Correlation: The Correlation tool measures the relationship between two data sets that are scaled to be independent of the unit of measurement. The population correlation calculation returns the covariance of two data sets divided by the product of their standard deviations.

One can use the Correlation tool to determine whether two data sets move together; that is, whether large values of one set are associated with large values of the other (positive correlation), whether small values of one set are associated with large values of the other (negative correlation), or whether the values in the two sets are unrelated (zero correlation - the correlation tends toward zero).

Sample 1	Sample 2	Sample 3
1.2	4.2	120
2.5	8.7	40
3.6	11.4	40
2.8	9.1	130
4.8	4.2	120
1.8	6.6	20
2.2	7.9	40
5.1	12.2	40

	Sample 1	Sample 2	Sample 3
Sample 1	1		
Sample 2	0.415546	1	
Sample 3	0.013415	-0.51417	1

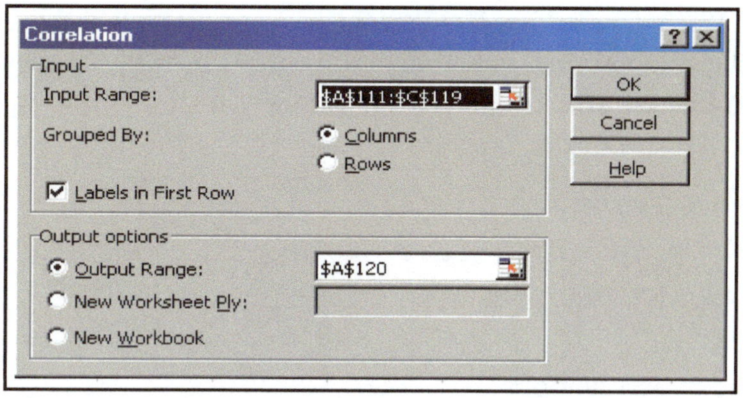

Regression : Performs linear regression analysis. Regression fits a line through a set of observations using the "least squares" method. Regression is used in a wide variety of applications that seek to analyze how a single dependent variable is affected by the values of one or more independent variables. For example, several factors affect a fishers' performance, including age, sex, height, and weight. Regression apportions shares in the performance measure to each of these four factors based on a set of performance data. One can then use the regression results to predict the performance of a new, untested fisherman.

Input Y Range: Type the reference for the range of dependent data you want to analyze. The dependent data should be typed in a single column. If you include a report label in the first row of the input y range, you must select the Labels check box or Microsoft Excel displays a message.

Input X Range: Type the reference for the range of independent data you want to analyze. Microsoft Excel orders independent variables in ascending order from left to right, using 1, 2, 3, and so on for the variable names in the summary output table. The maximum number of input ranges is 16.

Constant Is Zero: Select the Constant Is Zero check box to force the regression line to pass through the origin.

Labels: If the first row in the input range includes report labels, select the Labels check box. If your input range does not contain labels, clear the Labels check box. Microsoft Excel then generates the appropriate data labels for the summary output table.

Confidence Level: Select the Confidence Level check box if you want an additional level included in the summary output table. In the Confidence Level box, type an additional confidence level that you want Microsoft Excel to apply to the regression in addition to the default 95% confidence level.

Output Range: Type the reference for the upper-left cell of the range where you want the summary output table to appear. Allow at least seven columns for the summary output table. The summary output table includes the following:

ANOVA table

Coefficients

Standard error of y estimate

r^2 values

Number of observations

Standard error of coefficients

New Worksheet Ply: This option inserts a new ply into the workbook. The current ply resides in and pastes the results into cell A1 of the new ply. Use the box next to New Worksheet Ply to optionally name the new ply.

New Workbook: This option creates a new workbook, adds a new workbook ply in the new workbook, and pastes the results into cell A1 of the new ply.

Residuals: Select the Residuals check box if you want to include residuals in the residuals output table.

Standardized Residuals: Select the Standardized Residuals check box if you want to include standardized residuals in the output table.

Residual Plots: Select the Residual Plots check box if you want Microsoft Excel to generate a chart for each independent variable versus the residual.

Line Fit Plot: Select the Line Fit Plot check box if you want Microsoft Excel to generate a chart for predicted values versus the observed values.

Normal Probability Plot: Select the Normal Probability Plot check box if you want Microsoft Excel to generate a chart plotting normal probability plots.

Descriptive Statistics

The Descriptive Statistics tool generates a report of univariate statistics for data in the input range. The output values generated by the Descriptive Statistics tool are: Standard deviation of sample (sample variance), kurtosis, and skewness. These outputs are derived using the same algorithms used by the built-in functions STDEV, VAR, KURT, and SKEW, respectively.

Sample 1	Sample 2	Sample 3
1.2	4.2	120
2.5	8.7	40
3.6	11.4	40
2.8	9.1	130
4.8	4.2	120
1.8	6.6	20
2.2	7.9	40
5.1	12.2	40

Sample 1		Sample 2		Sample 3	
Mean	3	Mean	8.0375	Mean	68.75
Standard Error	0.4931	Stdd Error	1.0499	Stdd Error	16.1949
Median	2.65	Median	8.3	Median	40
Mode	#N/A	Mode	4.2	Mode	40
Standard Deviation	1.3948	Std Deviation	2.9698	Std Deviation	45.8062
Sample Variance	1.9457	Sample Variance	8.8198	Sample Variance	2098.214
Kurtosis	-0.9948	Kurtosis	-1.0764	Kurtosis	-2.07672
Skewness	0.4985	Skewness	-0.0317	Skewness	0.570952
Range	3.9	Range	8	Range	110
Minimum	1.2	Minimum	4.2	Minimum	20
Maximum	5.1	Maximum	12.2	Maximum	130
Sum	24	Sum	64.3	Sum	550
Count	8	Count	8	Count	8

t-Test: Paired Two-Sample for Means: Performs a two-sample paired student's t-Test. This form of the t-test tests whether a sample's means are distinct. It does not assume the variances of both populations from which the data sets are drawn are equal. A paired test is appropriate whenever there is a natural pairing of observations in the samples, such as when a sample group is tested twice, before and after an experiment. The two input ranges of data must contain the same number of data points.

Variable 1 Range: Type the reference for the first range of data you want to analyze. The data must occupy a single column or row on the worksheet. If any data points in the range of data are nonnumeric, Microsoft Excel displays a message.

Variable 2 Range: Type the reference for the second range of data you want to analyze. The data must occupy a single column or row on the worksheet. If any data points in the range of data are nonnumeric, Microsoft Excel displays a message.

Output Range: Type the reference for the upper-left cell of the range where you want the output table to appear.

If the first row or column of your input range contains labels, select the Labels check box.

If your input range does not contain labels, clear the Labels check box. Microsoft Excel then generates the appropriate data labels for the output table (Variable 1 and Variable 2).

Alpha: Type the confidence level for the test. The default alpha value is 0.05. If the alpha value is nonnumeric or outside the range 0...1, Microsoft Excel displays a message.

Hypothesized Mean Difference: Type the number that you want for the shift in sample means. A value of 0 indicates that the sample means are hypothesized to be equal.

Sample 1	Sample 2
1.2	4.2
2.5	8.7
3.6	11.4
2.8	9.1
4.8	4.2
1.8	6.6
2.2	7.9
5.1	12.2

t-Test: Paired Two Sample for Means

	Sample 1	Sample 2
Mean	3	8.0375
Variance	1.945714	8.819821
Observations	8	8
Pearson Correlation	0.415546	
Hypothesized Mean Difference	0	
df	7	
t Stat	-5.26532	
P (T<=t) one-tail	0.000583	
t Critical one-tail	1.894578	
P (T<=t) two-tail	0.001167	
t Critical two-tail	2.364623	

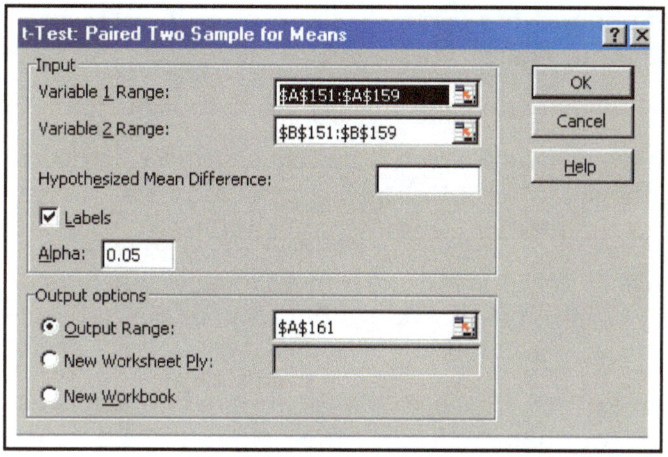

t-Test: Two-Sample Assuming Equal Variances: Performs a two-sample student's t-test. This form of the t-test assumes that the means of both data sets are equal and is referred to as a homoscedastic t-test. You use t-tests to determine whether the two samples' means are equal.

t-Test: Two-Sample Assuming Unequal Variances: Performs a two-sample student's t-test. This form of the test assumes that the variances of both ranges of data are unequal and is referred to as a heteroscedastic t-test. Use a t-test to determine whether two sample means are equal and when the groups under study are distinct. Use a paired test when there is one group before and after a treatment.

Forecasting and Showing Trends in Charts

Trendlines are used to graphically display trends in data and to analyze problems of prediction. Such analysis is also called regression analysis. By using regression analysis, you can extend a trendline in a chart beyond the actual data to predict future values.

Moving Average : You can also create a moving average (moving average: A sequence of averages computed from parts of a data series. In a chart, a moving average smooths the fluctuations in data, thus showing the pattern or trend more clearly.), which smoothes out fluctuations in data and shows the pattern or trend more clearly.

Chart types that support trendlines : You can add trendlines to data series (data series: Related data points that are plotted in a chart. Each data series in a chart has a unique color or pattern and is represented in the chart legend. You can plot one or more data series in a chart. Pie charts have only one data series.) in unstacked 2-D area, bar, column, line, stock, xy (scatter), and bubble charts. You cannot add trendlines to data series in 3-D, stacked, radar, pie, surface, or doughnut charts.

Types of trendlines : choosing the right one for your data

When you want to add a trendline to a chart in Microsoft Excel, you can choose any of the six different trend/regression types. The type of data you have determines the type of trendline you should use.

Trendline reliability : A trendline is most reliable when its R-squared value is at or near 1. When you fit a trendline to your data, Excel automatically calculates its R-squared value.

Linear : A linear trendline is a best-fit straight line that is used with simple linear data sets. Your data is linear if the pattern in its data points resembles a line. A linear trendline usually shows that something is increasing or decreasing at a steady rate.

Logarithmic **:** A logarithmic trendline is a best-fit curved line that is used when the rate of change in the data increases or decreases quickly and then levels out. A logarithmic trendline can use negative and/or positive values.

Polynomial : A polynomial trendline is a curved line that is used when data fluctuates. It is useful, for example, for analyzing gains and losses over a large data set. The order of the polynomial can be determined by the number of fluctuations in the data or by how many bends (hills and valleys) appear in the curve. An Order 2 polynomial trendline generally has only one hill or valley. Order 3 generally has one or two hills or valleys. Order 4 generally has up to three.

Power : A power trendline is a curved line that is used with data sets that compare measurements that increase at a specific rate — for example, the acceleration of a race car at 1-second intervals. You cannot create a power trendline if your data contains zero or negative values.

Exponential : An exponential trendline is a curved line that is used when data values rise or fall at increasingly higher rates. You cannot create an exponential trendline if your data contains zero or negative values.

Moving Average

A moving average trendline smoothes out fluctuations in data to show a pattern or trend more clearly. A moving average uses a specific number of data points (set by the period option), averages them, and uses the average value as a point in the line. If period is set to 2, for example, then the average of the first two data points is used as the first point in the moving average trendline. The average of the second and third data points is used as the second point in the trendline, and so on.

Add a trendline to a chart : The following example shows the trendline (linear) for the following data. The chart of XY scatter is created first and then right click on the graph line (data series) and click on add trend line. The regression equation and R^2 value showing options can be clicked in respective check boxes to get regression equation and R^2 value on chart itself. Likewise depending upon the data other types of trendlines can be chosen for logarithmic, polynomial, power and exponential.

Year	Fish Prod in tonnes
1997	1000
1998	1100
1999	1200
2000	1400
2001	1500
2002	1550
2003	1600
2004	1800
2005	2000
2006	2100

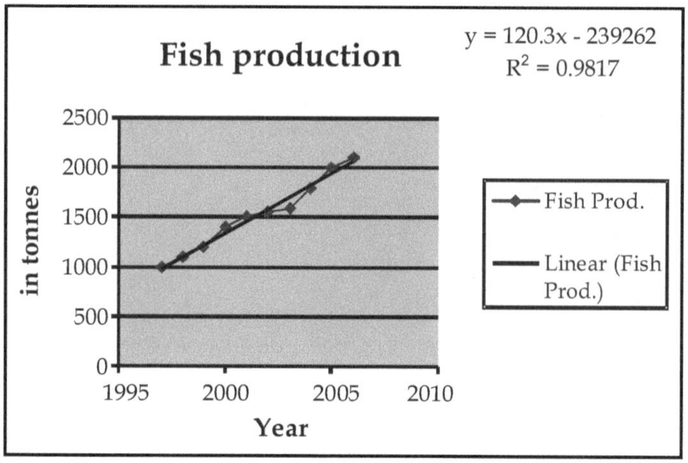

ECONOMICS

Chapter **32**

Fish Production Function and Environmental Sustainability

Kumar Das & S. T. Rout

Introduction

Fish is a renewable natural resource. It is a significant part of the domestic food supply which amount of foreign exchange. Export of fish and fish products has expanded very significantly. Fish earns increasing farming is an important component of India's rural economy. Inland fisheries including rivers and ponds produce about 35 to 40 percent of the total fish harvest of the country. Fresh water aquaculture belongs to the unorganised traditional sector, which consists of many socio economically weak fishermen and small-scale fish farming units that use traditional methods of fish culture. Fish farming faces dual problem of conservation and development. Conservation may be defined as an approach to resource management that seeks to leave fish resources relatively untouched. Development on the other hand amounts to the use and modification of fish resources. It may involve transformation of fish resources. It is considered as a renewable resources because with proper management it can be made to yield direct services indefinitely into the future. But its renewability does not imply that it is free from pressing problem of over exploitation and total depletion.

The characteristics of fish resources that make it different from other resources are those that relate to the problems of renewability, ownership and externalities. Despite these difference fish is subject to the operation of

Applied Bioinformatics, Statistics and Economics in Fisheries Research: pp. 473-480

New India Publishing Agency, New Delhi (India)

price mechanism in the same way as all other resources. So the question that arises here: Can the price system lead to a level of exploitation that will allow the maintenance of a desirable level of stock for the future? All renewable resources including fish are those resources that are subject to a natural rate of growth and therefore are capable of yielding a sustained level of production into the future. The important characteristic of fish is that the rate of its renewal directly depends on the amount of the resource left unharvested. The rate of rise of its population consists of its rate of growth less its mortality rate. The rate of growth of its population F can be considered to be directly proportional to the stock of fish. It gives a straight-line function, which is represented by line ON growth curve. In the absence of any catch fish population will expand exponentially. But expansion is arrested by mortality factors such as food shortage, predation and diseases. It can be assumed that there is a maximum fish population which the environment can support equal to F_m. With lower quantity of fish natural mortality rate is also low. As fish population increases mortality rate increases at an increasing rate. As the maximum fish quantity is approached food shortages and diseases etc become more important factors which cause high mortality rates. Mortality is represented by OM curve. Therefore the natural rate of growth of fish is obtained by subtracting both the curves. After doing this we get the natural growth curve of fish i.e. NG curve. It is concave downwards. It shows that a sustainable yield can be achieved for any level of fish quantity, provided the rate of harvest does not exceed the rate of change in its population. Sustainable yield therefore refers to the maximum amount that can be harvested from a given stock without decreasing its level in the next period or for future.

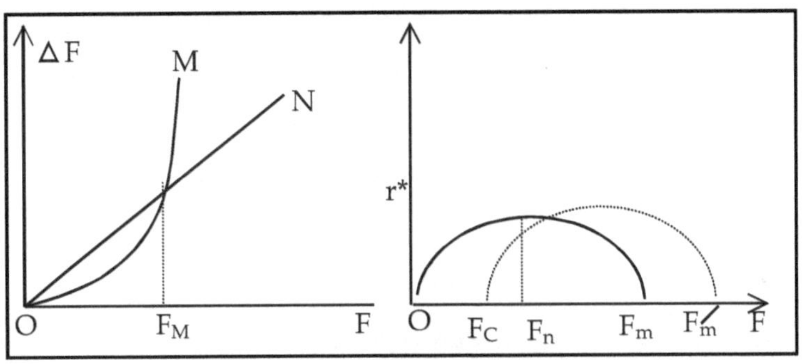

Fig. 1. Natural Growth Curve

Fish population increases above zero rate of growth. Its growth reaches maximum level at or* and fish increases to F_n. Beyond that level the rate of growth falls with further rise in fish population. At lower level of fish there is plenty of food and space for their growth. So there is rapid rise in fish stock.

With higher fish population the scarcity of food and space force down the rate of growth eventually to zero level at F_m. So ecological restraints automatically impose zero population growth on fish. If the fish resource fails to reproduce itself above the zero level than the natural growth curve will be the dotted curve. Here there is a critical minimum level of fish population below which it disappears .Here the critical minimum level of fish is of below this level fish can not maintain a viable rate of renewal. If man does not exploit any fish resource the natural equilibrium will be Fm which is a stable equilibrium point.

$$\text{If } F > F_m$$

Then ecological restraints will force the species back to Fm level

$$\text{If } F < F_m$$

Then the ecological factors will be conducive to sustain increase upto F_m.

The natural growth curve implies that with proper management policies it should be possible to harvest the resources annually in such a way that is subject to natural fluctuations, a constant level of population is maintained. The economic importance of the natural growth curve (ngc) lies in the fact that it can be used for determining the exact level of harvesting that will allow a sustainable yield and a constant level of population. It is perceived easily by understanding the relationship between growth rate and amount of harvesting of fish.

The fish Production function is denoted as:

$$H_t = f\ (F_t, E_t)$$

Where: H_t is the harvest level in t time, F_t is the fish stock in t time, E_t is the fishing effort in t time and $0 < H < H\ max$

But it should include another important factor i.e. the stock of the fish which is itself a function of past fish stocks, past fishing effort and past natural growth of fish stock.

$$\text{So } S_t = f\ (S_{t-1}, H_{t-1}, G_{t-1})$$

Where: H and G are harvesting and growth respectively. Harvesting reduces the stock and growth enhances the stock. When G_t outweighs the H_t stock rises. When H_t outweighs the G_t the stock declines.

So the original fish production function should be revised to include S_t.

Therefore more relevant fish Production function is:

$$H_t = f\ (F_t, E_t, S_{t-1}, H_{t-1}, G_{t-1})$$

Having the data on these variables one can estimate the fish production function. The results will be quite revealing. A Cobb Douglas production function can be fitted to estimate á and â value. A B values for both modern and traditional sectors can be estimated.

$$VH_T = A\, L_T^{\alpha}\, K_T^{\beta}\, D \text{ and}$$

$$VH_M = B\, L_M^{\alpha}\, K_M^{\beta}\, E$$

Where: VH_T = Value of harvested fish in the traditional sector

VH_M = value of harvested fish in the modern sector

D = Dummy variable for using or not using mechanised boat in the traditional sector

E = Dummy variable for using or not using propulsion power of the craft in the modern sector

$$F_t = F_{t+1} - F_t$$

$$= f(F_t) - H_t$$

$$\text{Thus } H_t = f(F_t) - F_t$$

So If $H_t > F_t$ fish resources declines

If $H_t < F_t$ fish resources grows

It simply states that the variation in the level of fish is equal to the growth of fish in period t i.e. $f(F_t)$ minus the amount of harvest (H_t) in period t. So the natural growth curve (ngc) can be used for derivation of a yield curve which help us to determine the optimum level of fish catching.

The rate of change in population is equivalent to sustainable yield. ie.

$$F = SY$$

With higher levels of fishing fish population falls and rate of change in fish population rises and thus yield rises. It eventually depresses the sustainable yield.

Harvesting involves money and effort. It has a positive cost. So the level of fish depends on the extent of fishing, which determines the level of cost. The cost curve is OC which shows the total cost with each harvesting drive. When it is superimposed on SYC the optimum level of output is determined. If it is like CPR, which is exploited by large number of individuals there will be externalities. There will be over exploitation of fish resources. The output will rise up to D_3 where average cost will be equal to average yield i.e.

$$AC = AY$$

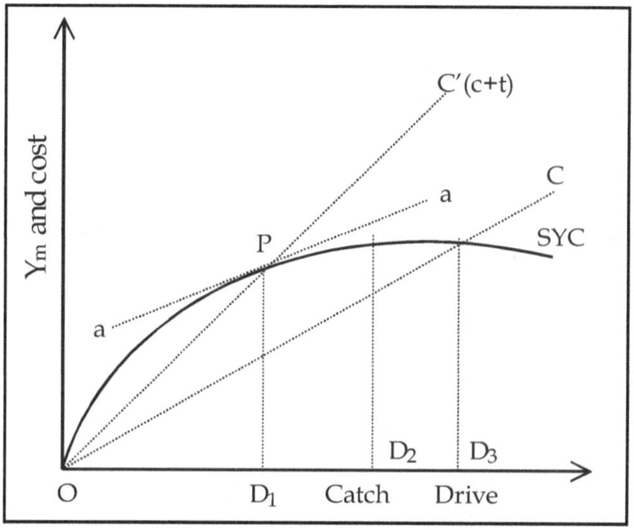

Fig. 2. Over Exploitation of Resources

Harvesting continues till there is profit. But the efficient level of exploitation will be that when marginal cost of fish harvesting is equal to the marginal yield i.e.

$$MC = MY$$

It occurs only at a level of output where the slope of SYC is equal to the slope of total cost curve. Line aa is parallel to OC curve which is tangent to SYC at point P.It implies that at lower level of catch drive D1 MC shown by the slope of total cost curve is equal to the marginal yield represented by the slope of the yield curve. In terms of money marginal yield is same as marginal revenue. It satisfies the necessary condition of profit maximising behaviour of any entrepreneur. From this it is very clear that maximum yield is not the optimum yield. It represents the non-economic solution to the economic question. It suffers from basic economic illiteracy. It results in misallocation of resources. It results in lower standard of living for the fishing community. Therefore economic solution ensures that cost of harvesting drive is taken into consideration. Maximum yield is economically inefficient and unjustified as it entails less earning and less production for each fish operator.

Efficiency Dimension

Efficiency means economic efficiency, which includes both technical efficiency and allocative efficiency. It is defined as the maximisation of profit for a given level of input or as the minimisation of cost for a given level of output. Efficiency in marine fishing is commonly measured by the catch per unit of effort. It is estimated in terms of weight, number or value of catch per fishing voyage. Efficiency issues arises from:

a. Inefficiency of traditional method or technology
b. Inefficiency due to under exploitation of deep sea zones
c. Inefficiency due to over exploitation of inshore zones

Marine fisheries are unmanaged involving open access. Every fisher has an incentive to catch as much fish as possible in the present regardless of its implication for the future fish harvest. This behaviour is optimal for the individual fisher but positively detrimental to all when practiced by everyone. This is tragedy of commons. Free riding permits the fisher to enter into physical competition with all other to derive the maximum harvest so long as his average cost is not prohibitive. The consequence is that the fish population is exploited in a reckless manner beyond its regenerative capacity. It causes permanent destruction and depletion of fish resources.

Equity Dimension

Equity concerns are closely related to that of efficiency, inefficiency and sustainability. Inefficiency and inequity are mixed together in fishing business. Equity problems are exacerbated by asymmetries among users, which creates opportunities for some at the expense of many. The modern sector has a clear edge over the traditional sector. The consequence is higher income and better means of institutional benefits, which in turn ensures continuous access to better technology. The lower income level of the traditional fishermen tends to prevent them from acquiring this improved technology. The existing market structure is also a source of inequality between both sectors. The traditional fishermen are scattered all along the coast and an oligopsonistic market structure prevails. On the other hand the primary market for the modern fisher is concentrated at a single point and is oligopolistic in nature. As a result the modern fishermen obtain better prices for their produce. So measure should be taken by the govt to narrow down the inequalities.

Technological Dimension

Technological polarisation is a serious phenomenon in the fishing business. It is closely linked with the issues of efficiency, sustainability and equity. There are two sectors: traditional sector and the modern sector. They represent different ends of a continuum, producing the same product, with different levels of technological sophistication. The traditional sector uses non-mechanised boats while the modern sector uses mechanised crafts. About 90 percent of the fishing boats are non-mechanised. Both the sectors are economically viable at their respective levels of technology. Traditional sector contributes about 60 percent of the total fish production of the country. But the level of income, fishing range, market outlet, employment, social mobility and financial dependence keep traditional fishermen subservient to the economic decisions and operating constraints placed on them by those who buy their product. Because their number is very large and their access to marine water is very limited (due to lack of technical sophistication) So the competition is very cut throat in the inshore waters. The income per person has been declining because of diminishing returns. This technological dualism accentuates the income inequality. This polarisation of fishing sector due to technology is a logical sequel to the developmental efforts undertaken in sixties to promote the fishing industry.

Public Policy

Without any regulation of the Common Property Resource(CPR) the optimal level of fish harvesting is not achieved and eventually fish is over exploited. In order to curb the over fishing practice several management alternatives can be adopted. The cost of harvesting can be raised by a suitable tax, so that the cost curve shifts. Therefore the equilibrium occurs at an earlier point. It gives optimum level of harvesting drive. The tax amount will be equal to the external cost involved. A tax on harvesting can also achieve the same result. Govt. could manage the resource and force the operation at optimum level, imposing quotas on harvest or number of operators. Or the public authority can make it similar to any other resource and place it under control of a profit maximising sole ownership who will automatically attempt to operate at the optimum level.

The intervention of the public authority is necessary to deal with the externalities of CPR. Apart from licensing and taxation regulation include the size of catch and limitation on seasons and equipments. These are the methods of internalising externalities of CPR. The public authority can regulate in three ways:

1. limiting catches or access by season and areas
2. regulating the kind of effort or technology used
3. controlling the amount of effort by licensing the trawlers or imposing taxes/fees

REFERENCES

Berkes, F.(1985). The Common Property Resources problem and the creation of limited property rights, *The Human Ecology*, Vol 13:187-208.

Jodha, N S.(1986). Common property resources and the rural Poor in dry regions of India, *Economic and Political Weekly* ,21 July 5: 1169-81

Kainath,G S.(1992). Strategy for Development of Marine Fishery, *Financial Express*, 4th Jan, 1992.

Singh, Kartar (1993). Marine Fishermen Cooperatives in Kerala, Working paper No.40, *Anand, Institute of Rural Management.*

Srivastav, UK and Dharma Reddy (1983). *Fishery Development in India*, Concept Pub, New Delhi.

Chapter **33**

Quantitative Methods in Aquaculture Economics Research

K. N. Sahoo

Introduction

Quantification of costs and revenues is necessary to evaluate farm performance in aquaculture at experimental station level. Such calculations aided by economic engineering methods can help in determining the most appropriate package of technologies for different categories of farmers in different geographical locations. The method described below is based on the works of T.V.R. Pillay (Aquaculture: Principles and Practices). The computation of capital cost (initial cost), annual operating cost (variable cost plus fixed cost), and gross revenue (income) shows the performance of aquaculture farms. In this chapter an attempt has been made to illustrate the above mentioned useful fundamental economic indicators along with the computational procedure of assessment of economic feasibility of aquacultural projects which is very important for any entrepreneur.

Cost and Return Analysis

Any business concern needs capital for a start. Economists define capital as "the produced means of production" (Bohm-Bawerk). Capital is man-made and productive. Capital is usually considered a stock, and increases by investment. In aquaculture cost incurred on pond construction, stores and workshop, water supply system, nets etc. are termed capital costs. Each of

Applied Bioinformatics, Statistics and Economics in Fisheries Research: pp. 481-488

New India Publishing Agency, New Delhi (India)

them has an economic life period expressed in years. Annual depreciation for any asset is calculated by dividing initial cost by economic life period.

Economists treat cost as either fixed or variable. But this is a distinction in the short run, when at least one factor is fixed in supply. In the long run, all factors are variable and therefore all costs. Variable and fixed costs are usually calculated annually. Variable cost and fixed cost together constitute annual operating cost. Variable costs are associated with variable factors of production which vary with the level of output. In aquaculture, cost incurred on seed, feed, hired labor, diesel, electricity etc. constitute variable cost. Total cost incurred on each item is calculated by multiplying the quantity used with unit price. Fixed costs are associated with fixed factors of production and do not vary with the level of output. Land lease, interest on loan, depreciation, maintenance, permanent labor (which includes self-labour) are considered fixed costs. Variable and fixed costs are usually calculated annually. Income or gross revenue is estimated by multiplying annual production with unit price of the product.

A hypothetical Example: would help us to clarify the concepts involved in cost and return analysis in aquaculture. The analysis is based on hypothetical data for one ha. pond area for one year operation. (Table I, II, III).

TABLE I. Initial costs

Item	Cost (in Rs.)	Economic life (Years)	Annual depreciation
Pond construction	1,00,000	NA	
Godown+ Labor hut	25,000	10	2,500
Water pump	25,000	10	2,500
Nets	10,000	5	2,000
Boat	5,000	5	1,000
Miscellaneous	5,000	NA	
TOTAL	**1,70,000**		**8,000**

TABLE II. Annual operating costs

Costs	Quantity	Price (in Rs.)	Total cost (in Rs.)
Variable costs			
Seed	5000	1	5,000
Weed control			2,000
Fertilizer			10,000

Contd...

Table II. Contd...

Feed			1,00,000
Casual labour			500
Electricity +Diesel			500
Sub total 1			1,18,000
Fixed costs			**10,000**
Land lease			
Interest on loan (10%)			11,800
Depreciation			8,000
Maintenance			5,000
Permanent labour			1,000
Water cess			1,250
Sub total 2			37,050
Total cost			1,55,050
SUB TOTAL (1+2)			

TABLE III. Income

Production (in tonne)	Unit price (in Rs.)	Income (in Rs.)	
Fish	10	20,000	2,00,000s

From the above hypothetical Example: several indicators of farm performance are estimated.

Profit

Economist's definition of profit is much wider than what accountants mean by profit. In economic theory, the difference between sales revenue and the full opportunity cost of resources involved in producing the goods is termed as profit (Samuelson & Nordhaus). In this hypothetical Example:

Profit = Annual income – Total annual operating cost
= 2,00,000 – 1,55,050
= Rs 44,950

Rate of Return on Capital Costs

Capital, once sunk in a particular activity is not available for other purposes, until and unless disposed off in the market. In aquaculture, return from capital is thought to be spread evenly over the years, excepting perhaps some initial years when the farmer/entrepreneur is not an expert in the science and art of aquaculture. It depends upon the level of profit, and profit in aquaculture mainly depends upon the level of production. In this hypothetical example:

Rate of Return on Initial Cost (expressed as a percentage)

= (Profit/Total initial cost) x 100

= (44950/1,70,000) x 100

= 26.44%

Rate of Return on Operating Cost

Similarly a rate of return on operating cost can be calculated. This rate increases with an increase in profit or with a decrease in the level of annual operating cost or both. In this Example:

Rate of return on operating cost

= (Profit/Annual operating cost) x 100

= (44950/1,55,050) x 100 = 29%

Pay Back Period

Pay back period is the number of years required to recover the initial investment. The less the number of years the more the likelihood that a particular investment would be carried out compared to others. But this is a crude measure. A short pay back period might be beneficial for the farmer/ entrepreneur, but in the long run thought must be expended on the sustainability of aquaculture projects . In this particular case

Pay back period = Total initial cost / (Profit + Annual depreciation)

= 170000/ (44950 + 8000) = 3.2 years

Concept of Break Even

Break even price/production ensures that all fixed and variable costs are covered. Break even ensures that an entrepreneur continues in his present business. In the given Example:

Break Even Price

= Total annual operating cost/Level of out put

= 1,55,050/10

= Rs 15,505/ tonne

Break Even Production

= Total annual operating cost/Unit price of product

= 1,55,050/20,000

= 7.75 tonne/ha.

Economic Feasibility of Aquacultural Projects

In the choice among alternative investments the objective is to make the best possible use of a limited resource, capital. An investment project can usually be written down as a series of monetary inflows and outflows (benefits and costs) expected over a number of years. Characteristically, there will be a major payment at the beginning of the project, a series of net receitpts over its useful life and perhaps a small receipt for scrap value at the very end. As people have a positive rate of time preference and attach higher values to money flows the earlier they are expected to occur, it is usual to convert the stream of dated sums of money to a *present value* by multiplying them by a *discount factor*. A discount factor is given by

$$D = 1 / (1 + i)^t$$

where i is the rate of interest or discount thought appropriate by the decision maker and t is the 'date' of the money flow. Suppose the discount rate to be 5%, then Rs 21 to be received next year has a present value of only Rs 20. Had the discount rate been 10% the present value would have been about Rs 19. Two things should be noticed

1. With a given discount rate the present value gets smaller the more distant the expected money flow, i.e., the distant future is less important than the near future.
2. The present value of a given stream of expected money flow is less the greater the rate of discount.

Net present value (NPV) criterion states that a project is acceptable if present value of receipts

- present value of operating and maintenance costs
- initial out lay exceeds zero or NPV>0

A closely related criterion uses the internal rate of return. NPV falls as the rate of discount increases. There must therefore, be some points at which NPV crosses over from being positive to negative. The rate of discount at which this crossing over occurs is defined as the internal rate of return. If the rate of return exceeds the rate of interest or discount (i) the project is acceptable. Thus $R > i$ is the appropriate criterion

IRR may be applied to the incremental cash flow from an incremental investment in a project as well as to an array of unrelated projects. The internal rate method postulates that the algebraic sum of the compound amounts of all cash flows for a project is zero at some internal rate of return are r found by trial and error. The decision criterion is the internal rate r, larger r being preferred. The present worth method also referred to as incremental present forth, is restricted to comparisons of projects that has identical lives L or that cover the same total life span.

If it can be assumed the costs and returns of replacements will repeat those of the original asset, multiple cycles of a short term project may be compared with a single long term project covering the same time span , e.g. three 5-year lives considered equivalent to one 15-year life . With this method present worth is the decision criterion, larger present worth being preferred . This method is suitable for projects that have no positive receipts, a situation in which IRR is indeterminate.

A hypothetical example: would clarify the concepts involved in IRR and NPV based on data in Table - IV and V.

Table IV : Computation of Net Present Value (NPV) (Hypothetical)

Year (c1)	Initial cost (c2)	Annual operatin g cost (c3)	Gross cost (c4) = c2 + c3	Gross benefit (c5)	Incremen tal Benefit (c6)	Discount factor (10%) (c7)	Present worth (c8)
1	1,70,000	1,35,250	3,05,250	2,00,000	-1,05,250	.909	-95,672
2	0	1,35,250	1,35,250	2,00,000	64,750	.826	53,483
3	0	1,35,250	1,35,250	2,00,000	64,750	.751	48,627
4	0	1,35,250	1,35,250	2,00,000	64,750	.683	44,224
5	0	1,35,250	1,35,250	2,00,000	64,750	.621	40,209
6	0	1,35,250	1,35,250	2,00,000	64,750	.565	36,583
7	0	1,35,250	1,35,250	2,00,000	64,750	.513	33,216
8	-	-	-	1,14,000 (a)	1,14,000	.467 (b)	53,238
Total	1,70,000	9,46,750	11,16,750	15,14,000	3,97,250	4.868	2,13,953

(a) = salvage value = Total initial cost - Total depreciation

= 1,70,000 - (8000* 7)

= 1,14,000

(b) = The value is omitted from total to avoid double counting.

NPV (10% d.f) over the eight years period = Rs.2,13,953

Table V : Computation of Internal Rate of Return (IRR) (Hypothetical)

Year	Initial cost	Annual operating cost	Gross cost	Discount factor (18%)	Present value of gross cost (18%)	Gross benefit	Present value of Gross benefit (18%)	Incremental Benefit (cash flow)	Persent Worth (18%)
C1	C2	C3	C4 = c2 + c3	C5	C6	C7	C8	C9	C10
1	1,70,000	1,35,250	3,05,250	0.847	2,58,546	2,00,000	1,69,400	-89,146	-75,506
2	0	1,35,250	1,35,250	0.718	97,109	2,00,000	1,43,600	46,491	33,380
3	0	1,35,250	1,35,250	0.609	82,367	2,00,000	1,21,800	39,433	24,014
4	0	1,35,250	1,35,250	0.516	69,789	2,00,000	1,03,200	33,411	17,240
5	0	1,35,250	1,35,250	0.437	59,109	2,00,000	87,400	28,296	12,365
6	0	1,35,250	1,35,250	0.370	50,042	2,00,000	74,000	23,958	8,864
7	0	1,35,250	1,35,250	0.314	42,468	2,00,000	62,800	20,332	6,384
8	-	-	-	0.266 (a1)	-	1,14,000 (b1)	30,324	30,324	8,066
Total	1,70,000	9,46,750	11,16,750		6,59,430	15,14,000	7,92,524	1,33,099	34,807

(a1) - omitted from total discount factor to avoid double counting
(b1) - salvage value

Benefit cost ratio (B/C) at 18% = 7,92,524 / 6,59,430 = 1.2

An investment with a benefit-cost ratio greater than one can be considered as feasible.

Net present worth at 18% = 7,92,524 - 6,59,430 = 1,33,094

For this hypothetical aquaculture project IRR would be much higher which brings net present worth equal to zero. This can be found by trial-error method.

REFERENCES

Sloman John, 1997. The theory of distribution of income. *Economics.* Third edition, Prentice Hall of India Pvt. Ltd., New Delhi. pp. 268-279.

Chaston Ian, 1988. Cost accounting. *Business management in fisheries and aquaculture.* Fishing News books Ltd., England. pp. 51-61

——— Investment appraisal and financial analysis. *Ibid.* 62-70.

Pillay, T.V.R. 1990. Economics and financing of aquaculture. *Aquaculture: Principles and practice.* Fishing News Books Ltd.,England. Pp. 256-257.

Lekhi, R.K. 1993. Project evaluation and cost benefit analysis.*Economics of Development and Planning* section V. Kalyani Publisher New Delhi. pp: 89-105

Collard David. 1972. Time. *Prices Markets and Welfare.* Faber and Faber London. pp: 22-29

Gittinger, J. P. 1972. Economic Analysis of Agriculture Projects. John Hopkins University Press, Baltimore.

Sahoo, K.N. Roy, A.K., Saradhi, K.P. and Saha, G.S. 2000. Break even analysis of carp culture in Eluru mandal- A case study. Paper presented in Fifth Indian Fisheries Forum held at Central Institute of Freshwater Aquaculture(CIFA), Bhubaneswar during 17-20 January, 2000.

Chapter **34**

An Econometric Analysis of Shrimp Farming

P. K. Katiha & Chinmoy Chakraborty

Introduction

In view of the existence of productive agro-climatic and soil-morphological conditions, a dynamic estuarine river system and unique network of productive brackishwater aquaculture has emerged in the coastal states of India. It is immensely suitable for the shrimp/shrimp+fish culture providing high valued export quality shrimp. The Indian contribution of world shrimp export is at 6% (Katiha *et al.*, 2003).Unfortunately, investment in shrimp farming has slowed down (Hempel, 2000) in recent years due to lack of interest among the entrepreneurs. Therefore, it is pertinent to work out and compare the economic feasibility of various shrimp farming system. Studies on cost and return for shrimp production are scanty, therefore, in depth analysis with identification of bottlenecks, yield gaps and constraints is the need of hour. It may lead to harness the potential for shrimp farming through inculcating entrepreneurs' interest. This may also help the financial institutions for reconsidering loan procedure and further investment in this enterprise. Considering, over one third of national potential area for the enterprise and its sustainability over past decade, the study mainly concentrates on shrimp farming in the state of West Bengal.

Methodology

Three types (*viz.* extensive, modified extensive and semi-intensive) of shrimp farming system prevailing in this state. Information regarding

Applied Bioinformatics, Statistics and Economics in Fisheries Research: pp. 489-499

New India Publishing Agency, New Delhi (India)

polyculture of *P. monodon* in case of extensive system and monoculture of *P. monodon* for modified extensive and semi-intensive systems were collected using a structured questionnaire. The data thus procured, were tabulated and analysed using simple average method. The cost and income analyses of different management systems were worked out to cater the need of the objectives of the study. Comparative economic analysis was done to identify the most profitable one.

Cost and Income Analysis

Capital cost included total cost incurred during farm design and construction and purchasing of assets *viz.* aerators, pumps etc. This was estimated for three different management systems. Fixed cost included land rent, cost of diesel used for farmhouses through out the year, cost of depreciation and interest. The depreciation cost was estimated at the rate of 10% per annum on capital cost. The interest on capital cost was calculated at the rate of 8% per annum. Farm specific cost included license fees, telephone bills, consultancy fees, farm maintenance charges and cost of food required for farm labour.

Shrimp and fish are the only output in brackishwater aquaculture. The total quantity of fish and shrimp harvested and sold were recorded from all the management systems. The cost and return components and their composition considered (Bhatta, 2004) in the present study are given below:

Total cost = Variable cost + Fixed Cost

Variable Cost = $A_1 + A_2 + A_3 + A_4 + A_5 + A_6 + A_7 + A_8 + A_9 + A_{10}$

Where, A_1= Cost of seed; A_2 = Feed cost; A_3= Inorganic fertilisers; A_4= Organic fertilisers; A_5=Growth promoter; A_6= Health management cost; A_7= Harvesting cost; A_8= Transportation cost; A_9= Diesel cost for pumping; A_{10} = Farm specific cost.

Fixed cost = Land rent (B_1) + Diesel cost for farm house (B_2)

Total revenue (R) = Return from fish sale (R_1) +Return from shrimp sale (R_2)

Gross profit (GP) = Total revenue (R) – [Variable (A) + Labour cost]

Net profit (NP) = Gross profit (GP) - [Fixed cost (B) + depreciation+ interest]

Comparative Economic Analysis

The comparative economic analysis was done to identify the most profitable shrimp culture system for adoption in farm level. Non-discounted measure (e.g. Pay back period) and discounted measures (Net present value,

Benefit cost ratio & Internal rate of return) were used to appraise the economic viability of different types of management systems. But for the purpose of simplicity only pay back period is mentioned.

Pay Back Period

It simply estimates time required to recover the initial investment out of expected earnings from the investment before any allowances for depreciation. It was estimated by following formula:

$$T = \frac{C}{E}$$

where, T = Pay back period

C = Initial investment cost

E = Average annual profits expected from the investment before depreciation.

Result and Discussions

Cost and income analyses

Variable cost : The important aspects of variable costs are availability of inputs. The stocking density has a major role in the production and it has a close bearing on returns to farmers. The growth rates of *P. monodon* vary depending on the rate of stocking and prevailing environmental conditions. Scarcity of *P. monodon* seed and exorbitant price hike of post larvae and outbreak of white spot disease resulted in under stocking in 'bhery'.

Good quality of seeds at reasonable price during stocking period is one of the key issues for success. Next to seed, feed is the second most important input in shrimp culture. Good quality of formulated feed is very essential to obtain faster growth of *P. monodon*. Brennan *et al.* (2000) investigated that investment in higher quality inputs had a positive effect on yield and income in the brackishwater region of the Mekong Delta, The cost of feed can be reduced by using feed with improved conversion ratio and by lowering the feed price. The feed conversion ratio can be improved by avoiding wastage of feed and uniform distribution of feed during feeding. If the molting stage is recognized, daily feed quantity can be reduced to minimize the feed cost. Shrimps have to be harvested immediately at marketable size, failure of which can lead to more expenditure on feed (Annamalai, 1994).

Fertilization and supplementary feeding both were used simultaneously to enhance the yield. The coastal water of West Bengal is highly productive and therefore cost of fertilisers can be reduced by proper water exchange in pond. The next important item is cost of diesel for pumping. Location of farms, biomass of the ponds and water quality fluctuations were the few direct or indirect regulating factors involved with the costs of diesel. The material input included inorganic and organic fertilisers, feed, growth promoters etc. Inorganic fertilisers commonly used in this farming system were SSP and Urea. Organic manure included cow dung.

The average quantity of inputs used under various shrimp production systems is mentioned in Table 1.

Table 1: Inputs used in different farming system in West Bengal.

ha^{-1} $crop^{-1}$

				Management system		
Purpose	**Item**	**Unit**	**Extensive**	**Modified extensive**	**Semi-intensive**	**Average**
Pond preparation	Labour	nos.	1.00	2.00	3.00	1.45
	Bleaching powder	kg	2.90	263.00	358.68	39.81
	Zeolite	kg		34.00	282.16	17.88
	Lime	kg	85.70	3029.00	2678.14	423.41
Grow out culture	Urea	kg	0.50			0.43
	Phosphate	kg	1.70			1.49
	Sulphate	kg	16.40			14.37
	Poultry excreta	kg	9.10			5.34
	Cow dung	kg	138.10			80.58
	MOC	kg	38.00			33.49
	Feed	kg	5.00	2585.00	4159.49	403.93
	Egg	nos.		25.00	311.00	18.95
	Vitamin powder	kg		1.42	12.79	0.79
	Vitamin tablet	nos.		143.03	1482.54	94.03
Health management	Protozoan	kg		7.99	12.00	1.21
	Fungal	kg		0.45	5.50	0.33
	Bacterial	kg		0.38		0.03
	Viral	kg		2.12	7.65	0.57
	Gas absorbent	kg		0.36		0.02
	Probiotics	kg		11.81	4.30	1.02
	Bottom sanitizers	kg	0.80	1.00		0.74
Harvesting	Labour	nos.	0.10	12.00	5.98	1.16
	Wooden box	nos.	1.00			0.02
	Sack	nos.		1.00		0.79
	Bag net	nos.		1.00	1.00	0.2
	Ice	kg		1803.00	1879.00	223.31
	Bucket	nos.	1.20	31.00	47.00	5.7

The cost structure for different grow out technologies of *P. monodon* farming indicated that cost of inputs fluctuated according to intensity of their use across different systems. The maximum cost was for high input semi-intensive system. The lowest cost was for low input extensive polyculture due to absence of major inputs. In this experiment, variable cost (Table 2) was estimated at Rs. 0.17 lakh ha^{-1} crop^{-1} for extensive system. It was observed Rs. 2.55 lakh ha^{-1} crop^{-1} for modified extensive and Rs. 3.82 lakh ha^{-1} crop^{-1} for semi-intensive system. The Average variable cost was estimated at Rs. 0.53 lakh ha^{-1} crop^{-1}.

Table 2: Variable costs of different management systems. (Rs ha^{-1} crop^{-1})

Item	Management system			
	Extensive	Modified extensive	Semi-intensive	Average
Seed value of *P. monodon*	5623.52	51535.58	79296.99	12682.88
Fish seed value	3723.99	-	-	3272.28
Total seed value	9347.51	51535.58	79296.99	15955.16
Feed value	207.81	135080.65	206585.37	20469.05
Cost of inorganic fertilizers	615.14	18495.49	29717.83	3407.80
Cost of organic fertilizers	458.02	-	-	401.96
Cost of growth promoter	-	977.43	5850.43	390.75
Health management cost	118.33	5619.92	14413.34	1278.88
Harvesting cost	178.60	3720.41	3582.02	601.31
Transportation cost	97.54	2432.88	5451.94	551.49
Diesel cost for pumping	563.68	23776.70	23752.59	3377.77
Other cost	5466.73	13785.90	13008.12	6435.02
Total Variable Cost	**17053.36**	**255424.96**	**381658.63**	**52869.19**

Fixed Cost

Brackishwater fish farming involves high capital investment (Ghosh, 1990). In this study, capital cost (Table 3) was estimated at Rs. 0.21 lakh ha^{-1} in extensive system. The capital cost in modified extensive system was Rs. 1.02 lakh ha^{-1} but in semi-intensive system, it was very high constituted about Rs. 4.30 lakh ha^{-1} primarily due to cost of aerators, more number of pumps, pipes etc. The Average capital cost for shrimp farming was estimated at Rs. 0.48 lakh ha^{-1}. The total fixed cost was Rs. 0.17 lakh ha^{-1} crop^{-1} in extensive system

(Table 3). It was observed Rs.0.36 lakh ha^{-1} $crop^{-1}$ in modified extensive and Rs. 0.97 lakh ha^{-1} $crop^{-1}$ in semi-intensive system respectively with an Average average at Rs. 0.22 lakh ha^{-1} $crop^{-1}$.

Table 3: Fixed costs of different management systems.

(Rs ha^{-1} $crop^{-1}$)

Item	Management system			
	Extensive	Modified extensive	Semi-intensive	Average
Capital cost	21317.29	102476.42	430416.07	48339.57
Land rent	12511.77	6037.22	8608.32	11872.06
Diesel cost for farm house	281.85	11888.35	11876.30	1688.88
Depreciation in capital cost (10%)	2131.73	10247.64	43041.61	4833.96
Interest on capital cost (8%)	1705.38	8198.11	34433.29	3867.17
Total	**16630.73**	**36371.32**	**97959.52**	**22262.07**

The land of all the 'bhery' in North and South 24 Parganas was privately owned with multiple ownership. Katiha (2002) studied that the number of landowners varied from 150 to 300. These were leased out through open auction. The auction was generally done in November to December. In few cases, although, the water body was leased out to individual, but the lease amount was shared by more than one person. The lease period varied from 1 to 4 years.

Valuation of Output

The price of fish and shrimp varied according to the grade and quality. The average price received by the farmers in extensive system was estimated at Rs. 105 Kg^{-1}, which included both fish and shrimp component. It was observed Rs. 277 kg^{-1} in modified extensive system and Rs. 287 kg^{-1} in semi-intensive system (Table 4). The range of price was found Rs. 110-340 kg^{-1} during the corresponding period depending on the count (no./kg), condition of shrimps and the method of selling. Total revenue generated from shrimp and fish in extensive system was Rs. 0.47 lakh ha^{-1} $crop^{-1}$. This was estimated at Rs. 4.85 lakh ha^{-1} $crop^{-1}$ in the modified extensive and Rs. 7.63 lakh ha^{-1} $crop^{-1}$ in semi-intensive system with an average at 1.15 lakh ha^{-1} $crop^{-1}$. The share of fish was 32% of total revenue in extensive system.

Table 4. Returns from different management systems. (Rs ha^{-1} $crop^{-1}$)

Item	Management system			
	Extensive	Modified extensive	Semi-intensive	Average
Price received (Rs kg^{-1})	105.13	276.84	286.83	143.29
Revenue from shrimp (Rs)	31913.36	484546.26	763305.83	101833.64
Shrimp yield (Kg ha^{-1} $crop^{-1}$)	303.56	1750.28	2661.18	710.68
Count (No. of pieces kg^{-1})	34.26	37.35	36.50	36.89
Revenue from fish (Rs)	14793.55	-	-	13304.78
Total revenue (Rs)	46706.91	484546.26	763305.83	115138.42
Share of shrimp in revenue (%)	68.33	100.00	100.00	88.44
Share of fish in revenue (%)	31.67	-	-	11.56

Cost, Return and Profitability of per kg Production in Different Management Systems

The cost and income analyses of different management systems are worked out in Table 5. In case of extensive system, the net profit was estimated at Rs. 0.07 lakh ha^{-1} $crop^{-1}$. The benefit cost ratio (BCR) was 1.17. The net profit for modified extensive system was estimated at Rs. 1.80 lakh ha^{-1}

Table 5: Profitability in different management systems (Rs ha^{-1} $crop^{-1}$)

Item	Management system			
	Extensive	Modified extensive	Semi-intensive	Average
1. Total revenue	46706.91	484546.26	763305.83	115138.42
2. Variable + labour cost	23286.50	268361.07	402318.60	60346.94
i. Variable cost	17053.36	255424.96	381658.63	52869.19
ii. Labour cost	6233.14	12936.11	20659.97	7477.75
3. Gross profit (1-2)	23420.41	216185.19	360987.23	54791.48
4. Fixed +depreciation+ interest	16630.73	36371.32	97959.52	22262.07
i. Fixed cost	12793.62	17925.57	20484.62	13560.94
ii. Depreciation +interest	3837.11	18445.75	77474.90	8701.13
5. Net profit (3-4)	6789.68	179813.87	263027.71	32529.41
8. B C Ratio	1.17	1.59	1.53	1.39

crop^{-1} with a BCR of 1.59. Among all the systems, highest net benefit was found Rs. 2.63 lakh ha^{-1} crop^{-1} in semi-intensive system with a BCR of 1.53. The net returns were higher for the management systems with higher investments. The Average net benefit in shrimp farming was estimated at 0.33 lakh ha^{-1} crop^{-1} with a BCR of 1.39. Profit was estimated at Rs. 15 kg^{-1}, Rs. 103 kg^{-1} and Rs. 99 kg^{-1} for extensive, modified extensive and semi-intensive systems respectively. Net profit (Rs. 6,790 ha^{-1} crop^{-1}) was very low in extensive system due to injudicious exploitation of natural fertility through the ages and lack of proper management know-how. This can be increased by stock entry control and intensive management care, which can consequently increase the survival rate at the time of harvest as well as the growth rate.

Miah & Islam (1992) projected the profitability of shrimp farming in Bangladesh. Although the yield of shrimps in Bangladesh was low compared with other shrimp producing countries, the results of the study indicated that it was a profitable business. Begum & Alam (2002) reported that an outbreak of viral disease in Bangladesh is crippling the shrimp farming and the economy. Pillai & Katiha (2004) opined that although semi-intensive shrimp farming is highly profitable, it is considered risky as well, because of overstocking and disease problems. Many companies went to the market to raise equity for shrimp farming business and there was a time when shares of such companies were considered as potential shares. Both the shrimp farming and the share market suffered serious setbacks in recent years and consequently, the situation is now considered highly volatile. High profitability and trade liberalization policies triggered a spurt in growth of semi-intensive shrimp farming in early nineties. But, it has not been put to rigorous financial and economic analyses so as to estimate the alleged trade-off and formulate sustainable policies for its development (Jayaraman & Selvaraj, 2000).

Comparative economic analysis (cash flow)

Various farming systems like extensive, modified extensive and semi-intensive have their own advantages, disadvantages and own implications with respect to sustainability, economics and equity (Rao & Krishnan, 2000). Israel *et al.* (1985) had done comparative economic evaluation of different pond culture systems for *P. monodon, P. indicus* and *P. merguiensis* using standard economic tools and methods of analysis. The extensive monoculture of shrimps (*P. monodon* or *P. indicus*) and the extensive polyculture of *P. monodon* and other shrimps with milkfish were found profitable. The semi-intensive culture system of *P. monodon* or *P. indicus* gave moderate results. This is largely due to higher capital requirements than the extensive systems (Anon., 1984).

In this study, cash flow is presented in Table 6. Pay back period was lowest in case of modified extensive (7 months) followed by extensive (24 months) and semi-intensive system (15 months). The Average pay back period was estimated at 15 months. Hence, considering pay back period modified extensive system was found profitable than other two systems due to lowest payback period.

Table 6: Comparative economic analysis of different management systems.

Management system	Year	Return	Cost	Net profit	Depreciation & interest	Net cash inflow	Payback period (approx)
Extensive	2000	46121	39392	6729	3837	10566	
	2001	47572	40692	6880	3837	10718	
	2002	46427	39667	6760	3837	10597	24 months
Modified Extensive	2000	468510	302607	165903	18446	184349	
	2001	443473	278703	164770	18446	183216	
	2002	541655	332887	208768	18446	227214	7 months
Semi intensive	2000	773350	507951	265399	77475	342874	
	2001	755732	494485	261247	77475	338722	
	2002	760836	498399	262437	77475	339912	15 months
Average	2000	111308	80827	30481	8701	39182	
	2001	111815	80757	31058	8701	39759	
	2002	122292	86243	36049	8701	44750	15 months

Conclusion

The benefit cost ratio was highest in modified extensive system (1.59) followed by semi-intensive system (1.53) and extensive system (1.17). Among all the systems, profit was highest in modified extensive system. Considering pay back period, modified extensive system was found profitable than other two systems, because this has lowest payback period.

The study revealed that modified extensive culture of *P. monodon* was comparatively profitable and commercially viable than other two systems. Based on these observations, it can be stated that modified extensive system may be adopted instead of traditional or extensive farming in this state and farmers may be encouraged to adopt this as a profitable enterprise. The study concluded that in West Bengal major emphasis may be given on sustainability of the system rather than on increasing fish yield. Following components should hold the key to ensure more sustainable production system in this state:

- More co-ordination with the financial institutions to be ensured with special emphasis on the Short Term Credits (STC).
- Similar to Kishan Credit Card (KCC) in agriculture sector, Fisheries Credit Card (FCC) may also be introduced to save their valuable time and assured access to finance in need.
- Formations of more self help groups (SHGs) are also required to exchange technology achieve cooperation, market information and post harvest activities – this will ensure higher yield, assured marketing, more income and ensured repayment.
- Considering the high risk factor in shrimp farming, a national level policy may be framed and implemented to cover shrimp farming under Aquaculture Crop Insurance to safeguard is from social problems like poaching, poisoning, disease as well as natural calamities like flood.

REFERENCES

Annamalai, N. 1994. Assessing the commercial viability of brackish water prawn farming. *In* Orientation course on brackish water shrimp culture for Indian entrepreneurs, 12-16 September, 1994. p xvii-1-11.

Anon. 1984. Economics of prawn culture systems. Annual report. *Tigbauan,* Iloilo City (Philippines), (1985). p 22.

Begum, Anwara and Alam, S.M.N. 2002. Social and economic impacts of shrimp disease among small-scale, coastal farmers and communities in Bangladesh. FAO Fisheries Technical Paper, Italy: Food and Agriculture Organization of the United Nations (FAO), No.406, p 191-199, Rome. Eds.Arthur, J.A., Phillips, M. J., Subasinghe, R. P., Reantaso, M. B., MacRae, I. H.

Bhatta, R. 2004. Socio-economic profile of fisheries and aquaculture in India. Draft final report of ICAR-ICLARM (WFC) project on 'Strategies and options for increasing and sustaining fisheries and aquaculture production to benefit poor households in Asia'. p 82.

Brennan, D., Clayton, H. and Thanh Be, Tran. 2000. Economic characteristics of extensive shrimp farms in the Mekong Delta. International Association of Aquaculture Economics and Management, *Aquaculture Economics & Management.* Vol. 4 (3-4). p 127-139.

Ghosh, A. 1990. Culture prospects of brackish water in fish and shellfish and their economics. *In*: Technologies for inland fisheries development, (eds.) Sugunan, V. V. & Bhaumik U., p111-122.

Hempel, E. 2000. Shrimp supplies now falling behind demand. *Seafood International.* London. Vol. 15 (10). p 25-32.

Israel, D., Apud, F. and Franco, N. 1985. The economics of different prawn and shrimp pond culture systems: A comparative analysis. Proceedings of the first international conference on the culture of penaeid prawns/ shrimps, Iloilo City, Philippines, 4-7 December 1984, Taki, Y.,Primavera, J.H.,Llobrera, J.A. (eds.). P 172-173,

Jayaraman, R. And Selvaraj, P. 2000. Constraints to Development of Shrimp Farming. National Centre for Agricultural Economics and Policy Research, Workshop Proceedings 7. (Eds. M. Krishnan and Pratap S. Birthal). P 65-73.

Katiha, P.K., Chakraborty Chinmoy, Chakraborty, I. and Toor, M.S. 2003. Trends, status and prospects for Indian fish exports in world fish trade. *J. Inland Fish. Soc. India,* Vol. 35 (1). p 85-93.

Katiha, P.K. 2002. Institutional arrangements in culture based fisheries management. *In* Culture based fisheries for inland fisheries development. Eds. V.V. Sugunan, B.C. Jha and M. K. Das. The Central Inland Fisheries Research Institute. Bull No. 113, P: 165-174.

Miah, M.T.H. and Islam, M.R. 1992. Shrimp farming in Satkhira district of Bangladesh: an economic analysis. *Economic Affairs* (Calcutta). Vol. 37 (2), P. 100-104.

Pillai, N.G.K. and Katiha, P.K. 2004. Aquaculture technologies and fishing practices in India. Draft final report of ICAR-ICLARM (WFC) project on 'Strategies and options for increasing and sustaining fisheries and aquaculture production to benefit poor households in Asia'. p149.

Rao, A., Srinivasa and Krishnan, M. 2000. Coastal aquaculture: issues of economics, equity and sustainability. *Fishing Chimes.* Vol. 20 (5), P: 13-15.

Chapter **35**

Econometric Approach for Estimation of Technical Efficiency of Aquaculture Farms

A. K. Roy & Nibedita Jena

Introduction

Aquaculture resources in India is vast in the form of ponds and tanks of 2.35 million ha, reservoirs of 3.15 million ha, oxbow lakes and derelict waters of 1.3 million ha and brakishwaters of 1.24 million ha. At present, the total fish production is 6.4 mmt. Out of which inland fish production is 3.4 mmt. The total aquaculture production of 2.47 million tonnes that constitutes 4.2% of global production was valued at US$ 2.9 billion (FAO, 2006). Ponds and tanks are the prime resources for aquaculture, however only about 0.8 – 0.9 million ha is used for aquaculture currently. Recently, National Fisheries Development Board with huge budget provision has been established to improve production, processing, storage, distribution and marketing of fisheries products (Anon, 2006 a, 2006 b). During 10th and 11th plan periods higher fish production targets have been kept evisiging a growth of about 8 percent from inland sector.

In Doha Development Agenda of WTO, fish and fisheries products are not covered by agreement in agriculture i.e. more stringent rules for subsidies in fisheries than for agricultural products. Fish was negotiated under market access for nonagricultural products. It was proposed to reclassify, eliminate or reduce capacity-enhancing subsidies. Therefore, it is presumed that due to imposed regulations, trade flow, economic efficiency and welfare may be

Applied Bioinformatics, Statistics and Economics in Fisheries Research: pp. 501-518

New India Publishing Agency, New Delhi (India)

affected posing a stiff competition of exporters in the international market.To face the international challenges from various regulations and domestic phenomenon of declining trend of marine fish production, emphasis has been given for efficient management of aquaculture resources. Therefore, measurement and analysis of production performance becomes critical. At the farm level, farmers must produce at maximun efficiency and productivity. Therefore, measuring Technical Efficiency (TE) at farm level, identifying factors associated with efficient production, assessing potential for and sources of future improvements are need of the hour for establishing sustainable aquaculture. It is generally suggested that instead of increasing the use of inputs to increase production, efforts should be made towards output growth through improved TE that means to produce more by utilizing inputs at hand more efficiently. Role of aquaculture in rural economic development of India is well established. For rational use of scarce resources, understanding of the pattern of input use and knowledge about impact of farm specific variables on proudctivity are necessary. This can be achieved through efficiency measurement of the aquaculture farms spread all over the country. Improvement in efficiency of resource management in aquaculture have been recognized as prime considerations by many aquaculturist, stakeholders and governments.

Empirical studies suggests that farmers in developing countries fail to exploit fully the potential of a technology making inefficient decision due to various reason of which management capacity is an important one. Various reports are available in the literature on potential enhancement of production through efficient use of resources. The profit of rice farmers in Pakistan's Punjab could be increased by 28% through improved efficiency, which is positively related with education and timeliness of input use (Ali and Flinn, 1989). Average technical efficiency for Nepalese agriculture is 84% for rice production suggesting that there is a potential of 16% increase in the firm output (Belbase and Grabowski, 1985). It is reported that the "poor but efficient hypothesis" may have overlooked important potential for increasing agricultural productivity (Battese, 1992 and Bravo-Ureta and Pinheiro 1993).

In the present communication, an attempt has been made to briefly define the concept of TE, various econometric models available for estimation of TE along with softwares, a brief literature survey on application of various models for TE estimation. Brief analytical procedure, output in the form of results and interpretation of input output variables and role of socioeconomic and farm specific variables on TE are also presented using an empirical aquaculture farm data.

Concept of TE

According to Greene, 1993, the level of technical efficiency of a particular firm is characterized by the relationship between observed production and some ideal or potential production. Another school of thought refers TE as the maximum attainable level of output for a given level of production inputs, given the range of alternative technologies available to the farmer. The measurement of farm specific technical efficiency is based upon deviations of observed output from the best production or efficient production frontier. If a farm's actual production point lies on the frontier, it is perfectly efficient. If it lies below the frontier then it is technically inefficient, with the ratio of the actual to potential production defining the level of efficiency of the individual farm. Farrell's (1957) definition of technical efficiency led to the development of methods for estimating the relative technical efficiencies of farms. The common feature of these estimation techniques is that information is extracted from extreme observations from a body of data to determine the best practice production frontier (Lewin and Lovell 1990). From this, the relative measure of technical efficiency for the individual firm can be derived. The approaches for estimating technical efficiency can be generally categorized under the distinctly opposing techniques of parametric and non-parametric methods (Seiford and Thrall 1990).

Economic, Allocative and Technical Efficiency

The concept of efficiency is decomposed into two components technical and allocative efficiency. Technical efficiency refers to the maximum attainable level of output for a given level of production inputs, given the range of alternative technologies available to the farmer. Allocative efficiency refers only to the adjustment of inputs and outputs to reflect relative prices, having chosen the production technology. Economic efficiency is the situation when technical and allocative efficiency is combined. Technical efficiency is just one component of overall economic efficiency. However, in order to be economically efficient, a firm must first be technically efficient. Profit maximization requires a firm to produce the maximum output given the level of inputs employed i.e. be technically efficient, use the right mix of inputs in light of the relative price of each input i.e. be input allocative efficient and produce the right mix of outputs given the set of prices i.e. be output allocative efficient (Kumbhaker and Lovell 2000). These concepts can be illustrated graphically using a simple example of a two input (x_1, x_2)-two output (y_1, y_2) production process (Fig 1).

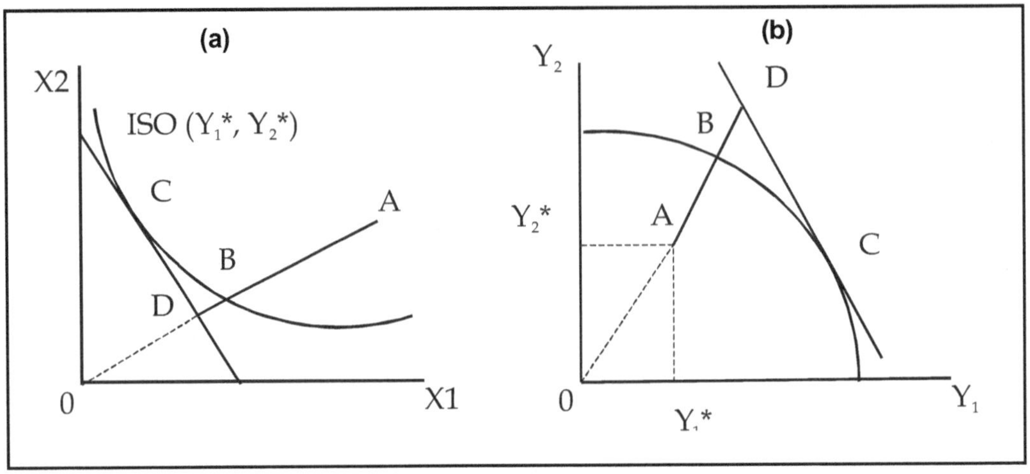

Fig 1: (a) Input Oriented & (b) Output Oriented efficiency measures
Source: Herrero and Pascoe, 2002

In Figure 1(a), the firm is producing a given level of output (y_1*, y_2*) using an input combination defined by point A. The same level of output could have been produced by radially contracting the use of both inputs back to point B, which lies on the isoquant associated with the minimum level of inputs required to produce (y_1*, y_2*). The input-oriented level of technical efficiency is defined by 0B/0A. However, the least-cost combination of inputs that produces (y_1*, y_2*) is given by point C i.e. the point where the marginal rate of technical substitution is equal to the input price ratio w2/w1. To achieve the same level of cost i.e. expenditure on inputs, the inputs would need to be further contracted to point D. The cost efficiency or economic efficiency is therefore defined by 0D/0A. The input allocative efficiency is subsequently given by 0D/0B in Figure 1.1(a) (Kumbhaker and Lovell 2000).

The production possibility frontier for a given set of inputs is illustrated in Figure 1(b) (i.e. an output-orientation). If the inputs employed by the firm were used efficiently, the output of the firm, producing at point A, can be expanded radially to point B. Hence, the output-oriented measure of technical efficiency can be given by 0A/0B. This is only equivalent to the input-oriented measure of technical efficiency under conditions of constant returns to scale. While point B is technically efficient, in the sense that it lies on the production possibility frontier, higher revenue could be achieved by producing at point C (the point where the marginal rate of transformation is equal to the price ratio p_2/p_1). In this case, more of y_1 should be produced and less of y_2 in order to maximize revenue. To achieve the same level of revenue as at point C while maintaining the same input and output combination, output of the firm would need to be expanded to point D. Hence, the revenue efficiency is given by 0A/0D. Output allocative efficiency is given by 0B/0D in Figure 1(b) (Kumbhaker and Lovell 2000).

For technical efficiency analysis, the two most popular methods are 1. Stochastic Frontier Production Function (SFP) and 2. Data Envelopment Analysis (DEA), which involve a parametric or econometric method and non-parametric or mathematical programming method, respectively. The two most commonly used packages for estimating of stochastic production frontiers and DEA are FRONTIER 4.1 (Coelli, 1996a) and DEAP (Coelli, 1996b) respectively. A brief outline of the models is furnished below.

Stochastic Frontier Production Function (SFP)

The Empirical Model: Farrell (1957) developed a conceptual model involving the contraction of inputs to an efficient frontier that laid the foundations for the development of both the parametric (SPF) and non-parametric (DEA) estimation of efficiency. The efficient frontier can be considered in terms of either the maximum output for a given set of inputs (an output orientation), or the minimum set of inputs required to produce a given set of output (an input orientation). Both relate to the assumption of profit maximization, the former approach representing the maximization of revenue for a given level of costs, while the later approach minimizes the costs for a given level of revenue (Tingley Diana et al., 2005). The stochastic frontier production function for farms is specified as follows taking into account the model developed and proposed by Aligner et al., (1977), Meioses, and van deer Brock (1977).

$$LnY_i = \beta_0 + \sum_{k=1}^{n} \beta_k \, LnX_{ki} + V_1 - U_i \qquad \text{...... (1)}$$

where subscript i refer to the i-th farm in the sample; Ln represents the natural logarithm; Y is output variable and Xs are input variables. βs are unknown parameters to be estimated; V_i is an independently and identically distributed N (0, σ_v^2) random error; and the U_i is a non-negative random variable associated with technical inefficiency in production, which is assumed to be independently and identically distributed and truncations (at zero) of the normal distribution with mean, m_i and variance, σ_u^2 ($|N(\mu_i, \sigma_u^2)|$). The maximum likelihood estimation of equation 1 yields estimators of β and γ where $\gamma = \sigma_u^2 / \sigma^2$ and $\sigma^2 = \sigma_u^2 + \sigma_v^2$ and σ explains the total variation of output from the frontier which can be attributed to technical inefficiency that lies between zero and one.

According to Battese and Coelli (1995), the technical inefficiency distribution parameter, m_i is defined as:

$$\mu_i = \delta_0 + \sum_{m=1}^{n} \delta_m \; Z_{mi} \qquad \text{...... (2)}$$

where Zs are various farm-specific variables and is are unknown parameters to be estimated. Since the dependent variable in Equation-2 is defined in terms of technical inefficiency, a farm-specific variable associated with the negative (positive) coefficient will have a positive (negative) impact on technical efficiency (Sharma and Leung, 2000a)

The technical efficiency of the i-th sample farm (TE_i) is given as:

$$TE_i = \exp(-U_i) \qquad \text{...... (3)}$$

The prediction of technical efficiencies is based on the conditional expectation of expression in equation-3, given the model specifications (Battese and Coelli, 1988). The parameters for the stochastic production frontier model in equation-1 and those for the technical inefficiency model in equation-2 can be estimated simultaneously using the maximum-likelihood estimation (MLE) program, FRONTIER 4.1 (Coelli, 1996a).

Data Envelopment Analysis (DEA)

DEA is a nonparametric approach to distance function estimation (Fa¨re et al.1994). The method involves the use of linear programming to construct a piecewise linear envelopment frontier over the data points such that all observed points lie on or below the production frontier. Let X be a K x N matrix of inputs, which is constructed by placing the input vectors, xi, of all N firms side by side, and Y denotes the M x N output matrix which is formed in an analogous manner. The output-oriented variable returns to scale (VRS) DEA frontier is defined by the solution to N linear programs of the form

$$\begin{aligned} &\min \ \theta \\ &\theta\lambda \\ &\text{subject to} \quad -y_i/\theta + Y\lambda \geq 0 \\ &\qquad\qquad\quad xi - X\lambda \geq 0 \\ &\qquad\qquad\quad N1'\lambda = 1 \\ &\qquad\qquad\quad \lambda \geq 0, \end{aligned}$$

where N1 is an N X 1 vector of 1s, k is an N X 1 vector of weights, and q is the output distance measure. We note that $0 \leq \theta \leq 1$ and that $1/\theta$ is the proportional expansion in outputs that could be achieved by the ith firm, with input quantities held constant.

In a similar manner, the input-oriented VRS DEA frontier is defined by the solution to N linear programs of the form

$$\begin{aligned} &\max_{\rho\,\lambda} \ \rho \\ &\text{subject to } -y_i + Y\lambda \geq 0 \\ &\qquad xi/\rho - X\lambda \geq 0 \\ &\qquad N1'\lambda = 1 \\ &\qquad \lambda \geq 0, \end{aligned}$$

where ρ is the input distance measure. We note that $0 \leq \rho \leq 1$

The technical efficiency measure under constant returns to scale (CRS), also called the 'overall' technical efficiency measure, is obtained by solving N linear programs of the form

$$\begin{aligned} &\min_{\theta_i^{CRS}} \ \theta_i^{CRS} \\ &\text{subject to } -Y\lambda + y_i \leq 0 \\ &\qquad \theta_i^{CRS} x_i - X\lambda \geq 0 \\ &\qquad \lambda \geq 0, \end{aligned}$$

where θ_i^{CRS} is a technical efficiency measure of the ith firm under CRS and

$$0 \leq \theta_i^{CRS} \leq 1.$$

The output and input-oriented models will estimate exactly the same frontier surface and therefore, by definition, identify the same set of firms as being efficient. The efficiency measures may, however, differ between the input and output orientations. Under the assumption of constant returns to scale, the estimated frontier and the efficiency measures remain unaffected by the choice of orientation (Coelli and Perelman, 1999).

The DEA model for computing technical efficiency may be solved using DEAP 2.1 (Coelli 1996b) and that for economic or 'revenue' efficiency may be solved using a general linear programming package, LINGO (1995).

Empirical Study on Estimation of Technical Efficiency (TE) using Aquaculture Farm Data

Background Information: Due to the methodological advances in efficiency and productivity analysis, many applications to aquaculture operations have been undertaken in the recent past abroad in carp farming (Sharma and Leung, 1998, 2000a, b); Linuma *et al.* 1999, Sharma *et al.* 1999; tilapia growout pond (Dey *et al.* 2000); tilapia hatchery operations (Bimbao *et al.*, 2000); mariculture of seabss and sea bream (Karagiannis *et al.* 2000) and salmon aquaculture (Tveteras and Battese, 2000, Vassdal and Roland, 1998). Some reports are available on the technical efficiency measurements in marine fisheries involving many factors and these include measures relating to both technology and skill, sonar and navigational aids, skipper education, experience, family history and age characteristics are incorporated as proxies for 'skipper skill' (Pascoe *et al.*, 1998, 2001, 2002, 2003, 2004) and verry few in India (Jayaraman, 1998, Roy *et al.* 2006a).

Data source : The present empirical study is based on a data set of 221 aquaculture farms that was collected using multistage stratified random sampling covering 73 revenue villages spread over 9 Mandals of West Godavari and Krishna districts of Andhra Pradesh during 1998-2002 under the AP-Cess fund project entitled "Technological innovations in aquaculture and its effects on sustainability of farming systems in Andhra Pradesh" funded by Indian Council of Agricultural Research (ICAR).

Model Used for TE Estimate : Of the two popular methods for estimation of TE, the parametric stochastic frontier production function involving econometric methods has been used for estimation of TE. In the present analysis the frontier production function defines potential output that can be produced by a farm with given level of inputs and technology. At first technical efficiency scores are estimated using stochastic frontier production function (Aigner, 1977). In the second stage, the influence of socioeconomic and farm specific variables on TE estimated using an established model proposed by Battese and Coelli, 1995 using DEAP (V 4.1).

Variables considered for the study: Input output Analysis and Technical Efficiency Estimate

Output : Yield (total quantity of fish harvested per hectare)

Input: Organic manure, inorganic fertilizer, lime, stocking biomass, Chemicals & Drugs, Feed, Labour and Water management cost.

Socio Economic Variables : Religion, caste, age, number of children in family, education, experience in carp culture, primary occupation, ownership type.

Farm Specific Variables : Pond size, pond renovation and asset cost, source of water, depth of water, source of seed, sources of advice taken, loan, harvesting technique, method of application of drug, method of application of feed, calamities faced, periodical netting for biomass checking, types of labour, types of organic manure, types of feed and disease encountered.

Results

The maximum likelihood estimates of the parameters for the stochastic frontier model and technical inefficiency model are reproduced in the following table 1.

Table 1: Maximum-likelihood estimates of stochastic production frontier and technical inefficiency models for carp production in Kolleru Lake, Andhra Pradesh, India.

	Parameter	Coefficient	Standard-error	T-ratio
Stochastic production frontier model				
Constant	β_0	1.2375***	0.1953	6.3358
Organic manure (X_1)	β_1	0.1411***	0.0377	3.7406
Inorganic fertilizer (X_2)	β_2	0.0240	0.0238	1.0094
Lime (X_3)	β_3	0.0330	0.0265	1.2825
Stocking weight (X_4)	β_4	0.0268*	0.0198	1.3549
Chemicals and Drugs (X_5)	β_5	-0.0146	0.0170	-0.8566
Feed (X_6)	β_6	0.2001***	0.0360	5.4119
Labor (X_7)	β_7	0.0117	0.0112	1.0472
Water management (X8)	β_8	-0.0226*	0.0117	-1.9281
Technical inefficiency model				
Constant	δ_0	0.5500	0.6035	0.9114
Religion (Z_1)	δ_1	-0.0853*	0.0508	-1.6771
Caste (Z_2)	δ_2	0.0136*	0.0098	1.3926
Age (Z_3)	δ_3	-0.0005	0.0014	-0.3748
No of children (Z_4)	δ_4	0.0041	0.0085	0.4839
Education (Z_5)	δ_5	-0.0016	0.0041	-0.4030
Experience (Z_6)	δ_6	0.0002	0.0022	0.1042
Primary occupation (Z_7)	δ_7	-0.0279*	0.0224	-1.2466
Pond size (Z_8)	δ_8	-0.0027**	0.0011	-2.4203
Renovation (Z_9)	δ_9	0.0236	0.0197	1.2025
Asset cost (Z_{10})	δ_{10}	0.0000	0.0000	0.4021
Ownership type (Z_{11})	δ_{11}	-0.0371	0.0332	-1.1183

Contd...

Table 1. Contd...

Source of water (Z_{12})	δ_{12}	-0.0975**	0.0400	-2.4363
Depth of water (Z_{13})	δ_{13}	-0.0009	0.0235	-0.0372
Source of seed (Z_{14})	δ_{14}	0.0300*	0.0164	1.8287
Sources of Advice taken (Z_{15})	δ_{15}	-0.0046	0.0103	-0.4458
Loan (Z_{16})	δ_{16}	0.0217	0.0245	0.8823
Harvesting technique (Z_{17})	δ_{17}	0.0593*	0.0409	1.4519
Method of application of drug (Z_{18})	δ_{18}	-0.0154*	0.0102	-1.5186
Method of application of feed (Z_{19})	δ_{19}	0.0033	0.0430	0.0770
Calamities faced (Z_{20})	δ_{20}	-0.0299*	0.0181	-1.6504
Periodical netting for biomass checking (Z_{21})	δ_{21}	0.0685**	0.0252	2.7139
Types of labour (Z_{22})	δ_{22}	-0.0424*	0.0305	-1.3880
Type of organic manure (Z_{23})	δ_{23}	0.0091	0.0132	0.6903
Type of feed (Z24)	δ_{24}	-0.0192	0.0163	-1.1807
Disease encountered (Z_{25})	δ_{25}	-0.0003	0.0041	-0.0664
	σ^2	0.0084	0.0008	10.8598
	γ	0.9992	1.3398	0.7458
	Log likelihood function	214.76039		
	LR test	63.053464		
	Mean TE	0.72608099		

(*Source*: Roy *et al.* 2006a). ***, **, * Significant at 1, 5 and 10 percent level of respectively.

Interpretation of econometric analysis of Input and output variables : From table 1 it reveal that the slope coefficients and output elasticities of inputs showed almost expected signs of the physical factors excepting water management cost used in the model. The elasticity of output for feed was found highest (0.2001) followed by organic manure (0.1411) and both coefficients are observed highly significant ($P<0.01$). This suggests the fact that both feed and organic manure have major influence on yield. Inorganic fertilizer, stocking weight, labour and lime have demonstrated positive impact but not proved to be statistically significant. Both chemicals and drugs and water management coefficient showed insignificant negative impact over yield.

Estimation of technical efficiency : TE estimates using the stochastic frontier model is presented in the form of frequency distribution in Table 2. The results derived from the econometric estimation indicate that technical efficiency (TE) scores for fish farmers of Kolleru Lake ranged from 0.5051 to 0.9899 with a mean of 0.7260 and modal class is 0.6 to 0.7 (Table-2).

Table 2: Frequency distribution of technical efficiency estimates for a stochastic frontier model.

Efficiency score class	Number of farmers	% of farmer	Cumulative frequency of farmer
0.50 – 0.60	6	2.71	17
0.60 – 0.70	90	40.72	125
0.70 – 0.80	88	39.82	199
0.80 – 0.90	30	13.58	218
0.90 – 1.0	7	3.17	221
Total	221	100.00	–

Source: Roy *et al.* 2006a

Scope for technical efficiency improvement : It is observed from table 2 that on average the farmers of Kolleru Lake area are able to obtain about 72.60% of potential output from the given set of production inputs. Therefore, in the short run there is a scope of increasing the productivity of carps by around 27.40% by adopting the technology and techniques used by best practice farms without adding any additional input. It is interesting to note that if the average farmer in the sample was to achieve the TE level of its most efficient farmer counterpart then the average farmer could realize 26.7 percent cost saving [1 - (72.60/98.99)]. Similarly the most technically inefficient farmer may achieve cost saving of 48.9 percent [1 - (50.51/98.99)].

Identification of determinants of Technical inefficiency : Socio economic and demographic variables like religion, caste, age, number of children, education, experience in carp culture, primary occupation, ownership type are taken into consideration with the expectations that these factors may have some impact on technical efficiency. Similarly farm specific variables like pond size, renovation, cost of assets, source of water, depth of water, source of seed, sources of advice taken, loan, harvesting technique, method of application of drug and feed, calamities faced, types of feed, types of organic manures applied, types of disease encountered, periodical biomass checking, types of labour engaged are also taken with the expectation to have impact over technical efficiency affecting carp yield. Results of maximum likelihood estimates of the technical inefficiency model and determinants of technical efficiency that are presented in Table-1. Twenty-five nonphysical socio-economic, demographic and farm specific factors were included in the model to test the possible impact and association with the technical efficiency of aquaculture farms operating in the Kolleru lake area. The inclusion of variables was based on the belief that these might be relevant in model estimations. A close look at Table-1 reveals that of the twenty-five factors considered for technical efficiency model, socio economic and farm specific factors like religion, age, education, primary occupation, pond size, ownership type, source of water, depth of water, sources of advice taken, method of application of drugs, calamities faced, types of labour, types of feed and disease encountered have positive impact on technical efficiency. On the other hand caste, number of children, experience, renovation, asset cost, source of seed, loan, harvesting technique, method of application of feed, periodical netting for biomass checking and types of organic manure showed negative association with technical efficiency. Of all the farm specific factors showing positive impact mentioned above only pond size and source of water were found significant at ($p<0.05$) and types of labour, religion, method of application of drug, primary occupation and calamities are found significant at ($p<0.10$). Likewise of the variables showing negative impact, periodical netting for biomass checking is found highly significant ($p<0.01$), but source of seed and harvesting technique and caste were found significant at lower level of significance ($p<0.10$) demonstrating the influence of these factors in efficiency of carp culture.

Interpretation : In depth analysis of impact of pond size on TE indicates that a consistent increase in mean TE with the size of the farm showing highest TE value of 0.7552 for aquaculture farms of size greater that 10 ha. This is at par with the popular belief that larger farms are capable to capture the economics of size and operates at higher efficiency levels compared to those of marginal and small farm. Pond area is estimated to have positive impact indicating large operations are technically more efficient than smaller ones (Iinuma *et al.*, 1999). Pond size is reported to have positive relationship on productivity (Roy *et al.*, 2002b). Water is the primary essential component of all aquacultural activities and its importance has been reflected through significant positive impact on TE. Farmers those who have set up their farms by the side of irrigation canals facilitating regular intake and periodical discharge of water are found to operate with significantly higher mean technical efficiency of 0.7287 compared to those using water from other sources with a mean TE of 0.6648 ($F_{cal}=5.57>F_{crit}=3.88$). Types of labour employed also showed positive impact on TE. Farms employing permanent labour could utilize their services for timely operation of various farm activities reflecting significantly higher mean TE of 0.7354 compared to those employed casual labour as and when needed reflecting mean TE of 0.6957 (F_{cal}=10.11>F_{crit}=3.88). Majority of the farm were operated by farmers of Hindu religion who are influential and economically sound being in an advantageous position for mobilizing resources for timely investment in various aquacultural activities resulting in significantly higher TE of 0.7275 compared to other religion showing 0.6207 (F_{cal}=5.33>F_{crit}=3.88). Method of application of drug has also shown positive impact over TE. To combat disease, drugs were used either diluting in water or spraying in water or mixing with feed demonstrating estimated TE of 0.7292, 0.7237 and 0.6284 respectively and the differences among treatment means were not found significant (F_{cal}=1.62<F_{crit}=3.04). Primary occupation was found to be significant at 10 percent level indicating that farmers doing fish farming as main activity showed significantly higher mean of 0.7359 compared to those of others showing mean TE of 0.6896 (F_{cal}=13.02>F_{crit}=3.88). This finding is at par with the observation of Sharma and Leung, 2000b. Impact of calamity on TE was found significant (p<0.10). One possible explanation may be that Kolleru lake is located in the flood and cyclone prone areas of Krishna and Godavari districts of Andhra Pradesh. As a result farms affected by flood or cyclone were automatically flushed out of organic load and metabolic wastes those used to be accumulated due to regular use of inputs like organic manure and feed resulting in marginally higher mean TE of 0.7276 compared to a mean TE of 0.7210 of farms not affected by calamities during last five years of operation although the difference is not proved statistically significant (F_{cal}=0.27<F_{crit}=3.88).

In Kolleru lake area, it is observed that about 86% of farmers resort to periodical netting for biomass checking that has shown significant negative impact over TE, may be due to the fact that as a result of frequent netting, the shallow pond bottom of farms were disturbed resulting in turbidity for a prolonged period due to presence of muddy substances creating unfavourable environmental condition for the growth of cultivated species. Mean TE of 0.7898 demonstrated by farms those are not doing periodical netting is observed significantly higher compared to TE of 0.7157 by farms resorting to periodical netting (F_{cal}=25.18 > F_{crit}=3.88). Contrary to priori expectation, source of seed showed negative impact on TE at 10 percent level of significance thereby indicating that the on-farm seed production system does not improve the efficiency of aquaculture farms. On the other hand, those purchased seed from outside may be of superior quality. From survey it is found that about 66% of sample farmers stocked ponds procuring seed from outside sources and the rest of the farmers reared seed in their farms for stocking the culture ponds. Mean TE of 0.7351 shown by farms stocked procuring seed from outside compared to mean TE of 0.7215 shown by farms those stocked seed rearing in their own farm. Analysis of another important factor harvesting technique showed negative impact on TE. It reveals that 95% of sample farmers are resorting to single stocking and single harvesting technique wherein mean TE of 0.7631 was obtained compared to mean value of 0.7241 demonstrated by farm practicing single stocking and multiple harvesting or multiple stocking and multiple harvesting techniques. However, the differences of TE between the two groups are not statistically significant (F_{cal}=2.47<F_{crit} = 3.38). In Kolleru lake general caste farmers used to be dependent on hired labour showed negative impact on TE. Mean TE of farms owned by SC, OBC and General caste farmers are 0.7333, 0.7195 and 0.7276 respectively. But differences among castes are found not statistically significant (F_{cal}=0.33<F_{crit} = 3.04). The results indicate that the farm specific variables included in the technical inefficiency model contribute significantly to the explanation of the technical inefficiencies among sample carp farms of Kolleru Lake area.

Conclusion and Recommendation

One of the most important aims of carp culture in India that contributes significantly towards food and nutritional security is enhancement of productivity. The present analysis of data has given attention and focused on the issues of productivity enhancement through improvement of TE of the carp farming system with available existing resources and technology of the carp farmers of Kolleru Lake area, popularly known as Carp Pocket of India for regularly supplying quality protein particularly eastern and northern states of India. The stochastic frontier estimate demonstrates the average technical efficiency of 72.60% indicating that there is scope for further increasing the output by 27.40% without increasing the levels of inputs. The use of stochastic frontier analysis enables ranking of the carp culture farms based on efficiency of yield of fish facilitating identification of the most efficient and inefficient farms across various carp farming system operating in the Kolleru lake area. The efficient farms may serve as a model farm for improving the efficiency of carp production in the kolleru lake area as well as throughout the whole country where the mean technical efficiency is reported to be lower (Sharma and Leung, 2000b) compared to the one reported in the present case. The empirical results suggest that there are immense possibilities to increase efficiency levels by extending culture operation in large size ponds, provision for supply of fresh water, application of drugs after diluting in water, employing labour on regular basis and leasing out the water bodies to farmers whose primary occupation is fish farming and educating the farmers to procure seed from reliable sources, proper harvesting technique and avoiding periodical netting. From the policy point of view, above mentioned farm specific variables are found to be most promising areas for action, suggesting that policy maker may look into the matter seriously for possible reforms to enhance the productivity without any additional input use.

REFERENCES

Aigner, D.J., C. A. K. Lovell, P. J. Schmidt, 1977. Formulation and estimation of stochastic frontier production function models. *J. Econometrics*, 6:21-37.

Anon, 2006a. Pamphlet, National Fisheries Development Board. Department of Animal Husbandry, Dairying and Fisheries, Ministry of Agriculture, Govt. of India, Block 401-402, Maitri Vihar, Huda Commercial Complex, Ameerpet, Hyderabad-500038, Andhra Pradesh, India.

Anon, 2006b. Inaugural address of Sri Sharad Pawar, Hon'ble Union Minister for Agriculture, Consumer Affairs, Food and Public Distribution. 9th September 2006, National Fisheries Development Board, Hyderabad, India.

Ali, M., J. C. Flinn, 1989. Profit efficiency among basmati rice producers in Pakistan Punjab. American *Journal of Agricultural Economics*, 71:303-310

Battese, G. E., T. J. Coelli, 1992. Frontier production function, technical efficiency and panel data: with application to paddy farmers in India. *J. Productiv. Anal.*, 3:153-169.

Battese, G. E., T. J. Coelli, 1988. Prediction of firm-level technical efficiencies with a generalised frontier production functions and panel data. *Journal of Econometrics*, 38: 387-399.

Battese, G. E., T. J. Coelli, 1995. A model for technical inefficiency effects in a stochastic frontier function for panel data. *Empirical Economics*, 20:325-332.

Belbase, K., and R. Grabowski, 1985. Technical Efficiency in Nepalese agriculture, *Journal of Developing Areas*, 19: 525-585.

Bravo-Ureta, E., E. Boris, Antonio, E. Pinheiro, 1993. Efficiency analysis of developing country agriculture: A review of the frontier function literature. *Agricultural and Resource Economics Review*, 22(1), 88-101.

Coelli, T.J., 1996 a. A guide to FRONTIER version 4.1: a computer program for frontier production function estimation. CEPA Working Paper 96/07, Department of Econometrics, University of New England, Armidale, Australia.

Coelli, T.J., 1996 b. A Guide to DEAP Version 2.1: A Data Envelopment Analysis (Computer) Program. CEPA working papers no. 8/96, Department of Econometrics, University of New England, Armidale, NSW 2351, Australia.

Coelli, T.J., S. Perelman, 1999. A comparison of parametric and non-parametric distance functions: with applications to European railways. *European Journal of Operational Research,* 117:326–339.

Dey, M. M., F. J. Paraguas, G. B. Bimbao, P. B. Regaspi, 2000. Technical efficiency of Tilapia grow out pond operations in the Philippines. *Aquaculture Economics and Management,* 4(1) & (2):33-47.

Färe, R., S. Grosskopf and E. C. Kokkelenberg, 1989. Measuring plant capacity, utilization and technical change: a non-parametric approach. *Int. Econ. Rev.,* 30(3): 655-666.

Färe, R., S. Grosskopf and C. A. K. Lovell, 1994. Production frontiers. United Kingdom, Cambridge University Press.

FAO, 2005. Aquaculture production, 2003. *Yearbook of fishery statistics* – Vol.96/2. Food and Agriculture Organization of the United Nations, Rome, Italy.

FAO, 2006. State of World Aquaculture: 2006. FAO Fisheries Technical Paper. FAO. Rome.

Farrell, M. J., 1957. The measurement of productive efficiency. *Journal of the Royal Statistical Society,* series A120, Part 3:253-281.

Greene, W. H., 1993. The econometric approach to efficiency analysis, In Fried, H.O., Lovell, C.A.K. and Schmidt, S.S. (Eds.), The Measurement of Productive Efficiency, Oxford University Press, New York, 68-119.

Iinuma, M., K. R. Sharma, P. S. Leung, 1999. Technical efficiency of carp pond culture in peninsula Malaysia: an application of stochastic production frontier and technical inefficiency model. *Aquaculture* 175:199-213.

Ines Herrero and Sean Pascoe, 2002. Estimation of technical efficiency: a review of some of the stochastic frontier and DEA software. CHEER, 15(1).

Jayaraman, R., 1998. Economics and technical efficiency in carp culture in Thanjavur District, Tamil Nadu State, India. *In: Eide, A., Vassdal, T. (Eds.), Proceedings of 9th International Conference of the International Institute of Fisheries Economics and Trade,* Tromso, July 8-11, 1998. Norwegian College of Fishery Science, Tromso, Norway, 71-82.

Karaginannis, G., katranidis, S. D., Tzouvelekas, V., 2000. Measuring technical, allocative and cost efficiencies of seabass and seabream farms in Greece. *Aquac. Econ. Manag.* 4, 191-207.

Kumbhakar, S. & C. A. K. Lovell, 2000. Stochastic frontier analysis. United Kingdom, Cambridge University Press.

Lingo, 1995. The modeling language and optimizer. Lindo Systems Inc.

Lewin, A. Y. and C. A. K. Lovell, 1990. Editors Introduction, *Journal of Econometrics,* 46:3-5.

Meeusen, W., J. Van der Broeck, 1977. Efficiency Estimation from Cobb-Douglas Production Functions with Composed Errors. *Int. Econ. Rev.*, 18: 435-444.

Neff, D. L., P. Garcia, C. H. Nelson, 1993. Technical efficiency: a comparison of production frontier methods. *J. Agric. Econ.* 44 (3), 479–489.

Pascoe, S., L. Coglan, 2002. Contribution of unmeasurable factors to the efficiency of fishing vessels. Am. *J. Agric. Econ.*, 84(3):45-57.

Pascoe, S., I. Herren, 2004. Estimation of a composite fish stock index using data envelopment analysis. *Fish Res.*, 69:91-105.

Pascoe, S., C. Robinson, 1998. Input controls, input substitution and profit maximization. *J. Agric. Econ.*, 49(1):16-33.

Pascoe, S., J. L. Andersen, J. W. de Wilde, 2001. The impact of management regulation on the technical efficiency of vessels in the Dutch beam trawl fishery. *Eur. Rev. Agric. Econ.*, 28(2):187-206.

Pascoe, S. D. tingley, S. Mardle, 2003. Single output measures of technical efficiency in EU fisheries. CAMARE Report 61. CEMARE, UK.

Roy, A. K., N. Sarangi, G.S. Saha and Nibedita Jena, 2006 a. Methods of estimation of technical efficiency of aquaculture farm. *In*: *Bioinformatics & Statistics in Fisheries Research*, Vol.III, A.K. Roy and N. Sarangi (Eds.), pp: 439-458 . CIFA, Bhubaneswar, India

Seiford, L. M. & R. M. Thrall, 1990. Recent developments in DEA: the mathematical programming approach to frontier analysis. *J. Econometrics*, 46: 7-38.

Sharma, K. R., P. S. Leung, 1998. Technical efficiency of carp production in Nepal: An application of Stochastic Frontier Production Function Approach, *Aquaculture Economics and Management.*, 2(3):129-140.

Sharma, K. R., and P. S. Leung, 2000 a. Technical efficiency of carp pond culture in South Asia: An Application of a stochastic meta-production frontier model, *Aquaculture Economics and Management.*, 4(3) & (4):169-189.

Sharma, K. R., P. S. Leung, 2000 b. Technical efficiency of carp production in India: A stochastic frontier production function analysis, *Aquaculture Research*, 31: 937-958.

Tingley, Diana, Sean Pascoe, Louisa Coglan, 2005. Factors affecting technical efficiency in fisheries: stochastic production frontier versus data envelopment analysis approaches, *Fisheries Research* 73: (2005) 363–376.

Tveteras, R., Battese, G. E., 2000. Technical change and productive efficiency change in Norwegian Salmon Farming: the influence of regional agglomeration externalities. Paper presented at the. *10th Biennial Conference of the International Institute of Fisheries Economics and Trade*, Oregon, July 10-14, 2000.

Vassdal, T., Roland, B.E., 1998. Technical change in the Norwegian salmon aquaculture sector: A Malmquist index approach. In: Eide, A., Vassdal, T. (Eds.), Proceedings of the 9th International Conference of the Internaational Institute of Fisheries Economics and Trade, Tromso, July 8-11, 1998, pp. 404-415.

www.banxia.com/frontier/glossary.html

Chapter **36**

Aquaculture Market Intelligence System

T. Ravisankar, A. G. Ponniah & M. Krishnan

Introduction

Market intelligence is a basic requirement for attaining global trade competitiveness in agriculture and allied sectors. In a free market economy, a commodity's current price and expected future price have consequences on its future supply as the entrepreneur's production plans are accordingly changed. Additionally market intelligence on non-price issues like changes in consumer perceptions, traceability and quality related issues of HACCP is also equally important. This can guide the farmers in deciding the quality and the quantity of a commodity to be produced for particular target market/market segment. Market analysis and quantity that is likely to be in demand and price forecasts for fish are important to producers for rational production planning and for processors/exporters for taking right marketing decisions. Domestic and international input suppliers, bankers, insurers and policy analysts need market intelligence to estimate the requirements of fishers at macro-level for their advance planning.

International Trade Scenario of Aquaculture and India's Position

India stands second in the world with an annual aquaculture production of about 2.94 million metric tones and about 4.2% of the world's aquacultured fish in both quantum and value (Table.1). The poor value realization by India is attributed to the low unit value of aquaculture production. Japan, on the other hand, accounted for 2.1% of total world aquaculture production by

Applied Bioinformatics, Statistics and Economics in Fisheries Research: pp. 519-528

New India Publishing Agency, New Delhi (India)

weight but for 6.0 % by value because of the high-value species cultured (e.g. amberjack, scallops and oysters).

Table 1: India's position among the top ten-aquaculture producers of the world (2004)

Country	Production Volume (tones)	Share in global production Volume (%)	Rank	Production Value (000 US$)	Share in global production value (%)	Rank	Unit value ($/ Kg)	Rank
China	41,329,608	69.6	1	35,997,253	51.2	1	0.87	9
India	2,472,335	4.2	2	2,936,478	4.2	3	1.19	8
Philippines	1,717,028	2.9	3	794,711	1.1	10	0.46	10
Indonesia	1,468,612	2.5	4	2,162,849	3.1	6	1.47	5
Japan	1,260,810	2.1	5	4,241,820	6.0	2	3.36	2
Viet Nam	1,228,617	2.1	6	2,458,589	3.5	5	2.00	3
Thailand	1,172,866	2.0	7	1,586,625	2.3	7	1.35	6
Korea, Republic of	952,856	1.6	8	1,211,741	1.7	9	1.27	7
Bangladesh	914,752	1.5	9	1,363,180	1.9	8	1.49	4
Chile	694,693	1.2	10	2,814,837	4.0	4	4.05	1

(Adopted from FAO, 2006a, State of World of Fisheries and Aquaculture)

In terms of increasing the value of aquaculture production, the eye-catching case of Chile, a developing country is noteworthy. Chile ranks 10th in production volume, but a rank 4th in total value, as it produces high value items with an average value of US$ 4.05 per kilogram. In 1995, the aquaculture production of Chile was estimated at 2, 06,000 tones with a value of US$584 million, which included 98 000 tones of salmon and 43,000 tones of rainbow trout. By 2004 the total production figure had increased to 6, 95,000 tones with an estimated value of US$2.8 billion. Of this value, 74 percent originated from Atlantic salmon and trout culture with an estimated value of over US$1.5 billion for Atlantic salmon and US$568 million for trout. Thus in one decade the production of salmon quadrupled to 4, 39,000 tones and that of trout, tripled to 1, 26,000 tones. Other products with estimated production values of over US$100 million in 2004 were Coho salmon (US$294 million), Peruvian calico scallop (US$240 million) and Chilean mussel (US$131 million).

The growth in Chilean aquaculture industry since 1995 can be traced to efficiency of the public institutions in organizing production systems, establishing market channels and establishing clear legislation (General fisheries and aquaculture law, 1989). Chile's economic policies in general and specifically the effective use of market intelligence with proactive involvement in signing bilateral and regional treaties that directly benefits aquaculture are other key drivers of aquaculture development in Chile.

One more reason for Chile's success is due to strong insurance support made available to aquaculture sector over last ten years. The Chilean subsidiaries of multi-national aquaculture companies are most likely to be insured under their groups' worldwide insurance policies. Relatively smaller national aquaculture entrepreneurs could also easily buy insurance. Confidential insurance industry estimates put the premium volume in the Chilean market at around US$8 million to US$9 million annually.

The picture of seafood exports from India is in contrast to Chile in that while there has been increase in the volume and value of seafoods over the years, there has been a drop in average unit value. Seafood exports data series of India reveal that the share of marine product exports has grown over the years from a mere Rs.3.92 crore in 1961-62 to Rs.6, 646.49 crore in 2004-05 accounting for approximately 2.35% of the total exports from India. Frozen shrimp continued to be the prime item exported, in terms of value, contributing 63.50% of the total value of export. Frozen fish continued to be the largest item in terms of volume with 34.62% of the total volume of marine products exported.

Table 2: Marine Product export scenario of India (2001-06)

Year	Quantity (metric tones)	Value (million US $)	Unit Value (US $/kg)
2001-02	424470	1253.35	2.95
2002-03	467297	1424.90	3.05
2003-04	412017	1330.76	3.23
2004-05	461329	1478.48	3.20
2005-06	512164	1644.21	3.21

(Adopted from Anon, 2006a)

Though total value realization has gone up from US$ 46.40 million in 1970-71 to US$ 1478 million in 2004-05, the unit vale realization has fallen from US$ 4.78/Kg in 1982-83 to US$ 3.21/ kg in 2005-06 (Table.2). The fall in unit value realization is mostly attributed to falling shrimp prices, anti dumping duty imposed by USA, and stringent quality standards enforced by European Union and USA in the past couple of years. Increase in exports of low value fishes is the other reason for falling unit value.

The persisting problems with Indian fishery trade are inadequate coordination among public institutions and actual players in the market channel, absence of a market intelligence system to drive the process to culture high value species, deficiency in policy formulation and implementation and lack of unity among corporate players. Diversification of fish culture with Asian sea bass and crab is mainly constrained due to low market price, apart from reasons of non-availability of adequate seed and suitable feed. The potential demand in domestic and international markets for shrimps other than tiger shrimp, sea bass and crab have not yet been explored in depth (Ravisankar *et al,* 2006).

More detailed market analyses are required for framing suitable policies for development of aquaculture in India. Rigorous economic analysis of various price policy issues requires advance information on the fish production, quantum of supply and demand, as well as regular monitoring of price movements in both the domestic and international markets. A strong database on different features of fish production and marketing is important for both the formulation of policy and the evaluation of its effectiveness with fish producers, intermediaries, and ancillary industries. Utilization of Information Technology for spreading the valuable market information could add in high unit value realization for Indian aquaculture sector.

It is inevitable that restructuring of monetary and fiscal policy climate pertaining to aquaculture is imperative to facilitate rapid development of Indian aquaculture and India can learn from experiences of other countries like Chile. Diversification into high aquaculture species (Ravisankar *et al,* 2006) and facilitating availability of accurate market information will make fisheries and aquaculture more profitable and India could increase its trade of aquaculture products from the present 4.2 % of global market.

Seafood Market Intelligence Abroad

Many lessons of market intelligence can be learnt without wasting resources based on experiences of other countries. Some noteworthy initiatives on seafood market intelligence are listed below:

- Economic Research Services of United States of America (ERS/USDA) publish 'Aquaculture Outlook' with market data for the past 2-3 years along with price estimates for current year for a dozen sea food commodities (UDDA, 2006).

- Economic and Social Department of Food and Agriculture Organisation circulates 'Commodity Market Review' in which demand and supply scenarios for major food items including sea food/ shrimp are analyzed and estimated/likely prices are indicated (FAO,2006b).
- Trade Research Institute of Viet Nam, based on market analysis, guides the Viet Nam Government on policy issues on seafood sector (Government of Viet Nam, 2006).
- 'FISHINFONETWORK', a web based market information system is run by a consortium of seven Inter Governmental Organizations like Globefish, Infofish and Eurofish, providing market analysis reports, item-wise and country wise on major fish items traded (Fishinfonet, 2006).

Except 'Aquaculture Outlook' of ERS/USDA, all others are mostly subjective expert estimates and no econometric/statistical analysis is reported to have been used.

Agriculture Market Intelligence in India

Aquaculture market intelligence set up could pick up few points on planning and implementation from similar experiences of other primary sectors of production listed below:

- Directorate of Economics and Statistics (DE&S) of Government of India collects, compiles and disseminates (Department of Agriculture and Cooperation, 2006) wholesale, retail and international prices, arrivals on weekly/monthly basis as weekly bulletins/ e-mails, providing market intelligence for agricultural commodities. The Directorate of Commercial Intelligence & Statistics, Kolkata is the official repository of all trade related statistics.
- 'Domestic and Export Market Intelligence Cells (DEMIC)' dealing with agricultural commodities, are functional in Division of Agricultural Economics, IARI, New Delhi and Department of Agricultural Economics, TNAU, Coimbatore (TNAU, 2006). National Informatics Centre, New Delhi is running AGMARKNET scheme for market intelligence on agricultural crops (NIC, 2006).
- Market intelligence to the closed user groups (e-choupal) is available from ITC. Commercial offers and farmers' price expectations are matched and the sale of soya, coffee, and prawns occur (ITC, 2006).

- Marine Product Export Development Authority (MPEDA) publishes *'Price indicators for marine product exports'*, a global market update on marine products (MPEDA, 2006).

The additional components required to strengthen MPEDA efforts are access to latest market information along with reliable price forecasts, which have immense practical utility to aquaculturists, intermediaries and ancillaries. One study along this line is the work of Sarada *et al* (2004) in which a model and forecast of frozen shrimp export values using Holt–Winters Exponential smoothing Approach is given. The weakness of existing fish market intelligence system are that they do not directly address the location specific and farm specific needs of Indian aquaculturists.

Market Intelligence System Requirements for Indian Aquaculture

The requirements of market intelligence system are governed by farming systems like group farming, contract farming and indebtedness of the fishers and aquaculturists. The market intelligence system could be more effective with aquaculture in India compared to capture fisheries, with the presence of more educated and resourceful aquaculture community.

The good market intelligence system required for Indian aquaculture producers should address the following requirements:

1. It should provide up to date domestic and international market intelligence for marketing fish/fishery products with a nation-wide network.
2. By providing market information on occurrences that influence the changes in demand patterns, it would increase the marketing efficiency of fish products, ensuring remunerative prices to the producers,
3. It may also consider publishing periodical 'market analysis reports' on various fish/fishery products and markets, ensuring a regular flow of critical market data, preferably through a web based database system, meant for domestic fish producers and related agencies, for production/ product diversification and domestic completion of value chain.
4. A web based set-up that provides prices and price forecast for major fish products, price behaviour, market trends, and arrival information and export-import details is what the farmers require.

5. The vital market information on export procedures, export standards for various products and regions, infrastructure facilities available in various ports, air cargo, railways, etc. should be made available in separate modules. Adequate web based resources for market information could be hyperlinked and details of key officials who could be contacted for clarifications may be made available.

Active involvement of corporate sector and farmers association in development of market intelligence system is essential for making it workable and self-sustainable.

Role of Market Intelligence in the Value Chain of Aquaculture

A schematic fish value chain drawn mainly based on shrimp and scampi farming in India is given in figure.1. The market intelligence flow requirements are given as curved arrows. This basic value chain applicable for other commodities cultured in independent fish farms in India could benefit a lot through market intelligence flow to fishers, feed millers, hatcheries, farmers and trading groups. The group farms and contract farming under corporate umbrella like seaweed farming in India under PEPSICO contract farming may have different value chains with different requirements of market intelligence flow (Anon, 2006b).

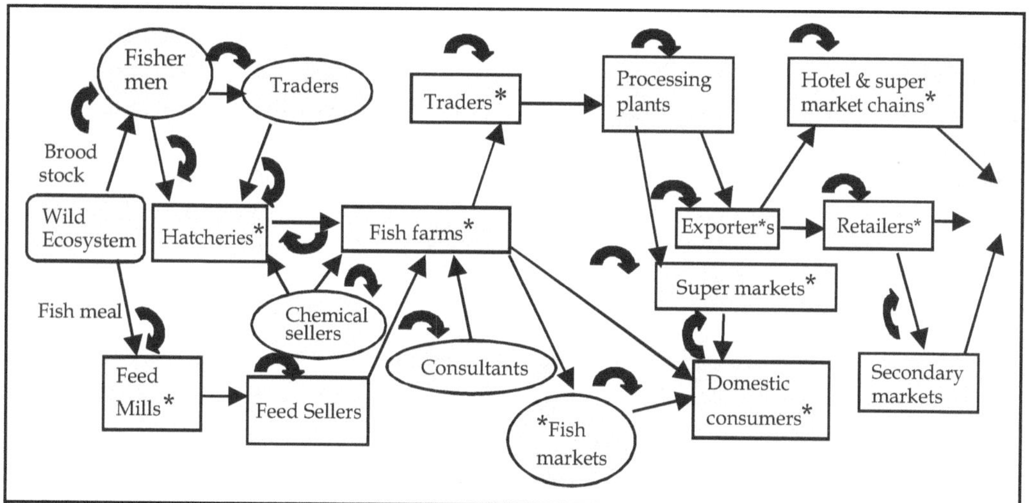

• Present Market intelligence flow information
* Key points to gather and monitor to build market intelligence systems
→ Flow of products or technical information

Fig. 1. Present Market intelligence flow along the schematic fish value chain in India

Based on informal and formal stakeholder consultations it is evident that the present flow of market information is generally limited to the next immediate step in the value chain and not from the export market to producers (Ravisankar *et al*, 2006). A field study done by CIBA (Mahalakshmi *et al.*, 2006) found only big farmers have benefited using market information provided by aquachoupals.

If we analyse the existing flow of market intelligence information among the different stakeholders it is evident that a fair share in consumer's rupee to the farmer could be ensured by reducing information asymmetries that exists among different value chain players like internet enabled corporate management and resource poor small farmers. The mass media like TV, Radio and Newspapers could be used to regularly pass on the market information to less enabled sections of the value chain.

Suggested Modeling Approaches

To forecast future prices, explanatory methods and/or the extrapolation methods and other forecasting tools, viz., best-fit, curve fitting/ regression, smoothing and seasonal smoothing methods, either as causal models employing smoothing techniques or as time series models (ARIMA, GARCH, and ARCH) could be used. Further fuzzy analysis, artificial neural networks could be used for refinement. As international markets are interrelated, any unforeseen change in a remote market will adversely affect the price in the other markets and the prediction models employed should be effective to capture this. Though improvement in modeling and updating is a continuous process, commodity market reports for major fish/fishery products, along with acceptably accurate price forecasts should be made available periodically. These forecasts should be regularly validated and models should be refined for reliability in a continuous manner to make the aquaculture market intelligence to be earnestly accepted by the actual users.

Strategic Action Points

The following strategic action points could help in establishing, managing and sustaining a good market intelligence system which is suitable for India:

1. Establishing a formal coordination network on market intelligence, trade promotion and policy initiatives among partner organizations such as Marine Product Export Development Authority (MPEDA), National Fisheries Development Board (NFDB) and Ministry of Agriculture.

2. Organizing and facilitating long-term market information provision activities and trade capacity building.

3. Increasing domestic awareness of export opportunities, export programs, and overseas market intelligence, with special emphasis on small and new-to export aquaculture species.
4. Utilize overseas field office staff of MPEDA and Indian embassies/high commissions to strengthen the market intelligence system.
5. Link the developed Market Intelligence System to specific promotional efforts to strengthen foreign market development and to target priority markets.
6. Develop human capital on trade analysis and market intelligence
7. Organising systematic and regular transmission of market information through web based system especially the Village Knowledge Centres, To overcome the 'digital divide', the less fortunate interest groups should be kept informed through TV, Newspapers and radio as done in agriculture sector.

Outcome and Impact of Market Intelligence System

Detailed and usable market analysis reports for fish and fishery products with short term and medium term price forecasts will help the aquaculture producers in developing their production and marketing plans. The increased awareness of changes in consumer preferences and production assessments of competing producers will lead to midcourse corrections in the product mix, product forms and destination markets. As outcome of above, aqua farmers in general and small scale operators, in particular will get adequate profits, country will get increased national income and aquaculture will again be considered forever as 'aqua gold' and 'sunrise enterprise'. The impact of this market intelligence system will ensure equitable distribution of benefits to all the players of value chain making aquaculture a sustainable enterprise.

Conclusion

Market intelligence system would reduce the information asymmetries among different value chain players making possible equitable benefit distribution. The development of an appropriate database and market intelligence facility, together with other policy initiatives is required to exploit the tremendous potential of Indian aquaculture sector. As discussed in length in the above sections, a web based Market Intelligence System for aquaculture with in built short range price forecasting will help in increasing the profitability for the people engaged in aquaculture directly and indirectly.

REFERENCES

Anon, 2006a. Export performance of marine products during 2005-06, *Seafood Export Journal* 36(6):11-19.

Anon, 2006b.http://www.thehindubusinessline.com/2006/09/29/stories/2006092904140300.htm as on 15-12-2006.

Department of Agriculture and Cooperation, 2006. http://dacnet.nic.in/ as on 15-12-2006.

Echoupal, 2006. www.echoupal.com/ as on 15-12-2006.

FAO, 2006. http://www.fao.org/ as on 15-12-2006.

FAO, 2006. State of World of Fisheries and Aquaculture 2006, FAO Fisheries Technical Paper 500, p.9.

Fishinfonet, 2006. http://www.fishinfonet.com/ as on 15-12-2006.

Government of Vietnam, 2006. http://www1.mot.gov.vn/en/ as on 15-12-2006.

Mahalakhsmi. P, D. Deboral Vimala and M. Krishnan, 2006. Web kiosks in aquaculture: A study of aquachoupal model in Prakasam District of Andhra Pradesh, Paper presented at *National Seminar on Extension Strategies for fostering knowledge centric agricultural growth, organized by Society of Extension Education*, TNAU, Coimbatore held at Pondicherry Dec 2-3, 2006.

MPEDA, 2006. www.mpeda.com/ as on 15-12-2006.

NIC, 2006. www.agmarknet.nic.in/ as on 15-12-2006.

Ravisankar T., C. Sarada and M.Krishnan, 2006. Diversification of Fish Culture and Exports among Major Shrimp-producing Countries of Asia: *A Spatial and Temporal Analysis, Agricultural Economics Research Review*, Vol. 18 July-December 2005, pp: 187-195.

Ravisankar.T, A.G.Ponniah, M.Krishnan and B.Shanthi, 2006. Vital missing links in coastal aquaculture-Workshop report of the stakeholder appraisal, CIBA Special publication No.28, Central Institute of Brackishwater Aquaculture, Chennai.

Sarada.C., M.Krishnan and T.Ravisankar, 2004. Modelling and forecasting of frozen shrimp export values- an application of Holt-Winters exponential smoothing approach, paper presented at *Annual Conference of Indian Society of Agricultural Statistics*, New Delhi, GBPUAT, 5-7 Feb 2004.

TNAU, 2006. www.tnau.ac.in/ as on 15-12-2006.

USDA, 2006. http//usda.mannlib.cornell.edu/as on 15.12.2006.

Chapter 37

Project Evaluation Techniques for Estimation of Research Benefits of Fisheries Research

A.K. Vasisht

Introduction

Project evaluation aims at analysing research and development projects, or activities or ideas, for any or all of the following purposes:

- Getting an overall understanding of the project.
- Making priorities among a set of projects.
- Taking a decision about whether or not to proceed with a project.
- Monitoring projects, eg. by following up the parameters analyzed when the project was selected.
- Terminating projects and evaluating the results obtained.

The project evaluation tools are used to provide information to assess the value of a potential project with particular reference to the estimation of costs, resources and benefits, in order to take a decision about whether or not to proceed with the project. It is an activity where money is spent to realize the expected returns and requires planning, financing and implementation as a unit. In fact, the whole complex of activities using resources to gain benefits constitutes the projects. Projects are of various kinds such as irrigation, dairy, poultry, piggery, rural credit, social forestry, land reclamation / leveling, rural development projects, aquaculture projects etc. A project can be either

Applied Bioinformatics, Statistics and Economics in Fisheries Research: pp. 529-539

New India Publishing Agency, New Delhi (India)

accepted or rejected once project costs and benefits are identified, priced and valued.

Aquaculture/ Fisheries Research

Research in fisheries is expected to provide higher genetic gains and productivity and contribute towards increase in fish supply, reduction in cost per kg of fish production, and increase in profitability of fish farm. The benefits, which accrue to fisheries research, are the social gains to consumer and producer. There are different aspects of costs of fishing and one can evaluate them in a cost and earnings analysis.

In order to assess the research impact, certain baseline information is also required such as, target domain; base year production, and productivity; research and development lag in years; assumption on the adoption and ceiling levels of genetically improved variety of fish; assumption on depreciation rate of technology for genetically improved variety; probability of success at the experiment stage for the genetically improved variety; own price elasticity of supply and demand by fish species; expected life of the project ; and research costs.

Broadly there are two methods of project appraisal, *viz;* undiscounted measures and discounted measures. In the undiscounted measures, payback period, ranking by inspection, proceeds per unit of outlay, etc; are important. In the discounted measures Net Present Worth (NPW), Internal Rate of Return, and Benefit-Cost Ratio are prominent.

Undiscounted Measures

Ranking by inspection

It is based on the size of costs and length of cash-flow stream. Suppose the two projects are with the same investment and the same net value of production, but with difference in the length of the period, then the project with longer duration is preferred to the one with shorter time period. This leads to bias in the choice obviously due to the absence of more elaborate and appropriate analysis.

Payback period

Another simple method of ranking a project is the length of time required to get back the investment on the project. The payback period of the project is estimated by using the straightforward formula:

$$P = \frac{I}{E}$$

Where, P= Payback period of the project in years, I= Investment of the project in rupees, and E= Annual net cash revenue in rupees.

The preference of a project is based on the lesser payback period.

Measurement of the Cash Flows of the Project

From the annual stream of gross benefits of the project, the capital invested and the other input costs like labour, machinery, fertilizers, pesticides, management etc; are deducted. From the residual, the return of capital and return on capital or return to capital, i.e., recovering investment made in the project (depreciation) and compensation for the use of money (interest) are computed. This residual is called cash flow of the project. In financial analysis the cash flow is the net incremental benefit of the project. The financial analysis aims at estimation of returns to all resources employed in the project. Hence, borrowed capital is considered as a benefit received, while its interest is considered as a cost and it is deducted from the gross returns. Whereas in economic analysis, this consideration is ruled out because of the assumption that all the resources employed in the project belong to someone or the other, within the society. In the economic analysis, it is important that the prices of some of the inputs must be shadow prices. On the other hand, in financial analysis all prices are market prices and they must include taxes and subsidies.

In its simplest form this approach requires the completion of a standard form, which asks for estimates of the expected cash outflows and inflows for the project. This can be done on a yearly basis for longer term projects or on a shorter time scale for shorter term projects, even down to weekly.

Cash outflows and inflows

	Cash Outflow	Cash Inflow	Net Cash Flow
1997	C0	B0	B1-C1
1998	C1	B1	B2-C2
1999	C2	B2	B3-C3
2000	C3	B3	B4-C4

A frequently used form of analysis applied to this data is to calculate the break-even point. This is the point at which the cumulative net cash flow is equal to zero and hence the cash inflow has covered all the cash outflows. The reason for this measure being seen as important is that the shorter the time scale to break-even, or payback, the less risk there is that the environment will change dramatically and hence seriously affect the estimates used in the cash flow projections. However, the analysis completely ignores the likely returns after the break-even point and therefore is not really a good indicator of the real potential of a project. In some cases, people choose to ignore the time value of money, although it is not difficult to handle this by including cost of capital payments on the net cash outflows as part of the costs.

Discounted Measures

The yearly net benefits accruing from a project are measured in terms of cash flows. If they are weighted by discount rate, they become discounted cash flows. These discounted cash flows are the best estimate to decide on the worth of the project. This approach will give the present net worth of the project.

Net Present Worth

This is the present worth of the cash flow stream. It is also referred to as Net Present Value (NPV). The present worth of the costs is subtracted from the present worth of the benefits in order to arrive at the Net Present Worth of the project every year. The choice of discount rate to be used in the measurement of Net Present Worth (NPW) poses some problems. NPW is helpful in working out benefit-cost ratio of the project. The selection criteria of the projects depend upon the positive value of the net present worth, when discounted at the opportunity cost of the capital. This could be satisfactorily done, provided there is a correct estimate of the opportunity cost of the capital. NPW is an absolute measure, but not relative. NPW of the project is estimated using the following equation:

$$NPW = \frac{P_1}{(1+i)^{t1}} + \frac{P_2}{(1+i)^{t2}} + \frac{P_3}{(1+i)^{t3}} + \text{------} + \frac{P_n}{(1+i)^{tn}} - C$$

Where,

P_i = Net cash flow in the ith year,
i = Discount Rate,
t = Time Period, and
C = Initial cost of the investment.

Projects with positive NPW are given weightage in the selection compared with those with negative present values, while zero NPW makes the investor indifferent.

The following example illustrates the steps in NPV calculations for two projects:

Estimation of NPW for two projects (Hypothetical)

Fisheries						Sericulture					
Year	Cost (in Rs.)	Returns (in Rs.)	Net income (in Rs.)	Discount factor at 12%	NPW (in Rs.)	Year (in Rs.)	Cost (in Rs.)	Returns	Net income (in Rs.)	Discount factor at 12%	NPW (in Rs.)
1	38900	-	-38900	0.8929	-34733.81	At the end of 6th year	25000	-	-25000	0.507	-12675
2	9239	28475	19236	0.7972	15334.94	7th year	4250	10260	6010	0.452	2716.52
3	10575	32550	21975	0.7118	15641.81	8th year	4792	12550	7758	0.404	3134.23
4	11952	35610	23658	0.6355	15034.66	9th year	5368	14530	9162	0.361	3307.48
5	12858	39802	26944	0.5674	15288.03	10th year	5975	16275	10300	0.322	3316.60
				NPW	**26565.63**	11th year	6456	19396	12940	0.287	3713.78
						12th year	7187	21470	14283	0.257	3670.73
										NPW	**7184.34**

Internal Rate of Return

IRR is a financial ratio expressing the net present value of expected future cash flows as a percentage of an investment. In financial analysis, IRR can be used for evaluating the return on an investment and to compare different investment options.

In the computation of Internal Rate of Return (IRR), the time value of money is accounted. The method of working IRR provides the knowledge of actual rate of return from the different projects. Thus IRR is known as 'marginal efficiency of capital or yield on the investment'. It is the discount rate at which the present values of the net cash flows are just equal to zero, *i.e*; NPW= zero.

$$\sum_{t=1}^{t=n} \frac{Bt - Ct}{(1+i)^t} = 0$$

Where,

Bt = benefit in each year

Ct = Cost in each year

t = 1,2,3, ..., n

n = number of years

i=interest (discount) rate

It is the discount rate that makes the net present worth of the incremental net benefit stream or incremental cash flow equal to zero. This discount rate is called the internal rate of return. It is the maximum interest rate that a project could pay for the resources used if the project is to recover its investment and operating costs and still break even. It is the "rate of return on capital outstanding per period while it is investment in the project". The IRR is also estimated through trial and error or interpolation method. The interpolation method is the rule for interpolating the value of the internal rate of return lying between discount rates too high on the one side and too low on the other side. The procedure is elucidated by some hypothetical examples below.

In the working procedure, an arbitrary discount rate is assumed and its corresponding NPW is arrived at. The positive NPW value of the project indicates that IRR is still higher and the next assumed IRR value must be comparatively higher than the initial level. This process is continued until NPW becomes negative. Then by interpolation method the exact IRR is found out using the following formula:

Internal rate of return	= Lower discount rate	+ Difference between the discount rates	Present worth of incremental net benefit stream (cash flow) at the lower discount rate / Sum of the present worths of the incremental net benefit streams (cash flows) at the two discount rates, Signs ignored

Steps in computation of IRR

Step I : Create the cash flow stream

Step II : Select a discount rate at which the NPV would be positive.

Step III : Using the discount factor for the above discount rate and compute the present value of cash flow stream.

Step IV : Select a slightly higher discount rate at which the NPV is expected to be negative.

Step V : Using the discount factor for the higher discount rate compute the present value of cash flow stream.

Step VI : Substitute the values of parameters generated in the formula for estimation of

IRR

Estimation of IRR for small-scale breeding and rearing unit for livebearers (Hypothetical Example)

Table 1

S. No.	Items	Amount (in Rs.)
I.	**Expenditure**	
A.	**Fixed Capital**	
1.	Low cost shed	10000
2.	Breeding tanks	10000
3.	Rearing tanks	5600
4.	Brood stock tanks	5600
5.	Larval tanks	9600
6.	Bore-well with pump	8000
7.	Oxygen Cylinder with accessories	5000
	Sub total	**53800**

Contd...

Table 1. Contd...

B.	**Variable cost**	
1.	800 female, 200 male @Rs.2.50/pc Guppy, molly, Sword tail & platy)	2500
2.	Feed (150 kg/yr @Rs.20/kg)	3000
3.	Different types of nets	1500
4.	Electricity/Fuel @Rs.250/month	3000
5.	Perforated plastic breeding basket (20 nos @Rs.30/pc)	600
6.	Wages (@Rs.1000/month)	12000
7.	Misc. Expenditure	2000
	Sub Total	**24600**
C.	**Total Cost**	
1.	Variable cost	24600
2.	Interest on fixed capital (15% per year)	8070
3.	Interest on variable cost (15% per year for 6 months)	1845
4.	Depreciation (20% of fixed cost per year)	10780
	Grand Total	**45295**
II	**Gross Income**	
	Sale of 76,800 nos fish @Re.1/pc, reared for one Month (@40 nos/female/cycle from 3 cycles / year, Survival 80%)	76800
III	**Net Income (Gross Income – Total Cost)**	**31505**

Table 2

Year	Costs	Income	Net Income	DF 40%	NPW	DF42%	NPW
1	53800	-	-53800	0.71	-38428.57	0.70	-37887.32
2	45295	76800	31505	0.51	16073.98	0.50	15624.38
3	36950	65650	28700	0.36	10459.18	0.35	10023.44
4	39256	67600	28344	0.26	7378.18	0.25	6971.20
5	41250	68140	26890	0.19	4999.78	0.17	4657.46
			61639		482.54		-610.84
				IRR =	40.89		

Benefit / Cost Analysis

Benefit /cost analysis is typically viewed as an alternative to program evaluation. However, it can also be seen as an extension of the evaluation process. As such, benefit - cost analysis provides a means to systematically quantify and compare program inputs to program outcomes in monetary terms. Valuing both benefits and costs in monetary terms allows them to be directly compared to determine the net impact of the program, make comparisons between alternative programs or projects, assist in program planning, advance organizational accountability and / or expedite program support.

In this technique, we compare the present worth of costs with present worth of benefits. Absolute value of the benefit-cost ratio will change based on the interest rate chosen. The higher the interest rate, the smaller the resultant benefit-cost ratio. The most common procedure of selecting / ranking the projects depending upon the B-C Ratio is to choose the projects having B-C ratio more than one, when discounted at opportunity cost of capital. The project having the highest B-C ratio is implemented among the alternatives. The following formula depicts the estimation of B-C ratio:

$$\textbf{B:C Ratio} = \frac{\sum_{t=1}^{t=n} \frac{Bt}{(1+i)^t}}{\sum_{t=1}^{t=n} \frac{Ct}{(1+i)^t}}$$

Where,

Bt = benefit in each year

Ct = Cost in each year

t = 1,2,3, ..., n

n = number of years

i = interest (discount) rate.

Example

The cash flow of an aquaculture project is as given below. Estimate the feasibility of the project given the opportunity cost of the capital as 10 %, 7 % and 3 %.

Year	BENEFITS					COSTS				Benefit-Cost Ratio
	Tourism	Fishing	Seafood	Recreation	Total	Abatement	Fishing	Lost Production	Total	
1	$ 5,000	$ -	$ 2,000	$ 2,500	$ 9,500	$ 8.000	$ 3,000	$ 20,000	$ 31,000	
2	5,350	-	2,060	2,580	9.998	8,000	2,500	15,000	25,500	
3	5,726	-	2,122	2,678	10.524	6,000	2,083	12,000	20,083	
4	6,126	-	2,185	2,772	11.082	4,000	1,736	10,000	15,736	
5	6,552	-	2,251	2,869	11.674	3,810	1,447	10,000	15,256	
6	7,013	-	2,319	2,969	12.301	3,628	1,206	10,000	14,834	
7	7,502	-	2,388	3,073	12.965	3,455	1,005	10,000	14,460	
8	8,029	1,000	2,460	3,181	14.669	3,291	-	10,000	13,291	
9	8,591	1,100	2,531	3,292	15.516	3,134	-	10,000	13,134	
10	9,192	1,210	2,610	3,407	16.419	2,985	-	10,000	12,985	
11	9,836	1,331	2,688	3,526	17.381	2,843	-	10,000	12,843	
12	10,522	1,464	2,768	3,650	18.407	2,707	-	10,000	12,707	
13	11,261	1,611	2,852	3,773	19.501	2,578	-	10,000	12,578	
14	12,049	1,772	2,937	3,910	20.668	2,456	-	10,000	12,456	
15	12,893	1,949	3,025	4,047	21.913	2,339	-	10,000	12,339	
16	13,795	2,144	3,116	4,188	23.243	2,227	-	10,000	12,227	
17	14,761	2,358	3,209	4,335	24.663	2,121	-	10,000	12,121	
18	15,792	2,594	3,306	4,487	26.180	2,020	-	10,000	12020	
19	16,900	2,853	3,405	4,644	27.801	1,924	-	10,000	11,924	
20	$ 18,083	$ 3,138	$ 3,507	$ 4,806	$ 29.534	1,832 $	- $	10,000	$ 11,832	
PV (10% Discount Rate)	$ 70,799	$ 6,065	$ 20,901	$ 27,086	$124,851	$ 36,993	$ 9,639	$ 99,861	$ 146,493	0.85
PV (7% Discount Rate)	93,458	8,979	26,663	34,700	163,801	43,288	10,473	121,288	175,047	0.94
PV (3% Discount Rate)	$ 142,819	$15,691	$ 38,835	$ 50,849	$248,193	$ 55,510	$ 11,793	$ 165,027	$ 232,330	1.07

Economic Surplus Approach to Evaluating Impact of Research

Economic Surplus Approach is widely used for evaluating the impact of technology which may be genetically improved strains of fish etc. on the economic welfare of households. The model is applied within the target domain of production environment.

Following the theory of demand and supply equilibrium, economic surplus (benefits) of research intervention is computed as:

Benefits = K * P * Q * (1+0.5 Z * PED), where K is supply shift, P is fish price, Q is production, Z is reduction in price due to technological change, and PED is price elasticity of fish demand

$$Z = K \times PES / (PES+PED), \text{ where}$$

PES: Price elasticity of fish supply & K is shift in supply curve

K = Net cost change * Probability of success in research * adoption rate * depreciation rate of technology

Net cost change = reduction in marginal cost - reduction in unit cost.

Reduction in marginal cost = Relative change in yield / PES

REFERENCES

Alston, J.M. 1991. Research benefits in a multi market setting: a review. *Review of Marketing and Agricultural Economics*, 59(1), 23-52.

Alston, J.M., Norton, G.W. and Pardey, P.G.*1995. Science Under Scarcity: Principles and Practice for Agricultural Research Evaluation and Priority Setting*. Cornell University Press, Ithaca, New York, (published in cooperation with the International Service for National Agricultural Research).

Gittinger. J.P., Comparing project cost and benefits. *In: Economic analysis of agricultural projects*, pp. 360.

Chapter **38**

Impact Assessment of Fisheries Research-Concepts and Measurement

G. P. Reddy, M. N. Reddy & B. S. Sontakki

Need for Impact Assessment

Fisheries Research is an economic activity. Like any other investment propositions, resource allocation to Fisheries Research needs to be justified. The society or the donors are always interested to know what happened to the money invested in Fisheries Research. It is important to document the returns and/ or benefits accrued from the research investment. Objective assessment of research investment helps in making decision and allocating resources in high returns research portfolio. It also helps to know which research areas and programs benefit the poor and regions. In the paradigm shift, the donors are seeking evidences on impact of past funding as a basis for future financial support. Systematic impact assessment forms the basis for efficient resource allocation in alternative research programs competing for financial support. Research programs demonstrate better historical performance, in terms of benefits generated for the society, and are rated higher for attracting required research resources. Impact assessment studies are also getting more prominence as the international environment is rapidly changing due to many emerging complex problems. Socioeconomic and environmental problems, like poverty, international trade, degradation of natural resources, are growing fast and the donors are looking for the research programs, which can overcome these challenges. Therefore, systematic

Applied Bioinformatics, Statistics and Economics in Fisheries Research: pp. 541-549

New India Publishing Agency, New Delhi (India)

impact assessment studies would form a strong base for higher research funding to overcome many regional, national and global problems.

Impact assessment of research is not a new phenomenon. Earlier, it was based on some partial evidences. Some times the changes in production, area and productivity enhancement were used as a proxy for contribution of research at regional or national levels. Other proxies used were increased export, import substitution, employment generation, and contribution towards improving nutritional security and conservation of soil and water resources. Such proxies were often questioned; as such changes were the result of numerous forces, including the research contribution.

Framework for Research Impact Assessment

The framework for research impact assessment is shown in Figure 1

Impact assessment is undertaken at three levels. First, ex-ante assessment, which is done to objectively assess the research portfolio and prioritize the research agenda. This is done to justify funding in different research options. In the figure, it is allocating resources in R&D for generating research outputs. The second is the concurrent evaluation, which is done to identify the impediments for larger adoption of the research outputs. The purpose is to correct the gaps and provide feedback for refining and tuning the technology as per the stakeholders' requirements. Often it is known as constraint analysis. In the figure, four circles are being shown. These are technology traits (e.g. duration, quality, etc.), policy environment (e.g. price support, procurement, etc.), institutional arrangements (e.g. Fish seed, credit availability, etc.), and infrastructure (e.g. markets, roads, power, processing facilities, etc.). All these four components determine the adoption of any technology. It is just like four wheels of a vehicle. When all the wheels have optimum air, pressing accelerator will pick-up the speed of the vehicle at desired speed. Less air even in one wheel will limit the speed. If one wheel is flat, the vehicle has to be dragged by other means. The same is true for adoption of research outputs or improved technologies. During the green revolution period, all the four circles were favorable, which resulted in fast adoption of improved technologies. On the contrary, many promising technologies were finding difficult to be largely adopted despite of favorable policy environment and infrastructure. The absence of appropriate institutional arrangement is hindering the speed of these promising technologies. Therefore, determining constraints for larger adoption forms a part of the impact assessment. Such studies are characterized as part of the early impact assessment. These provide useful information on conditions for larger returns and benefits of research investment.

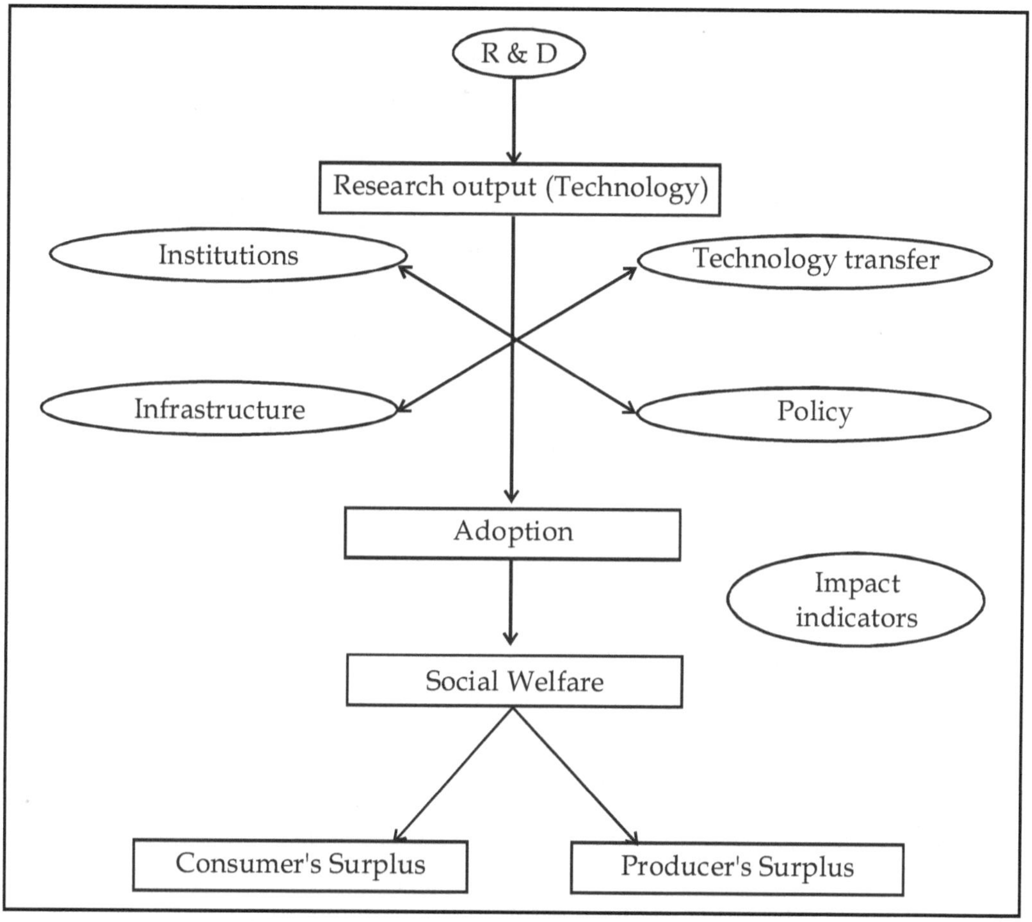

Fig. 1. Framework for assessing research impact

The third stage of impact assessment is known as ex-post assessment, which is done to validate past funding on research. These studies are being undertaken when the research outputs and technologies are largely adopted in the target domain, and assess their contribution to social welfare, resource conservation, trade, sharing of benefits of research outputs among different stakeholders (e.g. producers, consumers, industry), etc.

Impact Indicators

Impact indicators vary with technology and level of assessment. There are two types of benefits of research outputs: (i) tangible benefits are those, which can be assigned monetary values, and (ii) intangible benefits are those, which cannot be assigned monetary values but are important for the society.

Examples for the later type are improvement of environment, better health, and reduction in infant mortality, national defense, etc. These are important but difficult to assign any monetary value. These must be documented at least in physical terms.

The impact indicators will be different at farm, regional, national and global level. At the farm level, the direct beneficiaries are affected by adopting the technologies. At higher level, the society and the environment are being influenced and measured. Important farm-level and regional/ national-level indicators are listed in Box 1 and 2.

Box 1: Farm-level impact Indicators

- Efficiency
    - Income augmentation
    - Unit cost reduction
- Household food security
    - Nutritional security
- Poverty reduction
- Risk management
    - Improving yield or income stability in the absence of insurance
- Cropping intensity
- Gender related issues
- Natural resource conservation

Box 2: Regional/ National-level impact indicators

- Fish production
- Food self-sufficiency
- Employment generation
- Equity issues
    - Inter-regional
    - Inter-personal
- Poverty
- Trade
    - Prices
    - Export and/or import substitution
- Inter-sectoral linkages
    - Forward linkages (like markets, transport, processing, etc)
    - Backward linkages (like Fish seed, pesticide industry, machinery, etc.)
- Sustainability of natural resources

The emphasis of the listed indicators would vary with the type of research outputs. It is not necessary that all indicators would be applicable for any kind of technological change.

Measuring Efficiency Indicator

Following are the important methods for assessing the efficiency benefits of research impact:

Benefit cost analysis: The method compares the stream of benefits with that of stream of research cost. Following are the indicators for the benefit-cost analysis:

Benefit-cost ratio: It is the ratio of present worth of benefits stream and the present worth of cost stream.

Net present value: It is the present worth of the incremental net benefit stream.

Internal rate of return: It is the discount rate when net present worth of benefit and costs equal to zero.

Payback period: It is the period during which the entire research cost is recovered after the benefits are accrued.

Econometric approach: The approach assesses the changes in marginal productivity of research investment at macro-level. The econometric methods are powerful that can discern the contribution of research and other determinants in total change in output.

Total factor productivity: It is the ratio of total output and the whole set of inputs. It shows the residual left after incorporating the contribution of input quantities. The total factor productivity can be decomposed into the contribution of research resource allocation and other qualitative determinants.

Economic surplus approach: The approach estimates the economic surplus generated as a consequence of research outputs. The benefits can be decomposed into changes in the economic surplus to consumers and producers as a result of research success. The information derived through economic surplus approach is also be used to estimate benefit-cost ratio, internal rate of returns and net present value of research outputs.

In the Figure 2, the consumers' surplus as a result of fall in prices from F to P0 is the area beneath the demand curve (d) less cost of consumption. It is FP0a = FaQ0O-P0aQ0O.

The producers' surplus as a consequence of increase in prices from I0 to P0 is the area P0aI0. It is total revenue (P0aQ0O) less total cost of production (I0aQ0O).

The impact of research outputs or technological change is reflected by either unit cost reduction or yield enhancement. Cost reduction and adoption of technology would shift the supply function (from I0S0 to I1S1), which would form a new equilibrium price (from P0 to P1) and quantity (from Q0 to Q1) demanded and supplied. Figure 3 shows these shifts. The surplus generated as a result of shift in supply due to technological change will be as follows:

Change in consumers' surplus = P_0abP_1
Change in producers' surplus = $P_1bI_1 - P_0aI_0$
Change in economic surplus = I_0abI_1 or P_0abcd

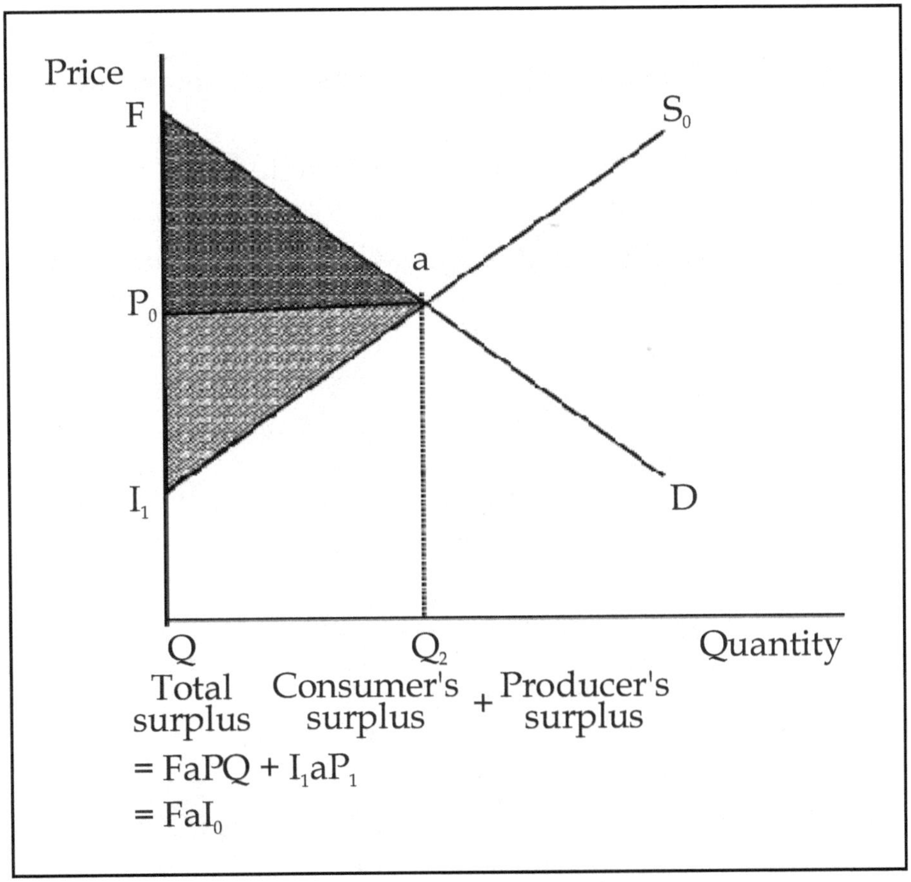

Fig. 2: Total economic surplus

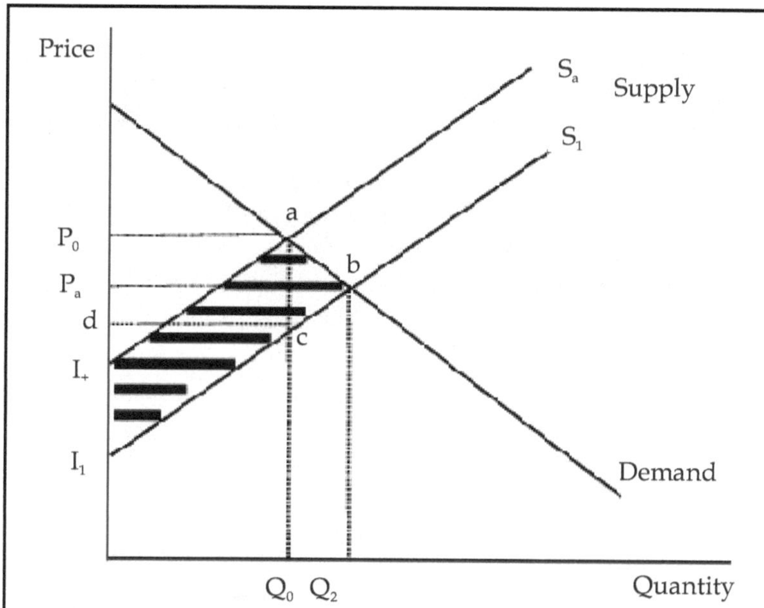

Change in consumer's surplus = P_0abP_1
Change in producer's surplus = $P_1bI_1 - P_0aI_1$
Change in economic surplus = I_0abI_1 or P_0abed

Fig. 3: Assessment of economic surplus

Research Cost

To develop the stream of cost, the research expenditure in developing specific technology need to be estimated. Under project-based budgeting, it is easy to have information on research cost. In the absence of such provisions, the research cost is to be estimated by consulting the scientists involved in the research.

Important elements of the research process, like (i) year of initiation of research, (ii) year when the final research output was identified, and (iii) year when the research output (or technology) was released. The time difference between starting year and research output release is known as the research lag. Shorter the research lag, early the research benefits reach the clients and vice-versa. In addition, the expenses incurred during the research lag period need to be estimated. These include the expenses on salary, operations, travel, equipment and overhead.

Another important component is the adoption of research outputs. At farm-level, adoption is the acceptance of new technology when farmer has full knowledge about its potential benefits. At aggregate-level, adoption is the process of spread of a new technology within a region. Estimates are needed on (i) year when the research output was adopted first time, (ii) the extent of adoption, and (iii) year when adoption reached to ceiling or maximum limit.

Figure 4 shows the research continuum, starting from research until the adoption started ceasing. Often, the agricultural technologies follow a sigmoid nature of adoption. Initially the technology is adopted at a slow pace, and then increases with increasing rate and finally reaches to ceiling (or maximum) level. After that when some new technology replaces the old one it starts descending.

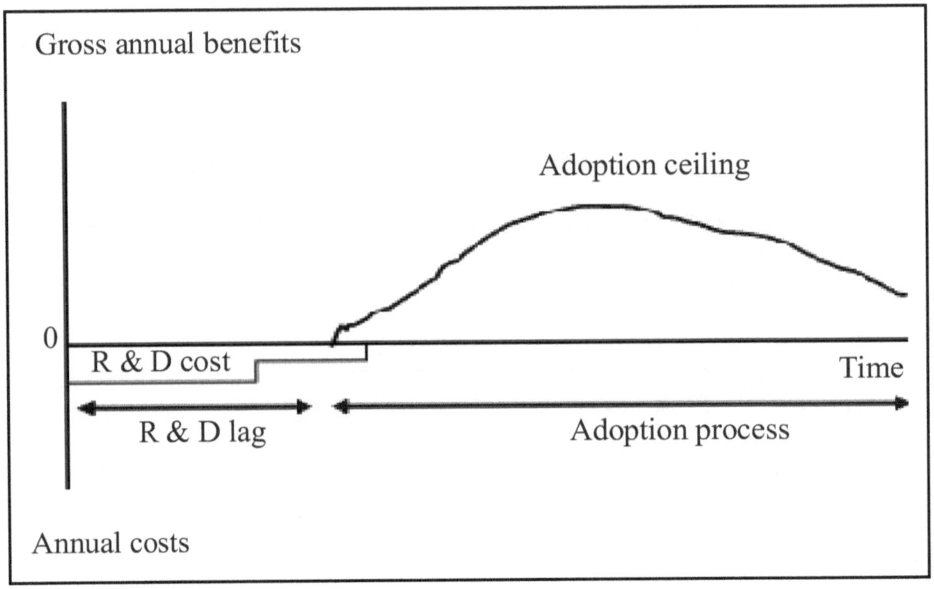

Fig. 4: Research continuum: research lag and adoption process

To develop the benefit stream, the change in economic surplus derived in Figure 3 is multiplied by the annual extent of adoption of the technology. The stream of benefits and stream of costs are used to estimate the internal rate of returns or discounted for computing the net present worth or benefit-cost ratio.

REFERENCES

Alston, Julian M., Norton, George W. and Pardy, Philip G. 1995. Science under scarcity: principles and practice for agricultural research evaluation and priority setting. The Hague, The Netherlands: International Service for National Agricultural Research.

Bantilan, M.C.S. and Joshi, P.K.1996. Returns to research and diffusion investments on wilt resistance in pigeonpea. Impact Series no.1. Patancheru, Andhra Pradesh, India: International Crops Research Institute for the Semi-Arid Tropics.

Bantilan, M.C.S. and Joshi, P.K. (eds.) 1998. Assessing joint research impacts. Patancheru, Andhra Pradesh, India: International Crops Research Institute for the Semi-Arid Tropics, Hyderabad.

Joshi, P.K. and Bantilan, M.C.S. 1997. Measuring the impact of resource management research. Impact Series no.2. Patancheru, Andhra Pradesh, India: International Crops Research Institute for the Semi-Arid Tropics, Hyderabad.

Chapter **39**

Priorities, Decisions and Evaluation of Fisheries Management and Governance of the Floodplain Wetlands in India

Nagesh K. Barik

Introduction

The floodplain wetlands or beels are used as multiple use and multiple users system. The large numbers of the stakeholders participate in the beels both within fisheries and across other sectors like forestry, animal husbandry, agriculture, flood control etc. (Sugunan & Sinha 2001, Ramsar Convention Secretariat 2004). The different set of stakeholders constitutes a diverse set of interests, goals and priorities for the fisheries. Therefore, under the present context, the fisheries managers face an extremely complex situation in governing fisheries of the wetlands. Such complexities lead to sub-optimal or unwise use of the wetlands. There is growing concern over the depletion of resources, poor quality of aquatic environment, poor potential realization, poor fisheries management, inequities, conflicts, over exploitation etc of the beels (CIFRI 2000, Pomeroy & William 1994, Welcomme & Bartley 1998). These issues are the outcome of the complex fisheries exploitation system and research efforts are being made to find suitable management regimes to address such issues. The evaluation of outcomes, identification of the shortcomings and understanding decision making process have been an important researchable areas (Pomeroy & William 1994, ICLARM & NSC 1998). Evaluation of a system should include depiction of its goals or objectives and

Applied Bioinformatics, Statistics and Economics in Fisheries Research: pp. 551-564

New India Publishing Agency, New Delhi (India)

an assessment of how well they are met as well as obstacles faced in meeting them. Accordingly, it is important to identify the management context of the beel fisheries management viz., the formal goals of a fishery management regime; the formal or informal goals of the various actors within it; the processes by which goals are identified and contested; processes through which they are linked to other aspects of the regime- including knowledge production and the implementation of action measures (ICES 2000). Therefore, it is imperative to understand the objectives, decision making process and criteria for evaluation of the fisheries management of the wetlands as a step towards further research in designing effective management regimes. The paper attempts to delineate the system in the simplistic way as well as review the relevant literature to map the objectives, decision making arrangements and evaluation criteria for assessing the fisheries management regimes of the beels. The present paper has been made based on the experience and understanding developed in the context of the beel fisheries management of Assam.

Methodology

The paper is based on the data collected from beels located in various parts of Assam during the period of 2000-2004 though Participatory Rural Appraisal (PRA) methods, focus group discussions, formal and informal communications with the state governments. The existing literatures were reviewed to develop theoretical background.

Theoretical Framework

Perspectives of management: There are many perspectives in which the fisheries of floodplain wetland operates. The important among them is conservation, development, social, economic and livelihood perspective. Each perspective set specific targets for the fisheries. The 'conservation perspective' is focused on the conservation of the ecosystem for providing various ecosystem services, biodiversity conservation from wildlife to fish to biotic communities (Ramsar Convention Secretariat 2004, Gowler *et al.,* 2002). These objectives has been attained through closure of wetland to fisheries uses (within national park, declaring sanctuary), restriction of the fisheries to the extent of providing non-threatening environment for them (Vani 2002). The perspective also attempts to restrict the severe alteration of the ecosystem, diversification to other uses, heavy water obstruction, pollution etc (V. K. Dewan 1998). The 'production or development perspective' see wetland in terms of the food, income and employment it provides (Welcomme & Bartley 1998). Therefore it seeks to optimize the fish production and productivity so as to generate greater food and income to the society. The 'social perspective' look for a system that is socially more acceptable and provide stable and sustainable relationship between the ecosystem and social system (Hanna & Munasinghe 1995). The 'Livelihood perspective' is predominantly gained importance as the national and international objective of eradiation of the poverty and livelihoods are being seen as the enforceable rights, therefore needs to be protected (Scoones 1998). The natural resources (so also fisheries resources) should aim at providing food and nutritional security and contribute in improving safety net to the fishers depending on them. These perspectives not only underpin equity consideration but also establish the rights of the primary stakeholders. The 'Economic perspective' is aimed at maximizing the benefits while minimizing the cost. It also prescribed for the management system that attain the objective at lowest cost without compromising on the quality (Carney & Farrington 1998).

The different perspective set up different goals for the fisheries which may be short-term or long-term. Important issues are the processes of goal formulation and prioritization, how conflicts between goals are dealt with, conditions under which coherency and consistency among goals is attempted, the degree of agreement about goals among different groups in a regime, and so forth. A critical question is of course which goals influence the decisions that are made and how (ICES 2000). The decision making process are variously termed as the fisheries governance (WRI 2000, WHAT 2000, Barik & Katiha 2003). The objectives set management to move to give rise to outcomes. The outcomes and objectives interrelated as dynamic process of mutual adjustments.

Outcome of an management arrangements can be evaluated on three related aspects viz, the meeting of management objectives, the impact on the resource and impact on its users (ICLARM & NSC 1998). The management objectives are sometimes differentiated into policy objectives and management objectives (Neiland & Bennette 2003). The first level of evaluation relates to overall institutional performance of management regimes versus other types of management arrangements. The second level of evaluation relates to performance in meeting specific management objectives and impacts at the operational level. Performance evaluation is conducted to determine how well the objectives are achieved and what the impacts of the management activities are on both the human and biophysical environment (ICLARM & NSC 1998).

Established Evaluation Criteria

The most common evaluative criteria are efficiency, equity and sustainability (Kerr, J. M. *et al.* 1996, ICLARM & NSC 1998). Some other criteria like effectiveness (Carney & Farrington 1998), livelihood and poverty

(Scoones 1998) derived respectively from efficiency and equity criteria but are given special attention.

i) *Efficiency* : It is also synonymous with productivity. This can be evaluated in term of technical and economic efficiency. Technical efficiency is achieved when with the given quantity of input, the resource yield highest possible physical product. The economic efficiency is achieved when total value of output is maximized for a given value of input. Therefore, the objective of attaining maximum productivity is a means of achieving efficiency (Kerr *et al.*, 1996). There are various measures of efficiency as applicable to fisheries. The first aspect of efficiency is whether fishers have achieved an optimal rate of use of the fishery (Oakerson 1992). A less rigorous criterion is that fishers are not exceeding the sustainable yield of the fishery. A second measure of efficiency has to do with the flow of benefits resulting from the management institutional arrangements and the costs (such as transaction costs) of establishing and maintaining such arrangements. A minimal efficiency criterion is that the benefits of operating and maintaining management arrangements exceed the full set of direct and indirect costs. A comparative efficiency criterion is that the difference between the benefits and the costs of management arrangements in one setting is the same as or greater than others (Ostrom 1992). In the marine fisheries sector, the inefficiency arise out of two things i. e. due to lack of proper fishing gear to fish (measured by catch per unit effort) and non utilization or improper utilization of the deep sea fisheries (resource use efficiency). The inefficiency can also arise out of conflicts of rights, technology constraints, sub-optimal investments in fisheries (Welcomme & Bartley 1998). The resource use efficiency increased though use of better fishing gears and enhancement measures like channeling the energy towards more useful species by rational choice of species spectrum, stocking, recapture options etc (Sugunan & Sinha 2001).

ii) *Equity* : It ensures that the benefit from using them are distributed fairly. However the meaning of the equity has been a matter of intense debate in the literature of the welfare economics. The Pereto-efficiency achieved when nobody can be better-off without some body else made worse off. Another criteria is that the equity is achieved when the benefit are skewed towards the poorer section of the society. The term social efficiency is also used to encompass objective like poverty reduction, responsiveness to local needs, facilitation of local initiatives, and respect or support for existing management practices (Carney & Farrington 1998). Equity (fair treatment for all people involved in managing, governing and using the resource) has four main components (Hanna 1995) i.e (a) Representation: a more equitable management regime should represent the range of interests in the fishery and accommodate the full diversity of those interests. (b) Process clarity: the management process should have a clear purpose and a transparent operation. (c) Homogenous expectations: the extent to which participants have similar expectations concerning the management process and its objectives. (d) Distributive effects: the management process should address the distributional changes embedded in the options under consideration.

Equity can be measured in several ways. First, is the distribution of an individual's return on contributions to the management and governance effort roughly similar to the benefits they receive? A second measure is to determine if there are patterns of redistribution that fishers wish to achieve at this level of institutional arrangements (ICLARM & NSC 1998).

iii) *Effectiveness* : Ability to meet the goals, objectives or needs. The management system is more effective when equitable access to resources, more economical in terms of administration and enforcement, increased sense of ownership of the resource by users, higher degree of acceptability and rule compliance, improved information about the resource, improved social cohesion in the community, and more participation (Pomeroy and Williams 1994). Therefore it incorporates much broader attributes than only cost benefit one. It also evaluates the effectiveness in achieving policy objectives (Neiland & Bennette 2003)

iv) Sustainability : In economic term sustainable allocation of the resource is efficient over time ensuring intergenerational equity. Sustainable utilization is defined as "human use of a wetland so that it may yield the greatest continuous benefit to present generations while maintaining its potential to meet the needs and aspirations of future generations" (Ramsar Convention Secretariat 2004). Sustainability can be divided into stewardship, resilience and governance (ICLARM & NSC 1998). Stewardship is the tendency for resource users to maintain productivity and ecological characteristics of the resource, is divided into three components: time horizons, monitoring and enforcement. To promote resources stewardship, the management process should expand time horizons beyond the short term. A sense of stewardship will be more likely if the effects of the management regime can be monitored and where necessary, enforcement measures taken. Resilience is the ability of the management system to absorb and deal with changes and shocks. The three components of resilience are rule flexibility, structural adaptation and market adaptation. Rules should be flexible enough to respond quickly to changing conditions. The management regime should be able to adapt to both changes in the structure of the industry as well as changes in the market (Hanna 1995). Governance includes the level of rule compliance as a major aspect. The willingness of fishers to regularly follow operational-level rules reflects the viability of the rules as coordinating devices. Other measures of governance may include overall reduction in conflict, existence of an effective conflict resolution mechanism, and existence of practical and implementable enforcement procedures.

Table 1. Criteria for evaluation of the fisheries in beels

Broad criteria	Attributes	Operational parameters
Efficiency	Technical efficiency	Productivity Output per unit input
	Economic efficiency	Benefit per unit cost low/high
	Low transaction cost	Whether Cost of management arrangements is higher/lower
	Resource use efficiency	What percentage of potential realized What percentage of fish caught Difficulties in catching fish
Equity	Distributions of benefits	What percentage of benefits goes to whom
	Participations	Whether all stakeholders participate Who participate in decision making
	Decision making process	Whether decision made transparently and consultatively
	Access to resources	Who access the resource for fishing What is the rule and mechanism of access
Sustainability	Maintenance of productivity	Whether the productivity declined over the period of time Whether important fishes availability declined Whether size of the fishes declined
	Time horizons	What is the tenure of lease Does the managers has long term plan

Contd....

Table 1. Contd....

	Monitoring trend	Are there is mechanism to monitor the trend Are there any trend data to monitor the fisheries
	Enforcement of corrective measures	Are there any mechanism to correct the problem of decline Are the mechanism responsive to the issues
	Rule Flexibility	Whether the rules of the managers flexible to adapt to any situation Are the rules of the government relaxed in the new situations
	Market adaptability	Whether the management are able to change the production as per market demand What adaptations followed for short term as well as long term demand
	Nature of conflicts	What are the major conflicts Whether conflicts affect sustainable use of fisheries What is the impact of the conflict
	Conflict resolution	What are the conflict resolution mechanisms How effective the mechanism Are the solutions are acceptable to parties
	Rule compliance	Whether the rules at the management and governance level are complied What is the level of non-compliance What is the enforcement mechanisms of the rules What are the incentives or disentwines of the rule compliance
Effectiveness	For policy objective	Whether the management arrangements achieve policy objective
	For management objective	How far the management able to attain the objectives of itself
Livelihood	Livelihood security	Whether system provided secured basis for livelihoods to fishers and other primary stakeholders Whether access to the resources are secured and institutionalized Whether fisheries provide livelihoo ds at the time of the need to the fishers
	Poverty alleviations	Does the poor are benefited from the system

Application of the Theoretical Understanding in the Fisheries Management of the Wetlands

The operational criteria for fisheries in beels : These abstract evaluative criteria must be converted into operational measures of value for use in appraising specific outcomes (Oakerson 1992). The operational measures changes with the resources as well as socio-economic context of the management. For the beel fisheries of Assam, the following operational measures can be used for the evaluations of the outcomes. These measures are the parameters for the evaluations of the outcome of the fisheries management of beels.

Governance and Decision Making Arrangements in Beels of Assam

The pertinent questions for the fisheries governance is who is setting objective? What are the determinants and mechanisms of setting objectives and prioritizing them. The complex process of governance is responsible for setting the objectives. The Governance is understood as a multidimensional decision-making process involving a wide range of actors interacting in various forms (Dore, 2001). Governance is defined as "The process of decision-making and the process by which decisions are implemented" (Petkova *et al.* 2002). Governance is the sum of the institutions, processes and traditions which determine how power is exercised, how decisions are taken, enforced and how citizens have their say. The elements of governance structure, process and mechanisms are *viz.,* Property rights and management regimes, rules, law and institutional environment, policies of the government, science and information, market and financial flow, participation and authority level (WRI 2003). Decision making arrangements are the interactions among the institutions and organization in an action situation (Oekersen 1986). Therefore the decision making arrangement study involves the actors, rules, institutions that govern the fisheries management decisions. Primary decision makers in case of the beels of Assam are government at various levels (Ministry, DoF, AFDC at state level, district fisheries officers, district revenue administration at operational level, and beel mangers, fisheries extension officers at field level); communities members *i.e* fishers, non-fishers, agriculturist, charitable institutions; managers as individuals, groups, cooperatives; scientific institutions; traders, marketing agents; local governance like Panchayat, community institutions, village institutions etc; international bodies as funding agencies like WorldBank (Table 2).

Table 2. Governance and decision making in the beels

Governance mechanism	Source of authority	Participants of decision making
Property rights	Law, lease arrangement, customary law, tradition	Managers, village communities, managing group, fishers
Rules and law	Constitution, legislations, judiciary, institutional arrangements	DoF, Panchayat, AFDC, officials, court, NGO
Policies	Government	Departments, AFDC, Revenue department, field functionaries, projects
Science and information	Research and knowledge system	Research organization, officials, NGO, communities, managers, fishers
Market and financial flow	Market, financial power	Traders, financiers, transporters,
Participation	Rights, law, rules, traditions, livelihoods	Fishers, communities, managers, filed functionaries,
Authority level	Hierarchy of authority	International bodies, national government, state government, district offices, Panchayat, communities

The existing structure of the management regimes of Assam is based on leasing and transferring the beels to the an autonomous body of either individual, groups or cooperatives (termed as beel managers). Many times the cooperatives further sub lease to an individual or group. In some cases, a power full group within cooperatives manage the beels. In the present framework, the beel managers are primary decision makers. The village communities negotiate to the extend to ensure better access to the fishers and non fishers, government plays the advisory roles, the scientific organization only provide technology information, NGO act as catalyst in community participation, court ensure access as per rule and adjudicate conflicts, market enter contact for supply of quality fish. In this framework, the beel managers are autonomous in taking management decisions while communities is important to allow the managers to enforce his right, state government lack mechanism to enforce its decision. Therefore, the beel managers, community governance, local governance, state governance, NGOs, market, fishers,

Table 3. Stakeholders and mechanism in the priority setting

Stakeholders/agents	Objectives	mechanism
Fishers	i. Maximization of catch ii. Minimization of cost of catch	i. Access to beels for fishing ii. Participation in decision iii. Providing information to management
Managers	i. Maximization of profit ii. Minimization of transaction cost iii. Minimizations of production cost	i. Taking operational decisions ii. Accessing information
Filed functionaries	i. Collection of data ii. Enforcement of regulation iii. Enforcement of rights	i. Monitoring and reporting to higher official ii. Advisory role
Communities	i. Maximization of fish to community members	i. Entering agreement with managers ii. Forceful access to resources
NGO	i. Equitable participation	i. Advisory role
Traders	i. Getting quality fish ii. More fish in scarcity season	i. Providing market information ii. Entering contract with managers for sale iii. Providing financial support
Local governance	i. Management of resource	i. Monitoring access rules
State governance	i. Maximization of lease rent ii. Enhancement of production	i. Monitoring production ii. Advisory role iii. Collecting rents Setting policy
Court	i. Equitable access as per rule	i. Enforcing rules ii. Adjudicating conflicts
Scientific institutions	i. Conservation of stock ii. Conservation of resources iii. Enhancement of production	i. Providing inforation on stock, depletions, practices, technology ii. Training
International bodies	i. Enhancement of production	i. Financing projects

scientific organizations are influencing power in order of effectiveness in setting management decisions. Based on the interactions of these actors, the following objectives are identified in order of priority for the fisheries management of the beels of Assam.

Priorities in Fisheries Governance

In the existing structure of the fisheries governance the various objectives can be categorized as high priority, medium priority and low priorities and non priorities from the management point of view. The high priorities objectives are implemented and enforced while medium are implemented to the extend not contradictory to the high priority one. The low priority objectives are attended as least important one. Non-priority are the objectives which are recognized but are not implemented and enforced. The non-priorities are the one which are prescribed by the scientific organization and government without the mechanism to implement it (Table 4).

Table 4 : Priorities in fisheries governance

	Objectives	**Reasons**
High priority	• Profit maximization of the managers • Lease rent maximization of the government • Maximum catch to the fishers • Maximization of the production	• Managers as lease owners for specific tenure (3-7yrs) • Govt lease given to highest bidders • Fishers catch fish for livelihood • Production maximization for profit and demand
Medium priority	• Increasing access of the local communities to the resources • Allocation of the fisheries lease as per rule	• Communities assets traditional rights • Court enforce the provisions
Low priority	• Monitoring of the fisheries trend and data collection • Enforcement regulations • Livelihoods	• Field officials were not given any resources • No legal power for enforcing regulations • Fishers negotiate with managers for access
Non priority objectives	• Conservation of stock • Conservation of resources quality	• No enforcement mechanisms • Managers have no long term incentives

The existing management regimes therefore inclined more towards short-term gain then over long-term one. The priorities are skewed towards profit maximization, higher lease rents that encourages over exploitations without proper corrective measures for sustainability.

Conclusion

The fisheries as multiple use and multiple users system operates in an environment of multiple and conflicting objectives. Actual objective is governed by the existing governance and management systems which decide on the priorities of the resources uses. The evaluations of the management system requires careful reading of the priorities as well as non-priority objectives. The evaluations should look at the areas which needs attention for better management. The system therefore close to the system of private management regimes with limited power of the state or governance to affect the decision of the managers. There is no credible mechanism to influence the decision of the managers.

REFERENCES

Barik, N. K. and P. K. Katiha. 2003. Management of Floodplain Wetlands: Institutional issues and options for Assam". *In* : Kumar, A. Katiha, P. K. and Joshi, P. K (eds.) *A Profile of people, technologies and policies in Fisheries sector of India,* National Centre for Agricultural Economics and Policy Research, New Delhi, pp. 141-157.

CIFRI. 2000. Ecology and Fisheries of beels in Assam. Bulletin, Central Inland Fisheries Research Institute, Barrackpore

Dewan V. K. 1998.North Eastern Region Local Acts and Rules Volume 9. Indian Law House. New Delhi

Diana, Carney and John Farrington 1998. Natural resource management and institutional change. Rutledge research/ODI Development Policy studies

Dore, John, 2001, *Environmental Governance in the Greater Mekong Sub-region,* Chiang Mai, Mimeo.

Edward B Barbier, Mike Acreman and Duncan Knowler 1997. Economic valuation of wetlands A guide for policy makers and planners, Ramsar Convention Bureau Gland, Switzerland

Gawler, M. (ed.) 2002. Strategies For Wise Use Of Wetlands: Best Practices In Participatory Management. Proceedings of a Workshop held at the 2nd International Conference on Wetlands and Development (November 1998, Dakar, Senegal). Wetlands International IUCN, WWF Publication No. 56, Wageningen, The Netherlands.

Hanna, S. 1995. Efficiencies of user participation in natural resource management. *In:* S. Hanna and M. Munasinghe, (eds.). *Property Rights and the Environment: Social and Ecological Issues*, pp. 59-68. Washington D.C., World Bank.

ICES 2000. Report of the Working Group on Fishery Systems. International Council for the Exploration of the Sea, Copenhagen, Denmark.

ICLARM and NSC. 1998 Analysis of Co-Management Arrangements in Fisheries and Related Coastal Resources: A Research Framework. International Centre for Living Aquatic Resources Management (ICLARM) and Institute of Fisheries Management and Coastal Community Development (IFM)

Neiland, A. E. and Elizabeth Bennett. 2003. Legal, Institutional and Policy Issues Affecting Access to Common Pool Resources in Developing Countries. The Case of Fisheries. IDDRA a paper prepared for the UK Department for International Development (DFID)/Hunting Technical Services (HTS/PASS) Project No. CP0034 'Initiate a learning process: Legal, institutional and policy issues affecting access to common pool resources', March 2003.

Oakerson, R.J. 1992. Analyzing the Commons: A Framework. pp. 41-59. *In:* D.W. Bromley, (eds.) Making the Commons Work: Theory, Practice and Policy. Institute for Contemporary Studies Press, San Francisco.

Ostrom, E. 1992. Crafting Institutions for Self-governing Irrigation Systems. Institute for Contemporary Studies Press, San Francisco.

Petkova E., Maurer C., Henninger N. and Irwin F. 2002. Closing the gap: Information, participation, and justice in decision-making for the environment. Washington, DC, World Resources Institute

Phukan. G. C. 2001. The Assam Fishery Rules, 1953. Assam Law House, Guwahati.

Pomeroy, R. S. and M. L. Williams. 1994. *Fisheries Co-management and Small-scale Fisheries: A Policy Brief.* Manila: International Center for Living Aquatic Resources Management.

Ramsar Convention Secretariat, 2004. *Ramsar handbooks for the wise use of wetlands.* 2nd Edition, Ramsar Convention Secretariat, Gland, Switzerland.

Scoones, Ian. 1998. Sustainable rural livelihoods: a framework for analysis ids working paper 72. Institute of development Studies, Sussex, UK.

Stuip, M.A.M., Baker, C.J. and Oosterberg, W. 2002. The Socio-economics of Wetlands, Wetlands International and RIZA, The Netherlands.

Sugunan V. V and M. Sinha 2001. Sustainable capture and culture -based fisheries in freshwaters of India in Pandian, T. J (eds.) Sustainable Indian Fisheries, National Academy of Agriculture Sciences New Delhi, 43-70

UNDP (1997). *Governance for Sustainable Growth and Equity:* Report of an International Conference, United Nations, New York, 28-30 July 1997

Vani M. S 2002. Customary law and modern governance of natural resources in India – conflicts, prospects for accord and strategies. Development Centre for Alternative Policies, New Delhi. Paper Submitted for the Commission on Folk Law and Legal Pluralism XIIIth International Congress , Chiang Mai University, Thailand.

WHAT. 2000. *Governance for sustainable future.II. Fishing for the future.* A report of Commission on Fisheries Resources. World Humanity Action Trust. London. UK

Welcomme R. L & D. M. Bartley 1998. An evaluation of present techniques for the enhancement of fisheries *In* : Petr, T. (ed). Inland fishery enhancements. Paper presented at the FAO/DFID Expert consultation on inland fishery enhancements. Dhaka, Bangladesh 7-11 April 1997. FAO Fisheries Technical paper. No 374. Rome, FAO 463 p.

WRI .2003. *Decisions for Earth: Balance, Voice and Power.* World Resources 2002–2004:. Washington, DC: World Resource Institute.

Chapter **40**

Linear Programming in Fish Diet Formulation

Nibedita Jena, Nirupama Panda & D.P. Rath

Introduction

Linear programming (LP) is one of the most widely applied operation research techniques and owes its popularity to George Dantzig's simplex method (Danzig 1963). It is an important field of optimization for several reasons. Many practical problems in operation research can be expressed as linear programming problems. Linear programming is heavily used in microeconomics and business management, either to maximize the income or minimize the costs of a production scheme. Some examples are food blending, inventory management, portfolio and finance management, resource allocation for human and machine resources, planning advertisement campaign etc.

In our economy, there are limited resources at our disposal. Our problem is to make such use of these resources to yield maximum quantity of production or to minimize the cost of production. Therefore, there arises the question of Linear Programming that is defined as a mathematical method of calculating the least cost combination of the factors of production, when there are many possible alternative combinations available to an enterprise.

Definition

The word linear means that the relationship existing between the variables are those represented by straight lines (relationship is in the form of y=mx+c,

Applied Bioinformatics, Statistics and Economics in Fisheries Research: pp. 565-574

New India Publishing Agency, New Delhi (India)

where y is dependent variable, x is independent variable, m is slope and c is y-intercept). In addition, the word programming means taking decision systematically. Thus, linear programming is a decision-making technique under given constraint on the assumption that the relationship amongst variables representing different phenomenon happen to be linear.

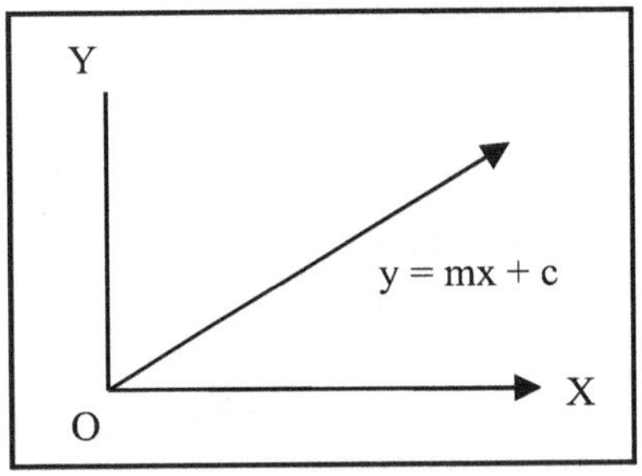

Fig. 1 Linear relationship

Linear Programming is essentially a time bound work. It has several aspects. A producer has an important aim, which is called objective function. He may like to maximize output, revenue or minimize cost.

Standard Form

Standard form is the usual and most intuitive form of describing a linear programming problem. It consists of the following three parts: [6]

- **A linear function to be maximized**
  Maximize $c_1x_1 + c_2x_2$
- **Problem constraints of the following form**
  $a_{11}x_1 + a_{12}x_2 \leq b_1$
  $a_{21}x_1 + a_{22}x_2 \leq b_2$
  $a_{31}x_1 + a_{32}x_2 \leq b_3$
- **Non-negative variables**
  $x_1 \geq 0$
  $x_2 \geq 0$

Example: Suppose that a farmer has a piece of pond, say *A* hectare area, to be stocked with either rohu or catla or some combination of the two. The farmer has a limited permissible amount *F* of fertilizer and *P* of insecticide which can be used, each of which is required in different amounts per unit area for rohu (F_1, P_1) and catla (F_2, P_2). Let S_1 be the selling price of rohu, and S_2 the price of catla.

If we denote the area stocked with rohu and catla with x_1 and x_2 respectively, then the optimal number of hectare to stock with rohu vs catla can be expressed as a linear programming problem:

Maximize $S_1x_1 + S_2x_2$

(maximize the revenue - revenue is the "objective function")

Constraints - Subject to $x_1 + x_2 \leq A$ (limit on total area)

$F_1x_1 + F_2x_2 \leq F$ (limit on fertilizer)

$P_1x_1 + P_2x_2 \leq P$ (limit on insecticide)

Non – negativity $x_1 \geq 0, x_2 \geq 0$ (cannot plant a negative area)

Augmented Form (Slack Form)

Linear programming problems must be converted into *augmented form* before being solved by the simplex algorithm. This form introduces non-negative *slack variables* to replace non-equalities with equalities in the constraints.

The example above becomes as follows when converted into augmented form:

Maximise $S_1x_1 + S_2x_2$ (objective function)

Subject to $x_1 + x_2 + x_3 = A$ (augmented constraint)

$F_1x_1 + F_2x_2 + x_4 = F$ (augmented constraint)

$P_1x_1 + P_2x_2 + x_5 = P$ (augmented constraint)

$x_1, x_2, x_3, x_4, x_5 \geq 0$

where x_3, x_4, x_5 are (non-negative) slack variables.

Simplex Method: When there are more than two variable, the graphical solution becomes extreamely complicated or impossible to draw. In this case, the algebraic solution (the simplex method) should be used.

Dual Problem: The basic problem whose solution is attempted by the linear programming technique is called the primal problem. To each primal problem corresponds a dual problem, which yields additional information to the decision maker. If the primal problem is the maximisation problem, its dual is a minimisation problem. Similarly, if the primal is a minimisation problem, its dual is a maximisation problem [1].

Assumptions of Linear Programming

The application of linear programming rests on several basic assumptions[3].

- *A well-defined objective function:* The decision maker wants to achieve certain goal. This may be maximization of net income, profit, sales or minimization of cost. The objective function should be expressed as a linear function of decision variables.
- *Activity or Process:* It is a combination of particular inputs to produce a particular output. It is a complex, large or small, of men and equipment.
- *Presence of constraints in the activities:* The linear programming technique can be applied to the problems, which involve certain kind of constraints in their activities. Such constraints are usually because of technological limitation or capacity and resource limitations. They exist primarily because resources are limited. These constraints are known as structural constraints.
- *Non-negativity of decision variable:* Along with these structural constraints, we have "Non-negativity" constraints which assume that negative value of the variable is not possible in the solution of linear programming problems.
- *Linear relations in the problem:* Linear programming problems are always stated in linear relations. The constraints along with the objective function are linear. Linearity means constant returns.
- *Feasible solution:* The feasible solutions can be explored only after the constraints have been duly established. Feasible solutions for the consumer are all of the possible combinations of commodities he can possibly or feasible buy, given his income and the prices of the commodities. With two commodities, feasible solutions for the consumer are all those combinations of the two goods that are on and to the left of the budget line. Likewise, one kind of feasible solution for the firm consists of all the combinations of the inputs that lie on or to the right of an iso-cost line.
- *Optimum Solution:* The optimum solution, of course is the best of the feasible solution. Sometimes, linear programming results in finding several feasible solutions, all equally good. In that case, there is no single optimum solution. *The simplex method is often used to find out the optimum solution among several feasible solutions. The simplex method consists of successively testing feasible solutions, successively eliminating the poorer ones, until finally the linear programmer is left with the optimum solution.*
- *Divisibility:* Activity levels are permitted to take fractional or integer values.

A Case Study

Least Cost Feed Formulation: Least cost feed formulation using linear programming methods is commonly used to derive minimum cost diets for fish and other food animals. The problem before the linear programmer is to minimize the cost of diet that meets minimum nutritional requirements of the fishes. Several feedstuffs are available in the market. Their selling prices are different per unit and possess different amounts of nutrients per kilogram, such as Vitamins, Minerals, and Proteins etc. The task of the linear programmer is to find out the cheapest combination of feedstuffs for the fishes. How he proceeds to do so?

Let us take a simple example to illustrate the methodology of the linear programming. A farmer let us suppose feeds fishes with two kinds of feedstuffs. The diet for the fish must give them each day certain minimum amounts of the above-mentioned three nutrients.

The objective function here is the minimization of the cost of the fish diet. Thus, there are two variables before the linear programmer, each with a different price. In addition, there are three constraints. Now, it is for the linear programmer to find out feasible solutions and from amongst them, the optimum solution. The methodology can be illustrated with the Fig.2 given below.

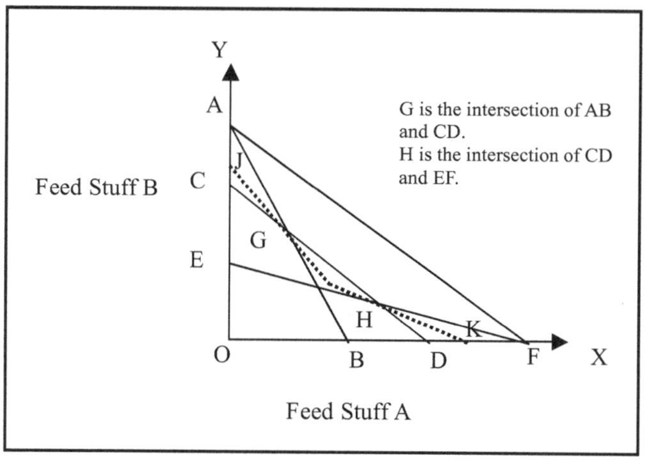

Fig. 2 Optimum Solution

In the above diagram, the OX axis shows kilogram of feedstuff A, while OY axis represents kilograms of feedstuff B. The minimum daily requirement per fish of nutrient I is satisfied by OB kilograms of feedstuff A or by OA kilograms of feed stuff B. Thus, the AB line shows how to combine feed stuffs A & B to meet requirements of nutrient I. It is evident from the diagram that feedstuff A is richer in nutrient I because fewer kilograms of it are needed. Likewise, the lines CD and EF show how the two feed stuff can be combined to provide the fish the minimum daily requirements of nutrients II & III. The line AGHF (with its segments AG, GH and HF) shows how to combine the two feedstuffs to meet all the three nutrients. These three nutrients are the 3 constraints. The line AGHF looks like an isoquant (derived from quantity and the Greek word iso [meaning equal], is a contour line drawn through the set of points at which the same quantity of output is produced while changing the quantities of two or more inputs), though it is not a smooth curve.

Any point on the line AGHF or to the right of it in the AGHF area represents feasible solutions of the diet problem. However, the problem now before the linear programmer is to find out the optimum solution among a number of feasible solutions. The optimum solution to the diet problem is represented by the discovery of the least-cost diet for the fishes. In order to arrive at the optimum solution, the programmer will have to bring into the picture the prices of the two feedstuffs. The ratios of the prices of the two feedstuffs A & B are represented by the iso-cost lines (which represents a combination of inputs which all cost the same amount). In the above diagram, the dashed lines J & K are the two possible iso-cost lines. In order to minimize cost, the linear programmer would like to get the lowest attainable iso-cost line. Let us take first the iso-cost line J. It occupies its lowest position, touching the point G. Thus, the G point represents the optimum diet when the price ratio is indicated by the line J. In this situation, B feed stuff is cheaper with a lower price per kg. The optimum solution, therefore, is to feed more kilograms of B than A. Let us now suppose that the iso-cost line K. indicates the prices of the two feedstuffs. In that case, the optimum solution would lie at H, because here A feedstuff is cheaper than B.

Solution Using Solver

Numeric Example

Nutrient	A	B	C	D
Weight (Gm)	90	50	20	02

The ingredients have the following nutrient values and cost

	A	B	C	D	Cost/kg
Ingredient 1 (gram/kg)	100	80	40	10	40
Ingredient 2 (gram/kg)	200	150	20	-	60

Solution

Variables

In order to solve this problem it is best to think in terms of one kg of feed mix. That kg is made up of two parts - Ingredient 1 and Ingredient 2

x_1= amount (kg) of Ingredient 1 in one kg of feed mix

x_2= amount (kg) of Ingredient 2 in one kg of feed mix

where $x_1 >= 0, x_2 >= 0$

Essentially these variables (x_1, x_2) can be thought of as the recipe telling us how to make up one kilogram of feed mix.

Constraints

Nutrient constraints

$100x_1 + 200x_2 >= 90$ (nutrient A)

$80x_1 + 150x_2 >= 50$ (nutrient B)

$40x_1 + 20x_2 >= 20$ (nutrient C)

$10x_1 >= 2$ (nutrient D)

$x_1, x_2 \geq 0$ (Non-negativity requirements)

Here we use of an inequality rather than an equality in these constraints, as we assume that the nutrient levels we want are lower limits on the amount of nutrient in one kg of feed mix.

Objective

Presumably to minimise cost, i.e

Minimize 40x1 +60x2 (objective function here is the cost function),

Which gives us our complete LP model for this feed problem.

In case you are interested the optimal solution to this LP (solved using the package i.e. using *Solver*)

There are of course numerous software packages, which are dedicated to solving linear programs (and other types of mathematical program), of which possibly LINDO, GAMS and XPRESS-MP are the most popular. All these packages tend to be DOS based and are intended for a specialist market, which requires tools dedicated to solving LPs. In recent years, however, several standard business packages, such as spreadsheets, have started to include an LP solving option, and Microsoft Excel is no exception. The inclusion of an LP solving capability into applications such as Excel is attractive for at least two reasons. Firstly, Excel is perhaps the most popular spreadsheet used both in business and in universities and as such is very accessible. Second to this, the spreadsheet offers very convenient data entry and editing features, which allows the student to gain a greater understanding of how to construct linear programs.

To use Excel to solve LP problems the *Solver add-in* must be included. Typically this feature is not installed by default when Excel is first setup on the hard disk. To add this facility to the Tools menu one need to carry out the following steps (once-only):

1. Select the menu option Tools | Add_Ins (this will take a few moments to load the necessary file).
2. From the dialogue box presented check the box for Solver Add-In.
3. On clicking OK, one will then be able to access the Solver option from the new menu option Tools | Solver.

Now Solver find the optimal values for Ingredient 1 and Ingredient 2. Generally, Solver generate three types of reports such as Answer Report, Sensitivity Report and Limits Report.

The answer report gives details of the solutions (in this case, cost is minimised when 0.37 gram of Ingredient 1 in one kg of feed mixed with 0.27 gram of Ingredient 2 and information concerning the status of each constraint with accompanying slack/surplus values is provided is shown in Table. 1. The Sensitivity Report, which provides information about how sensitive your solution is to changes in the constraints, is shown in Table 2. The Limits Report also provides sensitivity information on the RHS values is shown in Table 3.

The above sensitivity report (Table: 1) gives details of the solutions. In this case, total cost of feed is minimized when 0.37 gram of Ingredient 1 in one kg of feed mixed with 0.27 gram of Ingredient 2 [4].

Conclusion

With the help of the abovementioned methodology, various problems related to aquaculture may be solved to maximize return at minimum cost.

Table 1: Answer Report

Target Cell (Min)					
Cell	**Name**	**Original Value**	**Final Value**		
$B$5	Cost	0	30.66666667		
Adjustable Cells					
Cell	**Name**	**Original Value**	**Final Value**		
$B$10	Ingredient 1	0	0.366666667		
$B$11	Ingredient 2	0	0.266666667		
Constraints					
Cell	**Name**	**Cell Value**	**Formula**	**Status**	**Slack**
$B$14	Nutrient A	90	$B$14>=$C$14	Binding	0
$B$15	Nutrient B	69.33333333	$B$15>=$C$15	Not Binding	19.33333333
$B$16	Nutrient C	20	$B$16>=$C$16	Binding	0
$B$17	Nutrient D	3.666666667	$B$17>=$C$17	Not Binding	1.666666667
$B$18	Non-negativity 1	0.366666667	$B$18>=$C$18	Not Binding	0.366666667
$B$19	Non-negativity 2	0.266666667	$B$19>=$C$19	Not Binding	0.266666667

Table 2. Sensitivity Report

Adjustable Cells						
Cell	**Name**	**Final Value**	**Reduced Cost**	**Objective Coefficient**	**Allowable Increase**	**Allowable Decrease**
$B$10	Ingredient 1	0.366666667	0	40	80	10
$B$11	Ingredient 2	0.266666667	0	60	20	40
Constraints						
Cell	**Name**	**Final Value**	**Shadow Price**	**Constraint R.H. Side**	**Allowable Increase**	**Allowable Decrease**
$B$14	Nutrient A	90	0.266666667	90	50	26.36363636
$B$15	Nutrient B	69.33333333	0	50	19.33333333	1E+30
$B$16	Nutrient C	20	0.333333333	20	16	5
$B$17	Nutrient D	3.666666667	0	2	1.666666667	1E+30
$B$18	Non-negativity 1	0.366666667	0	0	0.366666667	1E+30
$B$19	Non-negativity 2	0.266666667	0	0	0.266666667	1E+30

Table 3 Limit Report

	Cell	Target Name	Value				
	$B$5	Cost	30.66666667				
	Cell	**Adjustable Name**	**Value**	**Lower Limit**	**Target Result**	**Upper Limit**	**Target Result**
	$B$10	Ingredient 1	0.366666667	0.366666667	30.66666667	#N/A	#N/A
	$B$11	Ingredient 2	0.266666667	0.266666667	30.66666667	#N/A	#N/A

REFERENCES

George B. Dantzig (1963). Linear programming and Extensions, Princeton University Press, Princeton University Press, Princeton, N.J.

http://people.brunel.ac.uk/~mastjjb/jeb/or/lp.html

Koutsoyiannis, A., 1991. Modern Microeconomics, "Linear Programming", Macmillan Education ltd, *London and Basingstoke*, pp: 414-434.

Linear programming - Wikipedia, the free encyclopedia.

MacDonald, Zyggy, 1995. Teaching Linear programming using Microsoft Excel Solver, University of Leicester, *CHEER*, 9:3.

Mehta, B.C. and G.M.K. Madnani, 1992. Mathematics for Economists, Linear Programming, Sultan Chand & Sons, Daryaganj, New Delhi, pp: 604-659.

Seth, M.L. 1983. Principles of Economics, "Linear Programming", Laxmi Narain Agarwal, Educational Publishers, Agra-3, pp: 316-321.

Chapter **41**

IPR Issues and Fisheries Research

Meenakshi Prajneshu & Abhinav Pragya

Introduction

A little information when shared can go a long way. Information patented can elevate India to great heights. Property (tangible) is associated with physical objects: house, land, etc., while intellectual property (intangible) includes Patents, Trade secrets, Copyrights, and Trademarks. Intellectual property rights (IPR) are legal rights which result from intellectual activity in industrial, scientific, literary, and artistic fields. Intellectual property protection (IPP) is right to protect intellectual property. Various types of IPR are:

(i) *Patent*: A newly invented product process is protected under it. Patentee has a legal right to make and sell

(ii) *Trademark*: It is a distinctive sign which identifies certain goods or services as those produced or provided by a specific person or enterprise

(iii) *Industrial Design*: It is ornamental or aesthetic aspect of an article

(iv) *Geographical Indications*: These are names of places used to identify origin, quality, reputation, or other characteristics of products

(v) *Copyright*: It is a legal term describing rights given to creators for their literary and artistic work

Applied Bioinformatics, Statistics and Economics in Fisheries Research: pp. 575-585

New India Publishing Agency, New Delhi (India)

Patenting Process

The patenting process involves the following steps:

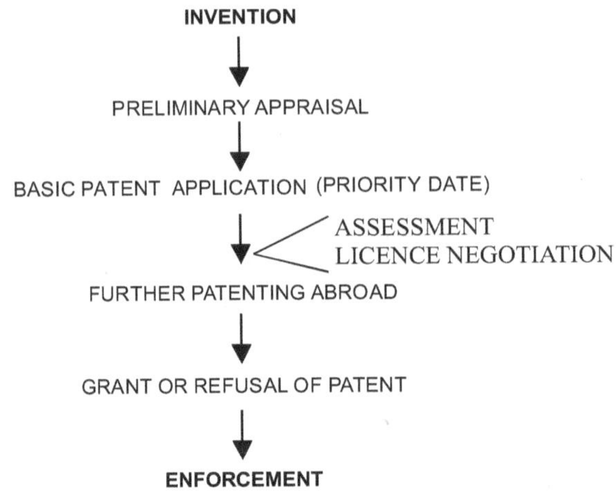

The Patent System in India is governed by the Patents Act, 1970 (No. 39 of 1970) as amended by the Patents (Amendment) Act, 2005 and the Patents Rules, 2003, as amended by the Patents (Amendment) Rules 2006 effective from 05-05-2006. The patent office is headquartered at Kolkata with branches at Chennai, New Delhi and Mumbai. The trade marks registry is located in Mumbai, and has branches in Kolkata, Chennai, Ahmedabad and New Delhi. The design office is located at Kolkata, while the office of the Patent Information System is in Nagpur. Intellectual Property Training Institutute (IPTI) is also at Nagpur. Patenting Information Centres (PICs) in India are in Andhra Pradesh, Assam, Chattisgarh, Gujarat, Madhya Pradesh, Manipur, Kerala, Punjab, Rajasthan, Sikkim, Tripura, Uttar Pradesh, Uttaranchal, West Bengal, and TIFAC (Delhi).

India is a member-state of World Intellectual Property Organisation *(WIPO)*, an International Organisation, responsible for promotion and protection of intellectual property throughout the world. India is a member of World Trade Organization with effect from 01-01-1995; Convention establishing World Intellectual Property Organisation; Paris Convention for protection of Industrial Property with effect from December 7, 1998; Patent Co-operation Treaty with effect from December 7, 1998; Budapest Treaty with effect from December 17, 2001.

A new product or process, involving an inventive step and capable of being made or used in an industry is a patentable invention. It means the invention to be patentable should be technical in nature and should meet the following criteria:

i) *Novelty:* The matter disclosed in the specification is not published in India or elsewhere before the date of filing of patent application in India.

ii) *Inventive Step:* The invention is not obvious to a person skilled in the art in light of prior publication/knowledge/ document.

iii) *Industrially applicable:* Invention should possess utility, so that it can be made or used in an industry.

Patents are granted for a period of twenty years if all requirements for first examination report (FER) are met or in case of opposition under Section 25(1), if opposition is decided in favour of the applicant, patent is granted after 6 months from date of publication. Under Section 11 A, the letter for patent is issued, entry is made in the register of patents and it is notified in the Patent Office, Journal; thereafter opening application, specification and other related documents for public inspection on payment of prescribed fee.

Where a patent covers a product, grant of patent gives patentee exclusive right to prevent others from performing, without authorisation, the act of making, using, offering for sale, selling or importing that product for above purpose. Where a patent covers a process, patentee has exclusive right to exclude others from performing, without his authorisation, the act of using that process, using and offering for sale, selling or importing for those purposes, the product obtained directly by that process in India. These rights created by statute are circumscribed by various conditions and limitations as provided in Patents Act, 1970 as amended by The Patents (amendment) Act, 2002. Registers of Patents are kept in the Patent offices where all its details are available. Patent has recently been granted for development of an innovative and less costly method for sewage treatment which is renewable (Tekes, 2006). Such types of patents can easily be obtained in fisheries. Patent analysis should be done to find out niche areas. Various issues relating to securing patents have been highlighted in clove (Prajneshu and Gupta, 2002a) and sugarcane (Prajneshu and Gupta, 2002b).

Opposition to a patent may be filed at concerned office but once a patent is granted, it takes a long time to oppose. Two of the examples in which patents granted have been opposed successfully are:

CASE STUDY 1: United States Patent Office (USPTO) granted patent on "Use of turmeric in wound healing" to University of Mississipi Medical Centre on 23 March, 1995.This generated debate on creating private ownership to traditional knowledge. C.S.I.R., New Delhi challenged it based on printed references, for example Indian Materia Medica, and Plants of India. Patent was ultimately revoked on 21 April, 1998. Some people even refer to it as the "Second battle of Haldighati".

CASE STUDY 2: Patent on "Basmati rice lines and grains" was granted to Rice Tec, Inc. Alvin Texas, U.S.A. on 2 September, 1997 by USPTO because it had produced a long grain rice with attributes similar to our own. Government of India challenged it on the ground that Basmati rice has been evolved by farmers of greater Punjab region and has been grown here for centuries. Rice Tec withdrew its claim in September, 2000. Had we lost that battle, we would not be allowed to produce, market, or sell Basmati rice anywhere in the World. Exports worth hundreds of crores of rupees were in jeopardy.

National and International Scenario

Indian Patent Databases are available as EKASWA - A and EKASWA - B. At the global level, a total of 609 biotech patents were taken on Rice during July, 1987 to September, 2000. Top 17 patent holders in the world are:

Company/Organisation	Country
DuPont	USA
Mitsui	Japan
Monsanto	USA
Aventis	France and Germany
Syngenta	Switzerland
Japan Tobacco	Japan
Cornell Research Foundation	USA
Mitsubishi	Japan
CSIRO	Australia
University of California	USA
Sumitomo	Japan
Purdue Research Foundation	USA
Dow Agrosciences	USA
US Department of Agriculture	USA
Advanced Technologies	USA
DNA Plant Technologies	USA
Okko	Japan

It may be pointed out that India figures nowhere in the above list. She started exhibiting its global presence only after 1 January, 1995 (Post-WTO period). Presently, leading public sector enterprises / other agencies in india are:

Name of the assignee	Main technology areas
Public sector enterprises:	
Indian Oil Corporation Ltd.	Detergent, catalysis
Indian Petrochemicals Ltd.	Chemicals, catalyst
Indian Explosives Ltd.	Explosive compositions
Oil and Natural Gas Commission	Pressure gauge inclinometer
S&T departments and agencies:	
Council of Scientific and Industrial Research	Chemicals, polymers, catalysts, biotechnology, natural products
National Institute of Immunology	Immunorepressive agent, vaccines
Defence R & D Organization	Alloys and superalloys
Indian Space Research Organization	Polyois
Sree Chitra Tirunal Institute	Heart valve
National Informatics Centre	Analog video add-on card
All India Institute of Medical Sciences	Medical technology
Delhi University	Microbiology and biotechnology
Jadavpur University	Drugs and pharmaceuticals

Further, leading private sector enterprises in India are:

Name of the assignee	Main technology area
Ranbaxy Pvt. Ltd.	Drugs and pharmaceuticals
Lupin Laboratories Ltd.	Drugs and pharmaceuticals
Dr. Reddy's Laboratories	Drugs and pharmaceuticals
Panacea Biotech Ltd.	Drugs and pharmaceuticals
Vijay Electricals Ltd.	Alloys
Cadbury India Ltd.	Vegetable fats
ICI India Ltd.	Chemicals
Hitek Fine Chemicals Pvt. Ltd.	Chemicals
Paramount Sintas Pvt. Ltd.	Chemicals

Living Organisms and Patenting

Although there was initial resistance to the patents of living organisms, this has now been overcome in most industrialized countries. In India, there is still some discussion and we are arguing against the patent of life forms, although eventually we may have to agree to the international practice.

If a substance found in nature has first to be isolated from its surroundings and a process is developed to obtain it from nature, then this process may be patentable as suggested by European Patent Office (EPO). Antibiotics and other microbial metabolites or a newly isolated microorganism are examples of such patentable substances. Several cases dealing with patents of products of nature have been recently contested in USA courts and a few are briefly discussed to illustrate the issue. Preparations of pure proteins or vitamin B_{12} by sophisticated purification techniques or by recombinant DNA methods are other examples. Before 1951, vitamin B_{12} had been available only as crude liver extracts, which was of no use for therapeutic application, but a patent claim for pure vitamin B_{12} has been allowed in USA. Microorganisms such as *E. coli*, in which human genes have been incorporated for production of human insulin, human growth, human tissue plasminogen activator (t-PA), etc. have been recognized for patents in USA. Microbial cells engineered to produce antigens and antibodies also qualified for patents. A protein may be modified by altering the gene sequences responsible for its synthesis. This alternation may or may not amount to infringement of a protein patent.

Patents on Higher Plants and Animals

A patentable invention must be capable of industrial application. Agriculture is also treated as an industry in this connection, so that a wide range of agricultural and horticultural methods and products can be patented. Biopesticides and bioinseticides are such examples. Other examples include:

(i) novel techniques of plant micropropagation, and

(ii) plant-cell and tissue-culture methods to prepare useful metabolites.

1. Among higher plants, an important example is the patent for a 'tryptophan - overproducing maize' obtained through tissue culture. The patent known as 'Hibberd Patent' was issued to 'Molecular Genetics Research and Development'.
2. A similar patent was allowed in USA for a transgenic animal, popularly described as 'oncomouse patent'.

3. Patent was also allowed for a pollyploid oyster produced by the application of hydrostatic pressure to zygotes.
4. In some cases, hybrid varieties were also considered patentable.

Sui Generis System and Fisheries

There are still no Animal Breeders' Rights similar to plant breeders' rights (PBR). There are, however, international processes considering such possibilities. The difficult question is how such a system could be designed. Major danger in this process is that such a system would borrow or use experiences from the plant sector without taking sufficiently into account special features of the fish breeding and farming sectors.

WHY ?

Most fish breeding systems are dependent on outbred, heterogeneous populations and are hence unsuited to fulfill the plant breeders' rights criteria of *new, distinct, uniform* and *stable.*

This reflects a need for a specially adapted type of intellectual property system for particular needs of aquaculture breeds. In legal parlance, this is called a sui generis system.

Patenting Transgenic Organisms

Transgenic plants and animals can be protected through patent claims in several countries now including USA, Japan, and Europe. For example, a boll worm resistant cotton may also be allowed patent protection. Transgenic Salmon has aqua-advantage, it produces high levels of hormone, matures in half the time of non-transgenic (A/F Proteins USA). Fish is genetically engineered. China is a big market for this fish. They have asked for a patent. There is no doubt, therefore, that in future patent claims for transgenic plants and animals fulfilling certain conditions will have to be allowed in several countries.

Fisheries and Patenting

A central socio-economic challenge in fish breeding arises from issues relating to access to and exclusive rights of genetic resources. A balance has to be created between the need for unencumbered and free access on one hand and, on other hand, need to ensure a right to the results from breeding and research (Rosendal *et al.,* 2006).

Fishery is one of the fastest growing sectors of food production, with an average increase of nine percent per year (FAO, 2004). There are great expectations that the aquatic Blue Revolution may constitute the next wave for enhanced food security in the World (Greer and Harvey, 2004). Similar to agriculture and livestock husbandry, fish farming involves environmental and socio-economic challenges along with its benefits (Prajneshu, 2002). Only 5-10% of total aquaculture production is based on genetically improved material (Gjedrem et al., 2005). Why the interest for investing in genetic improvement in aquaculture species is so low compared to other domesticated species, e.g. plants and livestock? Within the plant breeding sector, legal mechanisms for access and exclusive rights have been developed over a long period of time. The most important property rights to genetically improved plant varieties are based upon the plant breeders' rights.

Wild and improved materials are publicly owned by gene banks or common heritage of mankind. Plant breeders' rights and patent system is challenging it. The new biotechnologies have increasingly allowed for innovations in breeding and genetics fulfilling the criteria for patent protection (Prajneshu, 2005). The scope of Biological Diversity (CBD) covers wild species and improved breeding stocks as well as equitable sharing of benefits derived from use of the World's genetic resources.

Access or exchange of fish genetic resources and legal protection of investments and research in aquaculture have not been addressed extensively (Greer and Harvey, 2004). Fish biology shares some features with plants and some other features with terrestrial farmed animals. There are, however, crucial differences. Aquaculture species, such as carps, catfish, freshwater prawns, and mollusks have extremely high fecundity and reproductive capacity compared to other farm animals. This feature is more similar to the characteristics of plant species and provides a great potential for breeding and intensive selection. However, plant varieties are often formed as a result of homogenizing processes, like inbreeding and vegetative propagation. Accumulation of inbreeding does not seem to impair viability and performance of plants, as is normally seen in animals. The results are often highly uniform plant varieties, both phenotypically and genetically, that may be characterized, e.g. as in terms given by the UPOV (*new, distinct, uniform* and *stable*). Deliberate or unintended inbreeding has been practiced in fish breeding but often with severe inbreeding depression and reduced viability (Bentsen and Olesen, 2002). Consequently, genetic improvement programmes for fish will aim at minimizing inbreeding and maintaining as much genetic variation as possible within the population. The populations will then not be *uniform* and *stable*, but variable and evolving from generation to generation. Thus, there are important differences between plant and fish populations in terms of phenotypic and genetic characterization. The legal framework that has been developed for protection of plant varieties can therefore hardly be applied to populations of fish.

Nevertheless, the issues of access to, exchange of and exclusive rights to investment and breeding results are important also within the aquaculture sector.

In Europe, patents are presently granted to nearly all types of inventions including those that build upon biological material. Patent protection is available for fish breeding and fish farming. There are, however, only a few examples of patents in this sector. The need to protect genetic improvements in farmed species arises from ability of living organisms to pass on their performance traits to their progeny through improved composition of DNA. This means that a buyer/user of the organism may 'copy' performance by reproducing the organism. If public community does not pay costs of improvement efforts, these costs need to be covered by users or other beneficiaries in order to maintain improvement programmes or to stimulate new programmes.

Improved plant varieties may easily and quickly be propagated in large quantities. Since product sold to farmers is able to grow and multiply, it is easy to carry out on farm propagation of the seed. The main reason for farmers to return to the organized producer to buy new seed has been guarantees against contamination of the seed with diseases, pests or seed from weeds, guarantees regarding purity of origin of commercial seed and full heterosis effects of hybrid seeds. All these traits are similar in fish, but plant breeders' rights solution is not applicable due to the need for much higher genetic heterogeneity in most commercial fish populations. Fish populations can therefore not meet the criteria under UPOV convention.

The most common strategy in fish breeding programmes is continuous upgrading of the material. The sold material may, of course, be grown to sexual maturation and be illegally reproduced by buyer without knowledge of supplier, but reproduced material will then always lag behind by at least one generation compared to the material supplied directly from breeding programme.

In plant species, it is relatively easy to produce a large number of inbred lines and to test cross various combinations to identify hybrids that express significant heterosis effects. In fish breeding, applied hybrid programmes are scarce. As in plants, high female fecundity makes it feasible to produce and test cross large numbers of distinct parent lines and mass-produce commercial hybrids. However, cross-breeding experiments with stocks in aquaculture species have until now largely failed to detect major or applicable heterosis effects (Gjerde and Refstie, 1984; Gjerde, 1988; Gjerde et al., 2002; Bentsen et al., 1998). The gain from cross-breeding in aquaculture is typically lower than response after only one generation of selection in a simple selection programme. Possibly, deliberate inbreeding in parent lines may increase heterosis, but this strategy is expected to encounter same problems with inbreeding depression in parent lines as in other farm animals.

Traditionally, approach to seek legal property rights of genetically improved populations of fish has been to register product names and trademarks. Strictly speaking, this will not protect genetic material from being propagated and used by outsiders, but only prevent unauthorized use of registered name. There must be an access to and legal protection of aquaculture genetic resources and collaboration with selected, skilled multipliers through material transfer agreements (MTA).

Points to Ponder

President of the Guaymi General Congress, on learning of the patenting of a Guaymi woman's genes, said, "I never imagined people would patent plants and animals. It's fundamentally immoral, contrary to the Guaymi view of nature and our place in it and to patent human material or to take human DNA and patent its products. That violates the integrity of life itself, and our deepest sense of morality".

- ❐ Imagine that a medicinal plant your family and community have cultivated and used for as long as anyone can remember has been taken and patented by a multinational corporation.
- ❐ Imagine you are a medical researcher trying to find a cure for breast cancer, but you are blocked from using the genetic materials you need for your research because they have been patented.
- ❐ Imagine during a medical procedure that blood samples and scrapings are taken from your cheek without your consent, and from these, a research institute patents your cell lines.
- ❐ Imagine that you are a fish farmer who can no longer save your seed to re-grow next harvest, but must purchase it anew from the company each year because they have patented it.

Concluding Remark

Old proverb in academic circles, viz. "Publish or perish" has now been replaced by "Patent or perish"!

REFERENCES

Bentsen, H. B. and Olesen, I. (2002). Designing aquaculture mass selection programs to avoid high inbreeding rates. *Aquaculture*, 204, 349-59

Bentsen, H. B., Eknath, A. E., Palada-de Vera, M. C., Danting, J. C., Bolivar, H. L., Reyes, A., Dionisio, E. E., Longalong, F. M., Circa, A. V., Tayamen, M. M., and Gjerde, B. (1998). Genetic improvement of farmed tilapias: growth performance in a complete diallel experiment with eight strains of Oreochromis niloticus. *Aquaculture*, 160, 145-73

FAO (2004). *The State of World Fisheries and Aquaculture.* Sofia

Gjedrem, T., Gjerde, B. and Rye, M. (2005). Lessons from developing breeding programs for aquaculture species. *In* : Lessons from the Past to Optimise the Future. Extended abstracts and short communications". *Euro. Aqua. Soc. Spe.* Pub. No. 35. pp. 225-26

Gjerde, B. and Refstie, T. (1984). Complete diallel cross between five strains of Atlantic salmon. *Livest. Prod. Sci.*,11, 207-26

Gjerde, B. (1988). Complete diallel cross between six inbred groups of rainbow trout, Salmo gairdneri. *Aquaculture*, 75, 71-87

Gjerde, B., Reddy, P.V. G. K., Mahapatra, K. D., Saha, J.N., Jana, R. K., Meher, P. M., Sahoo, M., Lenka, S., Govindassamy, P., Rye, M. (2002). Growth and survival in two complete diallel crosses with five stocks of rohu carp (Labeo rohita). *Aquaculture*, 209, 103-15

Greer, D. and Harvey, B. (2004). *Blue Genes. Sharing and Conserving the World's Aquatic Genetic Resources.* Earthscan, London

Prajneshu, M. (2002). Aquaculture - The blue revolution and environment. *In: Sustainable Aquaculture.* Sahoo, D. and Qasim, S. Z. (Eds.). A. P. H. Pub. *Cor.*, New Delhi. pp. 85-99

Prajneshu, M. and Gupta, V. K. (2002a). Patenting on clove. *J. Intell. Prop. Rights*, 7, 24-32

Prajneshu, M. and Gupta, V. K. (2002b). Intellectual property rights in sugarcane research. *J. Intell. Prop. Rights*, 7, 416-22

Prajneshu, M. (2005). Intellectual property rights in biotechnological inventions. *In: Advances in Biotechnology*, Ed. Trivedi, P. C., Agrobios, Jodhpur. pp. 379-93

Rosendal, G. K., Olesen, I., Bentsen, H. B., Tvedt, M. W. and Bryde, M. (2006). Access to and legal protection of aquaculture genetic resources – Norwegian perspective. *J. Wor. Intell. Prop.*, 392-412

Tekes (2006). Treating sewage sludge from municipal waste water treatment plants. Israel

Chapter **42**

Fisheries Legislation and Policy in India

S. P. Singh

Introduction

States/organizations/agencies with statuary powers constitute Legislation, which are directly or indirectly related to their basic needs, smooth functioning of targeted system and to facilitate the safe survival of present as well as future generations of human being under jurisdiction. Legislation is a *Latin* word and customary that include all law constituted or in the process of constitution by empowered statuary houses/persons/group of people. Such Law has been categorized as Bill, Act, Rules and Regulations etc. Policy is a multipurpose dynamic tool, which activates different levels of administrative mechanism to perform various assignments systematically according to legislation systems. It includes collection of important information and statistics along with usefulness and validity of prevailing system of law. It brings required changes based on authentic and evaluated proper information. Legislation and policy and appropriate administrative mechanism are linked to each other in circular sequence.

Constitution of legal documents and its success is totally linked with administrative mechanism presumed to implement it. Basically administrative mechanism for implementation of legislation and policy is the responsibility of state/ government. To devise a good administrative mechanism to implement legislation successfully, is a complicated, dynamic and multidimensional process system which ensure the balance functioning of the different

Applied Bioinformatics, Statistics and Economics in Fisheries Research: pp. 587-605

New India Publishing Agency, New Delhi (India)

components with automatic feedback of the targeted output and effectiveness of the legislation system in place.

Fisheries legislations are enacted with the aim of managing our fisheries resources and dependent population. This area of public interest was addressed and studied in detail by Ancient Indian education system and the then local Institutions for its suitable and sustainable utilization. Various problems and issues were ably tackled at micro and macro level for whole Indian sub-continent irrespective of local states and small kingdom, till the emergence of British in India after first independence war of India during 1957. After successfully establishing their rule, British interfered to the long sustaining fishery legislation system in the area of fisheries of Indian sub - Continent, in principle, to raise their revenue from this state, where people became poorer and not able to bear the rising requirement of British empire from usual land revenue. This article aims to summarize the written fisheries legislation of India towards its different role for conservation of Fish Genetic Resources (FGR) and sustainable production from fisheries sector in general. Article contains six sections namely introduction, brief history and background, Fisheries Legislations of India, Legislations related to closely linked area to aquaculture and fisheries, Legislations in the indirectly related area to aquaculture and fisheries and conclusion. Efforts has been made to cover important basic law enacted by States and Union Government, but it may not be exhaustive and complete due to its vast nature and scattered availability. It is categorized in suitable groups and presented in the simple language. Currently in India, Union Government and State Governments are empowered to constitute law in their respective area assigned by law of the nation. Up to some extent, it is delegated by these statuary bodies to others by virtue of power vested to them. It is a dynamic activity and empowered states and Union Government are amending the existing law time to time or/and constituting new law for its compatibility to the system and the betterment of people therein. These tools are very useful to states/ nations /regions and people living therein for smooth functioning of different units, activities and better survival of generation over generation of human, resources and culture developed. Law making empowered agencies made their best efforts to constitute such law which is clear in all aspects like jurisdiction over time and space use of well defined words and match with other existing law of land.

Brief History and Background

India had a rich background towards development of fisheries legislation since time immemorial, although, it is available in the form of traditional practices, rituals, folksongs and festivals only, indicating on both the aspects *i.e.* conservation and production. Written evidences are not sufficienty available due to prolonged rule of foreigners and destruction of our old universities and libraries of these ancient Institutions. First documented evidence is the Edict, 246 BC of mighty king Ashoka of Maurya Dynasty. During British rein, a number of the enactment of fisheries law were initiated. Major one are (i)Fisheries Act,1875 (British (lower) Burma), (ii) Punjab State drafted Act and Rules in 1880, (iii) Bombay Bill for the preservation of game and fish in1881, (iv) Madras Fishery Bill in 1883 and (v) Bengal Private Fisheries Protection Act (Act 2 of 1889) for the prevention of poaching in private waters etc. To extend the jurisdiction of The Bengal Private Fisheries Protection Act 1889 with provision of prohibiting use of dynamite and poison in whole Indian territorial water, The Indian Fisheries Act, 1897 was enacted by British Government. This legislation is still in use by few states to manage their fisheries resources. It is working as supplemental fisheries legislation to other fisheries Law in position. After independence, The Constitution of India, is prime source to make provision of further legislating the required law in different area. Here , it is noteworthy that seventh schedule of The Constitution of India divided total fisheries sector into two portion i.e. inland and territorial water up to 12 Nautical Miles marine distance from baseline is included in the list II (State list) and territorial water from 12 to 200 Nautical Miles marine distance from baseline is in the List I (Union list). The constitution of fisheries legislation in India was accordingly divided. At present most of the major state governments of India had already constituted their own State Fisheries Acts followed by Rules, Regulation to manage and develop the fisheries sector of the states for betterment of territorial fishery resources under their jurisdiction. Theses Acts will be discussed in the coming section. List I also includes seafood exports trades, fishing harbors, fishing vessel industry, Marine and inland research, training etc. that is again jurisdiction of the Union government. Other related areas to fisheries sector like protection of wildlife, forest, endangered flora and fauna, coastal zone and marine biodiversity etc. are in the List III concurrent list which is joint responsibility of Union and States Government. All the Maritime States constituted Maritime Fishing regulation based upon the initiative taken by Union Government of Drafting a Bill of the Marine Fishing Regulation Act which was circulated among all the Maritime States during 1979. It was enacted by all States during 1980 to 2003. Fisheries legislation of India for Marine, Union Government is making effort consistently in this direction. A number of Acts and Rules in the area related to fisheries, sea food trades, port, harbor, vessel, coastguard, water, environment, forest, biodiversity and wildlife has been constituted and amended time to time. We can not claim all these legislation as fisheries law but at the same time, legislation constituted in these closely linked areas had its own importance. Name of fisheries legislation related to Marine and Coastal aquaculture effective at present, constituted by Union Government are depicted in the Table No. 6, other Acts which are closely associated with fisheries legislation and upliftment of the sector and its various stake holders are depicted in the table No. 7 and Acts which are indirectly associated with fisheries area are depicted in the table No. 8.

Fisheries Legislations of India

According to the provision of Article 246, Seventh Schedule of the Constitution of India, States have constituted different fisheries legislation for development of fisheries sector and management of the resources in sustainable manner. India is enriched with vast and diverse fishery resources spread from hilly upland coldwater to Exclusive Economic Zone (EEZ) along with many islands in the Bay of Bengal, Arabian Ocean and Indian Ocean. Resources under different categories under Indian territory are depicted in the table No. 3.To conduct research and frame the guidelines to ascertain conservation of Fish Genetic Resources (FGR) for sustainable production and equitable sharing of all stakeholder, legislation is a powerful tool. Hence, it is essential to know these legislation tools. For convenience of readers, existing important legislations of India can be divided into two categories, First is Marine water and second is Inland water. Further, for jurisdiction purpose inland water and marine water up to 12 nautical miles is under state government and EEZ area under the Union Government. There are three kind of water area for fisheries legislation, inland Water, Coastal water and Marine (EEZ) water. Clarity purpose, fisheries legislation constituted by State Government are divided further into two categories, legislation by (i) inland states and (ii) Maritime (coastal) states. Legislations of inland states are listed in table no.1 and Maritime (coastal) states are listed in table no.2.

Table 1: Name of the Inland State's fisheries legislation

S. No.	States	Name of the Legislation
1.	Arunachal Pradesh	The Arunachal Pradesh Fisheries Act, 2006
2.	Assam	The Assam Fisheries Rules, 1953
		The Assam Fish Seed Act,2005
		The Assam Private Fisheries Protection Act, 1935

Contd...

Table 1. Contd...

		The Assam Grant of Fishery Subsidy to Private Fish Farmers and Piscicultural Society Rules, 1971
		The Assam Grant of Fishery Subsidy to Fishermen/ Fishery Cooperative Societies Rules, 1971
3.	Delhi	Delhi Gazette Part IV
4.	Himachal Pradesh	The H.P. Fisheries Act ,1976
		Amended during1990
		The Himachal Pradesh Fisheries Rules,1979 Amended 1985, 1994
5.	Haryana	The Haryana Fisheries Rules, 1996
6.	Rajasthan	The Rajasthan Fisheries Act,1953
		The Rajasthan Fisheries Rules,1958 Amended 1981
7.	Madhya Pradesh	The Madhya Pradesh Fisheries Act, 1948
8.	Uttar Pradesh	The U.P. Fisheries Act , 1948
		The U.P. Fisheries Rules, 1954
9.	Nagaland	The Nagaland Fisheries Act ,1981
		The Nagaland Fisheries Regulation Rules,2006
10.	Panjab	The Panjab Fisheries Acts, 1914
		The Panjab Fisheries Rules,1985
11.	Pondicherry	The Indian Fisheries(Pondicherry) Amendment Act, 1965
12.	Meghalaya	The Indian Fisheries Act,1897
13.	Mizoram	Mizoram Fisheries Bill, 2002

Table 2: Name of the Maritime State's fisheries legislation

1.	Gujarat	The Gujarat Fisheries Act ,2003 The Gujarat Fisheries Rules, 2003
2.	Maharastra	The Maharashtra Fisheries Act , 1960 The Maharashtra Marine Fishing Regulation Act,1981

Contd...

Table 2. Contd...

		The Maharashtra Marine Fishing Regulation Rules,1982
3.	Goa, Daman, Diu	The Goa, Daman and Diu Marine Fishing Regulation Act, 1980
		Amended in 1989 and 1993
		The Goa, Daman and Diu Marine Fishing Regulation Rules, 1980
4.	Kerala	The Kerala Marine Fishing Regulation Act, 1980
		The Kerala Marine Fishing Regulation Rules, 1980
		The Kerala Marine Fishing Regulation (Appellate Board) Rules, 1980
	Kerala	Fisheries Development and Management Policy
5.	Lakshdweep	The Lakshdweep Mari1ne Fishing Regulation, 2000
		The Lakshdweep Marine Fishing Regulation Rules, 2001
6.	Orissa	The Orissa Marine Fishing Regulation Act, 1980
		The Orissa Marine Fishing Regulation Rules, 1980
7.	Karnataka	The Karanataka Marine Fishing(Regulation) Act, 1985
		The Karanataka Marine Fishing(Regulation) Rules, 1987
8.	Andhra Pradesh	The Andhra Pradesh Marine Fishing (Regulation) Rule ,1995
		The Andhra Pradesh Marine fishing (Regulation) Act, 1994
		The Andhra Pradesh Fisheries Conservation & Dev. of Fishery Resources (Bill)
9.	Tamil Nadu	The Tamil Nadu Marine Fishing (Regulation) Act, 1983, amended 2000
		Tamil Nadu Inland Fisheries Lease and License Rules, 1972
		The Kodaikanal Hills Fishing Rules,1969
		The Nilgiries Fishing Rules, 1966
		The Indian Fisheries Act, 1897
10.	West Bengal	The West Bengal Marine Fishing(Regulation) Act, 1993
		The West Bengal Marine Fishing(Regulation) Rules, 1995

Table 3. Total Fishery Resources of India

Resources	Measurement
Rivers & Canals (in km)	195210
Reservoirs(in Lakhs ha.)	21.60
Tanks &Ponds (in Lakhs ha.)	24.14
Floodplain lakes & derelict water (in lakhs ha.)	7.98
Brackish water(in lakhs ha.)	12.40
Continental shelf (in lakhs ha.)	530
Approx. length of coast line (kms)	8118
Exclusive Economic Zone (EEZ) (million sq. km)	2.02
Number of landing centers	1914
Number of fishing villages	3827

Source : Hand Book on Fisheries Statistics, 2004, Department of Animal Husbandry, Dairying and Fisheries, Ministry of Agriculture, Govt. of India

Fisheries Legislations of Inland States of India

The Indian Fisheries Act,1897 is the first fisheries act having jurisdiction over all Indian territory except the territories which immediately before the 1st November, 1956, were comprise within part B States subject to the provision of Section 8 and 10 of the General Clause Act, 1897. Now it is used as supplemental law to other existing fisheries legislation. Other Acts constituted by the state of Arunachal Pradesh, Uttar Pradesh, Manipur, Rajasthan, Madhya Pradesh, Punjab, Maharashtra, Nagaland and Himachal Pradesh are listed in table 1. Considering the importance of these Acts in conservation of fish biodiversity of country, provisions covered through different sections are summarized in brief and documented in table 4. This is only for purpose of introducing a preliminary study or research work. For any legislative matter, main documents must be consulted. These Acts are basis of making further amendment, rules, guideline and other legal policy. Freshwater is a multi-purpose commodity and demand is in rising trend. Its consumption and control is also with many departments of the government and other organizations. Hence a brief study of scenario by multi disciplinary team from policy planning, Industry, agriculture, commerce and science etc. is prerequisite for feedback to legislators to yield a suitable legislation which is suitable for our future requirement.

Delhi and Haryana made rule to regulate fishing under their territorial region based upon Punjab Fisheries Act, 1914 and Indian Fisheries Act, 1897.

Table 4. Important provision in sections in different State Fisheries Acts of inland States

		AR.P	UP		Raj. Man	M.P.	P	M	Naga.	HP
1.	Title and Extent	1	1	1	1	1	1	1	1	1
2.	Definition	2	2	2	2	2	2	2	2	2
3.	Prohibition and licencing of fishing	3,6	3	3	-	3	3	3	3	3
4.	Power to prohibit sale of fish	6	4		7	4	4	5	6	4
5.	Power to prohibit fishing by explosives/poison fish	4,5	-	-	-	-	-	-	-	-
6.	Arrest without warrant in some condition	7	6	7	9	8	6	7	8	-
7.	Power to fishery officer of search seizes and investigation as given to police sub-inspector S.I. in Criminal Procedure Code 1898	6	6	7	9	-	4	4	8	8
8.	Offence trail by magistrate not less than second class.	-	7	8	10	7	8	8	9	-
9.	Power to compound offences	-	8	9	11	8	3	9	10	10
10.	Penalties	6	5	5	8	4	5	6	7	5
11.	Power to make rules	6	3	12	5	3	8	4	3	3
12.	Repeal and savings	9	-	13	-	11	7	13, 14	-	12
13.	Officer deemed to be public servant		-	10	12	9	-	10	-	11
14.	Protection against action against good faith	-	11	13	10	-	11	-	-	-
15.	Total Section in the Act	9	8	13	12	11	8	14	10	12

NB: AR.P =Arunachal Pradesh, UP = Uttar Pradesh, Man = Manipur, Raj. = Rajasthan, M.P. =Madhya Pradesh, P= Punjab, M= Maharashtra, Naga= Nagaland and HP = Himachal Pradesh

Pondichery, made a slight amendment in Indian Fisheries Act, 1897 to make it suitable for its jurisdiction. It is known as Indian Fisheries (Pondicherry) Amendment Act, 1965 and 1996. Accordingly added section 7 provide right to make rule and get it approved by their Legislative Assembly. Section 8, determine the mode of recovery of license fee by regulation. It is similar to land revenue recovery. All the states having Acts had also constituted the various Rules to carry out the responsibility of the states provisioned in their Acts, except Arunachal Pradesh State. Arunachal Pradesh constituted the Act during 2006 with a provision of making Rules in the Section No.6. Respective Rules of different states are mentioned in above Table No. 1. Bengal and Assam inland fisheries is governed by their law relating to Fisheries; (i) Assam, The Assam Land and Revenue Regulation, 1886 and The Assam Private Fisheries Protection Act, 1935 and (ii) Bengal, The Fisheries Protection Act, 1889 of Bengal.

Fisheries Legislations of Maritime States of India

Fisher person of the coastal villages were earning their livelihood from fishing in the Marine since time immemorial. Due to rising competition and emergence of modern technologies in the Marine fishing, interest of poor fisher was adversely affected. To protect the interest of poor fisher and implement a homogeneous Rule, Union Government took lead and arrange a draft of fisheries legislation for all the Maritime States and Union Territories and circulated among them because State Government have jurisdiction under the fisheries up to specified distance in the Marine area. Legislators of the respective Maritime States considered the proposal of Union Government and process the bill through respective legislation houses. All the Government of Maritime States and Union Territories constituted said Acts during a time span of 1980-2003. Constitution of Marine Fishing Regulation Acts followed by constitution of the required Rules and Regulations exercising the provision of making Rules by State Government to carry out their duties assigned by the respective Acts. Marine Fishing Regulation Acts enacted by different States Government was result of the processing of similar draft for the purpose. Hence Provisions covered by fisheries legislation of different Maritime States were also in principle similar with differing in few sections, except Gujarat. Gujarat was the last State to enact Gujarat Marine Fishing Regulation Act, 2003. This State had given due emphasis to other aspects like biodiversity conservation and law and order in the Sea besides regulating the fishing. However Marine Fishing Regulation Acts of different States, Gujarat, Tamil Nadu, Andhra Pradesh, Karnataka, Orissa, Lakshdweep, Kerala, Goa, Daman, Diu, West Bengal and Maharashtra were summarized in brief and provisions covered by different sections are documented in the Table No. 5. It is simple to know the important provisions made by the different States in their Acts at a glance. The total No. of sections in different Maritime states is varying from 22 to 31.

Table 5: Items covered by sections in State Fisheries Act. Name of States with number of section

		Ka	M	Gu	W B	K	L	O	G	TN	AP
1.	Title and Extent	1	1	1	1	1	1	1	1	1	1
2.	Definition	2	2	2	2	2	2	2	2	3	2
3.	Authorization of Power to any Officer.	-	-	-	-	3	3	3	3	4	-
4.	Power to regulate and restrict fishing	3	4	3-7	-	4	4	4	4, 7-8	5	4
5.	Prohibition of use of fishing vessel	4	5	8	-	5	5	5	5	6	5
6.	License of fishing vessels	5	6	10	7	6	6	6	6	7	6
7.	Prohibition of unlicensed vessels	6	7	9	6	7	7	7	9	8	7
8.	Cancellation/ amendment of License	7	8	11	8	8	8	8	11	9	8
9.	Registration of vessel	8	9	12	9	9	9	9	12	10	9
10.	Information about movment Vessels is mandatory	-	10		10	10	10	10	13	11	10
11.	Information about return of Vessels is mandatory	-	11	14	11	11	11	11	14	12	11
12.	Finality of order	9	12		12	12		12	15	13	12
13.	Appeals against refusing license	10	13	16	13	13	12	13	16	14	13
14.	Power to search in fishing vessel	11	14	15	15	14	13	14	18	15	14
15.	Disposal of seized fish	11	15	15	16	15	14	15	19	16	15
16.	Arbitrative/Adjudicative	12	16	17	17	16	15	16	20	17	13
17.	Penalty imposition	13	17	21	18	17	16	17	21	18	17
18.	Constitution of Appellate Board	14, 16	18	-	19	18	17	18	22	-	18
19.	Appeal	14	18	17	19	18	17	-	22	19	18
20.	Power of Adjucating Officer and Appellate Authority	16	20	-	21	20	19	20	24	20	20
21.	Revision by Appellate Board	15	19	19	20	19	18	19	23	21	19
22.	Officer (Adju.) Appellate Board empowered as Civil Court	16	20	-	21	20	19	-	24	20	-
23.	Contravention by companies punishable	18	21	22	22	20	20	21	25	22	21

Contd...

Table 5. Contd...

24.	Cognizance of offence	18	21	23	22	21	20	21	25	22	21
25.	Offences as guilt	18	22	23	22	21	20	21	25	22	21
26.	Exemption	19	23	27	23	22	21	22	26	23	22
27.	Constitution and composition of Terminal Authority and its functions	-	-	24-26	14	-	-	-	-	-	-
28.	Protection against action in good faith	21	24	29	24	23	22	23	27	24	23
29.	Power to make Rule	22	25	30	25	24	23	24	28	26	24
30.	Repeal	-	-	31	-	25	-	-	-	27	-
31.	Savings	-	25	31	-	25	-	-	-	27	-
	Total Section in the Legislation	**22**	**25**	**31**	**25**	**25**	**23**	**24**	**28**	**27**	**24**

NB: G = Gujarat, TN = Tamil Nadu, AP= Andhra Pradesh, Ka=Karnataka, O= Orissa, L= Lakshdweep, K= Kerala, G= Goa, Daman, Diu, WB= West Bengal and M= Maharashtra

Eries Legislations and Policy of India Covering Marine Fisheries

Union Government is consistently making effort in this direction. Towards these efforts a number of the Legislation constituted to manage the fisheries resources under our federal system. Marine fisheries legislations and policy effective at present are described in the Table No. 6. To discuss all the legislations and policy is beyond the purview of this article. However few of them were discussed and other will be referring to the readers. Territorial Waters, Continental Shelf, Exclusive Economic Zone and Other Maritime Zones Act,1976 indicates farsightedness of policy planners which laid down the holistic approach to conservation, management of living resources and protection of the environment in the Indian Marine EEZ area with provision of constituting Rules and framing guidelines for this purpose. We have already discussed that India have vast Marine area including its 2.02 million sq. km. EEZ under the jurisdiction and function of the Union Government. Marine still have enormous potential to harness by this country to meet out its future requirement.

Maritime Zones of India (Regulation of Fishing by Foreign Vessels) Act, 1981 and Maritime Zones of India (Regulation of Fishing by Foreign Vessels) Rules, 1981 were constituted by Union Government for the regulation of fishing by foreign vessels in certain maritime zones of India and for matters connected therewith.

Act have 26 Section related all the aspects like jurisdiction, grant of license, prohibition of fishing in maritime zones, management set up, offences and

Table 6: Name of the Marine and coastal aquaculture fisheries legislation

S. No.	Name of the Legislation
1.	The Territorial Waters, Continental Shelf, Exclusive Economic Zone and Other Maritime Zones Act,1976
2.	The Maritime Zones of India (Regulation of Fishing by Foreign Vessels) Act, 1981
3.	The Maritime Zones of India (Regulation of Fishing by Foreign Vessels) Rules, 1981
4.	The Comprehensive Marine Fishing Policy, 2004
5.	The Coastal Aquaculture Authority Act, 2005
6.	The Coastal Aquaculture Authority Rules, 2005
7.	The Marine Products Export Development Authority Act, 1972
8.	The Marine Insurance Act, 1963

penalty etc. Rules were constituted for the discharge of the duties and functions of the Central Government of India.

Comprehensive Marine Fishing Policy, 2004 contain 9 section namely Fishery Resources, Harvest of Marine Fish Resources, Resource Management, Fishermen Welfare, Environmental Aspects, Infrastructure Development for Marine Fisheries, Legislative Support, Policy for Development of Fisheries in the Union Territories of Lakshadweep and Andaman & Nicobar Islands. This document is a land mark in the history of the constitution of Indian Fishery Legislation and Policy.

Through Comprehensive Marine Fishing Policy, 2004, Union Government planned strategically approach to manage the geographic base of Indian marine fisheries 8118 km. coastline, 2.02 million sq. km. of EEZ including 0.5 million sq. km. of continental shelf in the present scenario. Learning with the past deficiencies and criticism of the deep sea fishing Central Government included opinion and interest of all the stakeholders to manage the resources in sustainable manner. Environmental, ecological, biodiversity conservation, technical advancement and security in the sea etc. all possible aspects addressed.

The policy therefore planned well stringent management to further regulate fishing in the territorial waters with increasing deep sea fishing. Promoting exploitation in the deep sea and oceanic waters would be reducing fishing pressure in the traditional fishing areas and enhance the production and profitability. Income, employment, post harvest facilities, human resource development, fisher welfare measure and infrastructure for processing in sea itself will be certainly improved through its proper implementation.

Coastal Aquaculture Authority Act, 2005 and Coastal Aquaculture Authority Rules, 2005 was again a legislation to promote coastal aquaculture in well planned method. Central Government established Coastal Aquaculture Authority to regulate coastal aquaculture keeping environmental parameter within limit set by the The Environmental Protection Act, 1986 followed other environmental legislation. This Act is very useful to coastal aquaculture and made easy its establishment.

Legislation Related to Closely Linked Area to Aquaculture and Fisheries

Legislations in the area of Environment, forest, water, biodiversity, wildlife, port and ship/vessel etc. are closely linked with the area of fisheries and aquaculture and need due attention. Important legislations are depicted in the Table No. 7. The Environment (Protection) Act, 1986, Forest (Conservation) Act, 1972, Water (Prevention and Control of Pollution) Act, 1974, Major Port Trusts Act,1963, Inland Vessels Act , 1917 and followed amendment and Rules, Regulations are addressing more important issues related to the respective area which is closely associated with the fisheries, habitat and ecosystem. Contribution of these law towards development and manage the sustainable fisheries is highly significant.

Wildlife (protection) Act 1972 and its amendments is main document making provision of conservation of wild life and their habitat. Aquatic endangered animals and their habitat under its jurisdiction area are covered. Utilizing power of making rules under section 63 and 64 a number of rules has been constituted. This act was amended in 1982, 91, 93, 2002 and 2006. Latest amended act contain 66 sections divided in eight chapters which is very important to conserve the threatened species of animals including aquatic. Recent enactment The Biological Diversity Act, 2002 was enacted by the Union Government may be utilized to promote the fisheries in a sustainable manner and aquatic conservation can be integrated with biodiversity. Union Government constituted Rules and National Biodiversity Authority followed similar setup by States and local bodies provisioned in the Chapter VI and X of the Biological Diversity Act, 2002 respectively. This legislation with associated development will be unique tool to protect and conserve our genetic resources, indigenous knowledge and cultural diversity. This Act contains total 65 sections. This is the legislation which is capable to provide long run support to conservation and sustainable production from aquatic Genetic Resources after its successful implementation.

Table 7: Name of the legislation related to closely linked area to aquaculture and fisheries

S. No.	Name of the Legislation
1.	The Environment (Protection) Act, 1986
2.	The Environment (Protection) Rules, 1986
3.	The National Environment Appellate Authority Act, 1997
4.	The National Environment Tribunal Act, 1995
5.	The Forest (Conservation) Act, 1972
6.	The Biological Diversity Act, 2002
7.	The Biological Diversity Rules,2004
8.	The Destructive Insects and Pests (Amendment and Validation) Act, 1992
9.	The Inland Waterways Authority of India Act, 1985
10.	The Inland Vessels Act , 1917
11.	The Water (Prevention and Control of Pollution) Act, 1974
12.	The Water (Prevention and Control of Pollution) Cess Act, 1977
13.	The Wild Life (Protection) Act , 1972
14.	The Wild Life (Protection) Amendment Act, 1991
15.	The Wild Life (Protection) Amendment Act, 2002
16.	The Wild Life (Protection) Amendment Act, 2006
17.	The Major Port Trusts Act,1963
18.	The Live-stock Importation Act, 1898
19.	The Inter-State Water Disputes Act, 1956
20.	The Indian Ports Act, 1908
21.	The Inland Vessels Act , 1917

Legislation in the Indirectly Related Area to Aquaculture and Fisheries

Fishery legislation can not be utilized in isolation.

All the stakeholders including administrative mechanism assumed to implement it. There are legislation in the other area which is frequently used by concerned person for betterment of overall fishery system. It is listed in the Table No. 8

Table 8: Name of the legislation indirectly related area to aquaculture and fisheries

S. No.	Name of the Legislation
1.	The Coast Guard Act, 1978
2.	The Coast Guard(Amendment) Act, 2002
3.	The Coasting Vessels Act, 1838
4.	The Contract Labour (Regulation and Abolition) Act,1970
5.	The Co-operative Societies Act, 1950
6.	The Disaster Management Act, 2005
7.	The Dock Workers (Regulation of Employment) Act, 1948
8.	The Dock Workers (Regulation of Employment) (In applicability to Major Ports) Act, 1997
9.	The Dock Workers (Safety, Health and Welfare) Act Environment (Protection) Act, 1986
10.	The Goa, Daman and Diu Mining Concessions (Abolition and Declaration as Mining Leases) Act, 1987
11.	The Indian Carriage of Goods by Sea Act, 1925
12.	The Industrial Disputes Act, 1947
13.	The Industrial Disputes (Amendment and Miscellaneous Provisions) Act,1956
14.	The Industrial Employment (Standing Orders) Act, 1946
15.	The Insurance Act, 1938 amended 2002
16.	The Insurance Regulatory and Development Authority Act, 1999
17.	The Inter-State Corporation Act,1957
18.	The Inter-State Migrant Workmen (Regulation of Employment and The Conditions of Service) Act, 1979
19.	The Labour Laws (Exemption from Furnishing Returns and Maintaining Registers by certain Establishments) Act, 1988
20.	The Land Acquisition Act, 1894
21.	The Land Acquisition (Amendment) Act, 1962
22.	The Land Acquisition (Mines) Act, 1885
23.	The Merchant Shipping Act, 1958
24.	The Merchant Shipping (Amendment) Act, 1986
25.	The Merchant Shipping (Amendment) Act, 1987
26.	The Merchant Shipping (Amendment) Act, 2002
27.	The Mines Act, 1952
28.	The Mines and Minerals (Development and Regulation) Act, 1957
29.	The Minimum Wages Act, 1948
30.	The National Waterway (Sadiya-Dhubri Stretch of the Brahmaputra River) Act, 1988
31.	The Prevention of Cruelty to Animals Act, 1960
32.	The Trade Unions Act,1926
33.	The West Coast Canal and Champakara and Udyogmandal Canals) Act

Conclusion

India has strong fisheries legislations but its implementation and awareness is still weaker and needs improvement. Constitution of legislation, its amendment is recurring process. It must be based upon input of scientific facts and current reliable data. India need legislation in few area to properly address the issues and problems of fisheries resources, its unorganized sector like poor and land less fisher and low paid contract labors in the various associated small setup to fisheries industry. Inland fisheries need a fresh look for legislation towards developing and protecting our network of drainage system and other freshwater natural resources and genetic wealth therein due to rising population and reducing water resources. Other legislation in the priority area of fisheries and aquaculture are required to cover aquatic, exotic, quarantine, diseases (health aspects), quality seed production and habitat-specific ranching policy. Global development in legislation related to trade and commerce sector need to be studied with a long run approach to make competitive our legislation system. Proper linkages at different level must be developed for evolving good fisheries legislation system that should be dynamic, pro-environment, economic in implementation, socially viable and eco-friendly. To meet challenges of growing population, trade globalization and nutrition sufficiency, country must have vision to implement, update, co-ordinate legislative measure to enhance quality and production both. Administrative mechanism and policy must involve empowered village Panchayat and local Institutions/agencies to harness the benefit of linkages to make a system economic and efficient.

Self-built system of generating statistics through census or sampling is essential for planning, policy and research towards developing fisheries legislation because suitable and proper decisions are possible only based upon reliable current statistics. It may be concluded that development of the legislations must be initiated, processed and modified timely; it should reach to concern target people in brief and simple in their own languages with its specific merit. Policy makers and administrative machinery develop network for different purposes supported with strategies to implement it in true sense. Proper utilization of human resources for sustainable production in fisheries needs a dynamic legislation compatible to present scenario.

REFERENCES

The Andhra Pradesh Marine fishing (Regulation) Act, 1994

The Arunachal Pradesh Fisheries Act, 2006

The Assam Grant of Fishery Subsidy to Private Fish Farmers and Piscicultural Society Rules, 1971

The Assam Private Fisheries Protection Act, 1935

The Biological Diversity Act, 2002

The Biological Diversity Rules,2004

The Coast Guard Act, 1978

The Coast Guard(Amendment) Act, 2002

The Coastal Aquaculture Authority Act, 2005

The Coasting Vessels Act, 1838

The Comprehensive Marine Fishing Policy, 2004

The Constitution of India

The Contract Labour (Regulation and Abolition) Act,1970

The Co-operative Societies Act, 1950

The Destructive Insects and Pests (Amendment and Validation) Act, 1992

The Disaster Management Act, 2005

The Dock Workers (Regulation of Employment) Act, 1948

The Dock Workers (Regulation of Employment) (In applicability to Major Ports) Act, 1997

The Dock Workers (Safety, Health and Welfare) Act Environment (Protection) Act, 1986

The Environment (Protection) Act, 1986

The Forest (Conservation) Act, 1972

The Goa, Daman and Diu Marine Fishing Regulation Act, 1980 Amended in 1989 and 1993

The Goa, Daman and Diu Mining Concessions (Abolition and Declaration as Mining Leases) Act, 1987

The Gujarat Fisheries Act, 2003

The H.P. Fisheries Act, 1976 Amended during1990

The Haryana Fisheries Rules, 1996

The Indian Carriage of Goods by Sea Act, 1925

The Indian Fisheries Act, 1897

The Indian Fisheries (Pondicherry) Amendment Act,1965

The Indian Ports Act, 1908

The Industrial Disputes (Amendment and Miscellaneous Provisions) Act,1956

The Industrial Disputes Act, 1947

The Industrial Employment (Standing Orders) Act, 1946

The Inland Vessels Act , 1917

The Inland Vessels Act , 1917

The Inland Waterways Authority of India Act, 1985

The Insurance Act, 1938 amended 2002

The Insurance Regulatory and Development Authority Act, 1999

The Inter-State Corporation Act, 1957

The Inter-State Water Disputes Act, 1956

The Karanataka Marine Fishing(Regulation)_Act, 1985

The Kerala Marine Fishing Regulation Act, 1980

The Lakshdweep Mari1ne Fishing Regulation, 2000

The Land Acquisition (Amendment) Act, 1962

The Land Acquisition (Mines) Act, 1885

The Land Acquisition Act, 1894

The Live-stock Importation Act, 1898

The Madhya Pradesh fisheries Act, 1948

The Maharashtra Fisheries Act , 1960

The Maharashtra Marine Fishing Regulation Act,1981

The Major Port Trusts Act,1963

The Marine Insurance Act, 1963

The Marine Products Export Development Authority Act, 1972

The Maritime Zones of India (Regulation of Fishing by Foreign Vessels) Act, 1981

The Merchant Shipping (Amendment) Act, 1986

The Merchant Shipping (Amendment) Act, 1987

The Merchant Shipping (Amendment) Act, 2002

The Merchant Shipping Act, 1958

The Mines Act, 1952

The Mines and Minerals (Development and Regulation) Act, 1957

The Minimum Wages Act, 1948

The Nagaland Fisheries Act, 1981

The National Environment Appellate Authority Act, 1997

The National Environment Tribunal Act, 1995

The National Waterway (Sadiya-Dhubri Stretch of the Brahmaputra River) Act, 1988

The Orissa Marine Fishing Regulation Act, 1980

The Panjab Fisheries Acts, 1914

The Prevention of Cruelty to Animals Act, 1960

The Rajasthan Fisheries Act,1953

The Tamil Nadu Marine fishing (Regulation) Act , 1983, amended 2000

The Territorial Waters, Continental Shelf, Exclusive Economic Zone and Other Maritime Zones Act,1976

The Trade Unions Act, 1926

The U.P. Fisheries Act, 1948

The Water (Prevention and Control of Pollution) Act, 1974

The Water (Prevention and Control of Pollution) Cess Act, 1977

The West Begal Marine Fishing(Regulation) Act, 1993

The West Coast Canal and Champakara and Udyogmandal Canals) Act,

The Wild Life (Protection) Act, 1972

The Wild Life (Protection) Amendment Act, 1991

The Wild Life (Protection) Amendment Act, 2002

The Wild Life (Protection) Amendment Act, 2006

FSC
www.fsc.org
MIX
Papier | Fördert
gute Waldnutzung
FSC® C083411

Zeitfracht Medien GmbH
Ferdinand-Jühlke-Straße 7
99095 Erfurt, Deutschland
produktsicherheit@kolibri360.de